ENGINEERING METALLURGY

Part I   APPLIED PHYSICAL METALLURGY

ENGINEERING METALLURGY

Part I
APPLIED PHYSICAL METALLURGY

Sixth Edition

RAYMOND A. HIGGINS
B.Sc (Birm.), C.Eng., F.I.M.

Formerly Senior Lecturer in Metallurgy, West Bromwich College of Commerce and Technology; sometime Chief Metallurgist, Messrs Aston Chain and Hook Co., Ltd., Birmingham; and Examiner in Metallurgy to the Institution of Production Engineers, The City and Guilds of London Institute, The Union of Lancashire and Cheshire Institutes and The Union of Educational Institutes.

Edward Arnold
A member of the Hodder Headline Group
LONDON   SYDNEY   AUCKLAND

Edward Arnold is a division of Hodder Headline PLC
338 Euston Road, London NW1 3BH

© 1993 R A Higgins

First published in the United Kingdom 1957
Sixth edition 1993

8 7 6 5 4 3 2
99 98 97 96 95

British Library Cataloguing in Publication Data
Higgins, Raymond A.
Engineering Metallurgy.—Vol 1:
Applied Physical Metallurgy.—6 Rev. ed
I. Title
669

ISBN 0 340 56830 5

Typeset in 10/11 Linotron Times by
Rowland Phototypesetting Ltd, Bury St Edmunds, Suffolk
Printed and bound in the United Kingdom by
St Edmundsbury Press Ltd, Bury St Edmunds, Suffolk and
J W Arrowsmith Ltd, Bristol

PREFACE

To the First Edition

This text-book constitutes Part I of 'Engineering Metallurgy' and is intended primarily for students taking metallurgy as an examination subject for a Higher National Certificate in Mechanical or Production Engineering. The author hopes that it may also prove useful to undergraduates studying metallurgy as an ancillary subject in an Engineering Degree course. To students for whom metallurgy is a principal subject the book can offer a helpful approach to certain sections of the work in preparation for the Higher National Certificate and the City and Guilds Final Certificates in Metallurgy.

Comprehensive tables covering most of the alloys of importance to engineers are given in the appropriate chapters. In these tables an attempt has been made to relate British Standard Specifications to many commercially produced alloys. The author hopes that these tables will remain of use when the reader, no longer a student, finds it necessary to choose alloys for specific engineering purposes.

A generation ago much of a student's time was spent in dealing with the principles of extraction metallurgy. The widened scope of applied physical metallurgy has, however, in recent years, established prior claims upon the time available. Hence the brief survey in Chapter 2 of the production of iron and steel has to serve as a sufficient introduction to the methods of extraction metallurgy in general.

In the main, then, this book deals with the microstructural and mechanical properties of metals and alloys. Processes such as heat-treatment, surface hardening and welding are dealt with from the theoretical as well as the practical aspect. The author trusts that the treatment in Chapter 1 of the basic principles of chemistry will enable readers to follow the study of engineering metallurgy without being at a disadvantage if they have not previously studied chemistry as an independent subject.

It has been considered desirable to provide a basis for practical metallography; hence details of laboratory techniques are dealt with in Chapter 10.

At the end of each chapter will be found a selection of questions and exercises. Many of these have been taken from Higher National Certificate examination papers, and the author is greatly indebted to those authorities who have given permission for such questions to be used.

Although 'Engineering Metallurgy' Part II (Metallurgical Process Technology) is strictly speaking a sequel to the present volume, it may on occasion be read with advantage as a companion book—particularly when the approved syllabus for the Engineering Higher National Certificate is more than usually ambitious, or when direct contacts of students with metallurgical processes are limited by local circumstances.

In both Parts sections are numbered on the decimal system. In Part II frequent references to appropriate sections of Part I make it easy for the reader to look up the metallurgical principles governing any particular process under study.

Except where otherwise stated, the photomicrographs in this book are the work of the author or his students.

The author wishes to record his thanks to his wife for considerable help in producing the line diagrams; and to his friends J. H. Parry, Esq., FIM, of the School of Technology, Ipswich, and A. N. Wyers, Esq., AIM, of the Chance Technical College, for reading the original MS and making many helpful suggestions. He also wishes to record his appreciation of the generous assistance given to him at all stages in the production of this book by W. E. Fisher, Esq., OBE, DSc.

The author wishes to acknowledge the considerable help given by those connected with various industrial organisations, but in particular W. E. Bardgett, Esq., BSc, FIM (Messrs. United Steel Companies Ltd., Sheffield); J. F. Hinsley, Esq. (Messrs. Edgar Allen and Co. Ltd., Sheffield); Dr. J. R. Rait (Messrs. Hadfields Ltd., Sheffield); Dr. R. T. Parker and Dr. A. N. Turner (Messrs. Aluminium Laboratories Ltd., Banbury); Messrs. Samuel Osborn & Co. Ltd., and Prof. Dr. Fritz Gabler and Messrs. C. Reichert of Vienna.

R. A. HIGGINS

Department of Science,
  The Technical College,
    West Bromwich, Staffs.

PREFACE————————————
To the Sixth Edition

In 1937 I was a fledgling graduate with slim expectations of making a decent living as a scientist in British industry. The most exciting job the University Appointments Board was able to suggest for me involved the routine testing of aircraft carburettors on a twelve-hour shift basis (days and nights turn about) for the princely pittance of two guineas (£2.10) per week. Not unnaturally I spurned the offer and my name was summarily expunged from the files of the Appointments Board lest my 'unhelpful attitude' upset the delicate susceptibilities of their 'important clients' were I to be let loose near them.

There followed a dismal period when I eked out a precarious existence by twanging a Hawaiian steel guitar in a hula-hula band, performing in some of the more malodorous fleapits which served as variety theatres in those days of the nineteen-thirties. My lingering memory of that period of my life is not of the alleged glamour of showbiz but of the acrid smell of soft soap trapped in the wide cracks between the floorboards of the grubby little dressing rooms.

As war clouds gathered late in 1938 and the more enlightened sections of the metals industry anticipated the future need of scientists, I was offered a job at the then quite reasonable weekly wage of £3.10s (£3.50). This I grabbed with alacrity—after all it was a wage equal to that of a general shop-floor worker and slightly more than half that paid to the semi-skilled brass casters who were placed under my care, so who was I to grumble?

It will come as no surprise to the reader to learn that as soon as was possible I quit British industry for ever and sought employment in technical education, an occupation which for the next thirty-five years provided intellectual freedom, a decent standard of life, time to become a mountain-eer—and to write text books.

During the days which have followed the Second World War the functions of the scientist and engineer in industry have reputedly become increasingly important. But has status and remuneration improved proportionally? I suspect not. An examination of recent job advertisements leads me to believe that, allowing for some fifty years of inflation coupled with a higher proportion of salary lost to taxation as compared with pre-war days, *real* remuneration has changed little for the young graduate. Only the jargon of the advertisement is different. Now, instead of 'qualifications and experience', your 'CV' is required and the salary is quoted in £K—meant to impress I suppose. One advertisement I noticed recently preferred a 'Chartered Engineer or equivalent'. What, I wondered, would be regarded as an 'equivalent'? Applicants were asked to write to the 'Human Resources Department' which suggested to me that a prospective employee would be equated with so many tons of coal or some other expendable commodity. What's wrong with the old title 'Personnel Office' for God's sake? One wonders whether such an organisation has recruited Monty Python as its managing director.

It is now almost forty years since the late Dr W. E. Fisher, OBE, bullied me into producing the manuscript which became the First Edition of this book. Then in his late seventies and the dynamic Technical Editor of the then *English Universities Press*, he remains a great inspiration to me now that I in turn find myself at a similar age. Originally the book was written as a text for those student engineers taking metallurgy as a subject in the Higher National Certificate (Engineering) Courses. At the temporary demise of the Higher National Certificate some ten years ago this volume was largely rewritten to provide a treatment of general physical metallurgy at the elementary and intermediate levels.

When some years previously, 'metallurgy' had been replaced by 'materials science' in engineering syllabuses, many authors—attempting an adroit vault on to the bandwaggon—added a hurried chapter on 'plastics' to their existing texts. In many cases this served only to display a rather nebulous understanding of the true nature of the covalent bond. No mention was made of other non-metallic engineering materials. Obviously in almost forty years many new sophisticated metallic alloys have been developed whilst other metals, hitherto known only as symbols in the Periodic Classification of the Elements, have been drawn into the technology of the late twentieth century. Thus lithium, scandium, gallium, yttrium, indium, lanthanum, praseodymium, neodynium, samarium, gadolinium, dysprosium, erbium, thulium and ytterbium have all found uses during recent years in commercial alloys. They join boron, titanium, germanium, zirconium, niobium, cerium, hafnium and tantalum which had become metallurgically valuable during the immediately previous decades. Consequently this book has grown over the years so that it contains some 40% more pages than the first edition. Nevertheless it is still confined to a study of metallurgy and those who wish to study materials science for HNC or on a more general level should consult other titles.

R. A. HIGGINS

Walsall,
West Midlands.

CONTENTS

Dedicated to My Wife, Helen,
Who helped with the First Edition almost forty years ago—
and who still provides cups of tea whilst I scribble.

'The smith also sitting by the anvil, and considering the iron work, the vapour of the fire wasteth his flesh, and he fighteth with the heat of the furnace: the noise of the hammer and the anvil is ever in his ears and his eyes look still upon the pattern of the thing that he maketh; he setteth his mind to finish his work, and watcheth to polish it perfectly . . .'
Ecclesiasticus, c. 38; v. 28.

1

Some Fundamental Chemistry

1.10 Towards the end of the fifteenth century the technology of shipbuilding was sufficiently advanced in Europe to allow Columbus to sail west into the unknown in a search for a new route to distant Cathay. Earlier that century far to the east in Samarkand in the empire of Tamerlane, the astronomer Ulug Beg was constructing his great sextant—the massive quadrant of which we can still see to-day—to measure the period of our terrestrial year. He succeeded in this enterprise with an error of only 58 seconds, a fact which the locals will tell you with pride. Yet at that time only seven metals were known to man—copper, silver, gold, mercury, iron, tin and lead; though some of them had been mixed to produce *alloys* like bronze (copper and tin), pewter (tin and lead) and steel (iron and carbon).

By 1800 the number of known metals had risen to 23 and by the beginning of the twentieth century to 65. Now, all 70 naturally occurring metallic elements are known to science and an extra dozen or so have been created by man from the naturally occurring radioactive elements by various processes of 'nuclear engineering'. Nevertheless metallurgy, though a modern science, has its roots in the ancient crafts of smelting, shaping and treatment of metals. For several hundreds of years smiths had been hardening steel using heat-treatment processes established painstakingly by trial and error, yet it is only during this century that metallurgists discovered *how* the hardening process worked. Likewise during the First World War the author's father, then in the Royal Flying Corps, was working with fighter aeroplanes the engines of which relied on 'age-hardening' aluminium alloys; but it was quite late in the author's life before a plausible explanation of age-hardening was forthcoming.

Since the days of the Great Victorians there has been an upsurge in metallurgical research and development, based on the fundamental sciences of physics and chemistry. To-day a vast reservoir of metallurgical knowledge exists and the metallurgist is able to design materials to meet the ever exacting demands of the engineer. Sometimes these demands are over optimistic and it is hoped that this book may help the engineer to appreciate the limitations, as well as the expanding range of properties, of modern alloys.

1.11   Whilst steel is likely to remain the most important metallurgical material available to the engineer we must not forget the wide range of relatively sophisticated alloys which have been developed during this century. As a result of such development an almost bewildering list of alloy compositions confronts the engineer in his search for an alloy which will be both technically and economically suitable for his needs. Fortunately most of the useful alloys have been classified and rigid specifications laid down for them by such official bodies as the British Standards Institution (BSI) and in the USA, the American Society for Testing Materials (ASTM). Now that we are 'in Europe' such bodies as Association Française de Normalisation (AFNOR) and Deutscher Normenausschuss (DNA) also become increasingly involved.

Sadly, it may be that like many of our public libraries here in the Midlands, your local library contains proportionally fewer books on technological matters than it did fifty years ago, and that meagre funds have been expended on works dealing with the private life of Gazza—or the purple passion publications of Mills and Boon. Nevertheless at least one library in your region should contain, by national agreement, a complete set of British Standards Institution Specifications. In addition to their obvious use, these are a valuable mine of information on the compositions and properties of all of our commercial alloys and engineering materials. A catalogue of all Specification Numbers will be available at the information desk. Hence, forearmed with the necessary metallurgical knowledge, the engineer is able to select an alloy suitable to his needs and to quote its relevant specification index when the time comes to convert design into reality.

Atoms, Elements and Compounds

1.20   It would be difficult to study metallurgy meaningfully without relating mechanical properties to the elementary forces acting between the atoms of which a metal is composed. We shall study the structures of atoms later in the chapter but it suffices at this stage to regard these atoms as tiny spheres held close to one another by forces of attraction.

1.21   If in a substance all of these atoms are of the same type then the substance is a chemical *element*. Thus the salient property of a chemical element is that it cannot be split up into simpler substances whether by mechanical or chemical means. Most of the elements are chemically reactive, so that we find very few of them in their elemental state in the Earth's crust—oxygen and nitrogen mixed together in the atmosphere are the most common, whilst a few metals such as copper, gold and silver, also occur uncombined. Typical substances occurring naturally contain atoms of two or more kinds.

1.22 Most of the substances we encounter are either chemical *compounds* or *mixtures*. The difference between the two is that a compound is formed when there is a chemical join at the surfaces of two or more different atoms, whilst in a mixture only mechanical 'entangling' occurs between discrete particles of the two substances. For example, the powdered element sulphur can be mixed with iron filings and easily separated again by means of a magnet, but if the mixture is gently heated a vigorous chemical reaction proceeds and a compound called iron sulphide is formed. This is different in appearance from either of the parent elements and its decomposition into the parent elements, sulphur and iron, is now more difficult and can be accomplished only be chemical means.

1.23 Chemical elements can be represented by a *symbol* which is usually an abbreviation of either the English or Latin name, eg O stands for oxygen whilst Fe stands for 'ferrum', the Latin equivalent of 'iron'. Ordinarily, a symbol written thus refers to a single atom of the element, whilst two atoms (constituting what in this instance we call a *molecule*) would be indicated so: O_2.

1.24 Table 1.1 includes some of the more important elements we are likely to encounter in a study of metallurgy. The term 'relative atomic mass', (formerly 'atomic weight'), mentioned in this table must not be confused with the *relative density* of the element. The latter value will depend upon how closely the atoms, whether small or large, are packed together. Since atoms are very small particles (the mass of the hydrogen atom is 1.673×10^{-27} kg), it would be inconvenient to use such small values in everyday chemical calculations. Consequently, since the hydrogen atom was known to be the smallest, its relative mass was taken as unity and the relative masses of the atoms of other elements calculated as multiples of this. Thus relative atomic mass became

$$\frac{\text{mass of one atom of the element}}{\text{mass of one atom of hydrogen}}$$

Later it was found more useful to adjust the relative atomic mass of oxygen (by far the most common element) to exactly 16.0000. On this basis the relative atomic mass of hydrogen became 1.008 instead of 1.0000. More recently chemists and physicists have agreed to relate atomic masses to that of the carbon isotope (C = 12.0000). (See paragraph 1.90.)

1.25 The most common *metallic* element in the Earth's crust is aluminium (Table 1.2) but as a commercially usable metal it is not the cheapest. This is because clay, the most abundant mineral containing aluminium, is very difficult—and therefore costly—to decompose chemically. Therefore our aluminium supply comes from the mineral bauxite (originally mined near the village of Les Baux, in France), which is a relatively scarce ore. It will be seen from the table that apart from iron most of the useful metallic elements account for only a very small proportion of the Earth's crust. Fortunately they occur in relatively concentrated deposits which makes their mining and extraction economically possible.

In passing it is interesting to note that in the Universe as a whole hydro-

Table 1.1

Element	Symbol	Relative atomic weight (C = 12.0000)	Relative Density (Specific Gravity)	Melting point (°C)	Properties and Uses
Aluminium	Al	26.98	2.7	659.7	The most widely used of the light metals.
Antimony	Sb	121.75	6.6	630.5	A brittle, crystalline metal which, however, is used in bearings and type.
Argon	Ar	39.948	1.78×10^{-3}	−189.4	An inert gas present in small amounts in the atmosphere. Used in 'argon-arc' welding.
Arsenic	As	74.92	5.7	814	A black crystalline element—used to harden copper at elevated temperatures.
Barium	Ba	137.34	3.5	850	Its compounds are useful because of their fluorescent properties.
Beryllium	Be	9.012	1.8	1285	A light metal which is used to strengthen copper. Also used un-alloyed in atomic-energy plant.
Bismuth	Bi	208.98	9.8	271.3	A metal similar to antimony in many ways— used in the manufacture of fusible (low-melting-point) alloys.
Boron	B	10.811	2.3	2300	Known chiefly in the form of its compound, 'borax'.
Cadmium	Cd	112.4	8.6	320.9	Used for plating some metals and alloys and for strengthening copper telephone wires.
Calcium	Ca	40.08	1.5	845	A very reactive metal met chiefly in the form of its oxide, 'quicklime'.
Carbon	C	12.011	2.2	—	The basis of all fuels and organic substances and an essential ingredient of steel.
Cerium	Ce	140.12	6.9	640	A 'rare-earth' metal. Used as an 'inoculant' in cast iron, and in the manufacture of lighter flints.
Chlorine	Cl	35.45	3.2×10^{-3}	−103	A poisonous reactive gas, used in the de-gasification of light alloys.
Chromium	Cr	51.996	7.1	1890	A metal which resists corrosion—hence it is used for plating and in stainless steels and other corrosion-resistant alloys.
Cobalt	Co	58.933	8.9	1495	Used chiefly in permanent magnets and in high-speed steel.
Copper	Cu	63.54	8.9	1083	A metal of high electrical conductivity which is used widely in the electrical industries and in alloys such as bronzes and brasses.
Dysprosium	Dy	162.5	8.5	1412	Present in the 'rare earths', used in some magnesium-base alloys.
Erbium	Er	167.26	9.1	1529	A silvery-white metal. Present in the 'rare earths', used in some magnesium alloys and also used in cancer-therapy generators.

Table 1.1 *(continued)*

Element	Symbol	Relative atomic weight (C = 12.0000)	Relative Density (Specific Gravity)	Melting point (°C)	Properties and Uses
Gadolinium	Gd	157.25	7.9	1313	Used in some modern permanent magnet alloys. Also present in the 'rare earths' used in some magnesium-base alloys.
Gold	Au	196.967	19.3	1063	Of little use in engineering, but mainly as a system of exchange and in jewellery.
Helium	He	4.0026	0.16×10^{-3}	Below -272	A light non-reactive gas present in small amounts in the atmosphere.
Hydrogen	H	1.00797	0.09×10^{-3}	-259	The lightest element and a constituent of most gaseous fuels.
Indium	In	114.82	7.3	156.6	A very soft greyish metal used as a corrosion-resistant coating also used in some low melting point solders, and in semiconductors.
Iridium	Ir	192.2	22.4	2454	A heavy precious metal similar to platinum.
Iron	Fe	55.847	7.9	1535	A fairly soft white metal when pure, but rarely used thus in engineering.
Lanthanum	La	138.91	6.1	920	Used in some high-temperature alloys.
Lead	Pb	207.69	11.3	327.4	Not the densest of metals, as the metaphor 'as heavy as lead' suggests.
Magnesium	Mg	24.312	1.7	651	Used along with aluminium in the lightest of alloys.
Manganese	Mn	54.938	7.2	1260	Similar in many ways to iron and widely used in steel as a deoxidant.
Mercury	Hg	200.59	13.6	-38.8	The only liquid metal at normal temperatures —known as 'quicksilver'.
Molybdenum	Mo	95.94	10.2	2620	A heavy metal used in alloy steels.
Neodymium	Nd	144.24	7.0	1021	A yellowish-white metal. Used in some heat-treatable magnesium-base alloys and in some permanent magnet alloys.
Nickel	Ni	58.71	8.9	1458	An adaptable metal used in a wide variety of ferrous and non-ferrous alloys.
Niobium	Nb	92.906	8.6	1950	Used in steels and, un-alloyed, in atomic-energy plant. Formerly called 'Columbium' in the United States.
Nitrogen	N	14.0067	1.16×10^{-3}	-210	Comprises about 4/5 of the atmosphere. Can be made to dissolve in the surface of steel and so harden it.
Osmium	Os	190.2	22.5	2700	The densest element and a rare white metal like platinum.

Table 1.1   *(continued)*

Element	Symbol	Relative atomic weight (C = 12.0000)	Relative Density (Specific Gravity)	Melting point (°C)	Properties and Uses
Oxygen	O	15.999	1.32×10^{-3}	−218	Combined with other elements it comprises nearly 50% of the Earth's crust and 20% of the Earth's atmosphere in the uncombined state.
Palladium	Pd	106.4	12.0	1555	Another platinum-group metal.
Phosphorus	P	30.9738	1.8	44	A reactive element; in steel it is a deleterious impurity, but in some bronzes it is an essential addition.
Platinum	Pt	195.09	21.4	1773	Precious white metal used in jewellery and in scientific apparatus because of its high corrosion resistance.
Potassium	K	39.102	0.86	62.3	A very reactive metal which explodes on contact with water.
Rhodium	Rh	102.905	12.4	1985	A platinum-group metal used in the manufacture of thermocouple wires.
Samarium	Sm	150.35	7.5	1077	A light grey metal. Used in samarium-cobalt permanent magnets (electronic watches).
Selenium	Se	78.96	4.8	220	Mainly useful in the manufacture of photo-electric cells.
Silicon	Si	28.086	2.4	1427	Known mainly as its oxide, silica (sand, quartz, etc.), but also, in the elemental form, in cast irons, some special steels and non-ferrous alloys.
Silver	Ag	107.87	10.5	960	Widely used for jewellery and decorative work. Has the highest electrical conductivity of any metal—used for electrical contacts.
Sodium	Na	22.9898	0.97	97.5	A metal like potassium. Used in the treatment of some of the light alloys.
Strontium	Sr	87.62	2.6	772	Its compounds produce the red flames in fireworks. An isotope 'Strontium 90' present in radioactive 'fall out'.
Sulphur	S	32.064	2.1	113	Present in many metallic ores—the steelmaker's greatest enemy.
Tantalum	Ta	180.948	16.6	3207	Sometimes used in the manufacture of super-hard cutting tools of the 'sintered-carbide' type. Also unalloyed where high corrosion resistance is necessary (chemical plant).
Tellurium	Te	127.6	6.2	452	Used in small amounts to strengthen lead.
Thallium	Tl	204.37	11.85	303	A soft heavy metal forming poisonous compounds.
Thorium	Th	232.038	11.2	1850	A rare metal—0.75% added to tungsten filaments (gives improved electron emission).

Table 1.1 (*continued*)

Element	Symbol	Relative atomic weight (C = 12.0000)	Relative Density (Specific Gravity)	Melting point (°C)	Properties and Uses
Tin	Sn	118.69	7.3	231.9	A widely used but rather expensive metal. 'Tin cans' carry only a very thin coating of tin on mild steel.
Titanium	Ti	47.9	4.5	1725	Small additions are made to steels and aluminium alloys to improve their properties. Used in the alloyed and unalloyed form in the aircraft industry.
Tungsten	W	183.85	19.3	3410	Imparts very great hardness to steel and is the main constituent of 'high-speed' steel. Its high m.pt. makes it useful for lamp filaments.
Uranium	U	238.03	18.7	1150	Used chiefly in the production of atomic energy.
Vanadium	V	50.942	5.7	1710	Added to steels as a 'cleanser' (deoxidiser) and a hardener.
Ytterbium	Yb	173.04	7.0	817	Present in the 'rare earths' used in some magnesium-base alloys. A silvery-white metal.
Yttrium	Y	88.9	4.5	1522	Used in heat-treatable magnesium-base alloys. Also in some high-temperature alloys.
Zinc	Zn	65.37	7.1	419.5	Used widely for galvanising mild steel and also as a basis for some die-casting alloys. Brasses are copper-zinc alloys.
Zirconium	Zr	91.22	6.4	1800	Small amounts used in magnesium and high-temperature alloys. Also, un-alloyed, in atomic energy plant, and in some chemical plant because of good corrosion resistance.

gen is by far the most common element. In fact it accounts for some 90% of the total matter therein, with helium representing most of the remaining 10%; all other elements being restricted to no more than 0.2% in total. Magnesium, at about 0.003% is possibly the most abundant metal in the Universe. We are indeed fortunate in the variety and abundance of chemical elements on Earth.

Chemical Reactions and Equations

1.30 Readers will be familiar with the behaviour of the chemical substances we call *salts* during electrolysis. The metallic particles (or *ions* as they are termed) being positively charged are attracted to the negative electrode (or cathode), whilst the non-metallic ions being negatively charged are attracted to the positive electrode (or anode). Because of the behaviour of their respective ions metals are said to be electropositive and non-metals electronegative.

In general elements react or combine with each other when they possess opposite chemical natures. Thus the more electropositive a metal the more readily will it combine with a non-metal, forming a very stable compound. If one metal is more strongly electropositive than another the two may combine forming what is called an intermetallic compound, though usually they only mix with each other forming what is in effect a 'solid solution'. This constitutes the basis of most useful alloys.

1.31   Atoms combine with each other in simple fixed proportions. For example, one atom of the gas chlorine (Cl) will combine with one atom of the gas hydrogen (H) to form one molecule of the gas hydrogen chloride. We can write down a **formula** for hydrogen chloride which expresses at a glance its molecular constitution, viz. HCl. Since one atom of chlorine will combine with one atom of hydrogen, its **valence** is said to be one, the term valence denoting the number of atoms of hydrogen which will combine with one atom of the element in question. Again, two atoms of hydrogen combine with one atom of oxygen to form one molecule of water (H_2O), so that the valence of oxygen is two. Similarly, four atoms of hydrogen will combine with one atom of carbon to form one molecule of methane (CH_4). Hence the valence of carbon in this instance is four. However, carbon, like several other elements, exhibits a variable valence, since it will also form the substances ethene (C_2H_4), formerly 'ethylene', and ethine (C_2H_2), formerly acetylene.

1.32   Compounds also react with each other in simple proportions, and we can express such a reaction in the form of a chemical **equation** thus:

$$CaO + 2HCl = CaCl_2 + H_2O$$

Table 1.2   *The Approximate Composition of the Earth's Crust (to the Extent of Mining Operations)*

Element	% by mass	Element	% by mass
Oxygen	49.1	Nickel	0.02
Silicon	26.0	Vanadium	0.02
Aluminium	7.4	Zinc	0.02
Iron	4.3	Copper	0.01
Calcium	3.2	Tin	0.008
Sodium	2.4	Boron	0.005
Potassium	2.3	Cobalt	0.002
Magnesium	2.3	Lead	0.002
Hydrogen	1.0	Molybdenum	0.001
Titanium	0.61	Tungsten	9×10^{-4}
Carbon	0.35	Cadmium	5×10^{-4}
Chlorine	0.20	Beryllium	4×10^{-4}
Phosphorus	0.12	Uranium	4×10^{-4}
Manganese	0.10	Mercury	1×10^{-4}
Sulphur	0.10	Silver	1×10^{-5}
Barium	0.05	Gold	5×10^{-6}
Nitrogen	0.04	Platinum	5×10^{-6}
Chromium	0.03	Radium	2×10^{-10}

The above equation tells us that one chemical unit of calcium oxide (CaO or 'quicklime') will react with two chemical units of hydrogen chloride (HCl or hydrochloric acid), to produce one chemical unit of calcium chloride ($CaCl_2$) and one chemical unit of water (H_2O). Though the total number of chemical units may change due to the reaction—we began with three units and ended with two—the total number of atoms remains the same on either side of the equation. The equation must balance rather like a financial balance sheet.

1.33 By substituting the appropriate atomic masses (approximated values from Table 1.1) in the above equation we can obtain further useful information from it.

$$CaO + 2HCl = CaCl_2 + H_2O$$

$40 + 16$	$2(1 + 35\cdot5)$	$40 + 2(35\cdot5)$	$2(1) + 16$
56	73	111	18

Thus, 56 parts by weight of quicklime will react with 73 parts by weight of hydrogen chloride to produce 111 parts by weight of calcium chloride and 18 parts by weight of water. Naturally, instead of 'parts by weight' we can use grams, kilograms or tonnes as required.

We will now deal with some chemical reactions relevant to a study of metallurgy.

Oxidation and Reduction

1.40 Oxidation is one of the most common of chemical processes. It refers, in its simplest terms, to the combination between oxygen and any other element—a phenomenon which is taking place all the time around us. In our daily lives we make constant use of oxidation. We inhale atmospheric oxygen and reject carbon dioxide (CO_2)—the oxygen we breathe combines with carbon from our animal tissues, releasing energy in the process. We then reject the waste carbon dioxide. Similarly, heat energy can be produced by burning carbonaceous materials, such as coal or petroleum. Just as without breathing oxygen animals cannot live, so without an adequate air supply fuel cannot burn. In these reactions carbon and oxygen have combined to form a gas, carbon dioxide (CO_2), and at the same time heat energy has been released—the 'energy potential' of the carbon having fallen in the process.

1.41 Oxidation, however, is also a phenomenon which works to our disadvantage, particularly in so far as the metallurgist is concerned, since a large number of otherwise useful metals show a great affinity for oxygen and combine with it whenever they are able. This is particularly so at high temperatures so that the protection of metal surfaces by means of fluxes is often necessary during melting and welding operations. Although corrosion is generally a more complex phenomenon, oxidation is always involved and expensive processes such as painting, plating or galvanising must be used to protect the metallic surface.

1.42   It should be noted that, to the chemist, the term oxidation has a much wider meaning, and in fact refers to a chemical process in which the electronegative (or non-metallic) constituent of the molecule is increased. (1.74) For example, iron (II) chloride may be oxidised to iron (III) chloride—

$$2FeCl_2 + Cl_2 = 2FeCl_3$$
$$\text{Iron(II)} \qquad\qquad \text{Iron(III)}$$
$$\text{chloride} \qquad\qquad \text{chloride}$$

The element oxygen is not involved in this reaction, yet we say that iron (II) chloride has been 'oxidised' by chlorine since the chlorine ion is electronegative and so the electronegative portion of the iron (III) chloride molecule is greater than that of the iron (II) chloride molecule.

Since iron exhibits valences of both two and three this is indicated, in current chemical nomenclature, by writing either 'iron (II)' or 'iron (III)' as appropriate. Formerly the terms *'ferrous'* or *'ferric'* were used to describe these two series of compounds. In fact any metal exhibiting a variable valence gave rise to '-*ous*' and '-*ic*' series of compounds; -*ous* being used to describe that series in which the metal exhibits the lower valence and -*ic* that in which the metal exhibits the higher valence.

1.43   Whilst many metals exist in the Earth's crust in combination with oxygen as oxides, others are combined with sulphur as sulphides. The latter form the basis of many of the non-ferrous (that is, containing no iron) metal ores. The separation and removal of the oxygen or sulphur contained in the ore from the metal itself is often a difficult and expensive process. Most of the sulphide ores are first heated in air to convert them to oxides, eg—

$$2ZnS + 3O_2 = 2ZnO + 2SO_2$$
$$\text{Zinc sulphide} \qquad \text{Zinc oxide} \quad \text{Sulphur}$$
$$\text{('zinc blende')} \qquad\qquad\qquad \text{dioxide (gas)}$$

The oxide, whether occurring naturally or produced as indicated in the above equation, is then generally mixed with carbon in the form of coke or anthracite and heated in a furnace. In most cases some of the carbon is burned simultaneously in order to provide the necessary heat which will cause the reaction to proceed more quickly. Under these conditions carbon usually proves to have a greater affinity for oxygen than does the metal and so takes oxygen away from the metal, forming carbon dioxide and leaving the metal (often impure) behind, eg—

$$2ZnO + C = 2Zn + CO_2$$

This process of separating the atoms of oxygen from a substance is known as **reduction**. Reduction is thus the reverse of oxidation, and again, in the wider chemical sense, it refers to a reaction in which the proportion of the electronegative constituent of the molecule is decreased.

1.44 Some elements have greater affinities for oxygen than have others. Their oxides are therefore more difficult to decompose. Aluminium and magnesium, strongly electropositive metals, have greater affinities for oxygen than has carbon, so that it is impossible to reduce their oxides in the normal way using coke—electrolysis, a much more expensive process, must be used. Metals, such as aluminium, magnesium, zinc, iron and lead, which form stable, tenacious oxides are usually called *base* metals, whilst those metals which have little affinity for oxygen are called *noble* metals. Such metals include gold, silver and platinum, metals which will not scale or tarnish to any appreciable extent due to the action of atmospheric oxygen.

Acids, Bases and Salts

1.50 When an oxide of a non-metal combines with water it forms what we call an *acid*. Thus sulphur trioxide (SO_3) combines with water to form the well-known sulphuric acid (H_2SO_4), and sulphur trioxide is said to be the *anhydride* of sulphuric acid. Though not all acids are as corrosive as sulphuric, it is fairly well known that in cases of accident involving acids of this type it is necessary to neutralise the acid with some suitable antidote.

1.51 Substances which have this effect are called *bases*. These are metallic oxides (and hydroxides) which, when they react with an acid, produce water and a chemical compound which we call a *salt*. Typical examples of these acid–base reactions are—

$$H_2SO_4 + \quad CaO \quad = CaSO_4 + H_2O$$

Sulphuric   Calcium oxide   Calcium   Water
acid   ('quicklime')   sulphate
                               (a salt)

$$HCl \quad + NaOH = \quad NaCl \quad + H_2O$$

Hydrochloric   Sodium   Sodium
acid   hydroxide chloride
                ('caustic ('common
                 soda')   salt')

We can generalise in respect of equations like this and say—

Acid + Base = Salt + Water

Similarly, the acid *anhydride* will often combine with a base, forming a salt, eg—

$$SO_3 + CaO = CaSO_4$$

This type of reaction occurs quite frequently during the smelting of metallic ores.

1.52   One of the most common elements in the Earth's crust is silicon, present in the form of its oxide, silica (SiO_2). Since silicon is a non-metal, its oxide is an acid anhydride, and though all the common forms of silica, such as sand, sandstone and quartz, do not seem to be of a very reactive nature at normal temperatures, they are sufficiently reactive when heated to high temperatures to combine with many of the metallic oxides (which are basic) and produce neutral salts called silicates. Silica occurs entangled with most metallic ores, and although some of it is rejected by mechanical means before the ore is charged to the furnace, some remains and could constitute a difficult problem in that its high melting point of 1780° C would cause a sort of 'indigestion' in the furnace. To overcome this a sufficient quantity of a basic flux is added in order to combine with the silica and produce a slag with a melting point low enough to allow it to run from the furnace. The cheapest metallic oxide, and the one in general use, is lime.

$$SiO_2 + 2CaO = 2CaO . SiO_2$$

| Acid anhydride | Base | Salt (calcium silicate) |

The formula for the slag, calcium silicate, is generally written $2CaO.SiO_2$ rather than Ca_2SiO_4, since lime and silica will combine in other proportions. When the formula is written in the former manner, the reacting proportions of silica and lime can be seen at a glance.

1.53   On the other hand acid/basic reactions can constitute a problem when they involve similar reactions between slags and furnace linings. Thus, since we do not wish to liquefy our furnace lining, we must make sure that it does **not** react with the charge or the slag covering it. In short, we must make sure that the furnace lining is of the *same* chemical nature as the slag, ie if the slag contains an excess of silica, and is therefore acid, we must line the furnace with a similar silica-rich refractory, such as silica brick or ganister; whilst if the slag contains an excess of lime or other basic material, we must line the furnace with a basic refractory, such as burnt dolomite (CaO.MgO) or burnt magnesite (MgO). If the chemical nature of the furnace lining is the same as that of the slag, then, clearly, no reaction is likely to take place between them. Since silica and ganister, on the one hand, and dolomite and magnesite, on the other, all have high softening temperatures, they will also be able to resist the high temperatures encountered in many of the metallurgical smelting processes.

Atomic Structure

1.60   It was the Ancient Greek philosopher Leukippos and his disciple Demokritos who, during the fifth century BC, suggested that if matter were progressively subdivided a point would be reached where further subdivision was impossible. The Greek word for 'indivisible' is 'atomos'. More than 2000 years later—in 1808—a British chemist, John Dalton announced his *Atomic theory* which was based on the original Leukippos/

Demokritos idea. Dalton suggested that chemical reactions could be explained if it was assumed that each chemical element consisted of extremely small indivisible particles, which, following the Greek concept, he called 'atoms'. The Theory was generally accepted but before the end of the nineteenth century it was discovered that atoms were certainly not indivisible. Thus the 'atom' is an ill-described particle. But the title became so firmly established that it is retained to-day, though we usually modify our description to add that it is 'the smallest *stable* particle of matter which can exist'.

1.61   In 1897 the English physicist J. J. Thomson showed that a beam of 'cathode rays' was in fact a stream of fast-moving negatively charged particles—they were in fact *electrons*. Because the electron is negatively charged it can be deflected from its path by an electrical field and we make use of this feature in TV tubes where a beam of electrons, deflected by a system of electro-magnetic fields, builds up a picture by impingement on a screen which will fluoresce under the impact of electrons. Thomson was able to make only a rough estimate of the mass of the electron but was able to show that it was extremely small compared with a hydrogen atom. Thus Dalton's 'indivisible atom' fell apart.

1.62   Since atoms are electrically neutral—common everyday materials carry no resultant electrical charge—the discovery of the negatively charged electron stimulated research to find a positively charged particle. This was ultimately discovered in the form of the nucleus of the hydrogen atom, with a mass some 1837 times greater than that of the electron but with an equal but opposite positive charge. In 1920 the famous New Zealand born physicist Ernest Rutherford suggested it be called the *proton*.

So, atoms were assumed to be composed of equal numbers of electrons and protons, the electrons being arranged in 'shells' or 'orbits' around a bunch of protons which constituted the nucleus of the atom. But there was a snag: the true atomic weights of the elements, which had been derived by careful independent experiment over many years, were much greater than the atomic weights calculated from an assumption of the numbers of protons and electrons present in the atom of a particular element. Chemists explained this 'dead weight' in the atomic nucleus by suggesting that there were *extra* protons in the nucleus which had been 'neutralised' electrically by the presence of electrons also lurking there. Thus in the early 1930s electrons were classed as being either 'planetary' or 'nuclear.'

1.63   At about this time the English physicist Sir James Chadwick discovered the *neutron*, a particle of roughly the same mass as the proton but carrying no electrical charge. Its presence in the atomic nucleus made it easier to explain that part of the atomic mass not attributable to a simple electron/proton balance. Its electrical neutrality made it a useful particle in atomic research, since it could be fired into a nucleus without being repelled by like electrical charges. Chadwick's discovery did in fact alter the course of history since it made possible the development of the Atomic Bomb ten years later (18.74).

1.64   Since Chadwick's time the number of elementary particles has proliferated. These can be classified into three main groups:

1 *Baryons* (protons and other particles with a mass greater than that of the proton).
2 *Mesons* (any of a group of particles with a rest mass between those of the electron and proton, and with an integral 'spin').
3 *Leptons* (among which are the electron, positron or 'positive electron' and the neutrino which possesses neither charge nor mass but only 'spin').

The term 'quark' is used to describe any one of a number of hypothetical elementary particles with charges of $\frac{2}{3}$ or $-\frac{1}{3}$ of the electron charge, and thought to be fundamental units of all baryons and mesons. It is interesting to note that the word 'quark' was devised by James Joyce in *Finnegan's Wake*. Indeed, to date the existence of more than two hundred different subatomic particles has been reported. Of these only three—electron, proton and neutron—appear to have any substantial influence on the distinctive properties of each element. Consequently the rest will receive no further mention in this book. The essential features of electron, proton and neutron are summarised in Table 1.3. Since the real mass of these particles is inconveniently small for calculations their masses relative to that of the carbon atom (isotope C=12) are generally used.

Table 1.3

Particle	Actual rest mass (kg)	Relative mass ($^{12}C = 12$)	Charge (C)	
Electron	9.11×10^{-31}	0.000 548 8	-1.602×10^{-19}	Equal but opposite
Proton	1.672×10^{-27}	1.007 263	$+1.602 \times 10^{-19}$	
Neutron	1.675×10^{-27}	1.008 665	0	Zero

1.65 In the early days of the twentieth century Rutherford carried out a series of classical experiments in which he fired α-particles—the nuclei of helium atoms and therefore positively charged—at very thin gold foil. Most of these particles passed right through the foil whilst about one in 20 000 rebounded along its incident path. Others were deflected at various angles to the incident beam. From these experiments Rutherford concluded that the atom consisted of a comparatively small nucleus containing protons, around which circulated electrons. This concept of the atom was developed by the Danish physicist Niels Bohr. He proposed that the electrons were placed in a series of fixed orbits which varied in number with the complexity of the atom. The constitution of an atom was thus regarded as being similar to that of the solar system and containing about the same density of actual 'matter'. This model was later modified to the extent that the definite electron orbit was replaced by a mathematical function which represents the distribution of 'electrons' in the space occupied by the atom. This distribution is referred to as the *orbital* of the electron. In fact many visualise the electron as being in the nature of a 'cloud' of electricity rather than as a discrete particle, the orbital indicating the density of that cloud at any point within the atom. From the point of view of any diagram representing atomic structure it is more convenient to indicate the electron as being a definite particle travelling in a simple circular orbit round a nucleus consisting of protons and neutrons, but diagrams such as Fig. 1.1 should be studied with this statement in mind. Fig. 1.1 in no way represents what atoms 'look like'.

1.66 The most simple of all atoms is that of ordinary hydrogen. It consists of one proton with one electron in orbital around it. Since the positive charge of the proton is balanced by the equal but negative charge of the electron, the resultant atom will be electrically neutral. The mass of the electron being very small compared with that of the proton, the mass of the atom will be roughly that of the proton.

1.67 An atom of ordinary helium comes next in order of both mass and complexity. Here the nucleus contains two protons which are associated with two electrons in the same 'shell', ie similar orbitals surrounding the nucleus. The nucleus also contains two neutrons. However, in Table 1.4, which indicates the proton–electron make-up of some of the simpler atoms, neutrons have been omitted for reasons which will become apparent later (1.90). The number of protons in the nucleus, which is equal to the total number of electrons in successive shells, is called the *Atomic number* of the element.

In Table 1.4 it will be noted that with the metal lithium a new electron shell is formed and that this 'fills up' by the addition of a single electron with each successive element until, with the 'noble'* gas neon, it contains a total of eight electrons. With the metal sodium another new shell then begins and similarly fills so that with the noble gas argon this third shell also contains eight electrons. The next shell then begins to form with the metal potassium.

1.68 In the case of the elements dealt with in Table 1.4, this periodicity in respect of the number of electrons in the outer shell is reflected in the chemical properties of the elements themselves. Thus the metals lithium, sodium and potassium each have a single electron in the outer shell and all are very similar chemically. They will all oxidise very rapidly and react readily with water, liberating hydrogen and forming soluble hydroxides. Each of these elements has a valence of one. Physically, also, they are very similar in that they are all light soft metals, more or less white in colour.

In a similar way the gases fluorine and chlorine, with seven electrons in the outer shell in each case, have like chemical properties. Both are coloured gases (at normal temperatures and pressures) with strongly non-metallic properties.

The noble gases helium, neon and argon occur in small quantities in the atmosphere. In fact it is only there where they are likely to exist under natural conditions, since these noble gases are similar in being non-reactive and, under ordinary circumstances, unable to combine with other elements. Chemical combination between elements is governed by the number of electrons present in the outer shell of each atom concerned. When the outer shell contains eight electrons it becomes, as it were, 'satu-

---

* In the chemical sense the term 'noble' means that an element is not very reactive—thus, the 'noble' metals, gold, platinum, etc., are not readily attacked by other reactive substances, such as corrosive acids.

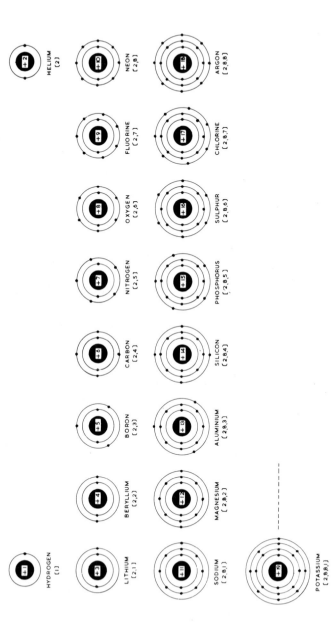

Fig. 1.1 This diagram indicates the electron–proton make-up of the nineteen simplest atoms, but it does not attempt to illustrate the manner in which they are actually distributed.

Table 1.4 *Electron Notation of the Elements in the First Four Periods of the Periodic Classification*

Element	Atomic Number (Z)	Shell 1 1s	Shell 2 2s	2p	Shell 3 3s	3p	3d	Shell 4 4s	4p	4d	4f
H	1	1									
He	2	2									
Li	3	2	1								
Be	4	2	2								
B	5	2	2	1							
C	6	2	2	2							
N	7	2	2	3							
O	8	2	2	4							
F	9	2	2	5							
Ne	10	2	2	6							
Na	11	2	2	6	1						
Mg	12	2	2	6	2						
Al	13	2	2	6	2	1					
Si	14	2	2	6	2	2					
P	15	2	2	6	2	3					
S	16	2	2	6	2	4					
Cl	17	2	2	6	2	5					
Ar	18	2	2	6	2	6					
K	19	2	2	6	2	6		1			
Ca	20	2	2	6	2	6		2			
Sc	21	2	2	6	2	6	1	2			
Ti	22	2	2	6	2	6	2	2			
V	23	2	2	6	2	6	3	2			
Cr	24	2	2	6	2	6	5	1			
Mn	25	2	2	6	2	6	5	2			
Fe	26	2	2	6	2	6	6	2			
Co	27	2	2	6	2	6	7	2			
Ni	28	2	2	6	2	6	8	2			
Cu	29	2	2	6	2	6	10	1			
Zn	30	2	2	6	2	6	10	2			
Ga	31	2	2	6	2	6	10	2	1		
Ge	32	2	2	6	2	6	10	2	2		
As	33	2	2	6	2	6	10	2	3		
Se	34	2	2	6	2	6	10	2	4		
Br	35	2	2	6	2	6	10	2	5		
Kr	36	2	2	6	2	6	10	2	6		
etc.											

Note: 4s begins to fill before 3d.

Note: 4f is not filled until element no.71 (Lu) by which point 6s has already been filled.

rated', so that such an atom will have no tendency to combine with others.

1.69 The periodicity of properties described above was noticed by chemists quite early in the nineteenth century and led to the advent of order in inorganic chemistry with the celebrated 'Periodic Classification of the Elements' by the Russian chemist Demitri Mendeleef in 1864. In more recent years this periodicity in chemical properties of the elements has been explained in terms of the electronic structure of the atom as outlined very briefly above.

In elements with atomic numbers greater than that of potassium, more complex shells containing more than eight electrons are present in the atom. These more complex shells are divided into sub-shells and some 'overlapping' of orbitals occurs so that new sub-shells tend to form before a previous shell has been 'filled' (Table 1.4). This gives rise to the 'transition metals' situated roughly in the centre of modern versions of the Periodic Table (Fig. 1.2). Nevertheless, elements in the vertical columns—or 'Groups'—have similar electron structures and therefore similar properties.

Chemical Combination and Valence

1.70 It was mentioned above that chemical combination between two atoms is governed by the number of electrons in the outer electron shell of each. Moreover, it was pointed out that those elements whose atoms had eight electrons in the outer shell (the noble gases neon and argon) had no inclination to combine with other elements and therefore had no chemical affinity. It is therefore reasonable to suppose that the completion of the 'octet' of electrons in the outer shell of an atom leads to a valence of zero. The noble gas helium, with a completed 'duplet' of electrons in the single shell, behaves in a similar manner.

As far as the simpler atoms we have been discussing are concerned the tendency is for them to attempt to attain this noble-gas structure of a stable octet (or duplet) of electrons in the outer shell. Their chemical properties are reflected in this tendency. With the more complex atoms the situation is not quite so simple, since these atoms possess larger outer shells which are generally sub-divided, to the extent that electrons may begin to fill a new outer 'sub-shell' before the penultimate sub-shell has been completed. As mentioned above this would explain the existence of groups of metallic elements the properties of which are transitional between those of one well-defined group and those of the next. The broad principles of the electronic theory of valence mentioned here in connection with the simpler atoms will apply. On these general lines three main forms of combination exist.

1.71 Electro-valent Combination   In this type of combination a metallic atom loses the electrons which constitute its outer shell (or sub-shell) and the number of electrons so lost are equivalent to the numerical valence of the element. These lost electrons are transferred to the outer electron shells of the non-metallic atom (or atoms) with which the metal is combining. In this way a complete shell of electrons is left behind in the metallic particle whilst a hitherto incomplete shell is filled in the non-metallic particle.

Let us consider the combination which takes place between the metal sodium and the non-metal chlorine to form sodium chloride (common salt). The sodium atom has a single electron in its outer shell and this transfers to join the seven electrons in the outer shell of the chlorine atom. When this occurs each resultant particle is left with a complete octet in the outer shell. (The sodium particle now has the same electron structure as the noble gas neon, and the chlorine particle has the same electron structure as the noble gas argon.) The balance of electrical charges which existed between protons and electrons in the original atoms is, however, upset. Since the sodium atom has lost a negatively charged particle (an electron), the remaining sodium particle must now possess a resultant positive charge. Meanwhile the chlorine atom has gained this electron so the resultant chlorine particle must carry a negative charge. These charged particles, derived from atoms in this manner, are called *ions*. In terms of symbols the sodium ion is written thus, Na^+, and the chlorine ion, Cl^-.

1.72   Since sodium ions and chlorine ions are oppositely charged they

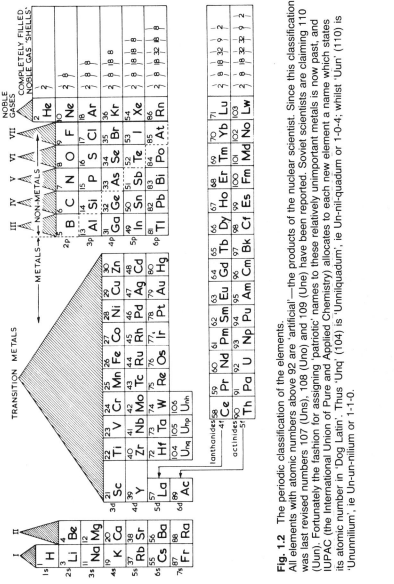

Fig. 1.2 The periodic classification of the elements.

All elements with atomic numbers above 92 are 'artificial'—the products of the nuclear scientist. Since this classification was last revised numbers 107 (Uns), 108 (Uno) and 109 (Une) have been reported. Soviet scientists are claiming 110 (Uun). Fortunately the fashion for assigning 'patriotic' names to these relatively unimportant metals is now past, and IUPAC (the International Union of Pure and Applied Chemistry) allocates to each new element a name which states its atomic number in 'Dog Latin'. Thus 'Unq' (104) is 'Unniliquadium', ie Un-nil-quadum or 1-0-4; whilst 'Uun' (110) is 'Ununnilium', ie Un-un-nilium or 1-1-0.

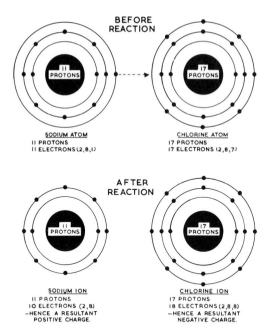

Fig. 1.3 The formation of the electro-valent bond in sodium chloride, by the transfer of an electron from the sodium atom to the chlorine atom.

will attract each other and the salt sodium chloride crystallises in a simple cubic form in which sodium ions and chlorine ions arrange themselves in the manner indicated in Fig. 1.4. Except for the force of attraction which exists between oppositely charged particles, no other 'bond' exists between sodium ions and chlorine ions, and when a crystal of sodium chloride is dissolved in water separate sodium and chlorine ions are released and can move as separate particles in solution. Such a solution is known as an electrolyte because it will conduct electricity. If we place two electrodes into such a solution and connect them to a direct-current supply, the positively charged sodium ions will travel to the negative electrode and the negatively charged chlorine ions will travel to the positive electrode. The applied EMF does not 'split up' the sodium chloride—the latter ionises as soon as it dissolves in water.

1.73   Thus, the unit in solid sodium chloride is the crystal, whilst in solution separate ions of sodium and chlorine exist. In reality there is no sodium chloride molecule and it is therefore incorrect to express the salt as 'NaCl'. Busy chemists are, however, in the habit of using symbols in this manner as a type of chemical shorthand. The author has in fact been guilty of this indiscretion earlier in this chapter when discussing formulae and equations in which electro-valent compounds are involved. For example, the equation representing the reaction between hydrochloric acid and caustic soda (1.51) would more correctly be written:

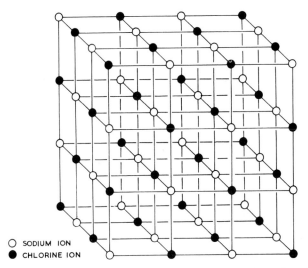

Fig. 1.4   A simple cubic crystal lattice such as exists in solid sodium chloride.

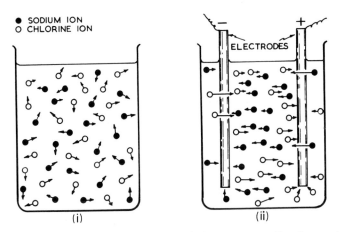

Fig. 1.5   (i) A solution of sodium chloride in which the separate sodium ions and chlorine ions are moving independently within the solution. Note that ionisation of the salt has taken place on solution and does not depend upon the passage of an electric current. (ii) When EMF is applied to the solution the charged ions are attracted to the appropriate electrode.

$$\underbrace{H^+ + Cl^-}_{\substack{\text{Hydrochloric}\\\text{Acid}}} + \underbrace{Na^+ + OH^-}_{\substack{\text{Sodium}\\\text{hydroxide}}} = \underbrace{Na^+ + Cl^-}_{\substack{\text{Sodium}\\\text{chloride}}} + H_2O$$

1.74   It was suggested earlier in this chapter (1.40) that the term 'oxidation' had a wider meaning in chemistry than the combination of an element with oxygen. Thus, when metallic iron combines with the gas chlorine to form iron(*II*) chloride, $FeCl_2$, the iron is said to have been oxidised, whilst when the iron(II) chloride so produced combines with still more chlorine to form iron(*III*) chloride, $FeCl_3$, the iron(II) chloride in turn has been oxidised. At each stage the 'electronegative portion' of the substance has increased. We can now relate this process of oxidation to a transfer of electrons. In being oxidised, an atom of iron has *lost* electrons to become an ion—first the iron(*II*) ion, Fe^{++}, at which stage it has lost two electrons and then the iron(*III*) ion, Fe^{+++}, when it has lost three electrons to the chlorine atoms:

$$Fe + Cl_2 \rightarrow Fe^{++} + 2Cl^- \xrightarrow{Cl} Fe^{+++} + 3Cl^-$$

iron(II) chloride   iron(III) chloride

Thus oxidation of a substance, in this case iron, involves a loss of electrons by atoms of that substance.

Conversely, reduction involves a *gain* of electrons. For example, when iron(III) oxide, Fe_2O_3, is reduced to metallic iron in the blast furnace the Fe^{+++} ion receives electrons and becomes an atom of iron:

$$Fe^{+++} + 3e^- \rightarrow Fe$$

1.75 Covalent Combination   In this type of chemical combination there is no 'loss' of electrons from one atom to another. Instead a certain number of electrons are 'shared' between two or more atoms to produce a stable particle which we call a *molecule*. In a molecule of the gas methane, four hydrogen atoms are combined with one carbon atom. The carbon atom has four electrons in its outer shell, but these are joined by four more electrons, contributed singly by each of the four hydrogen atoms (Fig. 1.6). Thus the octet of the carbon atom is completed and at the same time, by sharing one of the carbon atom's electrons, each hydrogen atom is able to complete its 'helium duplet'. This sharing of electrons by two atoms binds them together, and a molecule is formed in which atoms are held together by strong valence bonds. Each shared electron now passes from an orbital controlled by one nucleus into an orbital controlled by two nuclei and it is this control which constitutes the covalent bond. Chemists express the structural formula for the methane molecule thus:

$$\begin{array}{c} H \\ | \\ H-C-H \\ | \\ H \end{array}$$

Each co-valent bond is indicated so:—. Co-valent compounds, since they do not ionise, will not conduct electricity and are therefore non-electrolytes. They include many of the organic* compounds, such as benzene, alcohol, turpentine, chloroform and members of the 'alkane' series.

As the molecule size of co-valent compounds increases, so the bond

---

* Those compounds associated with animal and vegetable life and containing mainly the elements carbon, hydrogen and oxygen.

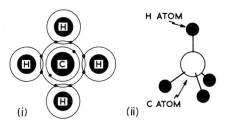

Fig. 1.6 (i) Co-valent bonding in a molecule of methane, CH_4. (ii) Spatial arrangement of atoms in the methane molecule.

strength of the material increases, as indicated in complex compounds such as rubber and vegetable fibres. Sometimes simple molecules of co-valent compounds can be made to unite with molecules of their own type, forming large chain-type molecules in which the bond strength is very high. This process is called 'polymerisation'. For example, the gas ethene C_2H_4 can be made to polymerise forming polythene:

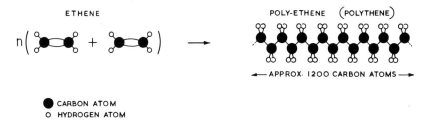

As the molecular chain increases in length, the strength also increases. Nylon (synthesised from benzene) and polychloroethene* (synthesised from ethine*) are both 'super polymers', the strength of which depends upon a long chain of carbon atoms co-valently bonded.

Forces of attraction (1.80), acting between points where these 'chain molecules' touch each other, hold the mass of them together. If such a substance is heated, the forces acting between the molecules are reduced and, under stress, the fibrous molecules will gradually slide over each other into new permanent positions. The substance is then said to be *thermoplastic*. Substances which (like water) are built up of simple molecules generally melt at a sharply defined single temperature since there is no entanglement, as exists with the ungainly molecules of the super polymers, and when the forces of attraction between simple molecules fall below a certain limit they can separate instantaneously. Super polymers, then, soften progressively as the temperature is increased rather than melt at a well-defined temperature.

Some polymers, when heated, undergo a chemical change which firmly anchors the chain molecules to each other by means of co-valent bonds.

---

* Formerly 'polyvinyl chloride' (PVC) and 'acetylene' respectively.

On cooling, the material is rigid and is said to be *thermosetting*. Bakelite is such a substance.

Rubber possesses elasticity (4.11) by virtue of the folded nature of its long-chain molecules. When stressed, one of these molecules will extend after the fashion of a spiral spring and, when the stress is removed, it will return to its original shape. In *raw* rubber the tensile stress will also cause the chain molecules to slide relative to each other, so that, when the stress is removed, some permanent deformation will remain in the material (Fig. 1.7). If the raw rubber is mixed with sulphur and heated it becomes 'vulcanised'. That is, sulphur causes the formation of co-valent links between the large rubber molecules which are thus held firmly together. Consequently, vulcanised rubber possesses elasticity due to the behaviour of its folded chain molecules, but it resists permanent deformation, since these molecules are no longer able to slide over each other into new positions.

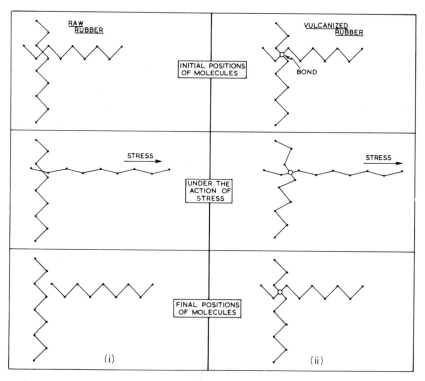

Fig. 1.7. Rubber molecules are of the long-chain type.
Due to their 'folded' form they become extended in tension, but return to their original shapes when the stress is removed. In raw rubber (i) a steady tensile force will cause separate molecules to glide slowly past each other into new positions, so that when the force is removed some plastic deformation remains, although the elastic deformation has disappeared. By 'vulcanising' the raw rubber (ii) the chain molecules are bonded together so that no permanent plastic deformation can occur and only elastic deformation is possible.

1.76 The Metallic Bond In most pure metals, atoms possess insufficient valence electrons to be able to form covalent bonds with each other. On the other hand any metallic ions which may be formed in a pure metal will carry like positive charges and so tend to repel each other so that electrovalent bonding is impossible. Yet we know that metals are crystalline in the solid state (3.10). How then is this situation achieved?

The explanation generally offered is that the valence electrons of each atom are donated to a common 'cloud' which is shared by *all* atoms present (Fig. 1.8). Thus, whilst the positively-charged ions which result, repel each other so that they arrange themselves in a regular pattern, they are held in these equilibrium positions by their mutual attraction for the negatively charged electron cloud which permeates them. Individual electrons no longer 'belong' to any particular atom but are the common property of all atoms present.

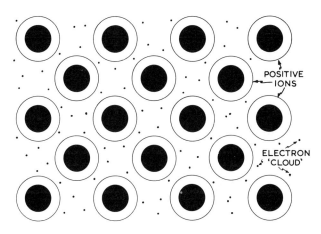

Fig. 1.8 Diagrammatic representation of the metallic bond.

A more detailed knowledge of the structure of the atom would indicate that the situation is not nearly so simple that metallic bonding can be explained in terms of this 'electron cloud' concept. However, for our present purposes it will be sufficient if we accept the results of this simplified interpretation, since it enables us to explain many characteristically metallic properties.

Since valence electrons in the common 'cloud' are able to travel freely among the positive ions this gives an explanation of the high electrical conductivity of metals, a current of electricity being nothing more than a movement of electrons in a particular direction. In a covalently-bonded compound on the other hand valence electrons are held captive in the chemical bond. Consequently most of the organic compounds—polythene, PVC and nylon—are insulators whilst liquids such as alcohols, benzene and oils are non-electrolytes.

The opaque lustre of metals is due to the reflection of light by free electrons. A light wave striking the surface of a metal causes a free electron to vibrate and absorb all the energy of the wave, thus stopping transmission. The vibrating electron then re-emits the wave from the metal surface giving rise to what we term 'reflection'.

The very important property of most metals in being able to undergo considerable plastic deformation is also due to the existence of the metallic bond. Under the action of shearing forces, layers of positive ions can be made to slide over each other without drastically altering their relationship with the shared electron cloud.

Secondary Bonding Forces

1.80 Stable atoms—and molecules—always contain equal numbers of protons and electrons. Consequently they will be electrically neutral and carry no resultant charge. Yet they must attract each other for how else can we explain the fact that all gases condense to form liquids and that all liquids either crystallise to form solids or else become so viscous that the strong *forces of attraction* between molecules make them behave almost like solids?

In short, whilst we have related the cohesion between metallic particles to the metallic bond we have not yet attempted to explain the forces which hold together covalently bonded molecules, or, for that matter, the single atoms in noble gases which, although they contain no valence electrons, must attract each other in some way since they ultimately liquify and solidify at very low temperatures. These weak secondary bonding forces are often referred to as van der Waals' forces since it was this Dutch physicist who first explained the deviations in the Gas Law (PV = RT) as being partly due to forces of attraction between molecules (or atoms in the case of noble gases) within the gas.

Consider two atoms of a noble gas such as argon. If these two atoms are in close proximity and their electrons happen to be concentrated as in Fig. 1.9, it is reasonable to suppose that mutual attraction will occur between the positively charged nucleus of atom X and the negatively charged electrons of atom Y at the moment when the nucleus of X is 'unshielded' by its own electrons. This situation will be continually changing as the distribution of electrons alters, but a resultant weak force of attraction exists.

Engineers will be familiar with the idea of 'centre of mass' in a solid body. In a similar way we can imagine electrical charges 'resolved' to give 'centres' of positive and negative charges respectively in a molecule. If

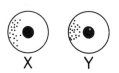

Fig. 1.9.

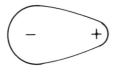

Fig. 1.10   A molecule with a resultant dipole moment.

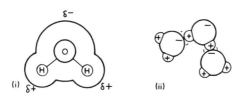

Fig. 1.11   The strong dipole moment of the water molecule (i) resulting in strong attraction between neighbouring molecules.

these centres do not coincide (Fig. 1.10) then the molecule will have a small *dipole moment* and will consequently attract (and be attracted by) other molecules with similar dipole moments.

In a molecule of water (Fig. 1.11 (i)) the two electrons which are contributed to the covalent bonds by the two hydrogen atoms tend to be drawn to the vicinity of the larger oxygen atom. Thus the centre of negative charge is shifted nearer to the oxygen atom leaving the positively-charged nuclei of the hydrogen atoms relatively 'exposed'. Consequently the water molecule has a very strong dipole moment and therefore a relatively strong force of attraction for its neighbours (Fig. 1.11 (ii)).

For this reason water has an abnormally high freezing point and boiling point as compared with other substances of similar molecular size. For example, methane melts at −183°C and boils at −162°C.

Particularly strong van der Waals' forces arise from the behaviour of the hydrogen atom in this way and are referred to as 'hydrogen bonds'.

When considered singly van der Waals' forces are very weak when compared with the forces acting within a single covalent bond. The combined effect, however, of van der Waals' forces acting at a large number of points between two adjacent chain-like polymer molecules such as those of polythene (1.75) can be very considerable. It also explains why polymer materials, though weaker than metals are highly plastic.

Isotopes

1.90   In the foregoing discussion of the mechanism of chemical combination no mention was made of the part played by the neutron. In fact the neutron, carrying no resultant electrical charge, has no apparent effect on ordinary chemical properties which are mainly a function of the *electron* structure of the atom. The principal role of the neutron is to increase the actual mass of an atom. Thus, the sodium atom with 12 neutrons in the nucleus, in addition to the 11 protons already mentioned, has a total nuclear mass of 23. The 11 electrons present are negligible in mass when compared with the massive protons and neutrons, so that the mass of the total atom is approximately 23 units, and it is from this value that the relative atomic mass is derived.

There are many instances, however, in which two atoms contain the same number of protons but unequal numbers of neutrons. Clearly, since they have equal numbers of protons, they will also have equal numbers of electrons and, chemically, such atoms will be identical. Differing numbers of neutrons, however, in respective atoms will cause these atoms to have unequal masses. An element possessing atoms which are chemically identical but which are of different mass is said to be *isotopic* and the different groups are known as *isotopes*.

1.91   Two such isotopes occur in the element chlorine. The chemical properties of these isotopes are identical because in each case an atom will contain 17 protons and 17 electrons. Only the relative masses of each atom will be different, since the nucleus of isotope II contains two more neutrons than that of isotope I (Fig. 1.12). Since there are about three times as many atoms of isotope I (usually written ^{35}Cl) as of isotope II (written ^{37}Cl) the relative atomic mass of chlorine 'averages out' at 35.45.

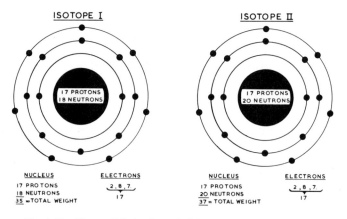

Fig. 1.12   The particle 'make-up' of the two isotopes of chlorine.

1.92   There are three isotopes of hydrogen of atomic masses one, two and three respectively (Fig. 1.13). The 'ordinary' hydrogen atom (now called 'protium') contains a single proton in its nucleus—and no neutrons. Its atomic mass is therefore one. However, approximately one hydrogen atom in every 6900 also contains a neutron in its nucleus and since the proton and neutron are roughly equal in mass then the atomic mass of the

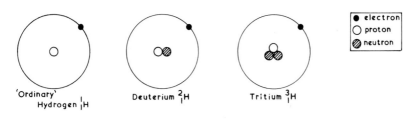

Fig. 1.13   The three isotopes of hydrogen.
Each isotope has the *same* electron/proton make-up so that the simple chemical properties of all three will be similar.

atom will be two. This atom, often known as 'heavy hydrogen', ^2_1H, is also given the special name *deuterium* and is the basis of the 'hydrogen bomb reaction'. A third, radioactive, isotope of hydrogen, called *tritium*, contains two neutrons along with the proton in the nucleus and so has an atomic mass of three, ie ^3_1H. It must be emphasised that deuterium and tritium are *not* different elements from hydrogen but only isotopes of that element. It is perhaps fortunate that separate isotopes of other elements are not given special names in this manner or confusion would be rife. The term 'isotope' tends to be associated in modern technology with the release of nuclear energy by suitable elements. The properties of the isotopes of uranium in this connection are dealt with later (18.75).

Exercises

1. If the valence of aluminium is three, write down the chemical formula of its oxide (1.31)
2. What mass of metallic copper would be deposited by the electrolysis of 10g of copper sulphate ($CuSO_4$) in water solution? (1.33)
3. Calculate the mass of iron obtained by reducing 10 tonnes of the ore hematite, assuming that hematite as mined contains 70% of the oxide Fe_2O_3. (1.33)
4. Why must aluminium be obtained from its ore by electrolysis instead of by the more usual process of reduction by carbon? (1.44)
5. In the chemical reaction: $Fe_3O_4 + C = 3FeO + CO$, has the Fe_3O_4 been oxidised or reduced? (1.44)
6. Complete (and balance) the chemical equation:

   HNO_3 + MgO = ?   (1.51)
   Nitric   Magnesium
   acid   oxide

7. The melting points of Li, Na, K and Rb are 179, 97, 62 and 39°C respectively. Estimate graphically the melting point of Cs. (1.69 and Fig. 1.2)
8. Without reference to any tables sketch (i) the electron structure of the element which has an atomic number of 14 (1.67); (ii) the electron structure of the magnesium ion. (1.71)
9. Show how far modern theory is successful in explaining not only many of the

mechanical properties of a metal but also that it is a conductor of electricity. (1.76).

10. Explain why water has abnormally high freezing point and boiling point as compared with substances of similar molecular mass such as ammonia. (1.80)

11. The metal copper (relative atomic mass—63.55) exists as two isotopes. 69.2% by mass of the metal consists of the isotope with a mass number 63. What is the likely mass number of the other isotope? (1.91)

Bibliography

Brown, G. I., *A New Guide to Modern Valency Theory*, Longman, 1971.

Cooper, D. G., *Chemical Periodicity*, John Murray, 1974.

Companion, A. L., *Chemical Bonding*, McGraw-Hill, 1979.

Cox, P. A., *The Elements, their origin, abundance and distribution*, Oxford Science Publications, 1989.

Hume-Rothery, W. and Coles, B. R., *Atomic Theory for Students of Metallurgy*, Institute of Metals, 1969.

Underwood, D. M. and Webster, D. E., *Chemistry*, Edward Arnold, 1985.

Wilson, J. G. and Newall, A. B., *General and Inorganic Chemistry*, Cambridge University Press, 1971.

2

The Physical and Mechanical Properties of Metals and Alloys

2.10 Of well over one hundred elements—if we include the increasing number of man-made ones—only eighteen have definite non-metallic properties. Six are usually classed as 'metalloids'—elements like silicon, germanium and arsenic—in which physical and chemical properties are generally intermediate between those of metals and non-metals, but the remainder have clearly defined metallic properties. Metals are generally characterised by their lustrous, opaque appearance and, in respect of other physical properties, metals and non-metals contrast strongly.

As we have seen (1.76) a metal consists of an orderly array of ions surrounded by and held together by a cloud of electrons. This is reflected in many of the physical properties of metals.

2.11 Melting point All metals (except mercury) are solids at ambient temperatures and have relatively high melting points (see Table 1.1) which vary between 234K ($-39°C$) for mercury and 3683K (3410°C) for tungsten. Non-metals include gases, a liquid (bromine) and solids. Their melting points vary much more widely: between 1K ($-272°C$) for helium and approximately 5300K (5000°C) for carbon.

2.12 Density The *relative density* (formerly *specific gravity*) of a material is defined as

$$\frac{\text{the weight of a given volume of the material}}{\text{the weight of an equal volume of water.}}$$

Metals generally have higher relative densities (Table 1.1) than non-metals. Values vary between lithium (0.534) which will float in water and osmium (22.5) which is almost twice the density of lead, which suggests that the simile 'as heavy as lead' needs revision.

2.13 Electrical conductivity   Non-metals are generally very poor conductors of electricity, indeed those where the bonding is entirely covalent will be insulators since all valance electrons are held captive in individual bonds and can move only in restricted orbits. By comparison in metals electrical conductivity arises from the presence of a sea or cloud of mobile electrons permeating the static array of ions. The electrons are able to flow through the ion framework when a potential difference is applied across the ends of the metal—which may be many miles apart as in the electric grid system. As indicated in Table 2.1 the electrical and thermal conductivities of metals follow roughly the same order. This is to be expected since both the flow of electricity and heat depend upon the ability of electrons to move freely within the metallic structure. For purposes of simple comparison Table 2.1 relates electrical and thermal conductivities of some important metals to those of silver (100). Although silver is marginally superior in terms of electrical conductivity to copper, the latter is used industrially because of relative costs. In fact for power transmission through the national grid aluminium lines are generally used for reasons given later (17.13). Electrical conductivity is reduced by alloying and the presence of impurities (16.21) as well as by mechanical straining.

Table 2.1   *Relative electrical and thermal conductivities of some metals*

Metal	Relative electrical conductivity	Relative thermal conductivity
Silver	100	100
Copper	96	94
Gold	69.5	70
Aluminium	59	57
Magnesium	41	40
Beryllium	40	40
Tungsten	29	39
Zinc	27	26.5
Cadmium	22	22
Nickel	(23)	21
Iron	16	17
Platinum	15	17
Tin	12.5	15.5
Lead	7.7	8.2
Titanium	2.9	4.1
Mercury	1.6	2.2

Electrical conductivity is measured in units Sm^{-1}, where the unit of conductance the *Siemen* (S), is equivalent to Ω^{-1}. Generally it is more convenient to consider the electrical *resistivity* (ϱ) of a material which is the inverse of its conductance and is of course measured in Ωm. Resistivity varies with temperature and over a limited temperature range a linear relationship of the form:

$$R_t = R(1 + \varrho t)$$

holds good. Here R_t is the resistance at the upper temperature, R the initial resistance and t the increase in temperature. ϱ is the temperature coefficient of resistance of the material.

2.14 Thermal conductivity   This arises in a similar way to electrical conductivity. Electrons pick up kinetic energy from the increased vibrations of the ions where the metal is hot. They pass rapidly through the ion framework where they collide with distant ions, causing them in turn to vibrate more rapidly. In this manner electrons behave as transporters of energy.

Metals are very good conductors of heat whereas most non-metals are not. The flow of heat in a conductor is governed by:

$$Q = -\lambda \frac{\partial T^*}{\partial x}$$

where Q is the heat flow across unit area, λ is the coefficient of thermal conductivity and T the temperature. $\frac{\partial T}{\partial x}$ will be the 'temperature gradient' at that unit area. Thermal conductivity is measured in units, $Wm^{-1}K^{-1}$.

2.15 Specific heat capacity   The specific heat capacity (C_p, C_v) of a substance is the quantity of heat required to raise the temperature of 1kg of the substance 1K. The units are $Jkg^{-1}K^{-1}$. The specific heat capacities of metals are low compared with those of non-metals so that it is less expensive to raise their temperatures.

Dulong and Petit's Law states that for all elements the product of the specific heat capacity and the atomic weight is approximately constant and this product is called the *Atomic heat*. This law was used more than a century ago to assess the atomic weights of many (then) new elements. It involves the relationship between heat capacity and the vibrational energy of atoms.

Table 2.2   *Thermal properties of some important metals*

Metal	Coefficient of thermal expansion ($\propto$) ($K^{-1} \times 10^{-6}$)	Specific heat capacity ($J\ kg^{-1}K^{-1}$)
Aluminium	23	913
Copper	17	385
Gold	14	132
Iron	12	480
Lead	29	126
Magnesium	25	1034
Nickel	13	460
Silver	19	235
Tin	23	226
Titanium	9	523
Zinc	31	385

2.16 Thermal expansion   As materials are heated the amplitude of atomic vibrations increases and this is evident as an increase in volume. The coefficient of cubic expansion (γ) is the increase in volume per unit

---

* This is basically similar to Ohm's Law governing the flow of electricity (electrons) through a conductor.

volume per unit rise in temperature (unit, K^{-1}). Similarly the coefficient of linear expansion (α) is the increase in length per unit length per unit rise

in temperature (unit, K^{-1}) or $\quad \alpha = \dfrac{l_t - l_o}{l_o t}$

where l_o = original length; l_t = length after a rise in temperature of t.

By suitable alloying additions to some metals it is possible to reduce α to low limits. Thus *Invar* (13.25) is used in long measuring tapes, pendulum rods for observatory clocks (before the days of electronic timekeeping), etc., whilst similar alloys are used in the delicate sliding mechanisms of instruments used under conditions of widely varying temperature, eg military rangefinders used in desert warfare. Further alloys are also used in bimetallic strips in small thermostats where the differential expansion of the two alloys of the strip leads to bending of the unit and a make/break contact.

2.17 Behaviour to light   Most metals reflect all wavelengths of light equally well for which reason they are white or nearly so. Notable exceptions are copper and gold whilst zinc is very faintly blue and lead slightly purple.

The reflecting capacity of metals is yet another aspect of the *mobility* of its electrons; an incident light wave causes the electrons near the surface of a metal to oscillate and as a result the incident wave is reflected back instead of being absorbed by the metal. Thus the reflection in a mirror is due to the oscillation caused in silver's mobile electron cloud.

2.18 Behaviour to short-wavelength radiations   Metals are transparent to γ-rays (2.93) and to those X-rays of short wavelength ('hard' X-rays) (2.91).

2.19 Magnetic properties   Most metals are magnetic to some slight extent but only in the metals iron, nickel, cobalt and gadolinium is magnetism strong enough to be of practical interest. The pronounced magnetism of this group is called 'ferromagnetism' (14.30).

Whilst many of these physical properties such as conductivity, magnetism and melting point dictate special uses for metals, it is mechanical properties such as strength, ductility and toughness which concern us principally in engineering design.

Fundamental Mechanical Properties

2.20   Whereas the *directional* nature of the covalent bond results in the extreme rigidity of substances like diamond and quartz, the *non-directional* nature of the metallic bond makes it relatively easy to bend a piece of metal. Moving groups of metallic ions through the electron 'sea' can be achieved in a number of ways such as hammering, rolling, stretching and bending. Fundamental mechanical properties of metals are related to the

amounts of deformation which metals can withstand under different circumstances of force application. *Ductility* refers to the capacity of a substance to undergo deformation under tension without rupture, as in wire-
or tube-drawing operations. *Malleability*, on the other hand, is the capacity
of a substance to withstand deformation under compression without rupture, as in forging or rolling. Substances which are highly ductile are also
highly malleable but the reverse may not be true since some extremely
malleable substances are weak in tension and therefore liable to tear.
Moreover, whilst malleability is usually increased by raising the temperature (for which reason metals and alloys are often hot-forged or hot-rolled),
ductility is generally reduced by heating, since strength is also reduced.

2.21 Toughness refers to a metal's ability to withstand bending or the
application of shear stresses without fracture. Hence, copper is extremely
tough, whilst cast iron is not. Toughness should not, therefore, be confused
with either strength or hardness, properties which will be discussed later.

2.22 Since these fundamental mechanical properties of ductility, malleability and toughness cannot be expressed in simple quantitative terms, it
has become necessary to introduce certain mechanical tests which are
related to these properties and which will allow of comparative numerical
interpretation. Moreover, the engineer is more concerned with the *forces*
which cause deformation in metals rather than with the deformation itself.
Consequently tensile tests and hardness tests correlate the amounts of
deformation produced with given forces in tension and compression
respectively, whilst impact tests are an almost direct measurement of
toughness. Such precise measurements of force-deformation values make
it possible to draw up sets of specifications upon which the mechanical
engineer can base his design.

Tenacity or Tensile Strength

2.30 The tensile strength of a material is defined as the maximum force
required to fracture in tension a bar of unit cross-sectional area. In practice
a test-piece of known cross-sectional area is gripped in the jaws of a testing
machine and subjected to a tensile force which is increased by suitable
increments. For each increment of force the amount by which the length
of a pre-determined 'gauge length' on the test piece increases is measured
by some device. The test piece is extended in this way to destruction.

A force-extension diagram can then be plotted (Fig. 2.1). At first the rate
of extension is very small and such extension as there is is directly proportional to the applied force; that is, OQ is a straight line. If the applied
force is removed at any point before Q is reached the gauge-length will return
to its original dimensions. Thus the extension between O and Q is *elastic*
and the material obeys Hooke's Law, which states that, for an elastic
body, the strain produced is proportional to the stress applied. The value
$\frac{Stress}{Strain}$ is constant and is equivalent to the slope of OQ. This constant value

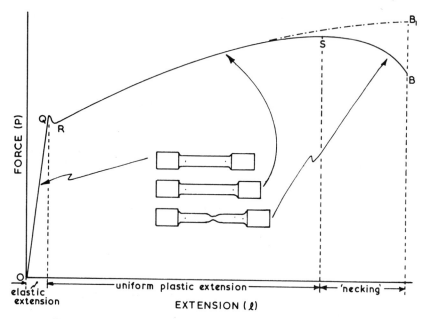

Fig. 2.1   The force-extension diagram for an annealed low-carbon steel.

is known as *Young's Modulus of Elasticity* (E) for the material. Consider a test piece of original length, L, and cross-sectional area, 'a', stretched elastically by an amount, 'l', under a force, P, acting along the axis of the specimen, then:

$$E = \frac{\text{Longitudinal Stress}}{\text{Longitudinal Strain}}$$

$$= \frac{P/a}{l/L}$$

$$= \frac{PL}{a\,l}$$

Young's Modulus is in fact a measure of the *stiffness* of the material in tension. This value and the stress range over which it applies are of great importance to the engineer. Young's modulus is measured in the same units as those of stress, since:

$$E = \frac{\text{Stress}}{\text{Strain}}$$

$$= \frac{[\text{Stress}]}{[\text{length}]/[\text{length}]}$$

$$= [\text{Stress}]$$

If at any point on the part of the curve under consideration the force is relaxed then the test piece will return to its original length, extension so far being entirely elastic.

2.31   If the test piece is stressed beyond the point Q the curve deviates from its straight-line characteristics. Q is therefore known as the *elastic limit* or *limit of proportionality* and if the force is increased beyond this point a stage is reached where a sudden extension takes place for no increase in the applied force (assuming that we are testing a specimen of annealed low-carbon steel as indicated in Fig. 2.1). An explanation of this phenomenon, known as the *yield point*, R, will be given later (8.61). If the force is now removed the elastic extension will disappear but a small permanent plastic extension or *permanent set* will remain.

As the force is increased beyond the point R the test piece stretches rapidly—first uniformly along its entire length and then locally to form a 'neck'. This 'necking' occurs just after the maximum force value has been reached at S, and since the cross-section decreases rapidly at the neck, the force at B required to break the test piece is *less* than the maximum force applied at S.

This might be an appropriate moment at which to mention the difference between a 'force/extension' diagram and a 'stress/strain' diagram since these terms are often loosely used by both metallurgists and engineers. Fig. 2.1 clearly represents a force/extension diagram since total force is plotted against total extension, and, as the force decreases past the point S, for reasons just mentioned, the decrease is indicated on the diagram. *Stress* however is measured as *force per unit area* of cross-section of the test-piece and if we wished to plot this we would need to measure the minimum diameter of the test piece at each increment of applied force. This would be particularly important for values of force after the point S, since from S onwards the effective cross-section is decreasing rapidly due to the formation of the neck. The test piece is only as strong as the force its minimum diameter will support.

If stress were calculated on this decreasing cross-section the resulting stress/strain diagram would follow a path indicated by the broken line to B_1 from S onwards. In practice, however, a *nominal* value of the *tensile strength* of a material is calculated using the maximum force (at S) and the *original* cross-sectional area of the test piece. Therefore:

$$\text{Tensile strength} = \frac{\text{Maximum force used}}{\text{Original area of cross-section}}$$

In this connection the term 'engineering stress' is often used; it implies the force at any stage of the loading cycle divided by the *original* area of cross-section of the material.

Although tensile strength is a useful guide to the mechanical properties of a material it is not of paramount importance in engineering design. After all, the engineer is not particularly interested in a material once it begins to stretch plastically—unless of course he is a production engineer engaged in deep-drawing or some other metal-forming process. In the case of structural or constructional engineering, the elastic limit, Q, will be of far greater significance than tensile strength.

2.32   The form of force/extension diagram described above is in fact a special case, obtained only for wrought irons and low-carbon steels in the soft condition (8.61). Most alloys, particularly if they have been heat-treated or cold-worked, show neither a definite elastic limit nor a yield point and give, on test, diagrams of the types shown in Figs. 2.2 and 2.3.

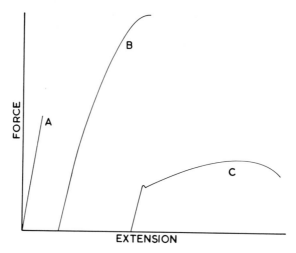

Fig. 2.2   The effects of heat-treatment on the force-extension diagram of carbon steel. (A) is in the quenched condition; (B) is quenched and tempered; and (C) represents the annealed condition.

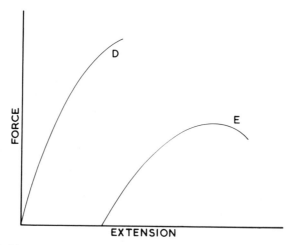

Fig. 2.3   Typical force-extension diagrams for a non-ferrous alloy, showing the absence of a well-defined yield point. (D) represents the cold-worked condition, and (E) the fully annealed condition.

Since the yield point is of greater importance to the engineer than the tensile strength itself, it becomes necessary to specify a stress which corresponds to a definite amount of permanent extension as a substitute for the yield point. This is commonly called the 'Proof Stress', and is derived as shown in Fig. 2.4. A line *BC* is drawn parallel to the line of proportionality, from a pre-determined point *B*. The stress corresponding to *C* will be the proof stress—in the case illustrated it will be known as the '0.1% proof stress', since *AB* has been made equal to 0.1% of the gauge length. The material will fulfil the specification therefore if, after the proof force is applied for fifteen seconds and removed, a permanent set of not more than 0.1% of the gauge length has been produced. Proof lengths are commonly 0.1 and 0.2% of the gauge length depending upon the type of alloy. The time limit of 15 seconds is specified in order to allow sufficient time for extension to be complete under the proof force.

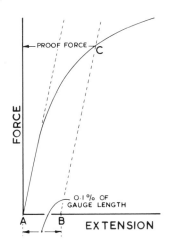

Fig. 2.4   Method used to obtain the 0.1% proof stress.

2.33   In addition to determining the tensile strength and the proof stress (or, alternatively, the yield stress), the percentage elongation of the test piece at fracture is also derived. This is an almost direct measure of ductility. The two ends of the broken test piece are fitted together (Fig. 2.5) so that the total extension can be measured.

In order that values of percentage elongation derived from test pieces of different diameter shall be comparable, test pieces should be geometrically similar, that is, there must be a standard relationship or ratio between cross-sectional area and gauge length. Test pieces which are geometrically similar and fulfil these requirements are known as *proportional* test pieces. They are commonly circular in cross-section. BSI lays down that, for proportional test pieces:

$$L_0 = 5.65 \sqrt{S_0}$$

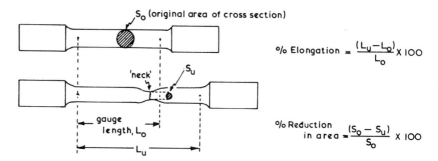

Fig. 2.5   The determination of percentage elongation and percentage reduction in area.

where L_0 is the gauge length and S_0 the original area of cross-section. This formula has been accepted by international agreement and SI units are used. For test pieces of circular cross-section it gives a value

$$L_0 \simeq 5d$$

where 'd' is the diameter at the gauge length. Thus a test piece 200 mm² in cross sectional area will have a diameter of 15.96 mm (16 mm) and hence a gauge length of 80 mm. Some old tensile testing machines may still be calibrated in 'tons force'. Since 10 kN ≡ 1.00361 tonf, dual value scales are not necessary, since, within the accuracy required, 1 tonf ≡ 10 kN.

2.34   The smallest diameter, S_0, of the neck is measured and from it the percentage reduction in area calculated (Fig. 2.5). Thus, from our complete set of observations we can derive the following values:

(a) Yield stress $= \dfrac{\text{Yield force}}{\text{Original area of cross-section}}$

$\left(\text{or Proof stress} = \dfrac{\text{Proof force}}{\text{Original area of cross-section}} \right)$

(b) Tensile strength $= \dfrac{\text{Maximum force}}{\text{Original area of cross-section}}$

(c) Percentage elongation $= \dfrac{\text{Increase in gauge length} \times 100}{\text{Original gauge length}}$

(d) Percentage reduction in area $=$

$$\dfrac{(\text{Original area of cross-section} - \text{Final area of cross section}) \times 100}{\text{Original area of cross-section}}$$

In terms of SI units stress is measured in N/m². However, since it is difficult to appreciate the very large force necessary to break a test piece

Plate 2.1 The Avery-Denison Servo-controlled Tensile Testing Machine, with an applied force capacity of 600kN.

The straining unit which is shown on the left embodies a double-acting hydraulic cylinder and ram. The force on the test piece is measured by load-sensitive cells and is indicated on the display panel of the control console shown on the right. The full load/extension cycle is electronically controlled and a permanent trace is produced. *(Courtesy of Messrs Avery-Denison Ltd, Leeds)*.

of one square metre in cross-section, most bodies, including BSI, quote tensile stress in metals in N/mm^2. This at least enables the student to relate the tensile strength of a steel to the force necessary to break in tension one of the thicker steel strings on his guitar.

2.35 Early tensile-testing machines were of the simple beam type in which the applied force was magnified by using a first-order lever system. With such machines an accurate evaluation of extension was possible up to the elastic limit by using a sensitive extensometer but beyond the maximum force, determinations of force/extension characteristics were impossible because the test piece fractured quickly as soon as necking began, there being no means of relaxing the applied force rapidly enough. Modern machines however are usually servo-hydraulically loaded (Plate 2.1) so that a complete force/extension relationship can be obtained. Since advanced computer control technology is now employed automatic calculation of proof stress, yield stress, ultimate tensile stress and percentage elongation are carried out; whilst software is available for cycling and data storage. Software programs can be written to meet other specific requirements. These machines can also be used for compression and transverse testing, and vary in size between large machines with a capacity of 1300 kN and small bench models having a capacity of only 20 kN.

2.36. In situations where a large amount of energy is being expended against gravity as in various types of aero-space travel—or even driving the humble 'tin Lizzie' up a hill—it becomes necessary to relate the tensile properties of a material to its relative density. Thus, what used to be called the 'strength-to-weight ratio' became important in the design of both land and air transport vehicles. In modern terminology this became 'specific strength'. Thus:

$$\text{Specific strength} = \frac{\text{Tensile strength of material}}{\text{Relative density of material}}$$

When *stiffness* is the prime consideration, however, Young's Modulus of Elasticity is a more appropriate guide to the required properties and a value termed *specific modulus* is now generally accepted as being relevant, ie—

$$\text{Specific modulus of elasticity} = \frac{\text{Young's modulus of elasticity}}{\text{Relative density}}$$

Hardness Tests

2.40 Classically, *hardness* could be defined as the resistance of a surface to *abrasion*, and early attempts to measure surface hardness were based on this concept. Thus in the Turner Sclerometer a loaded diamond point was drawn across the surface of the test piece and the load increased until a visible scratch was produced. In Moh's Scale—still used to evaluate the hardness of minerals—substances were arranged in order of hardness such that any material in the scale would *scratch* any material listed below it. Thus diamond (with a hardness index of 10) heads the list whilst talc (with an index of 1) is at the foot of the scale.

Whilst such methods undoubtedly reflect a true concept of the fundamental meaning of hardness, they have been abandoned in favour of methods which are capable of greater accuracy but in which the resistance of the surface layers to *plastic deformation* under static pressure is measured rather than true hardness. In most of these methods the static force used is divided by the numerical value of the surface area of the resulting impression to give the hardness index.

2.41 The Brinell Test, probably the best known of the hardness tests, was devised by a Swede, Dr. Johan August Brinell in 1900. In this test a hardened steel ball is pressed into the surface of the test piece using the appropriate specified force. The diameter of the impression so produced is then measured and the Brinell Hardness Number, H_B, derived from:

$$H_B = \frac{\text{Force, } P}{\text{Surface area of impression}}$$

It can be shown that the surface area of the impression is $\pi\dfrac{D}{2}\left(D - \sqrt{D^2 - d^2}\right)$ where D is the diameter of the ball and 'd' the diameter of the impression (Fig. 2.6). Since we are dealing with the actual area of the *curved surface* of the impression the derivation of the above expression is quite involved.

Hence,

$$H_B = \frac{P}{\pi\dfrac{D}{2}\left(D - \sqrt{D^2 - d^2}\right)}$$

and the units will be kgf/mm². To obviate tedious calculations H_B is found by reference to the appropriate set of tables.

2.42 It is obviously important that the stress produced by the indenter at the surface of the test piece shall suit the material being tested. If for example in testing a soft metal we use a force which is too -great relative to the diameter of the ball, we shall get an impression similar to that

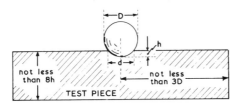

Fig. 2.6 The relationships between ball diameter, depth of impression and dimensions of the test piece in the Brinell-type test.

indicated in Fig. 2.7A. Here the ball has sunk to its full diameter and the result is obviously meaningless. The impression shown in Fig. 2.7B on the other hand would be obtained if the force were too small relative to the ball diameter and here the result would be likely to be very uncertain. For different materials then, the ratio P/D^2 has been standardised in order to obtain accurate and comparable results. P is measured in kgf and D in mm.

Material	Approximate H_B range	P/D^2 ratio used
Steel and cast iron	Over 100	30
Copper, copper alloys and aluminium alloys	30–200	10
Aluminium	15–100	5
Tin, lead and their alloys	3–20	1

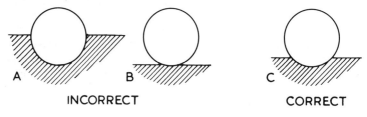

A — B   INCORRECT   C   CORRECT

Fig. 2.7   The influence of depth of impression on the accuracy of a Brinell determination.

Thus in testing a piece of steel we can use either a 10-mm ball in conjunction with a 3000 kgf load; a 5 mm ball with a 750 kgf load; or a 1 mm ball with a 30 kgf load. In the interests of accuracy it is always advisable to use the largest ball diameter that is possible. The limiting factors will be the width and thickness of the test piece, and the small ball would be used for thin specimens, since by using the large ball we would probably be, in effect, measuring the hardness of the table supporting our test piece. The thickness of the specimen should be at least eight times the depth, 'h', of the impression (Fig. 2.6). Similarly the width of the test piece must be adequate to support the applied force and it is recommended that the distance of the centre of the indentation from the edge of the test piece shall be at least three times the diameter of the indenting ball.

2.43 The Vickers Hardness Test—or Diamond Pyramid Hardness Test —uses as its indenter a diamond square-based pyramid (Fig. 2.8) which will give geometrically similar impressions under different applied forces.

This eliminates the necessity of deciding the correct P/D^2 ratio as is required in the Brinell test. Moreover, the diamond is more reliable for hard materials which have a hardness index of more than 500, since it does not deform under pressure to the same extent as a steel ball. Using the diamond point, however, does not eliminate the necessity of ensuring that

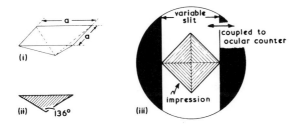

Fig. 2.8   The Diamond Pyramid Indentor and its resulting impression.

the thickness of the specimen is sufficient, relative to the depth of the impression.

In this test the diagonal length of the square impression is measured by means of a microscope which has a variable slit built into the eyepiece (Fig.2.8 (iii)). The width of the slit is adjusted so that its edges coincide with the corners of the impression and the relative diagonal length of the impression then obtained from a small instrument attached to the slit which works on the principle of a revolution counter. The ocular reading thus obtained is converted to Vickers Pyramid Hardness Number by reference to tables. The hardness index is related to the size of the impression in the same way as is the Brinell number.

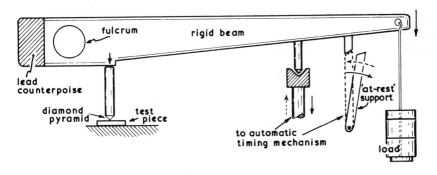

Fig. 2.9   Basic principles of the force application system in the Vickers Hardness Testing Machine.

2.44 The Rockwell Test   was devised in the USA, and is particularly suitable for rapid routine testing of finished material since it indicates the final result direct on a dial which is calibrated with a series of scales. A number of different combinations of indenter and indenting force can be used in conjunction with the appropriate scale:

Scale	Indenter	Total force (kgf)
A	Diamond cone	60
B	$^1/_{16}''$ steel ball	100
C	Diamond cone	150
D	Diamond cone	100
E	$^1/_8''$ steel ball	100
F	$^1/_{16}''$ steel ball	60
G	$^1/_{16}''$ steel ball	150
H	$^1/_8''$ steel ball	60
K	$^1/_8''$ steel ball	150

Of these, scale C is probably the most popular for use with steels.

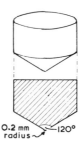

0.2 mm radius   120°

Fig. 2.10   The Rockwell Diamond Cone Indentor.

2.45 The Shore Scleroscope (Greek: 'skleros'—hard) tests the material very near to its surface. The instrument embodies a small diamond-tipped 'tup' which is allowed to fall from a standard height inside a graduated glass tube. The height of rebound is taken as the hardness index. Since the Shore Scleroscope is a small, portable instrument, it is very useful for the determination of hardness of large rolls, castings and gears, and other large components which could not easily be placed on the testing tables of any of the more orthodox testing machines.

The development of digital display units has enabled very small portable hardness testers of the indentation type to be manufactured. One of these consists of a small motorised probe which, when pressed against the surface of the test piece, makes a minute diamond impression using a force of only 8.4 N. Consequently such a test is virtually non-destructive and the instrument can be used in the most remote corners of the factory, hangar or repair yard. At the same time a high accuracy of ± 15 VPN is claimed over the hardness range of 50 to 995 VPN. Such instruments have largely replaced the Shore Scleroscope in terms both of accuracy and adaptability.

Table 2.3 gives representative hardness numbers, together with other mechanical properties, for some of the better-known metals and alloys.

Table 2.3 *Typical mechanical properties of some metals and alloys*

Metal or Alloy	Condition	0.1% Proof Strength (N/mm²)	Tensile Strength (N/mm²)	Specific Strength (N/mm²)	Young's Modulus (kN/mm²)	Specific Modulus (kN/mm²)	Elongation (%)	Hardness (Brinell)	Impact Value (Izod)(J)
Lead	Soft sheet	—	18	1.54	16	1.37	65	4	—
Aluminium	Wrought and annealed	—	60	21.8	70	25.9	60	15	27
Duralumin	Extruded and fully heat-treated	275	430	154	71	25.4	15	115	22
Magnesium-6Al/1Zn	Extruded bar	170	300	167	48	26.7	10	60	8
Copper	Wrought and annealed	46	216	24.1	130	14.5	60	42	59
70/30 Brass	Annealed	85	320	37.6			**68**	62	90
					100	11.7			
	Deep-drawn	370	465	54.6			19	132	—
Phosphor bronze (5% tin)	Rolled and annealed	120	340	38.1			66	72	—
					101	11.3			
	Hard-rolled	650	710	79.6			5	188	61
Mild steel	Hot-rolled sheet	270	400	50.8	210	26.7	28	100	75
0.45% carbon steel	Normalised	420	665	84.7			27	152	44
					200	25.4			
	Water-quenched and tempered at 600°C	540	780	99.4			25	200	65
4Ni/Cr/Mo steel	Air-hardened and tempered at 300°C	**1200**	**1550**	198	**225**	28.7	12	**444**	22
18/8 stainless steel	Softened	185	525	66.3	220	27.8	30	170	68
Grey cast iron	As cast	—	300	40.5	150	20.3	0	250	1
Titanium (commercially pure)	Annealed sheet	370	450	100	120	26.7	30	—	61
Titanium alloy (4Sn/4Al/4Mo 0.5Si)	Precipitation hardened	**1200**	1390	**309**	150	**33.3**	16	—	—

Bold type denotes maximum value in that property (where relevant).

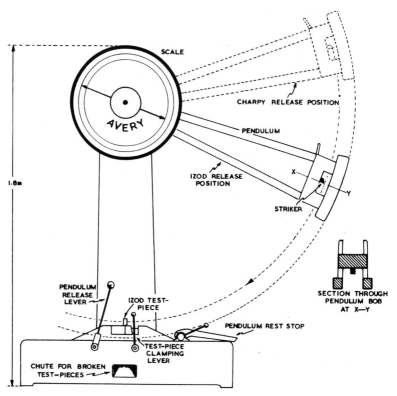

Fig. 2.11   The Avery-Denison Universal impact-testing machine.
This machine can be used for either Charpy or Izod impact tests. For Izod tests, the pendulum is released from the lower position, to give a striking energy of 170 J; and for the Charpy test it is released from the upper position, to give a striking energy of 300 J. (The scale carries a set of graduations for each test.) The machine can also be used for impact-tension tests.

Impact Tests

2.50   Impact tests indicate the behaviour of a material under conditions of mechanical shock and to some extent measure its toughness. Brittleness —and consequent lack of reliability—resulting from incorrect heat-treatment (13.42) or other causes may not be revealed during a tensile test but will usually be evident in an impact test.

2.51 The Izod Impact Test   In this test a standard notched specimen is held in a vice and a heavy pendulum, mounted on ball bearings, is allowed to strike the specimen after swinging from a fixed height. The striking energy of 167 J (120 ft lbf) is partially absorbed in breaking the specimen and, as the pendulum swings past, it carries a pointer to its highest point of swing, thus indicating the amount of energy used in fracturing the test piece.

2.52 The Charpy Test, developed originally on the Continent but now gaining favour in Britain, employs a test piece mounted as a simply-supported beam instead of in the cantilever form used in the Izod test (Fig. 2.12). The striking energy is 300 J (220 ft lbf).

2.53 To set up stress concentrations which ensure that fracture does occur, test pieces are notched. It is essential that notches always be standard, for which reason a standard gauge is used to test the dimensional accuracy of the notch. Fig. 2.12 shows standard notched test pieces for both the Izod and Charpy impact tests.

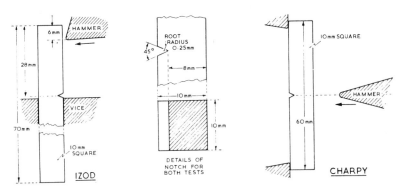

Fig. 2.12 Dimensions of standard test pieces for both Izod and Charpy tests. In the Izod test piece, notches 28mm apart may be cut in three different faces so that a more representative value is obtained.

2.54 The results obtained from impact tests are not always easy to interpret, and some metals which are ductile under steady loads behave as brittle materials in an impact test. As mentioned above, however, the impact test gives a good indication of how reliable the material is likely to be under conditions of mechanical shock. These tests are most likely to be specified for constructional steels of medium-carbon content.

Other Destructive Tests

2.60 These are often designed specifically for the measurement of some property peculiar to a single class of material or to assess the suitability of a material for a special purpose.

2.61 The Erichsen Cupping Test (11.54–Pt. II) is closely connected with the ductility of a material but is in fact designed to assess its deep-drawing properties.

2.62 Compression Tests are used to measure the capability of a cast iron to carry compressive loads. A standard test cylinder is tested in compression, usually employing a tensile testing machine running in 'reverse'.

2.63 Torsion Tests of various types are sometimes applied to materials in wire and rod form.

Non-destructive Tests

2.70 The mechanical tests already mentioned are of a destructive nature and are subject to the availability of separate test pieces which are reasonably representative of the production material. Thus, wrought products such as rolled strip, extruded rod and drawn wire are generally uniform in mechanical properties throughout a large batch and can be sampled with confidence. Parts which are produced individually, however, such as castings and welded joints, may vary widely in quality purely because they are made individually and under the influence of many variable factors. If the quality of such components is important and the expense justified—as in the case of aircraft castings—it may be necessary to test each component individually by some form of non-destructive test. Such tests seek to detect faults and flaws either at the surface or below it, and a number of suitable methods is available in each case.

The Detection of Surface Faults

2.80 It is often possible to detect and evaluate surface faults by simple visual examination with or without the use of a hand magnifier. The presence of fine hair-line cracks is less easy to detect by visual means and some aid is generally necessary. Such surface cracks may be associated with the heat-treatment of steel or, in a welded joint, with contraction during cooling.

2.81 Penetrant Methods   In these methods the surface to be examined is cleaned and then dried. A penetrant fluid is then sprayed or swabbed on the surface which should be warmed to about 90°C. After sufficient time has elapsed for the penetrant to fill any fissures which may be present the excess is flushed from the surface with warm water (the surface tension of the water is too high to allow it to enter the narrow fissure). The test surface is then carefully dried, coated with fine powdered chalk and set aside for some time. As the coated surface cools, it contracts and penetrant tends to be squeezed out of any cracks, so that the chalk layer becomes stained, thus revealing the presence of the cracks. Most penetrants of this type contain a scarlet dye which renders the stain immediately noticeable. Aluminium alloy castings are often examined in this way.

Penetrants containing a compound which fluoresces under the action of ultra-violet light may also be used. This renders the use of messy chalk unnecessary. When the prepared surface is illuminated by ultra-violet light, the cracks containing the penetrant are revealed as bright lines on a dark background. Penetrant methods are particularly useful for the examination of non-ferrous metals and austenitic (non-magnetic) steels.

2.82 Magnetic Dust Methods   consist in laying the steel component across the arms of a magnetising machine and then sprinkling it with a special magnetic powder. The excess powder is blown away, and any cracks or defects are then revealed by a bunch of powder sticking to the area on

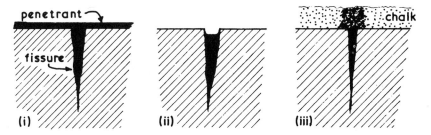

Fig. 2.13 The penetrant method of crack detection.
(i) The cleaned surface is coated with penetrant which seeps into any cracks present. (ii) Excess penetrant is removed from the surface. (iii) The surface is coated with chalk. As the metallic surface cools and contracts, penetrant is expelled from the crack to stain the chalk.

each side of the crack. Since the crack lies across the magnetic field, lines of force will become widely separated at the air gap (Fig. 2.14) and magnetic particles will align themselves along the lines of force.

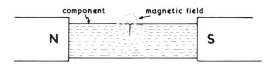

Fig. 2.14 The principles of magnetic particle crack detection.

The Detection of Internal Defects

2.90   Internal cavities in the form of blow holes or shrinkage porosity may be present in castings of all types, whilst wrought materials may contain slag inclusion and other subcutaneous flaws. Welded joints, by the nature of their production methods, may contain any of these defects. Since metals are opaque to light, other forms of electromagnetic radiation of shorter wavelength (X-rays and γ-rays) must be used to penetrate metals and so reveal such internal discontinuities. Although the railway wheeltapper was, for some obscure reason, always 'good for a laugh' at the mercy of the professional comedian, he was in fact using a 'sonic' method of testing the continuity of structure of the wheel and modern sophisticated methods of ultra-sonic testing use similar principles.

2.91 X-ray Methods   X-rays are used widely in metallurgical research in order to investigate the nature of crystal structures in metals and alloys. Their use is not confined to the research laboratory, however, and many firms use X-rays in much the same way as they are used in medical radiography, that is, for the detection of cavities, flaws and other discontinuities in castings, welded joints and the like.

X-rays are produced when a stream of high-velocity electrons impinges on a metal target. Fig. 2.15 illustrates the principle of an X-ray tube in which a filament supplies free electrons. Being negatively charged these electrons are accelerated away from the cathode towards the anode (sometimes called the 'anti-cathode') by the high potential difference between the electrodes. Collision with the anode produces X-rays. The containing tube is under vacuum, as the presence of gas molecules would obstruct the passage of relatively small electrons. Nevertheless only about 1% of the energy expended produces X-rays the remainder being converted to heat. Consequently the anode must be water-cooled. For greater output of X-rays (above 1 MeV) other types of generator such as the 'linear generator' or 'Linac' have to be used.

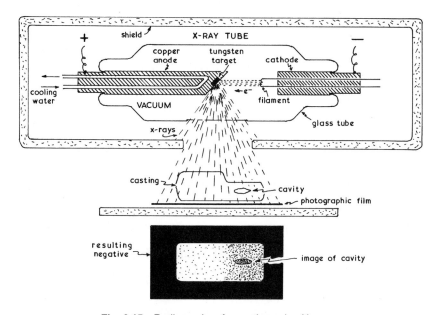

Fig. 2.15   Radiography of a casting using X-rays.

The penetrating power of electromagnetic radiation generally, depends upon its frequency. Thus radiation at the ultra-violet end of the visible spectrum will penetrate our skin to a depth of less than 1 mm but will nevertheless produce painful radiation burns (and possibly skin cancer) if we sunbathe carelessly. X-rays, having a much higher frequency than UV light, will penetrate more deeply and the 'harder' the X-rays (ie the higher the frequency) the greater the depth of penetration. X-rays used in metallurgical radiography are harder than those used in medicine, and are better able to penetrate metals. At the same time their properties make them much more dangerous to human body tissue, and plant producing radiation of this type needs to be carefully shielded in order to prevent the escape of stray radiations which would seriously impair the health of operators.

Like light, X-rays travel in straight lines, but whilst metals are opaque to light they are moderately transparent to X-rays, particularly those of high frequency. Fig. 2.15 illustrates the principle of radiography. A casting is interposed between a shielded source of X-rays and a photographic film. Some of the radiation will be absorbed by the metal so that the density of the photographic image will vary with the thickness of the metal through which the rays have passed.

2.92 X-rays are absorbed logarithmically—

$$I = I_o e^{-\mu d}$$

Where I_o and I are the intensities of incident and emergent radiation respectively, d the thickness and μ the linear coefficient of absorption of radiation. μ is lower for radiation of higher frequency.

A cavity in the casting will result in those X-rays which pass through the cavity being less effectively absorbed than those rays which travel through the full thickness of metal. Consequently the cavity will show as a dark patch on the resultant photographic negative in the same way that a greater intensity of light affects an ordinary photographic negative.

A fluorescent screen may be substituted for the photographic film so that the resultant radiograph may be viewed instantaneously. This type of fluoroscopy is obviously much cheaper and quicker but is less sensitive than photography and its use is generally limited to the less-dense metals and alloys.

2.93 γ-ray Methods can also be used in the radiography of metals. Since they are of shorter wavelength than are X-rays, they are able to penetrate more effectively a greater thickness of metal. Hence they are particularly useful in the radiography of steel, which absorbs radiation more readily than do light alloys.

2.94 γ-rays constitute a major proportion of the dangerous radiation emanating from 'nuclear waste' and from the fall-out of nuclear explosions. Initially naturally-occurring radium and radon (18.70) were used as a source of γ-rays but artificially activated isotopes of other elements are now generally used. These activated isotopes are prepared by bombarding the element with neutrons in an atomic pile. A nucleus struck by a neutron absorbs it and then contains an excess of energy which is subsequently released as γ-rays. Commonly used isotopes are shown in Table 2.4. Of these iridium-192 and cobalt-60 are the most widely used in industry.

2.95 Manipulation of the isotope as a source of γ-radiation in metallurgical radiography is in many respects more simple than is the case with X-rays, though security arrangements are extremely important in view of the facts that γ-radiation is 'harder' than X-radiation and that it takes place continuously from the source without any outside stimulation. All γ-ray sources are controlled remotely, generally using a manual wind-out system (Fig. 2.16). When not in use the isotope is stored in a shielded container of some γ-ray absorbent material such as lead. Because they are 'harder' than X-rays, γ-rays can be used to radiograph considerable thicknesses of steel. Since the radiation source is small and compact and needs no external

Table 2.4 *γ-ray sources used in industry*

Isotope	Symbol	Half-life	Relative energy output in terms of γ-radiation	Typical uses
Cobalt-60	$^{60}Co_{27}$	5.3 years	1.3	50–200mm of steel
Caesium-137	$^{137}Cs_{55}$	33 years	0.35	25–100mm of steel
Thulium-170	$^{170}Tm_{69}$	127 days	0.0045	2–13mm of aluminium
Ytterbium-169	$^{169}Yb_{70}$	31 days	0.021	2–13mm of steel
Iridium-192	$^{192}Ir_{77}$	74 days	0.48	10–90mm of steel

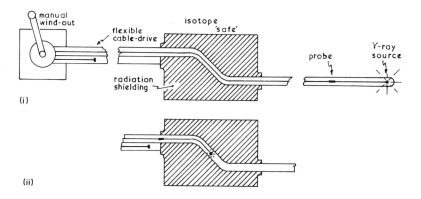

Fig. 2.16 γ-radiography manual remote wind-out system.
(i) γ-ray source exposed; (ii) γ-ray source stored.

power supply, γ-radiation equipment is very portable and can be used to examine materials *in situ*, eg welded joints in motor-way bridges.

All forms of ionising radiation such as X-rays and γ-rays are very harmful to all living tissue and their use in the UK is governed by the Factories Act—'The Ionisating Radiations (sealed sources) Regulations 1969'.

2.96 Ultra-sonic Methods In marine navigation the old method of 'Swinging the lead' was used to determine the depth of the water under the boat. This was replaced in the technological age by a 'sonic' method in which a signal was transmitted from the boat down through the water. The time interval which elapsed between transmission and reception of the 'echo' was a measure of the depth of the ocean bed. The ultrasonic testing of metals is somewhat similar in principle. Ordinary sound waves (of frequencies between 30 and 16 000 Hz) tend to bypass the small defects we are dealing with in metallic components and ultra-sonic frequencies (between 0.5 and 15 MHz) are used for metals inspection.

When an ultrasonic vibration is transmitted from one medium to another some reflection occurs at the interface. Any discontinuity in a structure will therefore provide a reflecting surface for ultrasonic impulses (Fig. 2.17). A probe containing an electrically-excited quartz or barium titanate

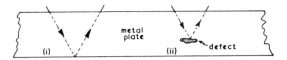

Fig. 2.17   The detection of a fault in plate material by ultrasonics.
In (i) the impulse is reflected from the lower surface of the plate; whilst in (ii) it is reflected from a defect. Measurement of the time interval between transmission of the impulse and reflection of the echo determines the depth of the fault.

crystal which can both transmit and receive high-frequency vibrations is used to traverse the surface of the material to be examined (Fig. 2.18.). The probe is coupled to a pulse generator and to a signal amplifier which transfers the resultant 'image' to a CRT (cathode-ray tube).

2.97   In satisfactory material the pulse will pass from the probe unimpeded through the metal and be reflected from the lower inside surface at A back to the probe, then acting as receiver. Both transmitted pulse and echo are recorded on the CRT and the distance, t_1, between peaks is proportional to the thickness, t, of the test piece. If any discontinuity is encountered such as a blow-hole, B, then the pulse is interrupted and reflected as indicated. Since the echo returns to the receiver in a shorter time an intermediate peak appears on the CRT trace. Its position relative to the other peaks gives an indication of the depth of the fault beneath the surface.

Fig. 2.18 shows separate crystals being used for transmitter and receiver but, as mentioned above, in many modern testing devices a single crystal fulfils both functions. Different types of probe are available for materials of different thickness and this method is particularly useful for examining material—such as rolled plate—of uniform thickness.

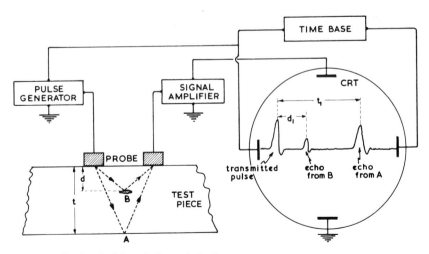

Fig. 2.18   Basic principles of ultrasonic testing.
The values of 'd' and 't' in the test piece are proportional to the values of 'd_1' and 't_1' shown on the CRT.

Exercises

1. Differentiate between:
   (i) Malleability and ductility;
   (ii) Toughness and hardness;
   (iii) Yield strength and tensile strength. (2.20)
2. An alloy steel rod of diameter 15 mm is subjected to a tensile force of 150 kN. What is the tensile stress acting in the rod? (2.30)
3. Fig. 2.19 represents the force-extension diagram for:
   (i) annealed copper;
   (ii) hard-drawn copper;
   (iii) annealed low-carbon steel; or
   (iv) cast iron? (2.32)

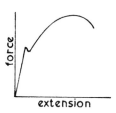

extension   **Fig. 2.19.**

4. When a steel wire 2.5 m long and of cross-sectional area 15 mm² was subjected to a tensile force of 4.0 kN, it stretched elastically by 3.2 mm. Calculate Young's Modulus of Elasticity for the wire. (2.30)
5. A low-nickel steel in the heat-treated condition had an 'engineering' tensile strength of 708 N/mm². The reduction in area of cross-section at the fracture was 44%. What was the true tensile strength of the steel? (2.31)
6. During a tensile test on a cold-worked brass the following figures were obtained for force and corresponding extension:

Extension (mm)	0.1	0.2	0.3	0.4	0.5	0.6	0.8	1.0
Force (kN)	23	46	69	82	89	94	102	110

Ext. (cont.)	1.5	2.0	2.5	3.0	4.0	4.3
Force (cont)	123	131	136	139	132	118 (Break)

The diameter of the test piece was 16 mm and the gauge length used was 80 mm. Draw the force-extension diagram on squared paper and determine:
   (i) Young's modulus of elasticity;
   (ii) the 0.1% proof stress;
   (iii) the tensile strength;
   (iv) The percentage elongation of the material. (2.30 and 2.32)
7. An aluminium alloy has a modulus of elasticity of 69 kN/mm² and a yield strength of 275 N/mm².
   (a) What is the maximum force which a wire 3 mm in diameter could support without suffering permanent deformation?
   (b) If a wire of this diameter and 25 m long is stressed by a force of 430 N what will be the elongation of the wire? (2.30)

8. What method of hardness determination would be suitable for each of the following components:
   (i) a small iron casting;
   (ii) a large steel roll *in situ*;
   (iii) small mass-produced finished components;
   (iv) a small hardened steel die.
   Justify your choice of method in each case. (2.40)
9. What important information is obtained from impact tests? (2.50)
10. What inspection techniques would be appropriate for detecting the following defects in cast products:
   (i) internal cavities in a large steel casting;
   (ii) surface cracks in grey iron castings;
   (iii) surface cracks in aluminium alloy castings;
   (iv) internal cavities in aluminium alloy casting?
   Give reasons for your choice of method in each case. (2.80–2.90)
11. What non-destructive testing methods would be applied to reveal the presence of:
   (i) subcutaneous slag inclusions in a thick steel plate;
   (ii) quench-cracks in a heat-treated carbon steel axle;
   (iii) surface cracks near to a welded joint in mild-steel plate?
   Give reasons for your choice of method in each case and outline the principles of the method involved. (2.80–2.90)

Bibliography

Bateson, R. G. and Hyde, J. H., *Mechanical Testing*, Chapman & Hall.

O'Neill, H., *Hardness of Metals and Its Measurement*, Chapman & Hall, 1967.

BS 18: 1987 *Methods for Tensile Testing of Metals (including aerospace materials)*.

BS 240: 1986 *Methods for Brinell Hardness Test*.

BS 427: 1981 *Methods for Vickers Hardness Test*.

BS 891: 1989 *Methods for Rockwell Hardness Test*.

BS 4175: 1989 *Methods for Rockwell Superficial Hardness Test (N and T Scales)*.

BS 131: 1989 *Methods for Notched-bar Tests (Part 1–Izod; Part 2–Charpy)*.

BS 3855: 1983 *Method for Modified Erichsen Cupping Test for Sheet and Metal Strip*.

BS 1639: 1983 *Methods for Bend Testing of Metals*.

BS 3889: 1983 and 1987 *Methods for Non-destructive Testing of Pipes and Tubes*.

BS 5996: 1980 *Methods for Ultrasonic Testing and Specifying Quality Grades of Ferritic Steel Plate*.

BS 4080: 1966 *Methods for Non-destructive Testing of Steel Castings*.

BS 4124: 1987 *Methods for Non-destructive Testing of Steel Forgings*.

BS 3923: 1972 and 1986 *Methods for Ultrasonic Examination of Welds*.

BS 6443: 1984 *Method for Penetrant Flaw Detection*.

BS 2600: 1973 and 1983 *Methods for Radiographic Examination of Fusion Welded Butt Joints in Steel*.

BS 3451: 1983 *Methods for Testing Fusion Welds in Aluminium and Aluminium Alloys*.

BS 709: 1981 *Methods for Testing Fusion Welded Joints and Weld Metal in Steel*.

BS 6072: 1986 *Methods for Magnetic Particle Flaw Detection*

BS 4331: 1987 and 1989 *Methods for Assessing the Performance Characteristics of Ultrasonic Flaw Detection Equipment*.

3

The Crystalline Structure of Metals

3.10 All chemical elements can exist as either solids, liquids or gases depending upon the prevailing conditions of temperature and pressure. Thus, at atmospheric pressure, oxygen liquifies at $-183°C$ and solidifies at $-219°C$. Similarly, at atmospheric pressure, the metal zinc melts at $419°C$ and boils at $907°C$. In the gaseous state particles are in a state of constant motion and the pressure exerted by the gas is due to the impact of these particles with the walls of the container. As the temperature of the gas is increased, the velocity of the particles is increased and so the pressure exerted by the gas increases, assuming that the container does not allow the gas to expand. If, however, the gas is allowed to expand the distance apart of the particles increases and so the potential energy increases.

Engineers will understand the term 'potential energy' as being that energy possessed by a body by virtue of its ability to do work. Similarly, matter possesses potential energy by virtue of its state. As the distance apart of particles increases, so the potential energy increases. In fact this is a simple application of the First Law of Thermodynamics which states that if a quantity of heat δQ is supplied to a system, part of that heat energy may be used to increase the *internal* energy of the system by an amount δU and part to perform *external* work by an amount δW. Thus:

$$\delta Q = \delta U + \delta W$$

In this case δW is the work done by the gas as it expands against some external pressure. Whatever changes occur, energy is conserved.

In a gas such as oxygen the 'particles' referred to are molecules, each of which consists of two covalently-bonded atoms but in a metallic gas these particles consist of single atoms since insufficient valence electrons are available for metallic atoms to be covalently bonded. Each atom has its own complement of electrons and in the gaseous state the metallic bond does not exist.

On condensation to a liquid the atoms come into contact with each other to form bonds (Fig. 3.1), but there is still no orderly arrangement of the atoms, though a large amount of potential energy is given up in the form of latent heat. When solidification takes place there is a further discharge of latent heat, and the potential energy falls even lower as the atoms take up orderly positions in some geometrical pattern which constitutes a crystal structure. The rigidity and cohesion of the structure is then due to the operation of the metallic bond as suggested in (1.76).

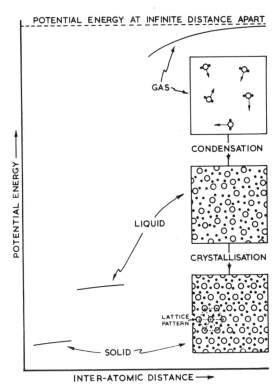

Fig. 3.1 Relative potential energy and atomic arrangements in the three states of matter. In the gaseous and liquid states these arrangements are disorderly, but in the solid state the ions conform to some geometrical pattern. (Note that in this diagram ions are indicated thus: ○, whilst *valence* electrons are indicated so: •, ie the metal is assumed to be bivalent).

3.11 Substances can be classified as either 'amorphous' or 'crystalline'. In the amorphous state the elementary particles are mixed together in a disorderly manner, their positions bearing no fixed relationship to those of their neighbours. The crystalline structure, however, consists of atoms, or, more properly, ions, arranged according to some regular geometrical pattern. This pattern varies, as we shall see, from one substance to another. All metals are crystalline in nature. If a metal, or other crystalline solid, is stressed below its elastic limit, any distortion produced is temporary and, when the stress is removed, the solid will return to its original shape. Thus, removal of stress leads to removal of strain and we say that the substance is elastic.

The amorphous structure is typical of all liquids in that the atoms or molecules of which they are composed can be moved easily with respect to each other, since they do not conform to any fixed pattern. In the case of liquids of simple chemical formulae in which the molecules are small, the forces of attraction between these molecules are not sufficient to prevent the liquid from flowing under its own weight; that is, it possesses high 'mobility'. Many substances, generally regarded as being solids, are amorphous in nature and rely on the existence of 'long-chain' molecules, in which all atoms are co-valently bonded, to give them strength as in certain super polymers (1.75). In these substances the large thread-like molecules, often containing many thousands of atoms each, are not able to slide over each other as is the case with relatively simple molecules in a liquid. The sum of all separate van der Waals' forces of attraction acting between these large molecules are much greater and, since there will be considerable mutual entanglement by virtue of their fibrous nature, the resultant amorphous mass will lack mobility and will be extremely viscous, or 'plastic'. An amorphous structure, therefore, does not generally possess elasticity but only plasticity. A notable exception to this statement is provided by rubber, in which the 'folded' nature of the long-chain molecules gives rise to elasticity (1.75).

A piece of metal consists of a mass of separate crystals irregular in shape but interlocking with each other rather like a three-dimensional jig-saw puzzle. Within each crystal the atoms are regularly spaced with respect to one another. The state of affairs at the crystal boundaries has long been a subject for conjecture, but it is now widely held that in these regions there exists a film of metal, some three atoms thick, in which the atoms do not conform to any pattern (Fig. 3.2). This crystal boundary film is in fact of an amorphous nature. The metallic bond acts within and across this crystal boundary. Consequently, the crystal boundary is not necessarily an area of weakness except at high temperatures when inter-atomic distances increase, and so bond strength decreases. Thus at high temperatures metals are more likely to fail by fracture following the crystal boundaries, whilst at low temperatures failure by transcrystalline fracture is common.

3.12   When a pure liquid solidifies into a crystalline solid, it does so at a fixed temperature called the freezing point. During the crystallisation process the atoms assume positions according to some geometrical pattern (Fig. 3.1), and whilst this is taking place, heat (the latent heat of solidification) is given out in accordance with the laws of thermodynamics, without any fall in temperature taking place. A typical cooling curve for a pure metal is shown in Fig. 3.3.

3.13   Atoms are very small entities indeed, and it has been calculated that approximately 85 000 000 000 000 000 000 atoms are contained in 1 mm^3 of copper. Thus an individual copper atom measures approximately 2.552×10^{-10} m in diameter, putting it far beyond the range of an ordinary optical microscope with its maximum magnification of only 2000. However,

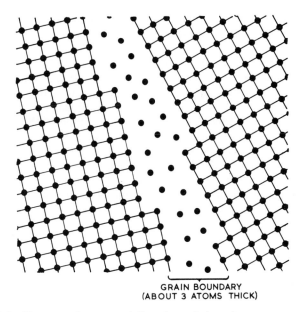

GRAIN BOUNDARY
(ABOUT 3 ATOMS THICK)

Fig. 3.2 Diagrammatic representation of a grain boundary.
The atoms (ions) here are farther apart than those in the crystals themselves.

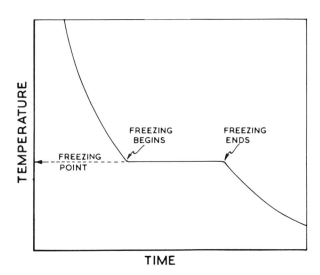

Fig. 3.3 Typical cooling curve of a pure metal.

modern high-resolution electron microscopes, capable of magnifications of a million or so, can show planes of atoms in metals; whilst field-ion microscopy producing magnifications of several millions reveals individual atoms in the structures of some metals.

We have said that the atoms in a solid metal are arranged according to some geometrical pattern. How, then, was this fact ascertained and what form do these patterns take?

Little work was possible in this direction until in 1911 Max von Laue employed X-rays in an initial study of the structures of crystals. Since then, X-rays have found an increasing application in the study of crystal structures, including those of metals. When a beam of monochromatic* X-rays is directed as a narrow 'pencil' at a specimen of the metal in question, diffraction takes place at certain of the crystallographic planes. The resultant 'image' is recorded on a photographic film as a series of spots, and an interpretation of the patterns produced leads to a reconstruction of the original crystal structure of the metal. One such method, that of 'back reflection', is shown in Fig. 3.4. Other methods are in use, but only this brief mention is possible here.

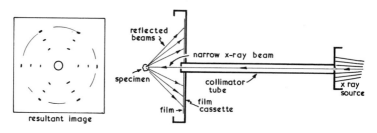

Fig. 3.4   The Laue back-reflection method used to determine the lattice structure of a metal by X-ray diffraction.

3.14   There are several types of pattern or space lattice in which metallic atoms can arrange themselves on solidification, but the three most common are shown in Fig. 3.5. Of these the hexagonal close-packed represents the closest packing which is possible with atoms. It is the sort of arrangement obtained when one set of snooker balls is allowed to fall in position on top of a set already packed in the triangle. (This is illustrated at the top right-hand corner of Fig. 3.5.) The face-centred cubic arrangement is also a close packing of the atoms, but body-centred cubic is relatively 'open'; and when, as sometimes happens, a metal changes its crystalline form as the temperature is raised or lowered, there is a noticeable change in volume of the body of metal. An element which can exist in more than one *crystalline* form in this way is said to be *polymorphic†*. Thus pure iron can exist in three separate crystalline forms, which are designated by letters of the Greek alphabet: 'alpha' (α), 'gamma' (γ) and 'delta' (δ). α-iron, which is body-centred cubic and exists at normal temperatures, changes to γ-iron, which is face-centred cubic, when heated to 910°C. At 1400°C the face-

---

* As in the case of light, the term monochromatic signifies radiation of a single wavelength.

† This term is now used to describe elements which occur in more than one *crystalline* form, whereas the term '*allotropic*' is used to describe those which occur in different forms which are not necessarily crystalline.

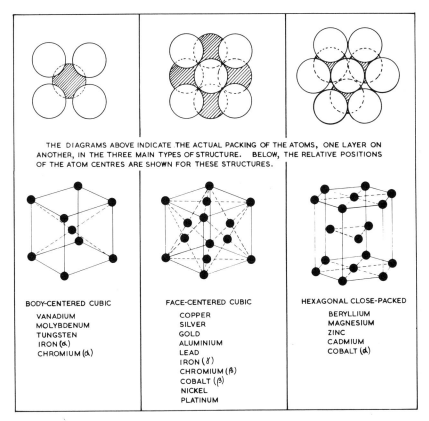

THE DIAGRAMS ABOVE INDICATE THE ACTUAL PACKING OF THE ATOMS, ONE LAYER ON ANOTHER, IN THE THREE MAIN TYPES OF STRUCTURE. BELOW, THE RELATIVE POSITIONS OF THE ATOM CENTRES ARE SHOWN FOR THESE STRUCTURES.

BODY-CENTERED CUBIC	FACE-CENTERED CUBIC	HEXAGONAL CLOSE-PACKED
VANADIUM	COPPER	BERYLLIUM
MOLYBDENUM	SILVER	MAGNESIUM
TUNGSTEN	GOLD	ZINC
IRON (α)	ALUMINIUM	CADMIUM
CHROMIUM (α)	LEAD	COBALT (α)
	IRON (γ)	
	CHROMIUM (β)	
	COBALT (β)	
	NICKEL	
	PLATINUM	

Fig. 3.5 The three principal types of structure in which metallic elements crystallise.

centred cubic structure reverts to body-centred cubic δ-iron. (The essential difference between α-iron and δ-iron, therefore, is only in the temperature range over which each exists.) These polymorphic changes are accompanied by changes in volume-contraction and expansion respectively as shown in Fig. 3.6(i).

The contraction which takes place as the body-centred cubic structure changes to face-centred cubic can be demonstrated with the simple apparatus shown in Fig. 3.6(ii). A wire is held taut under a steady load, and an electric current, sufficient to heat it above the α → γ change point, is passed through it. As the change point is reached, the instantaneous contraction of the wire is indicated by a sharp 'kick' of the pointer to the left. As the wire cools again, when the current is switched off, there is a kick to the right, accompanied by a brightening of the red glow emitted by the wire. This brightening is particularly noticeable when the experiment is made in a darkened room. This indicates that the γ → α change is accompanied by a liberation of heat energy, known as 'recalescence'. In the actual experiment a steel wire is used for the sake of convenience, since it is more easily

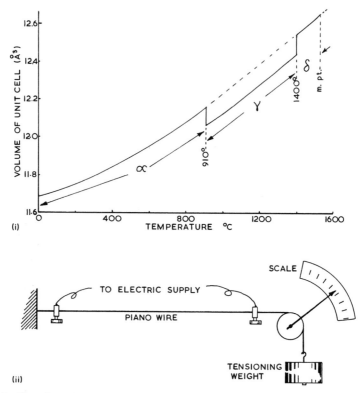

Fig. 3.6 The effect of polymorphic transformations on the expansion of iron.
(i) The 'close-packing' of the γ phase causes a sudden decrease in volume of the unit cell at 910°C (pure iron) and a corresponding increase at 1400°C when the structure changes to δ. (ii) The α→γ transformation (in steel) can be demonstrated using the simple apparatus shown.

obtainable than a pure iron wire. (The α → γ change will be exhibited in a similar way but at a lower temperature than if a pure iron wire were used.) It is this polymorphic change in iron which makes possible the hardening of carbon steels by quenching. Thus, if iron did not chance to be a polymorphic element, can one imagine that Man would have reached his present state of technological development? A world without steel is a prospect difficult to visualise.

3.15 Miller indices   The space lattices indicated in Fig. 3.5 represent the simplest units which can exist in the three main types mentioned. In actual fact a metallic crystal is built up of a continuous series of these units, each face of a unit being shared by an adjacent unit. Any atom will belong to several different crystallographic planes cutting in different directions through a crystal. In order to be able to specify these planes some system of reference is required. The system used is that of 'Miller indices'.

These are the smallest whole numbers proportional to the reciprocals of the intercepts which the plane under consideration makes with the three

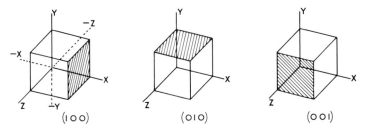

Fig. 3.7   The derivation of Miller indices.

crystal axes (X, Y and Z). Consider a simple cubic structure (Fig. 3.7). Each face intersects only one axis so that the intercepts are $(1, \infty, \infty)$, $(\infty, 1, \infty)$ and $(\infty, \infty, 1)$ respectively. The reciprocals of these numbers are $(1, 0, 0)$, $(0, 1, 0)$ and $(0, 0, 1)$, and these are the Miller indices of the three planes coinciding with the three faces of the cube under consideration. For quick reference they are usually written (100), (010) and (001). The three opposite faces in each case would have negative signs, eg ($\bar{1}$00). In Fig. 3.8(i) the plane indicated is represented by Miller indices (110) and that in Fig. 3.8(ii) by (111). Similarly the Miller indices of the plane shown in Fig. 3.8(iii) will be derived as follows-

Intercepts: 3, 1, 2
Reciprocals: ⅓, 1, ½
Smallest whole numbers
in the same ratio: 2, 6, 3

Hence the Miller indices are (263).

In order to define planes in hexagonal structures more simply a fourth index is introduced as compared with ordinary Miller indices and the resultant indices are termed Miller-Bravais indices. The system is indicated in Fig. 3.9 in which the axes w, x and y on the basal plane of the hexagon are at 120° to each other and normal to the z axis. The intercepts of the plane indicated are ∞, ∞, ∞ and 1 and the Miller-Bravais indices will therefore be (0001).

3.16 Coordination Number   As mentioned earlier and illustrated in Fig. 3.5 both hexagonal close-packed and face-centred cubic represent crystal structures in which atoms (or ions) are the most closely packed, whilst in the body-centred cubic structure atoms (or ions) are packed relatively loosely. If we refer to the upper part of Fig. 3.5 we see that the central atom in the HCP structure is touched by six nearest neighbours on the same plane, also by three (shown in broken line) on the plane above, as well as by three on a similar plane below; making a total of twelve nearest neighbours, ie all 'touching' the central atom. This total, twelve, is known as the *coordination number* of the HCP lattice.

In the FCC structure the central atom has four nearest neighbours on the same plane, four (shown in broken line) on the plane above and four

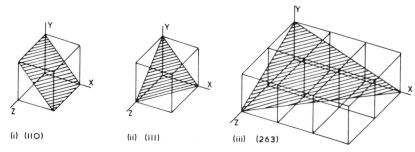

(i) (110)   (ii) (111)   (iii) (263)

Fig. 3.8.

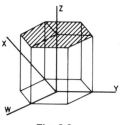

Fig. 3.9.

in similar positions on the plane below. Again the total, twelve, is the coordination number of the FCC lattice structure. In the BCC structure however the central atom (in broken line) has only eight nearest neighbours—four on the plane above and four on the plane below—so that its coordination number is eight.

Only one metal, polonium, is known to crystallise in simple cubic form, ie with one atom at each corner of a cube. Here any atom is touched by six nearest neighbours so that this very loose packing of atoms has a coordination number of only six. Hence it will be seen that the coordination number is an indication of how closely atoms (or ions) are packed in a crystal structure.

3.17   When a pure metal solidifies, each crystal begins to form independently from a nucleus or 'centre of crystallisation'. The nucleus will be a simple unit of the appropriate crystal lattice, and from this the crystal will grow. The crystal develops by the addition of atoms according to the lattice pattern it will follow, and rapidly begins to assume visible proportions in what is called a 'dendrite' (Gk 'dendron', a tree). This is a sort of crystal skeleton, rather like a backbone from which the arms begin to grow in other directions, depending upon the lattice pattern. From these secondary arms, tertiary arms begin to sprout, somewhat similar to the branches and twigs of a fir-tree. In the metallic dendrite, however, these branches and twigs conform to a rigid geometrical pattern. A metallic crystal grows in this way because heat is dissipated more quickly from a point, so that it will be there that the temperature falls most quickly leading to the formation of a rather elongated skeleton (Fig. 3.10).

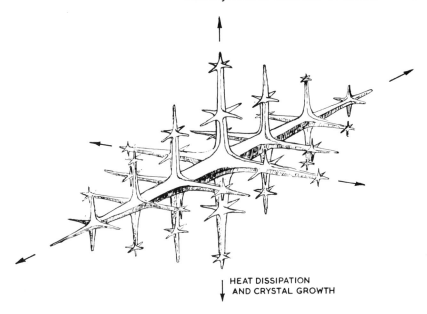

HEAT DISSIPATION
AND CRYSTAL GROWTH

Fig. 3.10 The early stages in the growth of a metallic dendrite.

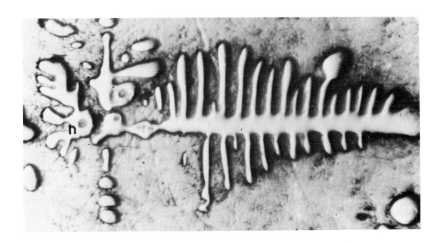

Plate 3.1 Dendritic growth.
This iron dendrite grew from a nucleus at 'n' in a molten mixture of iron and copper. After all the available iron had been used up the dendrite ceased to grow, and the molten copper solidified as the matrix in which the iron dendrite remains embedded. (In fact the iron dendrite will contain a little dissolved copper—in 'solid solution'—whilst the copper matrix will contain a very small amount of dissolved iron). × 300..

The dendrite arms continue to grow and thicken at the same time, until ultimately the space between them will become filled with solid. Meanwhile the outer arms begin to make contact with those of neighbouring dendrites which have been developing quite independently at the same time. All these neighbouring crystals will be orientated differently due to their independent formation; that is, their lattices will meet at odd angles. When contact has taken place between the outer arms of neighbouring crystals further growth outwards is impossible, and solidification will be complete when the remaining liquid is used up in thickening the existing dendrite arms. Hence the independent formation of each crystal leads to the irregular overall shape of crystals. The dendritic growth of crystals is illustrated in Fig. 3.11. In these diagrams, however, the major axes of the crystals are all shown in the same horizontal plane, ie the plane of the paper, whereas in practice this would not necessarily be the case. It has been shown so in the illustration for the sake of clarity.

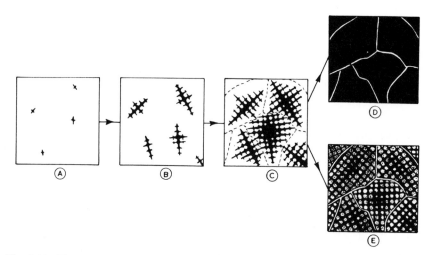

Fig. 3.11   The dendritic growth of metallic crystals from the liquid state.
A solid pure metal (D) gives no hint of its dendritic origin since all atoms are identical, but an impure metal (E) carries the impurities between the dendritic arms, thus revealing the initial skeleton.

3.18   If the metal we have been considering is pure we shall see no evidence whatever of dendritic growth once solidification is complete, since all atoms are identical. Dissolved impurities, however, will often tend to remain in the molten portion of the metal as long as possible, so that they are present in that part of the metal which ultimately solidifies in the spaces between the dendrite arms. Since their presence will often cause a slight alteration in the colour of the parent metal, the dendritic structure will be revealed on microscopical examination. The areas containing impurity will appear as patches between the dendrite arms (Fig. 3.11E). Inter-dendritic porosity may also reveal the original pattern of the dendrites to some extent. If the metal is cooled too rapidly during solidification, molten metal is often unable to 'feed' effectively into the spaces which form between the dendrites due to the shrinkage which accompanies freezing. These spaces then remain as cavities following the outline of the solid dendrite. Such shrinkage cavities can usually be distinguished from blow-holes formed by dissolved gas. The former are of distinctive shape and occur at the crystal boundaries, whilst the latter are quite often irregular in form and occur at any point in the crystal structure (Fig. 3.12).

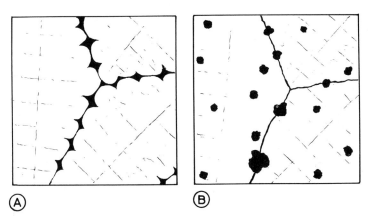

Fig. 3.12   Porosity in cast metals.
Shrinkage cavities (A) tend to follow the shape of the dendritic arms and occur at the crystal boundaries, whilst gas porosity (B) is usually of irregular shape and occurs at almost any point in the structure.

3.19   The rate at which a molten metal is cooling when it reaches its freezing point affects the size of the crystals which form. A slow fall in temperature, which leads to a small degree of undercooling at the onset of solidification, promotes the formation of relatively few nuclei, so that the resultant crystals will be large (they are easily seen without the aid of a microscope). Rapid cooling, on the other hand, leads to a high degree of undercooling being attained, and the onset of crystallisation results in the formation of a large 'shower' of nuclei. This can only mean that the final crystals, being large in number, are small in size. In the language of the foundry, 'chilling causes fine-grain casting'. (Throughout this book the term 'grain' and 'crystal' are used synonymously.) Thus the crystal size of a pressure die-casting will be very small compared with that of a sand-casting. Whilst the latter cools relatively slowly, due to the insulating properties of the sand mould, the former solidifies very quickly, due to the contact of the molten metal with the metal mould. Similarly, thin sections, whether in sand- or die-casting, will lead to a relatively quicker rate of cooling, and consequently smaller crystals.

In a large ingot the crystal size may vary considerably from the outside surface to the centre (Fig. 3.13). This is due to the variation which exists

Plate 3.2A   Dendrites on the surface of an ingot of antimony.
Antimony is one of the few metals which expand during solidification. Hence the growing dendrites were raised clear of the remaining liquid so that their growth could not be completed.

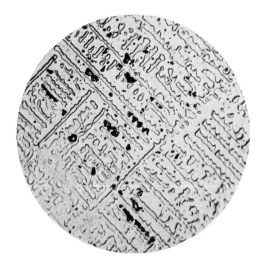

Plate 3.2B   Shrinkage cavities (black areas) in cast tin bronze.
These roughly follow the shape of the original dendrites and occur in that part of the alloy to solidify last. × 200. Etched in ammonia–hydrogen peroxide.

in the temperature gradient as the ingot solidifies and heat is transferred from the metal to the mould. When metal first makes contact with the mould the latter is cold, and this has a chilling effect which results in the formation of small crystals at the surface of the ingot. As the mould warms up, its chilling effect is reduced, so that the formation of nuclei will be retarded as solidification proceeds. Thus crystals towards the centre of the ingot will be larger. In an intermediate position the rate of cooling is favourable to the formation of elongated columnar crystals, so that we are frequently able to distinguish three separate zones in the crystal structure of an ingot, as shown in Fig. 3.13. More recent research into rapid solidification processes (RSP) has been carried out with the object of obtaining metals and alloys with extremely tiny crystals, and in some cases retaining the amorphous structure of the original liquid at ambient temperatures (9.110).

A number of defects can occur in cast structures. The more important are dealt with below.

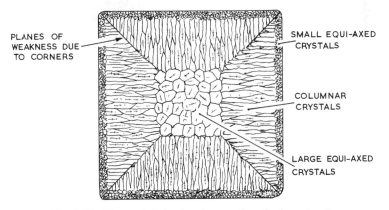

PLANES OF WEAKNESS DUE TO CORNERS

SMALL EQUI-AXED CRYSTALS

COLUMNAR CRYSTALS

LARGE EQUI-AXED CRYSTALS

Fig. 3.13 The crystal structure in a section of a large ingot.

Blow-holes

3.20 These are caused by furnace gases which have dissolved in the metal during melting, or by chemical reactions which have taken place in the melt. Gas which has dissolved freely in the molten metal will be much less soluble in the solid metal. Therefore, as the metal solidifies, gas will be forced out of solution. Since dendrites have already formed, the bubbles of expelled gas become trapped by the dendrite arms and are prevented from rising to the surface. Most aluminium alloys and some of the copper alloys are susceptible to 'gassing' of this type, caused mainly by hydrogen dissolved from the furnace atmosphere. The difficulty can be overcome only by making sure that there is no dissolved gas in the melt prior to casting (1.54 and 1.60—Part II).

3.21 Porosity may arise in steel which has been incompletely deoxidised prior to being cast. Any iron oxide (present as oxygen ions) in the molten steel will tend to be reduced by carbon according to the following equation:

$$FeO + C \rightleftharpoons Fe + CO$$

This is what is commonly called a *reversible* reaction and the direction in which the *resultant* reaction proceeds depends largely upon the relative concentrations of the reactants and also upon the temperature. When carbon (in the form of anthracite for example) is added to the molten steel the reaction proceeds strongly to the right and since carbon monoxide, a gas, is lost to the system the reaction continues until very little FeO remains in equilibrium with the relatively large amount of carbon present. As the ingot begins to solidify, it is almost pure iron of which the initial dendrites are composed. This causes an increase in the concentration of carbon and the oxide, FeO, in the remaining molten metal, thus upsetting chemical equilibrium so that the above reaction will commence again. The bubbles of carbon monoxide formed are trapped by the growing dendrites, producing blow-holes. The formation of blow-holes of this type is prevented by adequate 'killing' of the steel before it is cast—that is, by adding a sufficiency of a deoxidising agent such as ferromanganese. This removes residual FeO and prevents the FeO–C reaction from occurring during subsequent solidification. In some cases the FeO–C reaction is utilised as in the production of 'rimmed' ingots (2.21—Part II).

3.22 Subcutaneous blow-holes, ie those just beneath the surface may be caused in ingots by the decomposition of oily mould dressing, particularly when this collects in the fissures of badly cracked mould surfaces. The gas formed forces its way into the partially solid surface of the metal ingot, producing extensive porosity.

Shrinkage

3.30 The crystalline structure of most metals of engineering importance represent a close packing of atoms. Consequently solid metals occupy less space than they do as liquids and shrinkage takes place during solidification as a result of this decrease in volume. If the mould is of a design such that isolated pockets of liquid remain when the outside surface of the casting is solid, shrinkage cavities will form. Hence the mould must be so designed that there is always a 'head' of molten metal which solidifies last and can therefore 'feed' into the main body of the casting as it solidifies and shrinks. Shrinkage is also responsible for the effect known as 'piping' in cast ingots. Consider the ingot mould (Fig. 3.14A) filled instantaneously with molten steel. That metal which is adjacent to the mould surface solidifies almost immediately, and as it does so it shrinks. This causes the level of the remaining metal to fall slightly, and as further solidification takes place the process is repeated, the level of the remaining liquid falling still further. This sequence of events continues to be repeated until the metal is completely solid and a conical cavity or 'pipe' remains in the top portion of the ingot. With an ingot shaped as shown it is likely that a secondary pipe would be formed due to the shrinkage of trapped molten metal when it solidifies. It is usually necessary to shape large ingots in the way shown in Fig. 3.14A, that is, small end upwards, so that the mould can be lifted from the solidified ingot. Therefore various methods of minimising the pipe must be used (2.21—Part II). One of the most important of these methods is to pour the metal into the mould so that solidification almost keeps pace with pouring. In this way molten metal feeds into the pipe formed by the solidification and consequent shrinkage of the metal. Smaller ingots can be cast into moulds which taper in the opposite direction to that shown in Fig. 3.14A, ie large end upwards (Fig. 3.14B), since these can be trunnion-mounted to make ejection of the ingot possible.

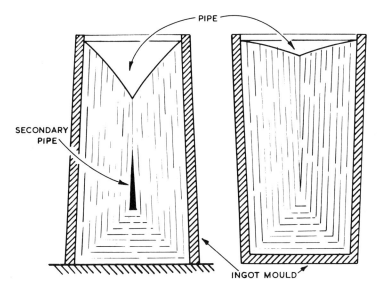

Fig. 3.14   The influence of the shape of the mould on the extent of piping in a steel ingot.

Segregation of Impurities

3.40   There is a tendency for dissolved impurities to remain in that portion of the metal which solidifies last. The actual mechanism of this type of solidification will be dealt with later (8.23), and it will be sufficient here to consider its results.

3.41   The dendrites which form first are of almost pure metal, and this will mean that the impurities become progressively more concentrated in the liquid which remains. Hence the metal which freezes last at the crystal boundaries contains the bulk of the impurities which were dissolved in the original molten metal. This local effect is known as *minor* segregation (Fig. 3.15A).

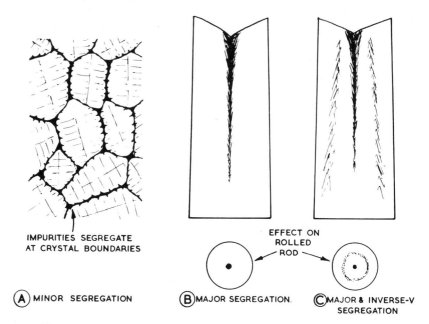

IMPURITIES SEGREGATE
AT CRYSTAL BOUNDARIES

EFFECT ON
ROLLED
ROD

(A) MINOR SEGREGATION   (B) MAJOR SEGREGATION.   (C) MAJOR & INVERSE-V
                                                                    SEGREGATION

Fig. 3.15 Types of segregation which may be encountered in steel ingots.

3.42 As the columnar crystals begin to grow inwards, they will push in front of them some of the impurities which were dissolved in the molten metal from which they themselves solidified. In this way there is a tendency for much of the impurities in the original melt to become concentrated in the central pipe. If a vertical section of an ingot is polished and etched, these impurities show as V-shaped markings in the area of the pipe (Fig. 3.15B). The effect is called *major* segregation.

3.43 With very large ingots the temperature gradient may become very slight towards the end of the solidification process, and it is common for the band of metal which has become highly charged with impurities, just in front of the advancing columnar crystals, to solidify last. Some impurities, when dissolved in a metal, will depress its freezing point considerably (similarly when lead is added to tin a low melting point solder is produced). Hence the thin band of impure metal just in advance of the growing columnar crystals has a much lower freezing point than the relatively pure molten metal at the centre. Since the temperature gradient is slight, this metal at the centre may begin to solidify in the form of equi-axed crystals, so that the impure molten metal is trapped in an intermediate position. This impure metal therefore solidifies last, causing inverted V-shaped markings to appear in the etched section of such an ingot. It is known as '*inverse-vee*' segregation (Fig. 3.15C). Rimming steels contain no heavily-segregated areas because of the mechanical stirring action introduced by the evolution of carbon monoxide during the FeO/C reaction (3.21).

3.44 Of these three types of segregation, minor segregation is probably the most deleterious in its effect, since it will cause overall brittleness of the castings and, depending upon the nature of the impurity, make an ingot hot- or cold-short, that is, liable to crumble during hot- or cold-working processes.

3.50 From the foregoing remarks it will be evident that a casting, suffering as it may from so many different types of defect, is one of the more variable and least predictable of metallurgical structures. In some cases we can detect the presence of blow-holes and other cavities by the use of X-rays, but other defects may manifest themselves only during subsequent service. Such difficulties are largely overcome when we apply some mechanical working process during which such defects, if serious, will become apparent by the splitting or crumbling of the material undergoing treatment. At the same time a mechanical working process will give a product of greater uniformity in so far as structure and mechanical properties are concerned. Thus, all other things being equal, a forging is likely to be more reliable in service than a casting. Sometimes, however, such factors as intricate shape and cost of production dictate the choice of a casting. We must then ensure that it is of the best possible quality.

Line and Points Defects in Crystals

3.60 In the foregoing sections we have been dealing with such defects as are likely to occur in cast metals. These defects may be so large that a microscope is not necessary to examine them. Others are small yet still within the range of a simple optical microscope. On the atomic scale however metallic structures which would be regarded as being of very high quality in the industrial sense nevertheless consist of crystals which contain numerous 'line' and 'points' defects scattered throughout the crystal lattice. These defects (Fig. 3.16) occur in wrought as well as in cast metals and though small in dimensions have considerable influence on mechanical properties.

3.61 The most important line defect is the *dislocation*. Here part of a plane of atoms is missing from the lattice, and the 'glide' of this fault through a crystal under the action of a shearing force is the mechanism by which metals can be deformed mechanically without fracture taking place (4.15).

3.62 Of the various points defects occurring in crystals the term *vacancy* describes a lattice site from which the atom is missing. When such a vacancy is formed by the resident atom migrating to the surface of the metal it is known as a *Schottky defect* (8.24). If a number of vacancies occur near to each other, stress within the lattice will be reduced if these vacancies diffuse together to form a void. Such voids are likely to precipitate the formation of cracks when the lattice is subjected to sufficient stress. A *Frenkel defect* is caused by the displacement of an atom from its lattice position into a nearby interstitial site.

To the mind not scientifically trained the term 'solution' suggests only

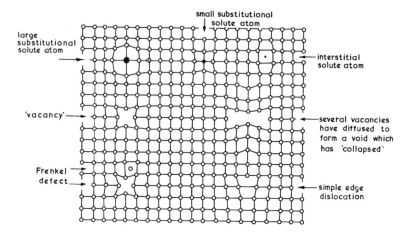

Fig. 3.16   The imperfect nature of a metallic crystal.

that some solid substance has been dissolved by a liquid, but in science a solution is a homogeneous mixture of two (or more) substances in which the atoms or molecules of the substances are completely dispersed—whether in the gaseous, liquid or solid states. Thus in metals a *solid solution* (8.20) consists of a crystal structure in which 'stranger' atoms take up positions within the lattice of the parent metal.

In a *substitutional* solid solution (8.21) some lattice sites are occupied by 'stranger' atoms (Fig. 3.16), that is, they are *substituted* for some of the atoms of the parent metal. Such stranger atoms are generally of a similar size—either larger or smaller—to those of the parent metal. If the stranger atoms are smaller than those of the parent metal they are quite likely to dissolve *interstitially*, that is they fit into the spaces (or interstices) between the parent atoms forming an *interstitial solid solution*. The very small atoms of hydrogen, carbon and nitrogen are able to dissolve interstitially even in solid γ-iron, the latter two elements making the processes of carburising and nitriding possible.

3.63   Whatever the nature of a points defect it is likely to stop or at least impede the smooth movement of dislocations through a crystal so that a greater force must be used to produce a new movement of dislocations. For this reason solid solutions are stronger than pure metals.

Exercises

1. Compare and contrast:
    (i) the properties of covalent and metallic bonding;
    (ii) atomic packing in FCC and HCP structures. (3.11 and 3.14)
2. What is meant by 'polymorphism'? Discuss the term with particular reference to iron, showing the connection which this property has with the engineering uses of the metal.

How could one of the manifestations of polymorphism be demonstrated in the laboratory? (3.14)
3. Sketch the *three* most important types of spatial arrangement encountered in the lattice structures of metals.
  Show how the density of packing of the atoms in each case affects volume changes in those metals which, like iron, are polymorphic. (3.14)
4. Derive Miller indices to represent the plane shown in Fig. 3.17. (3.15)

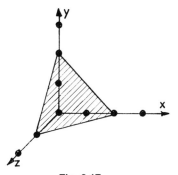

Fig. 3.17.

5. Why do metallic crystals generally have irregular boundaries? (3.16)
6. Explain the term 'dendritic solidification'. Show how certain mechanical properties of cast metals can be explained by reference to this type of crystal structure. (3.16)
7. Fig. 3.18 illustrates cross-sections of three castings of similar shape and composition but which have been cast under different conditions of pouring temperature and mould material.
  Account for the different structures produced. (3.18)

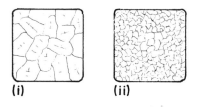

Fig. 3.18.

8. Discuss the formation and distribution of gas porosity in
    (i) cast steels;
    (ii) cast aluminium alloys.
  Show in each case how this form of defect may be minimised. (3.20)
9. Show how a 'pipe' tends to form during the solidification of a large ingot. What methods may be used to minimise the pipe?
  Why do impurities generally tend to segregate in the pipe? (3.30 and 3.42)

Bibliography

Barratt, C. S. and Massalski, T. B., *Structures of Metals*, Pergamon Press, 1980.
Brown, P. J. and Forsyth, J. B., *The Crystal Structure of Solids*, Arnold, 1973.
Cahn, R. W., *Physical Metallurgy*, North Holland, 1980.
Chadwick, C. A. and Smith, D. A., *Grain Boundary Structures and Properties*, Academic Press, 1976.
Hume-Rothery, W., Smallman, R. E. and Howarth, C. W., *The Structure of Metals and Alloys*, Institute of Metals, 1988.
Kennon, N. F., *Patterns in Crystals*, John Wiley, 1978.
Smallman, R. E., *Modern Physical Metallurgy*, Butterworths, 1985.
Woolfson, M. M., *An Introduction to X-ray Crystallography*, Cambridge University Press, 1978.

4

Mechanical Deformation and Recovery

4.10 Deformation in metals can occur either by elastic movement or by plastic flow (2.30). In elastic deformation a limited distortion of the crystal lattice is produced, but the atoms do not move permanently from their ordered positions, and as soon as the stress is removed the distortion disappears. When a metal is stressed beyond the elastic limit plastic deformation takes place and there must, clearly, be some movement of the atoms into new positions, since considerable permanent distortion is produced. We must therefore consider ways in which this extensive rearrangement of atoms within the lattice structure can take place to give rise to this permanent deformation.

4.11 Plastic deformation proceeds in metals by a process known as 'slip', that is, by one layer or plane of atoms gliding over another. Imagine a pile of pennies as representing a *single* metallic crystal. If we apply any force which has a horizontal component, ie any force not acting vertically, the pile of pennies will be sheared as one slides over another, provided that the horizontal component is sufficient to overcome friction between the pennies.

The results of slip in a polycrystalline mass of metal may be observed by microscopical examination. The direction of the slip planes is indicated in such a piece of metal after deformation by the presence of slip bands which form on the surface of the metal. If a piece of soft iron is polished and etched and then squeezed in a vice so that the polished surface is not scratched, these slip bands can be seen on the surface. Their method of formation is indicated in Fig. 4.2. Such slip bands are generally parallel in any individual crystal but differ in orientation from one crystal to another. It has been shown by electron microscopy that a single visible slip band consists of a group of roughly 10 steps on the surface, each about 40 atoms thick and approximately 400 atoms high. If the deformation has been excessive, the presence of slip bands is apparent even when a specimen is

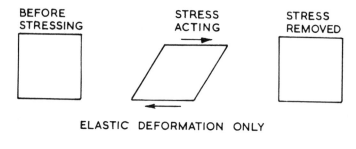

BEFORE
STRESSING

STRESS
ACTING

STRESS
REMOVED

ELASTIC DEFORMATION ONLY

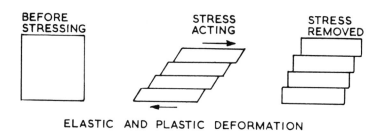

BEFORE
STRESSING

STRESS
ACTING

STRESS
REMOVED

ELASTIC AND PLASTIC DEFORMATION

Fig. 4.1 Diagrams illustrating the difference, in action and effect, of deformation by elastic and plastic means.

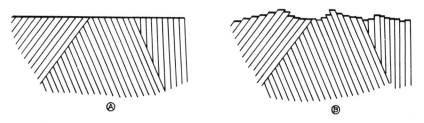

Fig. 4.2 The formation of slip bands.
(A) Indicates the surface of the specimen before straining, and (B) the surface after straining. The relative slipping along the crystallographic planes is apparent as ridges (visible under the microscope) on the surface of the metal.

polished and etched *after* deformation. Heavily stressed parts of the crystal —which contain considerable strain energy—dissolve more quickly during etching, revealing these so-called 'strain bands'.

4.12 All metals of similar crystal structure slip on the same crystallographic planes and in the same crystallographic directions. Slip occurs when the shear stress resolved along these planes reaches a certain value —the critical resolved shear stress. This is a property of the material and does not depend upon the structure. The process of slip is facilitated by the presence of the metallic bond, since there is no need to break direct bonds between individual atoms as there is in co-valent or electro-valent structures, nor is there the problem of repulsion of ions of like charge as

Plate 4.1 'Slip' in metallic crystals.
(i) Shows 'slip steps' (see Fig. 4.3) in a single crystal of cadmium approx 2mm in diameter. This was grown as a single crystal in the form of wire which was then stretched by hand. × 15. (ii) Illustrates 'slip bands' on the surface of annealed copper. The specimen was polished, etched and then squeezed gently in the jaws of a vice. × 200.

there is for an electro-valently bonded crystal. When slip occurs in co-valent or electro-valent structures it does so with much greater difficulty than in metals. Some types of crystal are more amenable to deformation by slip than others. For example, in the face-centred cubic type of structure there are a number of different planes along which slip could conceivably take place, whereas with the hexagonal close-packed structure slip is only possible on the basal plane of the hexagon. Thus, metals with a face-centred cubic structure, such as copper, aluminium and gold, are far more malleable and ductile than metals with a hexagonal close-packed lattice like zinc.

4.13 It is possible, under controlled conditions, to grow single crystals of some metals, and under application of adequate stress such crystals behave in a manner very similar to that of the pile of pennies. An offset on one side of the crystal is balanced by similar offset on the other side (Fig. 4.3), and both offsets lie on a single continuous plane called a slip plane. Whilst in the case of the pile of pennies the force necessary to cause slip was that required to overcome friction between them, in a single metallic crystal the force necessary to cause slip is related to that required to overcome the resistance afforded by the metallic bond.

For a single crystal the shear component, T (Fig. 4.4) of a tensile stress, σ, resolved along the direction of slip, OS, on a slip plane, P, can be calculated. ON is normal to P.

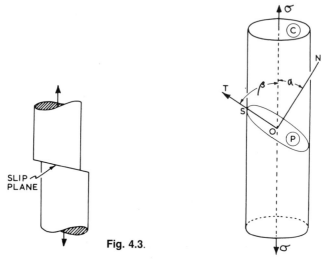

Fig. 4.3.

Fig. 4.4.

Component of σ along OS = σ.cos β
Area of projection on C of unit area of P = cos α
$$T = \sigma \cos \alpha . \cos \beta$$

Experiments show that different single crystals of the same metal slip at different angles and different tensile stresses, but if the stresses are resolved along the slip plane, all crystals of the same metal slip at the same critical value of resolved shear stress.

4.14   From a knowledge of the forces acting within the metallic bond it is possible to derive a *theoretical* value for the stress required to produce slip by the *simultaneous* movement of atoms along a plane in a metallic crystal. However, the stress, T, actually obtained practically in experiments on single crystals as outlined above, is *only about one thousandth of the theoretical value* assuming simultaneous slip by all atoms on the plane. Obviously then slip cannot be a simple simultaneous block movement of one layer of atoms over another. Nor does such a simple interpretation of the idea of slip explain the work-hardening which takes place during mechanical deformation. A perfect material in plastic deformation would presumably deform without limit at a constant yield stress as indicated in Fig. 4.5A, but in practice a stress-strain relationship of the type indicated in Fig. 4.5B exists.

4.15   Earlier theories which sought to explain slip by the simultaneous gliding of a complete block of atoms over another have now been discarded and the modern conception is that slip occurs step by step by the movement of so-called 'dislocations' within the crystal.

If the reader has ever tried his hand at paper-hanging he will know that wrinkles have a habit of appearing in the paper when it is laid on the wall. Attempts to smooth out these wrinkles by pulling on the edge of the paper would undoubtedly prove fruitless, since the tensile force necessary to cause the *whole sheet of paper to slide* would be so great as to tear it.

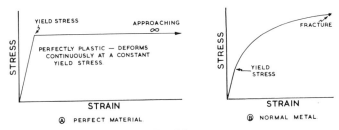

Fig. 4.5.

Instead, gentle coaxing of the wrinkles individually with the aid of a brush or cloth leads to their successful elimination, causing them to glide and 'pass out of the system', with the application of a force which is small compared with that necessary to slide the whole sheet of paper simultaneously. The movement of dislocations on a slip plane in a metallic crystal probably follows a similar pattern.

Dislocations are faults or distorted regions (Fig. 4.6) in otherwise perfect

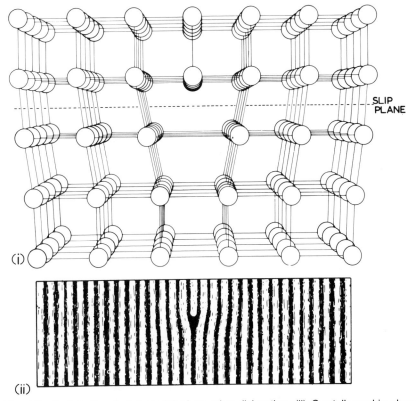

Fig. 4.6 (i) A 'ball-and-wire' model of an edge dislocation. (ii) Crystallographic planes containing an edge dislocation, as they appear in aluminium at a magnification of several millions. (A sketch of the structure as revealed by a high-resolution electron microscope).

crystals and the step-by-step movement of such faults explains why the force necessary to produce slip is of the order of 1000 times less than the theoretical, assuming simultaneous slip over a whole plane.

The fundamental nature of a dislocation is illustrated in Fig. 4.7 (in which for the sake of clarity only the centres of atoms are indicated). Assume that a shearing stress, σ, has been applied to the crystal causing the top half, $APSE$, of the face, $ADHE$, to move inwards on a slip plane, $PQRS$, by an amount PP_1 (one atomic step). There has been no corresponding slip, however, at face $BCGF$. Consequently the top half of the crystal contains an extra half-plane as compared with the bottom part of length P_1Q. In Fig. 4.7 this is the half-plane $WXYZ$ and the line XY is known as an *edge dislocation*. It separates the slipped part $PXYS$ of the slip plane from the unslipped part $XQRY$. During slip this 'front' XY moves to the right through the crystal.

The movement of such an edge dislocation under the action of a shearing force is illustrated in Fig. 4.8. Here an edge dislocation exists already (i), and the application of the shearing stress (ii) causes the dislocation to glide along the slip plane in the manner suggested. In this case it has been assumed that the dislocation glides out of the crystal completely, producing a slip step of one atom width at the edge of the crystal (iii). The dislocation can be moved through the crystal with relative ease, since only one plane is moving at a time and then only through a small distance.

Slip can also take place by the movement of 'screw' dislocations. These differ from edge dislocations in that the direction of movement of the dislocation is *normal* to the direction of formation of the slip step. The

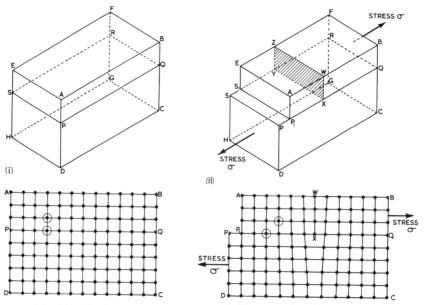

Fig. 4.7 The formation of an edge dislocation by the application of stress.

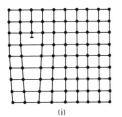 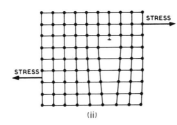 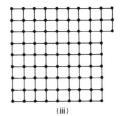

Fig. 4.8.

mechanism of the process is indicated in Fig. 4.9. Here a shear stress, σ, had displaced P to P_1 and Q to Q_1 on one face of the crystal so that P_1Q_1YX has slipped relative to $PQYX$. In this case the screw dislocation XY separates the slipped part $PQYX$ from the unslipped part $XYRS$ of the slip plane $PQRS$. It will be noted that XY is moving in a direction which is normal to the direction in which slip is being produced.

It is also possible for slip to take place by the combination of a screw dislocation with an edge dislocation. A curved dislocation is thus evolved which moves across the slip plane (Fig. 4.10).

4.16   Until comparatively recently much of the foregoing commentary concerning dislocations was of a speculative nature. Whilst many of the properties of a metal could be best explained by postulating the existence of dislocations, no one had actually seen a dislocation in a metallic structure since dimensions approaching atomic size are involved. True, dislocations had been observed in structures of crystalline complex compounds and it was reasonable to suppose that similar dislocations occurred in the crystalline structures of metals. During the last few years the rapid development of electron microscope techniques has made it possible to observe crystallographic planes within metallic structures. With the aid of high-resolution electron microscopy, photographs of edge dislocations in metals such as aluminium have been produced (Fig. 4.6(ii)).

Since the stress required to produce dislocations is great, it is assumed that they are not generally initiated by stress application but that the majority of them are formed during the original solidification process. During any subsequent cold-working process, dislocations have a method of reproducing themselves from what are called Frank–Read sources (Fig. 4.11), so that in the cold-worked metal the number of dislocations has greatly increased.

The relationship between the force necessary to initiate dislocations and to move those which already exist is notably demonstrated by the tensile properties of metallic 'whiskers'. These are hair-like single crystals grown under controlled conditions and generally having a single dislocation running along the central axis. If a tensile stress is applied along this axis the dislocation is unable to slip. As no other dislocations are available, the crystal cannot yield until a dislocation is initiated at E (Fig. 4.12). The stress then falls to that necessary to move the dislocation (Y), and dislocations then reproduce rapidly so that plastic flow proceeds.

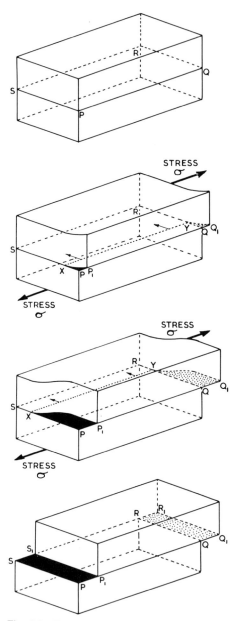

Fig. 4.9 The movement of a screw dislocation.

4.17 A brief mention of experimental evidence supporting the operation of slip in a polycrystalline mass of metal was made earlier in this
chapter (4.11), but so far we have been dealing mainly with the methods
by which slip can take place in individual crystals. If slip occurred com-

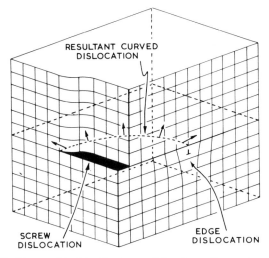

Fig. 4.10 The combination of a screw dislocation with an edge dislocation.

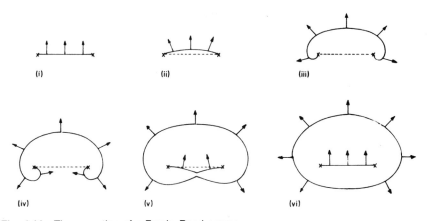

Fig. 4.11 The operation of a Frank–Read source.
The original source is visualised as a dislocation line anchored at its ends, possibly by other faults. The action of a suitable shear stress causes the line to bow outwards, ultimately turning upon itself (iii), and eventually forming a complete dislocation loop (v). Since the original dislocation line still remains, the process can repeat *ad lib* producing a series of dislocation loops or ripples flowing out from the original source.

pletely over a whole plane, as was assumed above, the dislocation would pass right out of the crystal and there would be no reason for any change in mechanical properties. Yet, as we know, metals, which have solidified under ordinary conditions and which consist of a mass of individual crystals, become harder and stronger as the amount of cold work to which they are subjected increases, until a point is reached where they ultimately fracture.

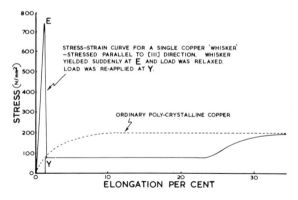

Fig. 4.12   Differences in tensile properties between a single copper 'whisker' and ordinary polycrystalline copper.

Consequently, it is assumed that many dislocations remain in every crystal and that increase in hardness results from their mutual interference and the building up of a transcrystalline 'traffic jam'. Increase in hardness and strength is due to the greater difficulty in moving new dislocations against the jammed ones, whilst a 'pile up' of jammed dislocations may propagate a fracture. Dislocations will be unable to escape at crystal boundaries by forming steps because of the adherence of crystals in terms of the amorphous film (3.11). Moreover, the amorphous film will act as an effective barrier preventing dislocations from passing from one crystal to another, even supposing that the lattice structures of two neighbouring crystals were suitably aligned to make this possible. Since individual crystals develop at random in a polycrystalline mass, this will rarely be so.

Individual crystals in a polycrystalline metal deform by the same mechanisms as do single crystals, but since they are orientated at random, shear stress on slip planes will attain the critical value in different crystals at different loads (acting on the metal as a whole). Thus there is considerable difference between the deformation of a single crystal and that of a polycrystalline mass. The yield stress of the latter is generally higher and glide occurs almost simultaneously on several different systems of slip planes. Dislocations cannot escape from individual crystals and so they jam. A 'pile up' on one slip plane will prevent dislocations in intersecting planes from moving, producing a situation not uncommon in the case of motor-traffic conditions at most crossroads near the South Coast on almost any Sunday in the summer.

For these reasons, strain-hardening proceeds much more rapidly in a polycrystalline metal than it does in a single crystal. A decrease in crystal size leads to strain-hardening taking place more quickly so that, though the material is stronger, it is less ductile. Strain-hardening, however, is supremely important. If a metal did not strain-harden but continued to slip, as would any perfectly plastic material (Fig. 4.5A), failure would be inevitable, immediately it were loaded above its yield point.

4.18 In addition to deformation by slip, some metals, notably zinc, tin and iron, deform by a process known as 'twinning'. The mechanism of this process is illustrated in Fig. 4.13. In deformation by slip all atoms in one block move the same distance, but in deformation by twinning, atoms in each successive plane within a block will move different distances, with the effect of altering the direction of the lattice so that each half of the crystal becomes a mirror image of the other half along a twinning plane. It is thought that twinning also proceeds by the movement of dislocations.

Twins thus formed are called 'mechanical twins' to distinguish them from the 'annealing twins' which are developed in some alloys—notably those of copper—during an annealing operation which follows mechanical deformation. The mechanical twins formed in iron by shock loading are known as 'Neumann bands'. Twin formation in a bar of tin can actually be heard as the bar is bent and used to be called 'The Cry of Tin'.

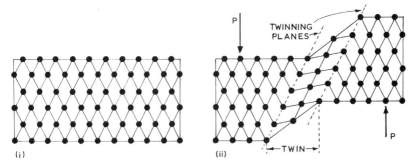

Fig. 4.13 The formation of 'mechanical twins' (i) before stressing, (ii) after stressing under shearing force *P*.

Energy of Mechanical Deformation

4.19 As deformation proceeds the metal becomes progressively harder and stronger and, whether by slip or by twinning, a point is reached when no more deformation can be produced. Any further increase in the applied force will lead only to fracture. In this condition, when tensile strength and hardness have reached a maximum and ductility a minimum, the material is said to be *work hardened*. As deformation proceeds the capacity for further deformation decreases and the force necessary to produce it must increase. All dislocations present at the start of stress application have moved into 'jammed' positions, as have all new dislocations generated by the progressive increase in stress, until no further slip by movement of dislocations is possible. At this point of maximum resistance to slip (maximum strength and hardness) further increase in stress would give rise to fracture. The material must then be annealed if further cold work is to be carried out on it.

During a cold-working process approximately 90% of the mechanical energy employed is converted to heat as internal forces acting within the metal are overcome. The remaining 10% of mechanical energy used is stored in the material as a form of potential energy. The bulk of this—about 9% of the energy originally used—is that associated with the number of dislocations generated. These have energy because they result in distortion of the lattice and cause atoms to occupy positions of higher-than-minimum energy. The remaining potential energy (1% of the energy originally used) exists as locked-up residual stresses arising from elastic strains internally balanced.

The increased-energy state of a cold-worked metal makes it more chemically active and consequently less resistant to corrosion. It was suggested earlier in this chapter that as deformation proceeds dislocations will tend to pile up at crystal boundaries. Consequently these crystal boundaries will be regions of increased potential energy due to the extra micro-stresses present there. For this reason the grain-boundary areas will corrode more quickly than the remainder of the material so that intercrystalline failure will be accelerated.

This stored potential energy is also the principal driving force of recovery and recrystallisation during an annealing process.

Annealing and Recrystallisation

4.20   A cold-worked metal is in a state of considerable mechanical stress, resulting from elastic strains internally balanced. These elastic strains are due to the jamming of dislocations which occurred during cold deformation. If the cold-worked metal is heated to a sufficiently high temperature then the total energy available to the distorted regions will make possible the movement of atoms into positions of equilibrium so that the elastic strains diminish and the 'locked-up' energy associated with them 'escapes'. Since dislocations will once more be in positions of minimum energy from which they can be moved relatively easily, tensile strength and hardness will have fallen to approximately their original values and the capacity for cold-work will have returned. This form of heat-treatment is known as *annealing*, and is made use of when the metal is required for use in a soft but tough state or, alternatively, when it is to undergo further cold deformation. Annealing may proceed in three separate stages depending upon the extent of the required treatment.

4.30 Stage I—The Relief of Stress   This occurs at relatively low temperatures at which atoms, none the less, are able to move to positions nearer to equilibrium in the crystal lattice. Such small movements reduce local strain and therefore the mechanical stress associated with it, without, however, producing any visible alteration in the distorted shape of the cold-worked crystals. Moreover, hardness and tensile strength will remain at the high value produced by cold-work, and may even increase as shown in the curve for cold-worked 70–30 brass (Fig. 4.14). It is found that a controlled low-temperature anneal at, say, 250°C applied to hard-drawn 70–30 brass tube will effectively reduce its tendency to 'season-crack' (16.33) without reducing strength or hardness.

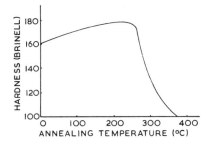

Fig. 4.14 The relationship between hardness and annealing temperature (cartridge brass).

4.40 Stage II—Recrystallisation Although a low-temperature annealing process intended to relieve 'locked-up' stresses may sometimes be used, annealing generally involves a definite and observable alteration in the microstructure of the metal or alloy. If the annealing temperature is increased a point is reached when new crystals begin to grow from nuclei produced in the deformed metal. These nuclei are formed at points of high potential energy such as crystal boundaries and other regions where dislocations have become entangled. The crystals so formed are at first

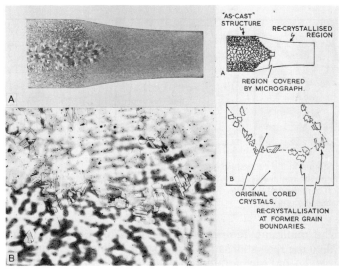

Plate 4.2 A cast 70–30 brass slab (38mm thick) was being cold-rolled when the mill was stopped. The partly rolled slab was then annealed at 600°C.
The photomicrograph (A) reveals two regions: (1) the coarse-grained region which suffered no cold-work and consequently did not recrystallise on annealing (hence it shows the original coarse as-cast structure); (2) the heavily cold-worked part of the slab which has recrystallised on annealing so that the crystals are much too small to be visible in the photomicrograph. The photomicrograph (× 100) (B) is taken from a region which suffered very little cold work. On annealing, recrystallisation has just begun at points on the original crystal boundaries where, presumably, a pile-up of dislocations had just commenced. Small twinned crystals have been formed at these points, whilst the remainder of the structure is still the original as-cast.

small, but grow gradually until they absorb the entire distorted structure produced originally by cold-work (Fig. 4.15). The new crystals are equi-axed in form, that is, they do not show any directional elongation, as did the distorted cold-worked crystals which they replace. They are, in fact, of equal axes.

This phenomenon is known as recrystallisation, and it is the principal method employed, in conjunction with cold-work, of course, to produce a fine-grained structure in non-ferrous metals and alloys. Only in rare cases —notably in steels and aluminium bronze, where certain structural changes take place in the solid state—is it possible to refine the grain size solely by heat-treatment.

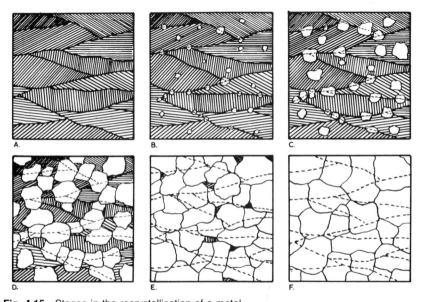

Fig. 4.15   Stages in the recrystallisation of a metal.
(A) Represents the metal in its cold-rolled state. At (B) recrystallisation had commenced with the formation of new crystal nuclei. These grow at the expense of the old crystals until at (F) recrystallisation is complete.

4.41   The minimum temperature at which recrystallisation will take place is called the recrystallisation temperature. This temperature is lowest for pure metals, and is generally raised by the presence of other elements. Thus, pure copper recrystallises at 200°C, whilst the addition of 0.5% arsenic will raise the recrystallisation temperature to well above 500°C; a useful feature when copper is to be used at high temperatures and must still retain its mechanical strength. Arsenical copper of this type was widely used in the boiler tubes and fire-boxes of the steam locomotives of former years. Similarly, whilst cold-worked commercial-grade aluminium recrys-tallises in the region of 150°C, that of 'six nines' purity (99.999 9% pure) appears to recrystallise below room temperature and, consequently, does not cold-work.

Other metals, for example lead and tin, recrystallise below room temperature, so that it is virtually impossible to cold-work them, since they recrystallise even whilst mechanical work is taking place. Thus, they can never be work-hardened at normal temperatures of operation. Most engineering metals have recrystallisation temperatures which are well above ambient temperatures, and are therefore hot-worked at some temperature above their recrystallisation temperature.

4.42 The recrystallisation temperature is dependent largely on the degree of cold-work which the material had previously received, and severe cold-work will generally result in a lower recrystallisation temperature, since the greater the degree of cold-work the greater the amount of locked-up potential energy available for recrystallisation. It is therefore not possible accurately to quote a recrystallisation temperature for a metal to the extent that it is for, say, its melting point. For most metals, however, the recrystallisation temperature is between one-third and one-half of the melting point temperature, T_m. (T_m being measured on the *absolute* scale, K.) Thus the mobilities of all metallic atoms are approximately equal at the same fraction of their melting point (K).

4.50 Stage III—Grain Growth   If the annealing temperature is above the recrystallisation temperature of the metal, the newly formed crystals will continue to grow by absorbing each other cannibal-fashion, until the structure is relatively coarse-grained, as shown in Fig. 4.15. Since the crystal boundaries have higher energies than the interiors of the crystals, a polycrystalline mass will reduce its energy if some of the grain boundaries disappear. Consequently, at temperatures above that of recrystallisation large crystals grow by absorbing small ones. As indicated in Fig. 4.16, a crystal boundary tends to move towards its centre of curvature in order to shorten its length. To facilitate this, atoms move across the boundary to positions of greater stability where they will be surrounded by more

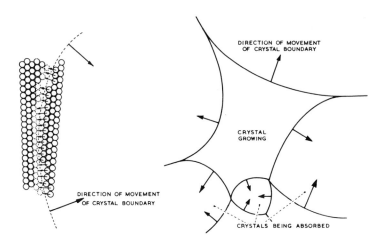

Fig. 4.16   The 'cannibalising' of small crystals by larger ones.

neighbours in the concave crystal face of the growing crystal. Thus the large grow larger and the small grow smaller. The same type of mechanism (in this case surface tension along a liquid film) causes small bubbles to be absorbed by the larger ones in the froth in a glass of beer. The extent of grain growth is dependent to a large degree on the following factors:

(*a*) The annealing temperature used—as temperature increases, so grain size increases (Fig. 4.17).

(*b*) The duration of the annealing process—grains grow rapidly at first and then more slowly (Fig. 4.17).

(*c*) The degree of previous cold-work. In general, heavy deformation will lead to the formation of a large number of regions of high energy within the crystals. These will give rise to the production of many nuclei on recrystallisation and consequently the grain size will be small. Conversely, light deformation will give rise to few nuclei and the resulting grain size will be large (Fig. 4.18). What is generally referred to as the *critical* amount of cold-work is that which is just necessary to initiate recrystallisation and so give rise to extremely coarse grain. This is likely to occur at lightly-worked regions in a deep-drawn component which is subsequently annealed (Fig. 4.19).

(*d*) The use of certain additives in a metal or alloy. Thus the presence of nickel in alloy steels limits grain growth during heat-treatment processes. Nickel is in fact the universal grain refiner and does much to increase the toughness of many alloys by limiting grain growth.

Insoluble particles are also thought to act as barriers to grain growth in some cases. For example small amounts of thoria (thorium oxide) are added to the tungsten used for electric lamp filaments. Here the films or particles of thoria prevent excessive grain growth which would otherwise result from maintaining the filament for long periods at temperatures well above that of recrystallisation for tungsten.

4.51 Of these factors the one requiring the most accurate control, in normal conditions, is the annealing temperature. Some alloys, particularly the brasses, are exceptionally sensitive to variations in the annealing temperature and an error of 100°C, on the high side, may increase crystal size

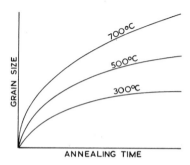

Fig. 4.17 The relationship between grain size and annealing time.

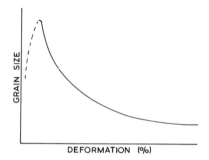

Fig. 4.18 The relationship between grain size and the degree of original deformation.

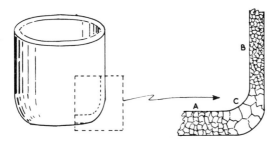

Fig. 4.19 Grain size in a deep-drawn component which has subsequently been annealed. The grain size at the base of the cup (A) will be that of the original material since there has been no cold-work, and consequently no recrystallisation during the annealing process. In the side walls of the cup (B) grain size is small since heavy cold-work during the drawing process has resulted in the formation of many nuclei during recrystallisation. In the region (C) cold-work has been slight—the 'critical' amount—so that, during annealing, few nuclei formed and the resultant grain is very coarse.

five-fold for a given annealing time. This coarse grain will, in turn, lead to loss in ductility and the formation of a rough, rumpled surface (known as 'orange peel') during a subsequent forming operation. When coarse grain has been thus produced in material of finished size, it cannot be refined, as is possible with steel, by heat treatment. The only remedy would be to remelt the brass and go through the complete stages of rolling and annealing, until once more the finished size was reached.

4.52 It follows that a reasonably accurate assessment of grain size is necessary when a material is destined for a pressing or deep-drawing process (6.43). Grain size is usually determined by counting the number of crystals in a known area during the microexamination of a prepared specimen. The result is expressed as the number of grains per mm^2. In the USA the ASTM index is often used and this is obtained by counting the number of grains per in^2 when viewed at a magnification of $\times$ 100. The ASTM index, G, is then given by:

$N = 2^{(G-1)}$ where N is the number of grains per square inch at $\times$ 100. Values of G between 1 and 8 cover normal grain size ranges.

A further method of estimating grain size involves measuring the Average Grain Diameter (mm). Here an image of the prepared surface is projected on to the reflex screen of a microscope at a magnification of $\times100$. A line AB, 100 mm long is previously drawn randomly on the screen and the number of crystals cut by the line counted. The 100 mm line on the screen represents a 1 mm line on the surface of the specimen, hence:

$$\text{Average Grain Diameter} = \frac{1}{\text{Number of crystals cut by line}} \text{ mm}$$

In Fig 4.20, 13 crystals have been intersected by the line AB, therefore the Average Grain Diameter = $^1/_{13}$ mm = 0.08 mm. (The optimum grain

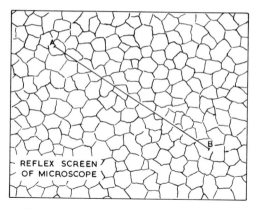

Fig. 4.20 A method of assessing average grain diameter.

size for most deep-drawing metals and alloys is between 0.04 and 0.06 mm AGD.)

Naturally a more representative value of Average Grain Diameter will be obtained if the specimen is moved several times relative to the line AB and an average of the counts used. BS 4490:1989 lays down that two 100 mm lines at right angles to each other be used along with two 150 mm lines at right angles to each other and diagonal to the 100 mm lines. In this way the effects of any directionality of grain shape are minimised.

Superplasticity

4.60 Superplasticity is a property of some alloys in which very great plastic elongations of up to 2000% or more can be obtained under the action of quite low tensile stresses. This behaviour is akin to that of heated glass where elongation is dependent upon the quantity of material to be thinned as the material 'flows'. Although the phenomenon has been known for more than sixty years it has only recently begun to be exploited.

4.61 When most metals are deformed plastically at temperatures well below that of recrystallisation (approximately 0.5 T_m*) then tensile strength increases with strain as work-hardening occurs (Fig 4.21(i)). Many metals and alloys have stress-strain curves which are roughly governed by the expression:

$$\sigma = k\varepsilon^{\eta} \tag{1}$$

where σ is the tensile stress, ε the strain and η the work-hardening exponent. If the temperature during deformation is now raised above that necessary for recrystallisation—say to 0.5 or 0.6 T_m—the stress required to produce plastic flow falls because metals are weaker when hot. Moreover,

---

* As mentioned earlier the melting point of a metal, T_m, is measured on the *absolute scale* (K), ie °C + 273, for these purposes.

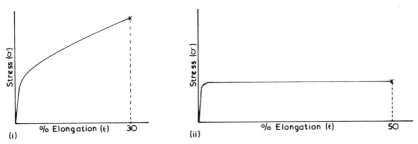

Fig. 4.21 The stress/strain characteristics of an ordinary metal (i) below $0.3T_m$ (ii) above $0.5T_m$.

work-hardening disappears since the metal is effectively being hot-worked so that recrystallisation and recovery accompany deformation.

Since work-hardening is no longer taking place, $\eta = 0$ and substitution of this into (1) gives:

$$\sigma = k \tag{2}$$

represented by a straight line parallel to the ε axis and giving a typical hot-working stress–strain relationship (Fig 4.21(ii)).

4.62 In both of the cases mentioned above the extent of elongation is limited by necking which ultimately occurs in the test piece. This necking, which leads to rapid failure, is due to the presence of some form of inhomogeneity in the metal giving rise to stress concentrations. In superplastic alloys, however, necking is inhibited and stress is not proportional to strain but to the *strain rate*, $\overset{\circ}{\varepsilon}$ (Fig 4.22). That is, the tensile stress required to produce plastic flow increases as the strain rate increases, in a way similar to which the viscous 'drag' of an oil increases as its rate of displacement is increased. Thus as soon as a neck tends to form the strain rate in that region increases and stress also tends to increase there. As a result deformation ceases there in preference to the lower-stressed regions. This stress–strain relationship is governed by:

$$\sigma = K\overset{\circ}{\varepsilon}{}^m \tag{3}$$

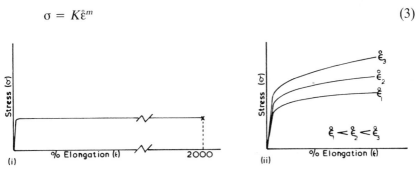

Fig. 4.22 (i) The stress/strain characteristics of a superplastic metal at temperatures above $0.5T_m$ and a *slow* strain rate, $\overset{\circ}{\varepsilon}$ (ii) the effect of strain rate on the stress required to cause flow in a superplastic metal.

where $\overset{\circ}{\varepsilon}$ represents the strain rate and m the strain-rate exponent. When $m = 1$ then:

$$\sigma = K\overset{\circ}{\varepsilon} \tag{4}$$

which indicates that stress is proportional to strain rate and necking will not take place. Most metals over a temperature range of, say, 0.1 to 0.9 T_m have 'm' values in the region of 0.1, whilst superplastic alloys usually have 'm' values within the range 0.4 to 0.8.

4.63 Hence certain alloys show large elongations at temperatures in the region of 0.5 T_m and at strain rates generally lower than are employed in industrial forming operations. In many of these alloys crystal structure is an important factor. Their crystals are generally very small—about 0.002 mm in diameter as compared with 0.05 mm average diameter for most common metals and alloys. Moreover these small crystals are very regular in shape. After large deformations have taken place grain size may have increased slightly but crystal shape will not have changed significantly and will remain non-directional.

4.64 Although opinions differ as to the reasons for superplasticity, it is probable that in fine-grained metals strained at a *controlled rate* as outlined above, the generation of new dislocations at grain boundaries is balanced by the disposal of existing dislocations as they pass across grains into other grain-boundary 'sinks' after a relatively unimpeded passage across the small grains.

Thus a superplastic metal is one which will have a very small grain size and will maintain it at temperatures in excess of 0.5 T_m. Whilst it is not difficult to produce small grains in most metals at temperatures below 0.5 T_m by means of heavy cold-working programmes, to maintain these small grains at 0.5 T_m and above is generally frustrated by the rapid grain growth which normally occurs. Some two-phase structures are superplastic because one phase obstructs the grain growth of the other. In fact a recent *Guinness Book of Records* quotes (as a world record for superplasticity) an elongation of 4850% attained with a tin–lead eutectic, ie 62Sn–38Pb.

4.65 In practice superplastic forming is a hot-stretching process whereby material is forced into or over a one-piece die usually by air pressure. Thus the majority of components are formed from sheet. One of the first alloys to be shaped superplastically on an industrial scale was Zn–22Al (again a eutectic alloy), at a temperature of 260°C.

Since necking does not occur during superplastic extension the dieless drawing of wire and tube is possible in much the same way as has been used for the manufacture of glass rod and tube for a very long time. Airframe manufacturers have pioneered the development of some titanium alloys exhibiting superplasticity. Thus fine-grained Ti–6Al–4V will deform superplastically under low strain-rate conditions around 900°C. This has led to the development of manufacturing processes for complex parts from thin sheet. The technique is also in use for an increasing number of airframe and missile parts. Some aluminium alloys (17.32) too are shaped using superplastic properties.

Exercises

1. What visual evidence is there to support the view that permanent deformation in metals takes place by a process of 'slip'?
   Outline the theory which seeks to explain the essential nature of slip. (4.11 and 4.14)
2. In a tensile test on a single metallic crystal the yield stress of the material was found to be 55 N/mm^2, and slip planes were formed at an angle of 40° to the axis of the specimen.
   Calculate the value of the stress which caused slip along these planes.
   This value will be found to be about 10^3 times less than the 'theoretical' value. Account for this (4.13)
3. Describe the nature of an 'edge dislocation' in a metallic structure.
   Explain with sketches how plastic deformation proceeds by the movement of such faults in a crystal. (4.15)
4. Discuss the relationship between yield stress of a metal and the presence of dislocations in the structure. Illustrate your answer by reference to metallic 'whiskers'. (4.16)
5. Why do ductile metals work-harden progressively during plastic deformation? (4.17)
6. Discuss the effects of (a) cold working (b) subsequent heat-treatment, on the structure and mechanical properties of a pure metal.
   Explain the terms *stress relief, recrystallisation* and *grain growth*. (4.19 and 4.20)
7. Given that the melting point of nickel is 1458°C make a rough estimate of its recrystallisation temperature. (4.42)
8. The melting point of pure tin is 232°C. Show why it is not possible to work-harden the metal at ambient temperatures. (4.42)
9. Describe the mechanism of grain growth in metals and show how the rate of grain growth is related to the annealing temperature. (4.50)
10. A heavily cold-worked copper bar is placed with one end in a furnace at 900°C whilst the other end reaches a temperature of 25°C. Assuming that there is a uniform temperature gradient along the whole length describe the structures you would expect to find along the bar after about two hours of such treatment. (4.30 to 4.50)
11. A nail is driven through a piece of sheet metal which has previously been cold-rolled and annealed under conditions which produced a small grain size. After puncture the specimen is annealed just above its recrystallisation temperature. Sketch and describe the type of crystal structure which would now be found in the region of the nail hole and account for the results you describe. (4.50)
12. What is 'superplasticity' as applied to metals? To what extent is this phenomenon related to microstructure. (4.60)

Bibliography

Cahn, R. W., *Physical Metallurgy*, North Holland, 1980.
Cottrell, Sir Alan, *Theoretical Structural Metallurgy*, Arnold, 1965.
Cottrell, Sir Alan, *Dislocations and Plastic Flow in Crystals*, Oxford University Press, 1979.

Honeycombe, R. W. K., *The Plastic Deformation of Metals*, Arnold, 1984.
Martin, J. W. and Doherty, R. D., *Stability of Microstructures in Metallic Systems*, Cambridge University Press, 1976.
Smallman, R. E., *Modern Physical Metallurgy*, Butterworths, 1985.
BS 4490: 1989 *Micrographic Determination of the Grain Size of Steel.*

5

Fracture of Metals

5.10 In an 'ideal' metal, that is one containing no flaws or defects, fracture will occur when *atomic bonding* is overcome across an atomic plane which is perpendicular to the tensile force; but in practice metals fail at much lower stresses than the theoretical (4.14). This is due partly to the presence of impurities and other discontinuities in the structure (Fig. 3.16), and partly to the polycrystalline nature of metals which in itself leads to the 'pile-up' of dislocations at or near crystal boundaries. Thus fracture is a common cause of failure in metals and whilst it will always occur when a metal is stressed beyond its tensile strength, under certain conditions it can also occur at stresses even below the elastic limit. Thus, some metals may *creep* during long periods of time, particularly at high temperatures, until failure occurs; whilst fracture which is the result of repeated cyclic stress is termed *fatigue* failure. Some metals fail by a form of *brittle fracture* which is influenced by low temperature conditions.

5.11 The nature of fracture differs from one metal to another. The type of fracture which follows large amounts of plastic deformation is generally referred to as *ductile* fracture whilst that which occurs after little or no plastic deformation is called *brittle* fracture (Fig. 5.1).

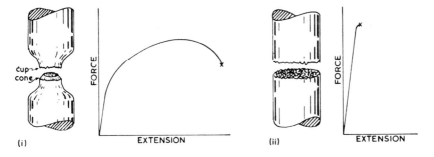

Fig. 5.1 Types of fracture (i) ductile—showing the well-known cup and cone; (ii) brittle—with a strongly 'crystalline' appearance.

It was stated in the previous chapter that during plastic deformation slip takes place along some crystallographic planes more readily than along others. Similarly, metallic fracture—or *cleavage*—tends to follow preferred crystallographic planes. Thus FCC metals are likely to suffer cleavage along the {111} planes (3.15), whilst BCC metals cleave along the {100} planes. In CPH metals cleavage occurs along the basal planes of the hexagon (designated the {0001} planes).

Brittle Fracture

5.20 Whilst plastic deformation takes place in a ductile metal by the 'rippling' of dislocations along slip planes (4.16), brittle fracture occurs as a result of complete and sudden separation of atoms as indicated in Fig. 5.2. The theoretically-calculated stress necessary to achieve such separation is considerable, yet in practice, the actual stress required to cause brittle fracture is often relatively small. Moreover, the results obtained for tensile strength measurements on a large number of apparently identical test pieces of a single brittle metal are often very variable (Fig. 5.3(i)) compared with similar determinations made for a ductile metal (Fig. 5.3(ii)). Such results suggest that in brittle materials some factor is present which gives rise to the variability of results.

5.21 In 1920, A. A. Griffith postulated that fractures in brittle solids were propagated from minute flaws in the material. He demonstrated that the strength of freshly-drawn glass fibres often approached the 'theoretical'

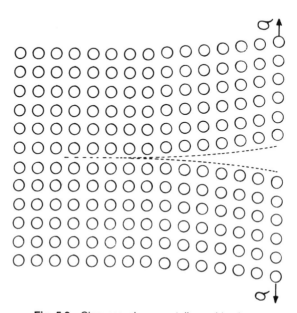

Fig. 5.2 Cleavage along crystallographic planes.

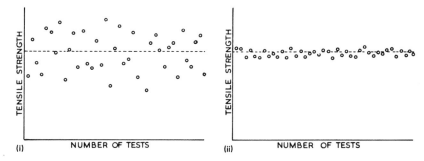

Fig. 5.3 The variability of results during the tensile testing of (i) a brittle material; (ii) a ductile material.

value, but if these fibres were allowed to come into contact with any other substance, including the atmosphere, even for short periods then the strength was considerably reduced. This suggests that the strength of glass fibre was very dependent upon surface perfection and that anything likely to initiate even minute surface irregularities would weaken it. In certain respects these principles may also be applied to brittle metals. Engineers will already be familiar with the concept of 'stress raisers' and stress concentrations associated with the presence of sharp in-cut corners and the necessity of eliminating these in engineering design whenever possible. Thus in iron castings in-cut corners are rounded by using leather 'fillets' on the wooden pattern.

5.22 Griffith's Crack Theory. Whilst the presence of a small fissure (Fig. 5.4) will obviously reduce the effective cross-section of the material, the reduction in breaking stress is very much greater than can be accounted for by this reduction in cross-sectional area. This is because an applied stress, S, generates stress concentrations at the tip of the fissure.

Griffith concluded that the concentrated stress, S_c, is related to the applied stress, S, the width of the crack, c, and the radius of curvature of the tip of the crack, r, by:

$$S_c = 2S \sqrt{\frac{c}{r}} \quad \text{or} \quad \frac{S_c}{S} = 2 \sqrt{\frac{c}{r}}$$

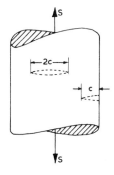

Fig. 5.4.

If we assume that the tip radius of such a crack is of the order of 10^{-10} m (roughly one atomic radius) and that such cracks are about 10^{-4} m in width, then the value S_c/S (which we can call the *factor of stress concentration*) is of the order of 10^3. This indicates that the actual stress, S_c, operating at the tip of the crack is some thousand times greater than the applied stress, S. In this case fracture is taking place at a localised stress nearer to the theoretical value. When a crack progresses through a brittle material under the action of a constant applied stress, S, the concentrated stress, S_c, at the tip increases since an increase in c gives rise to an increase in the term c/r. The speed of crack propagation therefore increases and failure is certain.

5.23 So far we have considered only the effects of an applied *tensile* stress on a brittle material. If however the applied stress is *compressive* this stress will be transmitted across existing micro-cracks without causing any stress concentrations—that is, a compressive stress will tend to 'close up' existing fissures. For this reason many brittle metallic materials such as cast iron are relatively weak in tension but strong in compression. Failure in compression will ultimately take place when compressive forces are so high that they produce *tensile* components of sufficient magnitude along crystallographic planes in the region of a fissure tip.

5.24 Intercrystalline Brittle Fracture   In the foregoing paragraphs we have been dealing with brittle fracture in terms of cleavage along *trans*crystalline planes. Brittle fracture also occurs by the propagation of cracks along grain boundaries—that is, by *inter*crystalline fracture. In some cases this type of fracture is due to the presence of grain-boundary films of a hard, brittle second phase. Such films may be formed by segregation during solidification (3.41) so that fracture of this type is more commonly found in cast materials. Brittle films of bismuth segregate at the crystal boundaries of copper in this way so that very small quantities of bismuth (less than 0.01%) may cause excessive brittleness in copper.

Impurities present in solid solution may also segregate at grain boundaries during the normal process of coring (8.23). A high concentration of the solute atoms in the grain-boundary region may give rise to brittleness. This is probably due to the pegging of dislocation movements and the consequent initiation of micro-cracks at the grain boundaries.

5.25 Temper Brittleness   This occurs in some low-alloy steels when tempered in the range 250–400°C (13.42), and is probably due to the precipitation of films or particles of carbides at grain boundaries. Although tensile strength, and even ductility, are not seriously reduced a very large reduction in impact value is experienced and fracture is intercrystalline.

Ductile Fracture

5.30   As was indicated in Fig. 5.3 the stress at which a ductile metal is likely to fail is much more predictable than that stress in a brittle metal and this fact alone makes ductile metals more suitable for fail-safe engineering design. Moreover, as plastic deformation begins, the tip radius of any micro-crack present is likely to increase and this will automatically reduce the concentrated stress, S_c, as the value of c/r (in the Griffith equation) decreases. As a result of this stress relaxation propagation of the crack may cease. The fracture which ultimately occurs in ductile materials is due partly to strain hardening which progressively reduces the ductility which is necessary for further stress relaxation.

5.31   It is difficult to differentiate implicitly between brittle and ductile fracture since some brittle metals undergo some ductile deformation before fracture, whilst many ductile metals exhibit final brittle-type cleavage. However, whilst a *brittle* fracture is one in which the progress of the crack involves very little plastic deformation of the neighbouring metal, a *ductile* fracture is one which proceeds as a result of high localised plastic deformation of the metal near the tip of the crack. Thus, whilst the two extremes of these methods of failure are easily recognised, there can be no sharp division between brittle and ductile fracture. A fracture which is almost completely of the brittle variety will exhibit a mass of minute facets which reflect light strongly and reveal the crystalline nature of the metal. An almost completely ductile fracture, on the other hand, shows a rough, dull dirty-grey surface because much of that surface has been deformed along the general plane of fracture. When viewed under the scanning electron microscope (10.33) however the surface of a ductile fracture has a 'dimpled' appearance, caused by the presence of a multitude of minute cavities accompanying the rupture process. Closer examination reveals that the formation of these minute cavities is due to the presence of very small particles of impurity and other phases. This confirms the observed fact that as the purity of a metal increases so does its ductility.

5.32   Most ductile polycrystalline metals fail with the well-known *cup-and-cone* fracture. Fracture of this type follows the formation of a *neck* in a tensile test piece. Crack formation begins at the centre of the neck on a plane that is roughly normal to the applied stress axis. Small cavities form near the centre of this cross section (Fig. 5.5(i)) and these coalesce to form a visible fissure (Fig. 5.5(iii)). As deformation proceeds the crack spreads outwards towards the edges of the test piece. The final fracture then occurs rapidly along a surface which makes an angle of approximately 45° with the stress axis. This leaves a circular lip on one half of the test piece and a corresponding bevel on the surface of the other half, leading to the formation of a cup on one half and a cone with a flattened apex on the other half (Fig. 5.1(i)).

Factors Leading to Crack Formation

5.40   Very pure metals are much more ductile than those of slightly lower purity and will often draw down to a 'chisel point' where the cross-section at failure is approaching zero. It is therefore reasonable to suppose that even minute inclusions in the microstructure play an important part in nucleating cracks. When slip in a metal takes place dislocations will tend to pile up at the metal/inclusion interface. Assuming that the inclusion

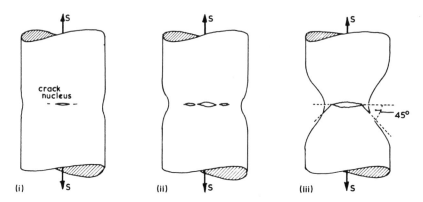

Fig. 5.5 The nucleation and development of a crack in a ductile material.

itself is strong and does not shear, a minute fissure will develop at the interface (Fig. 5.6). Of course, the development of the fissure will also depend upon the degree of *adhesion* between the surface of the inclusion and that of the metal. Thus, if there is little adhesion as, for example, at the interface between copper and cuprous oxide particles, then fissures will form easily, but if adhesion is high then the particles may indeed have a strengthening effect. Thus sintered aluminium powder (SAP), which contains a high proportion of aluminium oxide particles, is stronger than pure aluminium because of the high adhesion at the Al/Al_2O_3 interface.

5.41   Other barriers to the movement of dislocations can also initiate micro-cracks in a similar manner. Thus Fig. 5.7 suggests how a grain boun-

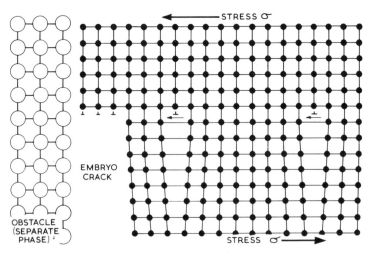

Fig. 5.6 A pile-up of dislocations at some obstacle leading to the formation of a crack which will be propagated if the stress, σ, is increased.

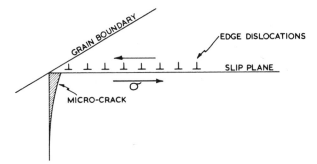

Fig. 5.7   The nucleation of a grain-boundary fissure.

dary can act as a barrier to the movement of dislocations so that a pile-up occurs and nucleates a micro-crack. Such a crack continues to grow as further dislocations, possibly generated from the same Frank–Read source, move to join it. Dislocations will not cross grain boundaries since lack of alignment and some disorder will always be present there.

Crack initiation can also be caused by movement of dislocations along close-packed or other planes within a crystal (Fig. 5.8(i)). These dislocations then pile up at the intersections of the slip planes and so form a crack (Fig. 5.8(ii)).

5.42   It was shown earlier (5.22) that, arising from Griffith's Crack Theory, at the tip of the crack:

$$\text{Stress concentration} \propto \sqrt{\frac{\text{Crack length}}{\text{Crack-tip radius}}}$$

Thus the spread of a crack in a plate glass window is arrested by drilling a hole in front of the tip of the advancing crack. Similarly, catastrophic brittle fracture in welded-steel ships (5.50) was prevented by including dummy rivet holes to arrest the progress of cracks already initiated at structural faults in the steel.

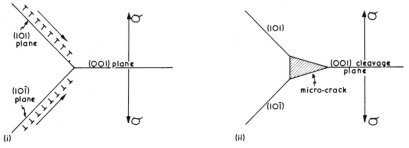

Fig. 5.8   The initiation of a micro-crack by a running together of dislocations (after A. H. Cottrell).

Ductile-Brittle Transition in Steels

5.50 There have been many instances in the past of failure of metals by unexpected brittleness at low temperatures. That is, metals which suffered normal ductile fracture at ambient temperatures would fail at low temperatures by sudden cleavage fracture and at comparatively low stresses. The failure of the motor sledges during the very early stages of the British South Pole Expedition of 1912–13 may well have been due to fracture of this type and contributed towards the final disaster which overtook the Polar party. In more recent years similar failure was experienced in the welded 'liberty ships' manufactured during the Second World War for carrying supplies from America to Europe and was unexpected and dangerous.

Under normal conditions the stress required to cause cleavage is higher than that necessary to cause slip, but if, by some circumstances, slip is suppressed, brittle fracture will occur when the internal tensile stress increases to the value necessary to cause failure. This situation can arise under the action of bi-axial or tri-axial stresses within the material. Such stresses may be residual from some previous treatment, and the presence of points of stress concentration may aggravate the situation. The liberty ships mentioned above were fabricated by welding plates together to form a continuous body. Cracks usually started at sharp corners or arc-weld spots and propagated right round the hull, so that, finally, the ship broke in half. Had the hull been riveted, the crack would have been arrested at the first rivet hole it encountered. Some riveted joints are now in fact incorporated in such structures to act as crack arresters.

5.51 The relationship between mechanical properties and the method of stress application has already been mentioned (2.52) with particular reference to the impact test. Plastic flow depends upon the movement of dislocations and this occurs in some finite time. If the load is applied very rapidly it is possible for stress to increase so quickly that it cannot be relieved by slip. A momentary increase of stress to a value above the yield stress will produce fracture.

5.52 As temperature decreases, the movement of dislocations becomes more difficult and this increases the possibility of internal stress exceeding the yield stress at some instant. Brittle fracture is therefore more common at low temperatures. This is supported by the fact that the liberty ships were in service in the cold North Atlantic.

5.53 Those metals with a FCC structure maintain ductility at low temperatures, whilst some metals with structures other than FCC tend to exhibit brittleness. BCC ferrite is particularly susceptible to brittle fracture, which follows a transcrystalline path along the (100) planes (see 4.12). This occurs at low temperatures as indicated in Fig. 5.9 and the temperature at which brittleness suddenly increases is known as the *transition temperature*.

Other things being equal, as the carbon content of a steel increases so does the transition temperature, making steel more liable to brittle fracture near ambient temperatures. Phosphorus has an even stronger effect in

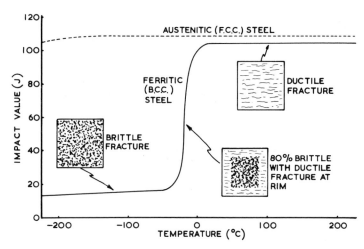

Fig. 5.9 The relationship between brittle fracture and temperature for ferritic and austenitic steels.

raising the transition temperature in steel and this is one reason why phosphorus is one of the least desirable impurities in ordinary carbon steels. Some elements, notably manganese and nickel, have the reverse effect in that they depress the transition temperature. Thus, for applications involving atmospheric temperatures the transition temperature can be reduced to safe limits by increasing the manganese-carbon ratio of the steel, whilst at the same time controlling the grain size by small additions of aluminium. A suitable steel contains 0.14% carbon and 1.3% manganese. Where lower temperatures are involved it is necessary to use low-nickel steels.

Fatigue

5.60 Engineers have long been aware that either 'live' loads or alternating stresses of relatively small magnitude can cause fracture in a metallic structure which could carry a much greater static or 'dead' load. Under the action of *repeated or fluctuating stresses* a metal may become *fatigued*. Some of the earliest quantitative research into metal fatigue was carried out in 1861 by Sir William Fairbairn. He found that raising and lowering a 3-tonne mass onto a wrought iron girder some 3×10^6 times would cause the girder to break, yet a static load of 12 tonnes was necessary to cause failure of a similar girder. From further investigations he concluded that there was some load below 3 tonnes which could be raised and lowered an infinite number of times without causing failure. Some ten years later Wöhler did further work in this direction and developed the fatigue-testing machine which bears his name.

Many engineering metals are subjected to fluctuating stresses during service. Thus the connecting rods in a piston engine are successively pushed and pulled; contact springs in electrical switch gear are bent to and fro; rotating axles suffer a reversal of stress direction through every 180° of rotation; and aircraft wings are continually bent back and forth as they pass through turbulent air. Often a structure will vibrate in sympathy with some external vibration emanating from equipment such as an air compressor. This too may initiate fatigue failure so that fracture by *fatigue is probably the most common cause of engineering failure* in metals.

5.61   The yield strength of a material is a measure of the *static* stress it can withstand without permanent deformation, and is applicable only to components which operate under static loading. Metals subjected to fluctuating or repeated forces fail at lower stresses than do similar metals under the action of 'dead' or steady loads. A typical relationship between the number of cycles of stress (N) and the stress range (S) for a steel is shown in the *S–N* curve (Fig. 5.10 (ii)). This indicates that if the stress in the steel is reduced then it will endure a greater number of stress cycles. The curve eventually becomes almost horizontal indicating that, for the corresponding stress, the member will endure an infinite number of cycles. The stress born under these conditions is called the *fatigue limit*, S_D. For many steels the fatigue limit is approximately one half of the tensile strength as measured in a 'static' test.

5.62   Most non-ferrous metals and alloys and also some steels operating under conditions of corrosion, give *S–N* curves of the type shown in Fig. 5.10(iii). Here there is no fatigue limit as such and a member will fail ultimately if subjected to the appropriate number of stress reversals even at extremely small stresses. With such materials which show no fatigue

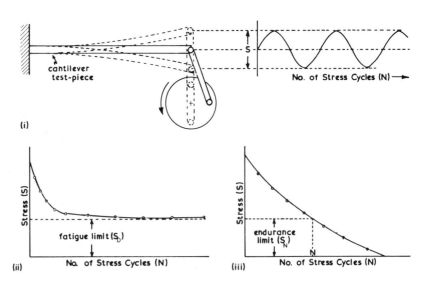

Fig. 5.10 (i) Represents a test piece suffering alternating stress of range S about a mean value of zero; (ii) a typical *S–N* curve for steel; (iii) an *S–N* curve typical of many non-ferrous alloys and some steels, particularly those operating under conditions promoting corrosion.

limit an *endurance limit*, S_N, is used instead. This is the maximum stress which can be sustained for a stated number, N, of cycles of stress. Components made from materials of this type must therefore be designed with some specific life (in terms of stress cycles) in mind and then 'junked' —as our American friends put it—after an appropriate working life, that is, before the number of cycles (N) for the corresponding stress (S) has been reached.

It should be noted that many authorities now use the terms 'fatigue limit' and 'endurance limit' to mean the same, but the above distinction still seems valid in differentiating between the two classes of *S–N* curve obtained for different materials.

5.63 Fatigue-testing machines vary in design but many are based on the original Wöhler concept (Fig. 5.11). Here the test piece is in the form of a cantilever which rotates in a chuck, the load, W, being applied at the 'free' end. For every rotation of 180° the stress will change from W in one direction to W in the opposite direction. Thus the stress-range will be $2W$ with a mean value of zero. To determine the fatigue limit (or endurance limit) a number of test pieces are tested in this way, each at a different value of W, until failure occurs, or alternatively, until an infinite number of stress reversals have been endured—since it is somewhat impracticable to apply an infinite number of reversals, a large number (say 20×10^6) is used instead.

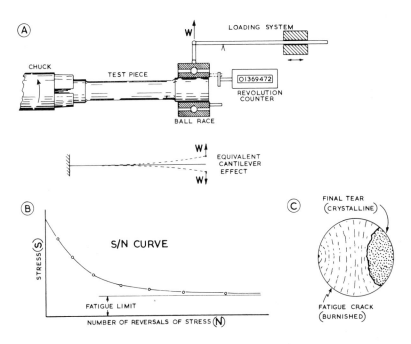

Fig. 5.11 The principles of a simple fatigue-testing machine in which the stress range, $S = 2W$.

5.64 The Mechanism of Fatigue Failure Fatigue failure begins quite early in the service life of the member by the formation of a small crack, generally at some point on the external surface. This crack then develops slowly into the material in a direction roughly perpendicular to the main tensile axis. Ultimately the cross-sectional area of the member will have been so reduced that it can no longer withstand the applied load and ordinary tensile fracture will result. A fatigue crack 'front' advances a very small amount during *each* stress cycle and each increment of advance is shown on the fracture surface as a minute ripple line. These ripple lines radiate out from the origin of fracture as a series of approximately concentric arcs. The individual ripples are far too small to be visible on the fractured surface except by using very high-powered metallographic methods, but under practical conditions a few ripples much larger than the rest, probably corresponding to peak stress conditions, are produced and these are visible on the fractured surface showing the general path which the crack has followed (Fig. 5.12).

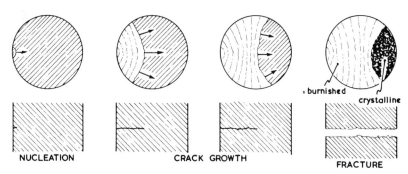

Fig. 5.12 The progress of fatigue failure.
A fatigue fracture is often easy to identify because the crack-growth region is burnished by the mating surfaces rubbing together as stresses alternate. The ultimate fracture is strongly 'crystalline' in appearance.

A fatigue fracture thus develops in three stages—nucleation, crack growth and final catastrophic failure. Since the crack propagates slowly from the source, the fractured surfaces rub together due to the pulsating nature of the stress and so the surfaces become burnished whilst still exhibiting the conchoidal markings representing the large ripples. Final fracture, when the residual cross section of the member is no longer able to carry the load, is typically crystalline in appearance. Fatigue failures in metals are therefore generally very easy to identify.

Fatigue cracks are not the result of brittle fracture but of plastic slip. During cyclic stressing, at stresses above the fatigue limit, plastic deformation is produced continuously and alternately positive and negative. It is this continuous to-and-fro plastic deformation in localised regions which ultimately propagates and spreads a fatigue crack. The continual plastic oscillation of metal layers along slip planes in and out of the surface causes some of the metal to produce ridges as work hardening sets in to resist 'back slip'. These ridges are forced up at the surface and are termed *extrusions* (Fig. 5.13). Narrow fissures or *intrusions* are formed in a similar manner. Several of these intrusions may then interconnect to initiate the start of a fatigue crack.

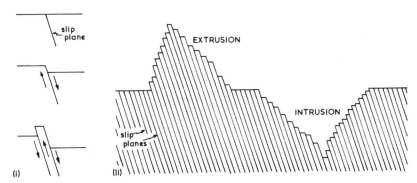

Fig. 5.13 (i) Stages in the formation of an extrusion by movements of blocks of atoms along slip planes; (ii) the nature of extrusions and intrusions.

5.65 Some Causes of Fatigue Failure   Since the fatigue characteristics of metals can be measured accurately, it seems reasonable to suppose that fatigue failure due to a lack of appropriate allowances in design should not occur. This is of course true, yet fatigue failure *does* continue to happen even in the most sophisticated items of engineering equipment. The catastrophic fatigue failure in Comet airliners some years ago is a case in point. In many such instances a member may be designed to carry a *static* load (well above S_D or S_N), yet it may be suffering undetected vibrations which give rise to reversal of stress at a value *above* S_D (or S_N). Such vibrations are often sympathetic.

Other design faults include the presence of such stress raisers as a sharp-cornered key way in a shaft or an unduly sharp radius in an in-cut corner. Poor workmanship in the shape of surface roughness, scratches or careless toolmarks can also cause stress concentrations. Minute quench cracks in heat-treated steels are also a source of fatigue failure as are the presence of small cavities, inclusions or other discontinuities just below the surface of the material. A surface weakened by decarburisation or other types of deterioration which have led to the softening and weakening of the surface layers also favours the initiation of micro-cracks via the formation of intrusions.

Corrosive conditions in the environment can also provide stress-raisers in the form of etch-pits and other surface intrusions such as grain boundary attack. Ordinary surface oxidation may have a similar effect and consequently nucleate a fatigue crack.

5.66 Methods of Improving Fatigue Strength   In order to maintain fatigue strength it follows from the above that surface finish should always be good. Engineer craftsmen of the past were not wasting their time when producing a high surface polish on many of their items of equipment.

Since the fatigue strength of a metal is approximately proportional to its tensile strength, any method used to increase the tensile strength of the material will correspondingly improve the fatigue strength. As fatigue failure nearly always commences at the *surface*, methods of surface hardening will be most effective in limiting the initiation of fatigue cracks. Thus, work-hardening of the surface by shot-peening is beneficial, whilst the carburising and nitriding (19.10) of steels will improve fatigue strength. A case-hardened axle provides a good all-round combination of core toughness, surface wear-resistance and *fatigue resistance*.

Creep

5.70 So far in this description of plastic deformation and ultimate failure of metals little mention has been made of the part played by *time*. Nevertheless, metals in service are often required to withstand steady stresses over long periods of time and it has been shown that under these conditions gradual deformation of the metal may occur at *all* temperatures and stresses. Metals can therefore fail in this way at a stress well below the tensile strength at that particular temperature. This phenomenon of continuous gradual extension under a steady force is known as *creep*.

The effects of creep in most engineering metals are not serious at ambient temperatures though some metals of low melting point (T_m) such as lead exhibit noticeable creep under these conditions. Lead sheeting which may have been protecting a church roof for centuries is sometimes found to be slightly thicker near the eaves than at the ridge. The lead has 'crept' under its own weight over the centuries.

Creep becomes very serious with many metals at temperatures above $0.4\ T_m$ where T_m, as noted previously (4.42) is measured on the Kelvin (or 'absolute') scale. Thus T_m for lead is equivalent to $327 + 273$ or 600 K. Hence $0.4\ T_m$ (for lead) is 240 K or $-33°C$. This explains why lead sheeting on a church roof will be likely to creep appreciably at ambient temperatures between, say, $-10°C$ and $30°C$. Similarly the effects of creep must be very seriously considered in the design of gas and steam turbines, steam and chemical plant and furnace equipment. Creep is thus associated in practical terms with both *time* and *temperature* and is basically due to slip by the movement of dislocations.

5.71 Fig. 5.14 shows the type of creep curve obtained when a metal is suitably stressed. This curve indicates that, following the initial elastic strain, the plastic strain associated with creep occurs in three stages:

(i) Primary, or *transient* creep, *EP*, beginning at a fairly rapid rate which then decreases with time because work-hardening sets in.

(ii) Secondary, or *steady-rate* creep, *PS*, in which the rate of strain is uniform and at its lowest value.

(iii) Tertiary creep, *SX*, in which the rate of strain increases rapidly

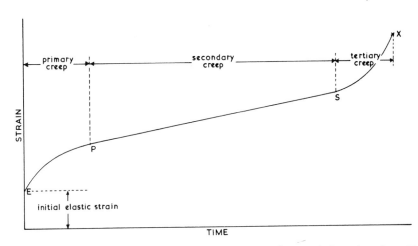

Fig. 5.14 A typical creep curve, showing three stages of creep during a long-time, high temperature creep test.

until fracture occurs at X. This stage coincides with necking of the test piece.

The form of relationship which exists between stress, temperature and the resultant creep rate is shown in Fig. 5.15. At a low stress and/or a low temperature (curve A) some primary creep occurs but this falls to a negligible value as strain-hardening prevents further slip from taking place. With increased stress and/or temperature (curves B and C) the rate of secondary creep also increases leading to tertiary creep and ultimate failure.

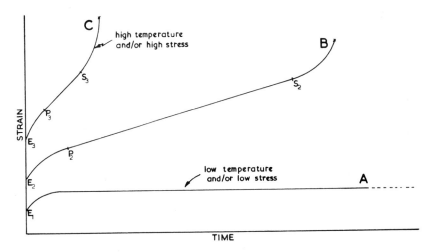

Fig. 5.15 Variations of creep rate with stress and temperature.
In curve A the creep rate soon becomes negligible as work-hardening sets in. In curve C the creep rate is higher than in curve B because of the use of either higher stress or higher temperature.

5.72 In creep testing, the specimen, in form similar to a tensile test piece, is enclosed in a thermostatically controlled electric tube furnace which can be maintained with accuracy over long periods at any given temperature up to 1000°C or more. A simple lever system is often used to load the test piece, and some form of delicate extensometer or strain-gauge system employed to measure the resultant extension at suitable time intervals. The extensometer is sometimes of the optical type, using mirrors, scales and telescopes.

Creep values were originally assessed in terms of the *limiting creep stress*. This was defined as that stress at any given temperature, below which no *measurable* creep takes place. Obviously this value depends upon the sensitivity of the measuring equipment and in any case it is now known that some creep occurs at all combinations of stress and temperature. Moreover, since formal creep tests involve long periods of time sometimes running into many months, other related tensile values are now used instead of the old concept of 'limiting creep stress'.

Most tests involve measuring the stress which will give rise to some predetermined *rate* of creep during the secondary stage of uniform deformation. The creep rate will therefore be equivalent to the slope of the curve during the secondary stage of creep. The Hatfield 'time-yield value' is derived from such a short-term creep test. It determines the stress at a given temperature which will produce a strain of 0.5% of the gauge length in the first twenty-four hours and a further strain of not more than one part per million of the gauge length in the next forty-eight hours. Creep tests of this type do not attempt to derive a 'creep limit' but are practical values upon which engineering design can be based.

5.73 The Mechanism of Creep Creep is a deformation process in which three main features appear to be involved:

(i) the normal movement of dislocations along slip planes;

(ii) a process known as 'dislocation climb' which is responsible for rapid creep at temperatures above $0.5 \ T_m$;

(iii) slipping at grain boundaries.

In the primary stages of creep, dislocations move quickly at first but soon become piled-up at various barriers. Nevertheless, thermal activation enables them to surmount some barriers, though at a decreasing rate so that the creep rate is reduced. At temperatures in excess of $0.5 \ T_m$, thermal activation is sufficient to promote a process known as 'dislocation climb' (Fig. 5.16). This would bring into use *new* slip planes and so reduce the rate of work hardening. Hence creep is a process in which work hardening is balanced by thermal softening which allows slip to continue. At low temperatures recovery does not take place due to lack of thermal activation (Curve A—Fig. 5.15) and so unrelieved work-hardening leads to a reduction in the creep rate almost to zero.

In addition to plastic deformation by dislocation movement, deformation by a form of slip at the grain boundaries also occurs during the secondary stage of creep. These movements possibly lead to the formation of 'vacant sites', that is lattice positions from which atoms are missing, and this in turn makes possible 'dislocation climb' (Fig. 5.16). The relationship

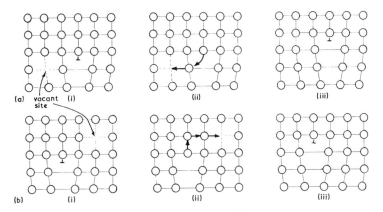

Fig. 5.16 Two examples illustrating 'dislocation climb'.
In each case this is made possible by the presence nearby of a 'vacant site' (or 'vacancy'), ie a lattice position not occupied by an atom. In both (a) and (b) the dislocation has 'climbed' on to a new slip plane.

between grain boundaries and creep is indicated by the fact that at high temperatures fine-grained metals creep more than coarse-grained metals of the same compositions, presumably because the fine-grained metals contain a higher proportion of grain-boundary region per unit volume of metal. At low temperatures, where the grain-boundary material is more 'viscous', fine-grained metals are more creep-resistant and generally tougher.

In the tertiary stage of creep micro-cracks are initiated at grain boundaries due largely to the movement of dislocations, but in some cases to the migration of vacant sites there. Necking and consequent rapid failure follow.

5.74 Creep Resistance   This can be increased by impeding the movement of dislocations in a metal and also by inhibiting the formation of new ones. Thus the presence of solute atoms (8.21) which do not diffuse rapidly will 'pin down' dislocations effectively; whilst the presence of small dispersed particles of a hard, strong constituent will have a 'particle-hardening' effect (8.62) by acting as barriers to the movement of a dislocation front. The high-temperature 'Nimonic' series of alloys (18.32) rely on this type of strengthening mechanism.

Exercises

1. Distinguish, with the aid of sketches, between *ductile* and *brittle* fracture. (5.20 and 5.30)
2. Show how Griffith's Crack Theory explains the considerable effects which the presence of microstructural faults have as stress raisers in a metal in tension. Why is cast iron weak in tension but strong in compression? (5.22)
3. What are the principal causes of intercrystalline brittle fracture? (5.24)

4. What are the main factors leading to crack formation in metals under the influence of mechanical stress? (5.40)
5. Discuss the relationship between impact value and service temperature for a 0.2% carbon steel. How may the properties of such a steel be improved for service at sub-zero temperatures? (5.50)
6. Write a short account of the importance of a consideration of fatigue in engineering design. (5.60)
7. Define the term *fatigue limit* and show how the character of this value varies as between ferrous and non-ferrous materials. Describe a method used to determine the fatigue limit of a given steel. (5.61 and 5.63)
8. Why is the concept of *limiting creep stress* no longer used as a criterion of creep phenomena? What other values are used to quantify creep and how are they derived? (5.72)
9. Creep is a high-temperature phenomenon. Show by what stages and mechanisms it proceeds. (5.71 and 5.73)

Bibliography

American Society for Metals, *Case Histories in Failure Analysis*.
Biggs, W. D., *The Brittle Fracture of Steel*, Macdonald and Evans, 1960.
Chell, G. G., *Developments in Fracture Mechanics*, Applied Science, 1979.
Colangelo, V. J. and Heiser, F. A., *Analysis of Metallurgical Failures*, John Wiley, 1974.
Duggan, T. V. and Byrne, J., *Fatigue as a Design Criterion*, Macmillan, 1977.
Fuchs, H. O. and Stephens, R. I., *Metal Fatigue in Engineering*, John Wiley, 1980.
Kennedy, A. J., *Processes of Fatigue and Creep in Metals*, Oliver and Boyd.
Osborne, C. J., *Fracture*, Butterworths, 1979.
Smallman, R. E., *Modern Physical Metallurgy*, Butterworths, 1985.
Sully, A. H., *Creep*, Butterworths.
BS 3518: 1984 *Methods of Fatigue Testing*.
BS 3500: 1987 *Methods for Creep and Rupture Testing of Metals*.

6

The Industrial Shaping of Metals

6.10 The chance discovery of small quantities of copper in the uncombined state undoubtedly led to the cold-forging of this malleable metal into various tools and weapons some seven thousand years ago. During the millennia which followed the craft of the metalworker developed so that, by the time of Tubal Cain, he was a well-established member of society. The melting and smelting of metals was probably achieved by chance in a primitive camp fire and the possibilities of metal casting ultimately recognised from the observation that solid metal retained the shape of the depression in the earth where it solidified.

What we now refer to as 'investment casting' was being used in Mesopotamia as long as five thousand years ago. A rough shape of the proposed casting was modelled in clay and this was coated with several layers of beeswax. The desired shape of the article was then carefully carved in the surface of the wax and this, in turn, was enclosed in a clay 'cocoon'. After allowing the clay to dry the wax was carefully melted out leaving a cavity between the outer clay shell and the inner clay core. Into this cavity molten copper or bronze was poured. The clay core was generally supported in position by being attached to the outer mould at some point. This left a hole in the final casting, generally arranged in some inconspicuous position.

The ancient metal worker was an artist rather than a scientist though he soon learned the value of cold forging in hardening both copper and bronze. Most of the empires of the ancient world relied for their military prowess on swords and spears fashioned from these metals.

6.11 In modern times metals and alloys may be shaped into something approaching the final form by one of the following programmes of operations:

(i) casting into some form of shaped mould;
(ii) casting as an ingot followed by a hot-working process;
(iii) casting as an ingot followed by a cold-working process;
(iv) compacting and sintering from a metallic powder.

Other processes such as electro-deposition and the condensation of metal vapours are sometimes used but these operations are usually confined to the surface treatment of metallic components rather than to their actual shaping. Thus, electro-deposition is sometimes used to build up parts which have become badly worn. Any one of the shaping methods mentioned above may be followed by some form of metal-cutting process, referred to generally as 'machining'. The study of this wide field of operations is the province of the production engineer but a very brief treatment of the metallurgical aspects of machinability is given later in this chapter. The industrial processes catalogued in this chapter are dealt with in greater detail in Part II but it was thought desirable to give a brief outline of the more important of them at this stage to help those readers who will not be studying metal shaping at a more detailed level.

Sand casting, die casting and allied processes

6.20 Large numbers of engineering and domestic components are produced by casting the molten metal or alloy into some form of mould cavity which will give to the component its final shape. The best known of these processes is sand casting. Here a suitable foundry sand is rammed around a wooden pattern, the two halves of the moulding box being capable of separation (Fig. 6.1) so that the pattern can be withdrawn leaving a mould cavity into which metal can be poured.

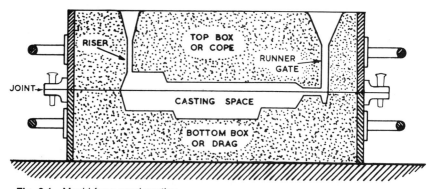

Fig. 6.1 Mould for a sand-casting.
A wooden pattern is used, the top box being raised so that the pattern can be removed.

If large numbers of castings are required and the process proves technically suitable a far superior product is obtained by casting into a metal mould. In permanent-mould casting (also known as *gravity* die casting) the molten alloy is allowed to run into the metal mould under gravity, whilst in die casting (also called *pressure* die casting) the charge is forced into the mould cavity under considerable pressure (Fig. 6.2). The use of die casting is confined mainly to zinc- or aluminium-base alloys and the product is

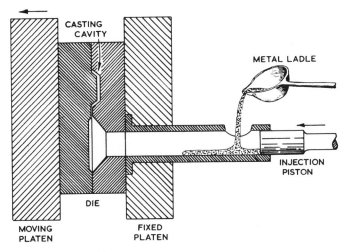

Fig. 6.2 A die-casting machine.
The molten metal is forced into the die by means of the piston. When the casting is solid the die is opened and the casting moves away with the moving platen, from which it is ejected by 'pins' passing through the platen.

metallurgically superior to a sand casting in that the internal structure is more uniform and the grain much finer because of the rapid cooling rates which obtain due to the metal mould. Moreover, output rates are much higher when using a permanent mould than they are when using sand moulds. Greater dimensional accuracy and a better surface finish are also obtained by die casting. However, some alloys which can be sand cast cannot be die cast because of their high shrinkage coefficients during solidification. Such alloys would inevitably crack due to their high contraction during solidification in the rigid metal mould.

Obviously the cost of producing metal dies is high so that die casting is essentially a process suitable for providing large numbers of castings of the types used in domestic washing machines, vacuum cleaners, automobile wind-screen wiper motors and the like. If only small numbers of castings are required then sand casting may have to be used since the cost of making a wooden pattern is small compared with that for a metal die. Moreover, many intricate shapes can be produced only by sand casting because of the possibility of using destructible sand cores.

6.21 Shell Moulding (5.30—Pt. II) is similar in principle to sand casting in that a two-part mould is used but, whilst a small proportion of clay acts as a natural bond in foundry sand, in shell moulding a fine clay-free silica sand with added bonding material of the thermosetting resin type (phenol- or urea-formaldehyde) is used. As is indicated in Fig. 6.3 each half of the shell mould is produced when the heated metal pattern is covered with the sand/resin mixture in a rotatable 'dump box'. The resin hardens on the hot pattern, binding together the sand particles to produce a strong tough half-mould. Though Fig. 6.3 illustrates a mould made from

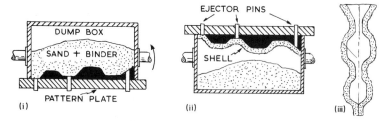

Fig. 6.3 Shell moulding.
(i) The heated pattern plate is attached to the dump box, which is inverted so that the pattern is covered with the sand/binder mixture; (ii) the dump box is returned to the charging position leaving a hardened shell of sand/resin mixture which is then detached from the pattern plate using the ejector pins; (iii) two halves, constituting the mould, are joined to receive their charge of molten metal.

two similar half-moulds the complete mould need not be symmetrical about the parting line so long as the halves 'match' at the parting line. The finished halves are joined together by adhesives or mechanically, and are usually supported in coarse sand or gravel to receive their metal charge. The main advantages of this process over sand casting are greater dimensional accuracy and a better surface finish.

6.22 Investment casting (5.40—Pt. II). As mentioned at the beginning of this Chapter the 'lost wax' process had its origins in ancient history. It was rediscovered in the sixteenth century by Benvenuto Cellini who used the process to produce many works of art in gold and silver, but the development of investment casting for engineering purposes took place more recently during the Second World War. In this process accurately dimensioned wax patterns are produced in a metal master mould. A number of these patterns are then attached to a central wax 'runner' (Fig. 6.4(ii)) and the resultant 'tree' invested with a mixture usually consisting of fine sillimanite sand and ethyl silicate. This, in the presence of moisture, hydrolyses to form a solid bond between the sillimanite particles. The wax pattern is then melted out and the mould baked, partly to harden it but

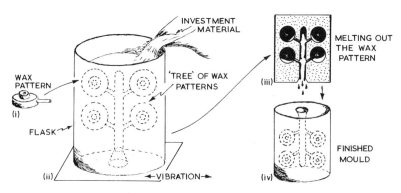

Fig. 6.4 The production of a mould in the investment casting process.

also to prevent chilling of the metal charge in thin sections. The main feature of the process is that small castings of very high dimensional accuracy can be produced. Moreover the casting has no 'parting line' and surface finish is excellent.

6.23 The Replicast ceramic shell process The 'lost wax' process described briefly above is limited generally to the manufacture of small precision castings since the cost of large amounts of investment material would make it uncompetitive for large castings. The Replicast process virtually replaces the wax pattern with one produced in *high-density* expanded polystyrene. Complex patterns can be made by sticking together several polystyrene units. Since this is a very light-weight material only a thin (4–6mm) shell of the investment material is necessary because of reduced stresses during the removal of the pattern. The thin shell mould is then supported by loose sand in a box prior to pouring the charge. The loose sand is vibrated so that it enters all recesses and so provides even support to the thin mould.

Even when modern technical controls are introduced during its production a casting is in many respects one of the least reliable or predictable of metallurgical structures because of the many variable factors which operate during the melting of the metal and its subsequent casting and solidification in the mould. Blow-holes, shrinkage cavities and oxidation defects may be present in a sand casting due to poor melting and/or casting practice. Moreover, the inherent characteristics of a cast metal, particularly in respect of coarse grain and segregation of impurities are such that its mechanical properties are always inferior to those of a wrought product.

If, therefore, mechanical strength and toughness are prime factors, a casting should properly only be used if the component is of such intricate shape that it cannot be produced by any other means, or if the alloy used is not amenable to any working processes, as, for example, some of the modern 'super alloys' (18.32 and 18.86) which are so strong and hard at high temperatures that they cannot be forged satisfactorily. In commerce the cost of manufacture must always be considered however, so that where only a few components are required, the choice of operation will frequently fall upon sand casting, notwithstanding its shortcomings.

6.24 The reader may be familiar with those 'non-drip' household paints which are said to be *thixotropic*. They are firm jelly-like materials which, when subjected to shear stress in the form of brush strokes, immediately become much less viscous and flow freely. When brushing ceases the paint viscosity increases again so that it does not drip. A metal shaping process known as *thixomolding* is based on this principle and is mentioned at this point as a process which in a sense is intermediate between casting and hot working.

The thixomolder consists essentially of a screw pump into which solid metal is fed in pellet form. As the metal is forced along the barrel it passes into a heating zone where it reaches a part-liquid/part-solid state. This slurry, containing up to 65% solid, is then forced at considerable pressure into a die somewhat similar in design to that used in diecasting. High pressure results in thixotropic flow of the alloy. The process is particularly useful in shaping magnesium alloys where foundry melting is hazardous and melting losses high.

Hot-working Processes

6.30 An increase in the temperature of a metal leads to an *increase* in inter-atomic spacings so that the bond strength will *decrease*. Moreover, the strain caused by the presence of a dislocation will also decrease due to the increase in inter-atomic distances. Thus the dislocation can be moved more easily through the crystal. As a result yield strength falls as temperature rises.

Since most metals become considerably softer and more malleable as temperature rises less energy is needed to produce a given amount of deformation. In fact, hot-working processes are invariably carried out above the recrystallisation temperature of a metal or alloy. Deformation and recrystallisation therefore take place simultaneously so that in addition to a saving of energy a considerable speeding up of the process is possible with no tedious inter-stage annealing operations such as attend cold-working processes. Some alloys can only be shaped by hot-working since they are hard or brittle when cold due to the presence of a hard microconstituent which is absorbed at the hot-working temperature.

Whilst malleability increases with rise in temperature ductility generally decreases because the material becomes less strong as the temperature rises and so tends to tear apart in tension. Consequently *hot-working processes* invariably involve the use of *compressive* forces as in forging, rolling and extrusion, whilst drawing processes which employ *tensile* forces are essentially *cold-working* operations. The main hot-working processes are dealt with briefly below.

6.31 Hot-rolling   The rolling mill was adapted by Henry Cort in 1783 as a simple 'two-high' non-reversing mill for the production of wrought-iron bar. The advent of the Bessemer process in 1856 made necessary the development of rolling-mill technique so that it could keep pace with the large amount of steel made by this new mass-production method. Thus in 1857 the 'three-high' mill was introduced so that the work piece could be rolled on the 'return pass' (Fig.6.5(i)). Unfortunately support of the bearings for the centre roll was generally found to be difficult so Ramsbottom developed the fore-runner of the modern two-high reversing mill (Fig.6.5 (ii)) at Crewe in 1866.

Hot-rolling is universally employed in the reduction of large steel ingots to sections, strip, sheet and rod of various sizes. In fact the only conditions under which cold-work is applied to steel are when the section is too small to retain its heat, or when a superior finish is required in the product.

A steel-rolling shop consists of a powerful two-high reversing mill to 'breakdown' the white-hot ingots, followed by trains of rolls (Fig. 6.5(iii)) which will be either plain or grooved according to the type of product being manufactured. Hot-rolling is similarly applied to most non-ferrous

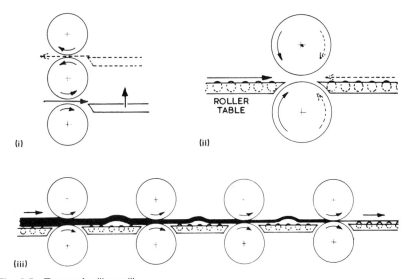

Fig. 6.5 Types of rolling mill.
(i) The three-high mill; (ii) the two-high reversing mill for 'breaking down' ingots; (iii) a train of non-reversing rolls for 'finishing'.

alloys in the initial breaking-down stages, but the finishing operations are more likely to involve cold work.

6.32 Forging   Tubal Cain (Genesis IV, 22) was forging metals by hand at least six thousand years ago but there is evidence that pre-historic man was using this type of process to fashion copper some two thousand years earlier. Hand forging is still used to a limited extent by the smith but most modern forging is power-assisted. Wrought iron was the traditional material of the smith but many ferrous and non-ferrous alloys are now shaped by forging processes and wrought iron has long become obsolete as an engineering material.

During forging the coarse 'as-cast' structure is broken down and is replaced, as recrystallisation proceeds, by one which is of relatively fine grain. At the same time impurities are redistributed in a more or less fibrous form. Therefore it is more satisfactory, all other things being equal, to forge a component than to cast it to shape.

6.33 Drop-forging   If a large number of identical forged components are required, then it is economically preferable to make them by a drop-forging process. In this process (Fig. 6.6) a shaped die is used, one half being attached to the hammer and the other half to the anvil. With complex shapes a series of dies may be used.

The hammer, working between two vertical guides, is mechanically lifted some distance above the anvil and allowed to fall under gravity on to the work piece which consists of a heated bar or billet of the metal placed on the anvil half of the die. As the hammer falls it forges the metal between the two halves of the die. A modification of drop-forging employs either mechanical or steam power to force the hammer downwards, thus increas-

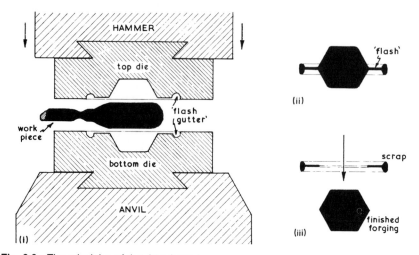

Fig. 6.6 The principles of the drop forging process.
In order to fill the die cavity completely, excess metal is required in the work piece. This excess is forced into the 'flash' gutter and the 'flash' produced (ii) is trimmed off in a suitable die to give the finished forging (iii).

ing the power of the blow. 'High-energy rate' forging in which the machines are generally operated by pneumatic pressure is now also used (8.37—Part II).

6.34 Heading 'Heading' or 'up-set' forging is employed extensively for the manufacture of bolts, rivets and other components, where an increase in diameter is necessary without loss of strength and shock-resistance. The stock bar is heated for a portion of its length and then forged, as shown in Fig. 6.7, in a machine which also parts off the component. It will be obvious that a bolt head forged in this way will be much stronger than one which has been machined from a bar, since forging does not cut into the fibrous structure of the material and thus introduce planes of weakness into the finished bolt head.

6.35 Hot-pressing Hot-pressing is a development of the drop-forging process, but is generally applied in the manufacture of more simple shapes. The hammer of drop-forging is replaced by a hydraulically driven ram, so that, instead of receiving a rapid succession of hammer blows, the metal is gradually squeezed by the static pressure of the ram. This downwards thrust is sometimes as great as 100 MN.

The main advantage of hot-pressing over drop-forging is that working is no longer confined to the surface layers, as it is with drop-forging, but is transmitted uniformly to the interior of the metal being shaped. This is particularly important when forging very large components, such as marine propeller shafts, which would otherwise suffer from having a non-uniform internal structure.

6.36 Extrusion The extrusion process is now used for shaping a variety of ferrous and non-ferrous metals and alloys. Its most important feature is

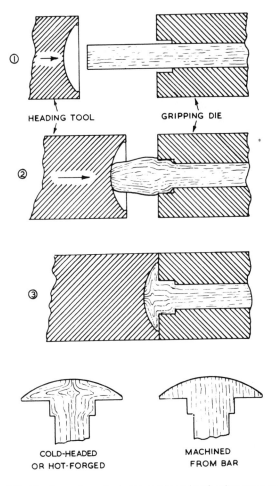

①

HEADING TOOL   GRIPPING DIE

②

③

COLD-HEADED   MACHINED
OR HOT-FORGED   FROM BAR

Fig. 6.7 The formation of a bolt head by a hot-forging process.

that we are able to force the metal through a die, and, in a single process from the cast billet, to obtain quite complicated sections of tolerably accurate dimensions. The metal billet is heated to the required extrusion temperature (350–500°C for aluminium alloys; 700–800°C for brasses; 1100–1250°C for steels) and placed in the container of the extrusion press (Fig. 6.8). The ram is then driven hydraulically with sufficient pressure to force the metal through a hard alloy-steel die. The solid metal section issues from the die in a manner similar to the flow of toothpaste from its tube.

Using this process, a wide variety of sections can be produced, including round rod, hexagonal brass rod (for parting off as nuts), brass curtain rail, small-diameter rod (for drawing still further to wire), tubes in many alloys including stainless steels; and many hollow stress-bearing sections in aluminium alloys (mainly for aircraft construction).

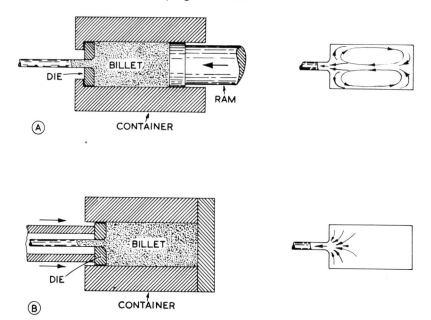

Fig. 6.8 Extrusion by (A) the 'direct' and (B) the 'indirect' process. The 'flow' of metal in the container is indicated in each case.

The most serious defect from which extruded sections suffer is called the 'extrusion defect' or 'back-end defect', because of its occurrence in the last part of the section extruded. It is caused by some of the surface scale of the original billet being drawn into the core of the section by the turbulent flow of metal in the container of the extrusion press. A number of methods are effective in reducing it, including the 'indirect' process of extrusion, where, in effect, the die is pushed into the billet, instead of, as in the ordinary direct process, the billet being forced through the die. In the latter process there will be relative movement between the billet and the walls of the container, setting up a turbulent flow of metal due to frictional forces, and so drawing surface scale into the main stream of extruded metal. Since relative movement between billet and container is reduced to a minimum in the indirect process, frictional forces are much smaller. Consequently, not only is less turbulence produced in the surface layers of the metal being extruded but also the power requirements are less than in the direct process where much more energy is lost in overcoming friction between the skin of the billet and the surface of the container.

Cold-working Processes

6.40 Cold-working from the ingot to the finished product, with, of course, the necessary intermediate annealing stages, is applied only in the case of a few alloys. These include both alloys which are very malleable in the cold and, on the other hand, those which become weak and brittle when heated.

Cold-working is more often applied in the finishing stages of production. Then its functions are:

(*a*) to enable accurate dimensions to be attained in the finished product;

(*b*) to obtain a clean, smooth finish;

(*c*) in some cases to straighten the work piece;

(*d*) by adjusting the amount of cold-work in the final operation after annealing, to obtain the required degree of hardness, or 'temper', in alloys which cannot be hardened by heat-treatment.

Raising the temperature of an alloy generally increases its malleability, but reduces its ductility because tensile strength falls as the temperature increases. This causes the metal to tear apart in tension before appreciable extension has occurred. Thus in hot-working processes we are always *pushing* the alloy into shape, whilst in cold-working operations we frequently make use of the high ductility of some alloys when cold, by *pulling* them into their required shapes. Therefore, processes involving the pulling or 'drawing' of metal through a die are always cold-working operations.

There are many cold-working processes, but the principal ones used in metallurgical industries are discussed below.

6.41 Cold-rolling   Cold-rolling is applied during the finishing stages of production of both strip and section and also in the production of very thin materials such as foil. The type of mill used in the production of the latter is shown in Fig 6.9. Here small diameter rolls must be used in order to produce any reduction in the already thin material. Moreover, since the sheet may be a metre or more in width the rolls must consequently be long and therefore liable to bend under the considerable roll pressures which

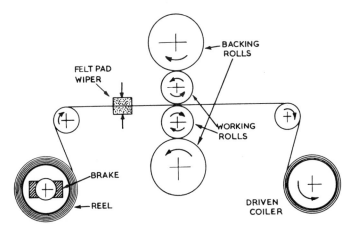

Fig. 6.9 The type of rolling mill used in the production of thin strip and foil. It is a four-high mill, the large-diameter backing rolls supporting the working rolls.

are necessary. This bending is minimised by using some system of backing rolls as in the four-high mill illustrated. In most other cases cold-rolling mills are similar in design to those used for hot-rolling.

The production of mirror-finished metal foil is carried out in rolls enclosed in an 'air-conditioned' cubicle, and the rolls themselves are polished frequently with clean cotton wool. Only by working in perfectly clean surroundings, with highly polished rolls, can really high-grade foil be obtained.

In addition to quality of finish, the objects of cold-rolling are accuracy of dimensions and the adjustment of the correct temper in the material.

Fig. Plate 6.1 A four-high non-reversing mill for the production of stainless-steel sheets. *(Courtesy of Messrs W. H. A. Robertson & Co Ltd, Bedford).*

6.42 Drawing of Solid and Hollow Sections Drawing is exclusively a cold-working process, since it relies entirely on the high ductility of the material being drawn and ductility is low at high temperatures. Both solid sections and tubes are produced by drawing through dies, and all wire is manufactured by this process. In the manufacture of tubes the bore may be maintained by the use of a mandrel, as shown in Fig. 6.10. Rods and tubes are drawn at a long draw-bench on which a power-driven 'dog' pulls the material through the die. Wire, and such material as can be coiled, is pulled through the die by winding it on to a rotating drum or 'block'. In each case the material is lubricated with oil or soap before it enters the die.

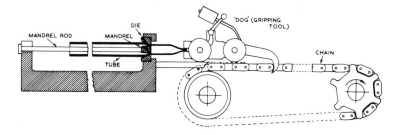

Fig. 6.10 A draw-bench for the cold-drawing of tubes.

Drawing dies (Fig. 6.11) are made, for different purposes: from high-carbon steel; from tungsten and molybdenum steels; from tungsten carbide (a tungsten carbide 'pellet' enclosed in a steel case); and, for very fine gauge copper wire, from diamond.

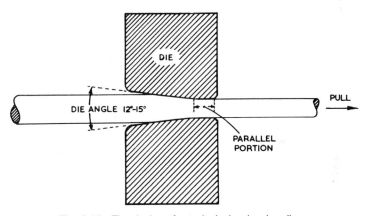

Fig. 6.11 The design of a typical wire-drawing die.

6.43 Cold-pressing and Deep-drawing These processes are so closely allied to each other that it is very difficult to define each separately. The operations range from making a suitable pressing in one stage to cupping followed by a number of re-drawings, as shown in Fig. 6.12. In each case the components are produced from sheet stock, and range from mild-steel motor-car bodies to cartridge cases (70–30 brass), bullet envelopes (cupro-nickel) and aluminium milk churns.

Deep-drawing demands very high ductility in the sheet stock, and only a limited range of alloys are, therefore, available for the process. The best known are 70–30 brass, cupro-nickel, pure copper, pure aluminium and some of its alloys, and some of the high-nickel alloys. Mild steel can be pressed and, to a limited extent, deep-drawn. It is used in the manufacture of innumerable motor-car and cycle parts by these methods.

Typical stages in a deep-drawing process are shown in Fig. 6.12, but it should be noted that wall-thinning may or may not take place in such a process. If wall-thinning is necessary, then one of the more ductile alloys must be used. Though Fig. 6.12 shows the processes of shearing and primary cupping as being effected in different machines, usually a combination tool is used, so that both processes take place in one machine.

Cold-pressing is very widely used, and alloys which are not quite sufficiently ductile for deep-drawing are generally suitable for shaping by simple press-work. Of recent years high-energy rate forming by the use of high explosives has also been developed (12.91—Part II).

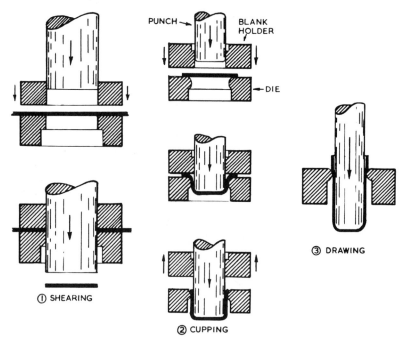

Fig. 6.12 Stages in the cupping and deep-drawing of a component.

6.44 Coining Coining, or embossing, is really a cold-forging process and, as its name implies, it is used for the production of coinage, medallions and the like. It has also been used on an experimental scale for producing engineering components to exact dimensions. A malleable alloy is essential for this type of process if excessive wear on the dies is to be avoided. *Embossing* is similar in principle except that no variation in thickness of the work piece is produced. It involves only light press work as in the forming of military buttons and badges.

6.45 Spinning Spinning is a relatively simple process in which a circular blank of metal is fixed to the spinning chuck of a lathe. As the blank rotates, it is forced up into shape by means of hand-operated tools of blunt steel or hardwood, supported on a hand-rest. The function of the hand-tool is to press the metal into contact with a former of the desired shape. This former is also fixed to the rotating chuck.

The operation is in some respects similar to that of a potter's wheel in which the lump of clay is replaced by a flat disc of metal. Large reflectors, aluminium teapots and hot-water bottles, and other domestic hollow-ware are frequently produced by spinning.

Sintering from a Powder

6.50   This method of producing metallic structures has become increasingly important in recent years, and is particularly useful when there is a big difference in the melting points of the metals to be alloyed, or when a metal has an extremely high melting point. As an example, products containing a high proportion of tungsten (m.pt 3410°C) are usually formed by sintering, since this metal is difficult and expensive to melt on a commercial scale.

6.51   The metal to be sintered is obtained as a fine powder, either by grinding; by volatilisation and condensation; or by reduction of its powdered oxide. Any necessary mixing is carried out, and the mixed powdered metals are then placed in a hardened steel die and compressed. The pressures used vary with the metals to be sintered, but are usually between 70 and 700 N/mm^2. As the metal particles rub together under these high pressures a degree of cold welding occurs between them at some points on the surfaces of contact. The brittle compressed mass is then heated in an electric furnace to a temperature which will cause sintering to take place and produce a mechanically satisfactory product. This sintering is dependent upon grain-growth taking place *across* the incipient cold-welds which occurred during pressing.

6.52   Tungsten is compacted and sintered in this way and the resulting sintered rod is drawn down to wire for the manufacture of electric-lamp filaments. The demand by engineers for tools which would be superior to high-speed steels for cutting and machining metals led to the development of sintered tungsten carbide products. Tools of this type are made by mixing tungsten carbide powder with up to 13% of powdered cobalt; the mixture is pressed into blocks and heated in a hydrogen atmosphere to a temperature of 700–1000°C. After this treatment the blocks are still soft enough to be cut and ground to the required shape. They are then heated in hydrogen to a temperature of 1400–1500°C, when sintering takes place, producing a material harder than high-speed steel. The function of cobalt is to provide a tough, shock-resistant, bonding material between the tungsten carbide particles. In addition to their use in cutting tools, cemented carbides of this type are employed in dies, other wear-resistant parts, percussion drills and armour-piercing projectiles. When making tools for machining steels, particularly with fine cuts or at high speeds, up to 20% titanium carbide may be incorporated.

6.53   Sintering is also used to produce an even distribution of some insoluble constituent in a metallic structure. In so-called oil-less bearings powdered graphite is compressed with powdered copper and tin to make a bronze bearing which is, as a result, impregnated with graphite. This is then sintered at about 800°C which is of course well above the melting point of tin which consequently melts and infiltrates the copper particles effectively 'soldering' them together. Thus the manufacture of bronze bearings differs from a true powder metallurgy process in this respect. Whilst still hot the bearings are quenched in lubricating oil. Such treatment results in a structure resembling a metallic sponge, which, when saturated with lubricating oil, produces a self-oiling bearing. In some cases the amount of oil held in the bearing is sufficient to last the lifetime of the machine. These self-lubricating bearings are used in the automobile industry, but find particular application in many domestic articles, such as vacuum cleaners, electric clocks and washing-machines; in all of which, long service with a minimum of maintenance is necessary.

The Machinability of Metals and Alloys

6.60   The mechanical forming operations already mentioned in this chapter are often classified by engineers as 'chipless-forming' processes. Many metal-shaping procedures however, rely on metal cutting. The considerable technology of these cutting processes is too vast a field to be described here and is properly the province of the production engineer. Nevertheless microstructure has considerable influence upon the ease with which a material can be machined and that aspect will receive a brief mention.

Machinability, or the ease with which a material may be machined, may be measured in a number of ways. It can be assessed as the energy required to remove unit volume of the work piece or the thrust required to remove material at a given rate. It may also be measured in terms of tool life. In each case standard conditions must be prescribed.

Machining is really a cold-working operation in which the cutting edge of the tool forms chips or shavings of the material being machined, and the process will be facilitated if minute cracks form just in advance of the cutting edge, due to high stress concentrations set up by the latter. Very ductile alloys do not machine well, because local fracture does not occur easily under the pressure of the cutting tool. Instead, such alloys will spread under the pressure of the tool and 'flow' around its edge (Fig. 6.13(i)), so that it becomes buried in the metal. Friction then plays its part, leading to the overheating and ultimate destruction of the cutting edge.

6.61   It follows, therefore, that a brittle alloy will be far more suitable for machining than would a ductile one. On the other hand, a brittle alloy will generally be unsuitable in ultimate service, particularly under conditions of shock. However, we can compromise by introducing what is, in effect, *local* brittleness in an alloy, whilst at the same time producing little or no deterioration in the impact toughness of the material *as a whole*. Work-hardening of the surface as in the case of 'bright-drawn steel' rod will produce this local brittleness but a more effective way of increasing

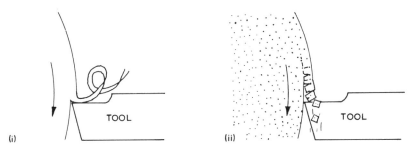

Fig. 6.13 The influence on machinability of isolated particles in the microstructure.

machinability is to introduce a suitable concentration of small isolated particles into the microstructure.

6.62 Such particles, whether of hard or soft material, have the effect of setting up stress concentrations locally, as the cutting edge approaches them. A minute fracture will therefore travel from the cutting edge to the particle in question, (Fig. 6.13(ii)), thus reducing friction between the tool and the material being cut. In addition to the reduction of wear on the tool, there will also be a reduction in the overall power required. Moreover, due to the discontinuity introduced by the particles, the swarf produced will be in small, conveniently sized pieces, instead of the long, curly slivers obtained when a ductile material is machined. In fact, from some of the free-cutting materials, such as leaded brasses, the swarf produced is more in the nature of a powder.

6.63 Many alloys having a duplex structure will fall naturally into the free-cutting or semi-free-cutting class. The slag fibres in wrought iron, the graphite in cast iron and particles of a hard compound (16.42) in a high-tin bronze are examples in which the presence of inclusions improves machinability. In many cases, however, a deliberate addition must be made in order to introduce such isolated particles. In the cheaper free-cutting steels the presence of sulphur is utilised by ensuring that sufficient manganese is also present to combine with the sulphur and form manganese sulphide. Manganese sulphide exists as isolated globules in the microstructure, whereas the iron (II) sulphide, which would form if manganese were absent, would exist as brittle intercrystalline films. Thus, whilst in a good-quality steel of the ordinary type sulphur is usually present only in amounts less than 0.06%, in a free-cutting steel as much as 0.20% sulphur may be present, the manganese content being raised to between 0.90 and 1.20% to ensure that all the sulphur is combined with manganese.

6.64 Some alloys, whether in the liquid or solid state, will not dissolve lead. It therefore exists as isolated globules scattered at random in the microstructure instead of being distributed as intercrystalline films, as is often the case when *liquid solubility* followed by *solid insolubility* leads to the segregation of a constituent at the crystal boundaries (4.41). Particles distributed in this globular form cause a relatively small deterioration in the mechanical properties.

Table 6.1 *The Effects of Lead on the Free-cutting and Mechanical Properties of Various Steels*

Type of steel	Condition	Composition (%)						Mechanical properties			Comparative machinability constants	Improvement in machinability when lead is added (%)
		C	Mn	S	Ni	Cr	Pb	Tensile strength N/mm^2	Elongation (%)	Izod (J)		
Free-cutting	6.35 mm diameter drawn	0.13	0.95	0.23	—	—	—	568	24.0	—	12.5	24.0
Leaded free-cutting	6.35 mm diameter drawn	0.12	0.98	0.22	—	—	0.20	571	23.0	—	15.5	
0.40% carbon	39.7 mm hexagonal normalised	0.40	0.61	0.03	—	—	—	567	26.0	32	6.2	29.0
Leaded 0.4% carbon	39.7 mm hexagonal normalised	0.42	0.61	0.03	—	—	0.20	602	25.0	34	8.0	
Nickel-chromium	31.75 mm diameter heat-treated	0.41	0.74	0.03	1.73	0.65	—	1043	18.0	75	4.75	35.0
Leaded nickel-chromium	31.75 mm diameter heat-treated	0.42	0.25	0.02	1.74	0.65	0.17	981	18.0	71	6.4	

Lead is used in this way to improve the machinability of steels as well as of non-ferrous alloys (in particular, brasses and bronzes), but whereas the amounts used are in the region of 1.0–3.0% in brass, in steels the amounts are only of the order of 0.15–0.35%, so that the particles of lead are so small that they are not generally visible using a simple microscope. These 'Ledloy' steels can be either of the plain carbon or alloyed type and, as shown in Table 6.1, machinability is improved by up to 35% (in terms of relative machinability constants) for alloy steels. The use of lead is particularly advantageous in introducing free-cutting properties into such alloys as the 13% chromium stainless steels, because the alternative method of utilising the presence of sulphur and manganese may lead to a reduction in the impact value by half, and a general reduction in other mechanical properties.

6.65   Elements other than manganese are sometimes added in the presence of sulphur to improve machinability. Molybdenum or zirconium, together with 0.1% sulphur, will produce non-abrasive sulphide globules. Selenium is sometimes used to replace sulphur, particularly in the austenitic stainless steels. Selenium is chemically very similar to sulphur, and forms selenide globules which act as chip breakers in the same way as sulphide globules. Some typical steels containing the elements mentioned are shown in Table 6.2.

6.66   As has been mentioned, free-cutting properties are imparted to the copper alloys by the addition of about 2.0% lead. The machinability

Table 6.2 *Typical Free-cutting Steels*

Type of steel	BS 970 designation	Composition	
Free-cutting mild steel	220M07	Carbon	0.1%
		Manganese	1.0%
		Sulphur	0.25%
Semi-free cutting '35–45' carbon bright drawn	216M28	Carbon	0.28%
		Manganese	1.3%
		Sulphur	0.16%
Free-cutting '40' carbon	225M44	Carbon	0.44%
		Manganese	1.5%
		Sulphur	0.25%
Chromium rust-resisting steel (free-machining)	416S21 and 416S41	Carbon	0.12%
		Chromium	13.0%
		Sulphur	0.3% max
		Manganese	1.5% max.
		Selenium	0.3% max.
		Molybdenum	0.6% max.
		Optional—Lead	0.35% max.
Free-cutting 18–8 stainless steel	303S21 303S41 325S21 and 326S26	Carbon	0.1%
		Chromium	18.0%
		Nickel	10.0%
		Manganese	1.5%
		Sulphur	0.25%
		Molybdenum	3.0% max.
		Titanium	0.6%

of brasses, bronzes and nickel silvers is improved in this way. The lead exists as tiny globules, plainly visible in the microstructure (Plate 6.2B). Since lead is insoluble in the liquid alloys, as well as in the solid, care has to be exercised to prevent segregation during the melting and casting of such alloys.

Although the initial cost of leaded brasses is high compared with some free-cutting steels, it must be realised that the *very fine powdery swarf* produced from leaded brasses gives a high tool life. Moreover, since the

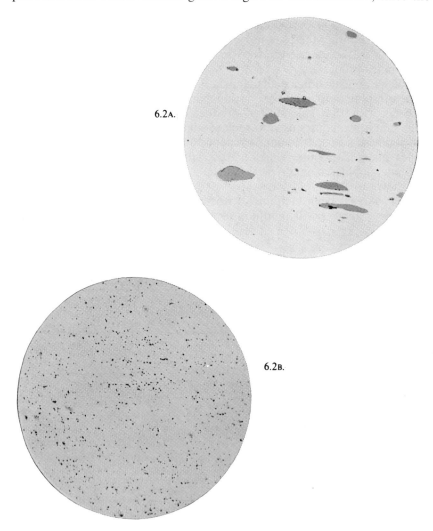

6.2A.

6.2B.

Plate 6.2   6.2A Free-cutting sulphur steel.
Longitudinal section showing inclusions of manganese sulphide. × 250. Unetched.
6.2B Free-cutting extruded brass rod (58% copper; 2.5% lead; balance zinc).
Longitudinal section showing globules of undissolved lead. × 100. Unetched.

swarf is compact it can be handled easily, economically transported and re-melted, so that its 'scrap-return value' is high. These factors mitigate the initial high cost of leaded brass.

6.67   Many heat-treatment processes (11.52 and 11.60), the objects of which are to improve machinability, also rely on the isolation of constituents in globular form so that they act as chip formers in this general manner.

Exercises

1. What casting processes would be likely to be used for the following:
   (a) the zinc-base alloy body of an automobile windscreen-wiper motor;
   (b) a cast iron water pipe approximately 150 mm in diameter;
   (c) the cast steel housing of a rolling mill.
   Give reasons for your choice in each case along with a brief outline of the process. (6.20)
2. Discuss the factors involved when choosing between the available methods of casting a component. (6.20)
3. Compare and contrast the effects of cold-working and hot-working on metals. (6.30 and 6.40)
4. Compare forging and casting as processes of production; with special reference to metallurgical effects. (6.32)
5. (a) Describe the effects of forging on the structure and properties of a metal;
   (b) Why is 'grain flow' important in a forged material?
   (c) Make a sketch to illustrate the correct 'flow lines' in the horizontal section of a crankshaft. (6.32 and 6.34)
6. Describe, briefly, suitable fabrication processes for the following:
   (a) a shell or cartridge case;
   (b) an aluminium saucepan;
   (c) a carburettor body;
   (d) a metal H-channel for a curtain runner. (6.20, 6.30 and 6.40)
7. What are the principal reasons for using a *powder-metallurgy* process? (6.50)
8. Show how microstructure and physical properties affect the suitability of
   (i) pure copper;
   (ii) grey cast iron, for machining operations. (6.60)
9. Discuss the factors which influence the machining properties of metals and alloys. Show how free-cutting properties are obtained in:
   (i) 'bright-drawn' steel rod;
   (ii) extruded brass rod;
   (iii) grey cast iron;
   (iv) free-cutting quality 18–8 stainless steel. (6.60)

Bibliography

Avitzur, B., *Handbook of Metal Forming Processes*, John Wiley, 1984.
Higgins, R. A., *Engineering Metallurgy (Part 2)*, Hodder and Stoughton, 1985.
Rowe, G. W., *Principles of Industrial Metalwork Processes*, Arnold, 1977.
Smart, R. F., *Developments in Powder Metallurgy*, Mills and Boon, 1973.
BS 970: 1973 and 1988 *Wrought Steels in the Form of Blooms, Billets, Bars and Forgings* (Part 1 includes free-cutting steels).

7

An Introduction to Steel

7.10 The military or industrial power of a nation has long been associated with its ability to produce steel. The might of sixteenth-century Spain was not unconnected with the quality of Toledo steel blades whilst in Britain at about that time the introduction of laws limiting the felling of trees for charcoal production, testified to the quantity of steel then being manufactured. Later, the growth of the Industrial Revolution in Britain gave her world supremacy in steel production as a result of the development of iron and steel making by Dud Dudley and Abraham Darby and, subsequently, by Huntsman, Bessemer, Gilchrist and Thomas. Towards the end of the great Victorian era Britain was manufacturing most of the world's steel and ruling an Empire on which the 'sun never set'.

During the present century depletion—and eclipse—of our home ore supplies combined with the development of vast ore deposits overseas has completely altered the situation. The USA and the CIS (formerly the USSR) owe their material power largely to the presence of high-grade iron ore within, or near to, their own considerable territories putting them among the world leaders in steel production. Britain currently occupies eleventh position behind the CIS, Japan, the USA, PR China, Germany, Italy, Republic of Korea, Brazil, the Benelux Group and France having fallen from the third position she held in the 'steel league' at the end of the Second World War. Many other countries too are increasing their steel-making capacities, generally from ore deposits within their own territories, so that outputs are approaching that of Britain. Of the big steel producers West Germany and Japan, as well as Britain, have to import a large proportion of their ore requirements.

7.11 The highest quality iron ores are the oxides magnetite, Fe_3O_4, and heamatite, Fe_2O_3, some deposits of the former containing almost 70% of the metal. British home-produced ore is low-grade material of the carbonate or hydroxide type, high in phosphorus and containing as little as 20% iron. Fortunately it occurs near to the surface as 'sedimentary' deposits which can be mined by open-cast methods. For this reason three-quarters of the ore used in British blast furnaces has to be imported in the form of

high-grade concentrates from Swedish Lapland, Africa, Canada or Vene-
zuela. The principal *ore* producing countries are (in order of production):
CIS, PR China (and DPR Korea), Brazil, Australia, the USA, India,
Canada, South Africa, Sweden, Venezuela, Liberia, Mauretania and
France.

Pig Iron Production

7.20   The smelting of iron ore takes place in the blast furnace (Fig. 7.1).
This is a shaft-type furnace some 60 m or more in height and having an
output capacity of up to 10 000 tonnes per day. The refractory lining of
such a furnace is designed to last for several years since once the furnace
goes 'on blast' it would not be economical to shut down for re-lining.
Production takes place on a 365 days-a-year basis, though 'damping down'
—with the blast turned off or reduced—has to be used for interruptions
such as strikes.

Ore, coke and limestone are charged to the furnace through the double
bell-and-cone gas-trap system, whilst a pre-heated air blast is blown in
through tuyères near the base of the furnace. In order to reduce coke
consumption still further and also increase furnace output fuel oil is some-

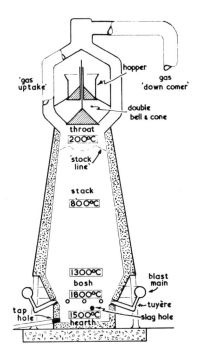

Fig. 7.1   The iron blast furnace. The complete structure may be sixty or more metres in
height.

times injected with the air blast. At regular intervals of several hours both tap hole and slag hole are opened in order to run off, first the slag and then the molten pig iron. The holes are then re-plugged with clay.

7.21   The smelting operation involves the following main chemical reactions:

(i) Coke in the region of the tuyères burns completely—

$$C + O_2 \rightarrow CO_2 + \text{Heat } (\textit{Exo}\text{thermic})$$

Just above the tuyères the carbon dioxide is reduced by the white-hot coke to carbon monoxide—

$$CO_2 + C \rightarrow 2CO - \text{Heat } (\textit{Endo}\text{thermic})$$

On balance the overall reaction is strongly exothermic, so that the temperature remains high.

(ii) As the carbon monoxide, which is a powerful reducing agent, rises through the charge it reduces the iron(III) oxide (in the ore)—

$$Fe_2O_3 + 3CO \rightarrow 2Fe + 3CO_2$$

(The reduction occurs in three stages: $Fe_2O_3 \rightarrow Fe_3O_4 \rightarrow FeO \rightarrow Fe$.) This reaction takes place in the upper part of the furnace where the temperature is too low for the iron so formed to melt. It therefore remains as a spongy mass until it moves down into the lower part of the furnace where it melts and runs down over the white-hot coke dissolving carbon, sulphur, manganese, phosphorus and silicon as it goes. Apart from carbon—which is absorbed from the coke—these elements are dissolved following their reduction from compounds present as impurities in the original ore.

(iii) At the same time as this reduction is taking place, the earthy waste or 'gangue' associated with the ore combines with lime (formed by the decomposition of limestone added with the charge) to produce a fluid slag:

$$CaCO_3 \rightarrow CaO + CO_2$$
Limestone   Lime

$$2CaO + SiO_2 \rightarrow 2CaO.SiO_2$$
Calcium silicate (slag)

The function of the lime here is to liquefy the gangue. The latter, being composed largely of silica, would not *melt* at the blast-furnace temperature and must therefore be attacked chemically to form a low melting-point slag which will run from the furnace.

7.22   The furnace is tapped at regular intervals, the iron generally being stored in the molten state in a 'mixer' prior to transfer to the steel-making plant. Some may be cast as 'pigs' for subsequent re-melting.

In the early days of this century an average blast furnace produced about 100 tonnes of pig iron per day but modern furnaces with outputs of 10 000 tonnes per day are in operation both here and abroad. The approximate quantities of materials in both charge and products for the daily 'throughput' of a medium sized blast furnace would be:

Charge (tonnes)		Products (tonnes)	
Ore (say 50% iron)	4 000	Pig iron	2 000
Coke	1 800	Slag	1 600
Limestone	800	Blast furnace gas	10 800
Air	8 000	Dust	200
Total	14 600	Total	14 600

The slag produced is of low value and is used mainly for 'filling' purposes —railway ballast, road making, concrete aggregate and for the manufacture of slag wool (for thermal and accoustic insulation). It should not be confused with the 'basic slag' used in agriculture which is a by-product of steel-making processes.

One item in the above table which may surprise the reader is the enormous volume of both air and resultant blast-furnace gas involved in the production of a tonne of pig iron. An upwards flow of gas on this scale demands a very *porous* charge in the furnace to allow the passage of gas. For this reason the bulk of the ore used and particularly powdery imported 'concentrates', have to be 'agglomerated' before being charged to the furnace. This involves sintering the fine material to produce strong large lumps. Any fine ore or dust is blown out at the top of the furnace by the upwards rush of gas. The coke too must be strong enough so that it does not crush under the enormous pressure of the charge. Hence special coking-coal is used in its production.

7.23   The blast-furnace gas contains considerable amounts of carbon monoxide which remain unused during reduction of the ore. Since the gas has a useful calorific value the blast furnace performs a secondary role as a giant gas producer and all of this gas is utilised as fuel. After being cleaned of dust much of it is burned in the Cowper stoves which, in turn, pre-heat the in-going air blast. Two such regenerator stoves are required for each blast furnace. One is being re-heated by the burning blast-furnace gas whilst the other is pre-heating the in-going air. The surplus gas is utilised in many ways in an integrated steel-making plant, eg for raising electric power or for firing different types of re-heating furnace used around the plant.

7.24 Pig iron   is a complex alloy. In addition to iron it contains up to 10% of other elements, the chief of which are carbon, silicon, manganese, sulphur and phosphorus. These elements are absorbed as the reduced iron melts and runs down through the white-hot coke and slag. The total amount of carbon is usually 3–4% and it may be present either as iron carbide, Fe_3C (also called *cementite*) or as un-combined carbon (*graphite*). A high silicon content of 2.5% or more favours the formation of graphite during solidification so that a fractured surface of the solid iron is grey and the iron is called a 'grey iron'. A low silicon content on the other hand, say 0.5%, will favour the formation of iron carbide, Fe_3C, during solidification and the resultant iron will be a 'white iron' since the fractured surface will show white iron carbide. However, the *rate* of solidification of the iron also affects the formation of either graphite or iron carbide as it does in cast iron (15.30). Some grades of pig iron—generally those low in sulphur and phosphorus—are used for the manufacture of iron castings but the bulk of pig iron produced is transferred, still molten, to the steel-making plant.

7.25 Direct Reduction Processes   The increasing scarcity of coke suitable for use in the blast furnace has led to the development of alternative methods of iron production. Whilst the bulk of iron produced still comes from the blast furnace, these new methods of iron production will doubtless become much more important during the next two decades.

In some of these processes 'pelletised' high-grade ore is fed continuously into a slowly rotating kiln which is fired by natural gas or other hydrocarbons. The fuel provides the necessary heat and also carbon monoxide which effects chemical reduction of the ore:

$$Fe_2O_3 + 3CO \rightarrow 2Fe + 3CO_2$$

In a similar process the ore is contained in enclosed retorts through which a mixture of carbon monoxide and hydrogen—derived from natural gas—circulates. Having reduced the ore in the first retort, the 'diluted' gas is then burned to heat the second retort, and so on. Moving-grate furnaces are also used to carry the ore through a reducing flame provided by the partial combustion of natural gas.

The processes mentioned briefly above employ gaseous fuels to heat and reduce the charge, but in other methods cheap solid fuels such as coke breeze or even lignite are used. The product of these processes is known as 'sponge iron'. It is reasonably pure iron but unfortunately is still mixed with the gangue present in the original ore. Since this gangue is mainly silica it must be fluxed with lime when the sponge iron is subsequently remelted for steel making.

7.26 Electric Iron Smelting Processes   These are also used where coke is expensive, relative to the cost of electricity, as in Scandinavia. These processes more nearly resemble blast-furnace smelting except that heat is supplied by an electric arc rather than by coke, though low-grade coke is used as the reducing agent. The coke is mixed with the ore and fed into a reaction hearth where carbon electrodes provide the heating current. The ore is reduced by the hot coke:

$$Fe_2O_3 + 3C \rightarrow 2Fe + 3CO$$

Unlike the direct reduction processes mentioned above where sponge iron is produced, the end-product here is a molten pig iron. The gas leaving the reduction chamber is very rich in carbon monoxide so it is cleaned and used as an energy source in the same way as blast-furnace gas.

Electric smelting of iron has the advantage that it produces less carbon dioxide, whereas for every tonne of pig iron tapped from the blast furnace some 3.5 tonnes of CO_2 finds its way—by whatever route—into the atmosphere. Thus a large blast furnace producing 10 000 tonnes of pig iron per day is responsible for discharging into the environment about 12 million tonnes of CO_2 every year. This joins the outflow of all other carbon-burning enterprises in promoting the 'greenhouse effect'.

An alternative method of smelting iron would seem to be by using hydrogen as a reducing agent:

$$Fe_2O_3 + 3H_2 \rightarrow 2Fe + 3H_2O \text{ (which ultimately falls as rain)}$$

The hydrogen would be provided by the electrolysis of water, using electricity from nuclear power. This is very expensive, and the anti-nuclear lobby would react; but do we have any long-term choice if the Earth itself is not to 'go critical' in a rather different sense? In the short term a much more efficient reclamation of scrap steel would help. Instead we allow millions of tonnes to rust and 'escape' into the environment every year.

The Manufacture of Steel

7.30   Prior to the introduction of the first blast furnaces during the fourteenth century, iron had always been reduced as a *solid* metal by heating a mixture of the ore and charcoal in a hearth-type furnace. Quite small hand-powered bellows were used to provide the air blast and the temperature attained was not high enough to melt the iron released by chemical action from the ore. Instead particles of iron, mixed with large amounts of slag, collected at the bottom of the hearth. These porous masses, called *blooms*, were then hammered so that much of the molten slag was expelled and a bar of relatively pure iron containing some remnant slag particles was the result. By repeatedly forge-welding such bars together and rehammering into a new bar, the slag particles were elongated into fibres (Pl. 11.1B) and *wrought iron* was produced. Later the reduction hearth was replaced by a small shaft-type furnace and much larger bellows were driven by water power. The temperature reached was sufficient to melt the reduced iron and so the first pig iron ran from a blast furnace. This pig iron was then remelted in an open hearth so that the impurities were oxidised and relatively pure iron, of consequently higher melting point, crystallised out. This was then forged to produce wrought iron as before. Oxidation of the remelted pig iron in this way was later known as 'puddling'.

Being of quite high chemical purity wrought iron had a good resistance to corrosion as is shown by examples of nineteenth-century shipbuilding. Thus the hull of the four-masted sailing ship *Munoz Gamero* built in 1875 but now lying at Punta Arenasin, Chile, shows a remarkably high state of preservation of her puddled iron plates.

7.31   By the Middle Ages wrought iron had been manufactured, by one method or another, for some four thousand years and it is reasonable to suppose that sooner or later some ancient craftsman would have noticed that 'wrought iron' could be hardened by cooling it in water provided that it had been heated in a charcoal fire for a sufficiently long time. This ultimately led to the manufacture of steel by what was later called the *Cementation Process*. Bars of wrought iron were packed into stone boxes along with charcoal and heated at about 900°C for a week. Carbon diffused into the solid wrought iron* and the product was forged to give a more homogeneous steel. Such production methods were expensive and steel was used only as a tool material. In the meantime wrought iron continued to be used for structural and constructional work and was not finally abandoned until the Tay Bridge disaster of 1879 when it was probably quite wrongly blamed for the collapse of the railway bridge.

7.32 In 1742 Benjamin Huntsman, a Sheffield clock maker, decided that his clock springs were breaking because they lacked homogeneity, due largely to the presence of slag in the wrought iron from which the cementation steel had been manufactured. He therefore melted bars of cementation steel so that most of the slag was lost. In this way he ultimately perfected *crucible cast steel* for which Sheffield became justly famous.

7.33 Modern mass-production methods of steel manufacture began in 1856 when Henry Bessemer attempted to speed up wrought-iron manufacture by blowing air *through* a charge of molten pig iron contained in a pear-shaped 'converter'. Due to the high rate of the chemical reactions involved in the oxidation of impurities such as carbon, silicon and manganese in the pig iron, the temperature ran so high that, instead of solid pure iron crystallising out, the final product remained molten in the converter and had to be cast. After preliminary difficulties had been overcome, *low-carbon steel* suitable for constructional purposes became available for the first time and soon the Bessemer process was established for the mass-production of steel.

Since the Bessemer process was a very rapid production method—the complete 'blow' lasted some half-hour—there was little time available to control the composition and quality of the product. This led to the introduction of the open-hearth process by Siemens-Martin in 1865. In this process pig iron could be melted and refined along with large quantities of steel scrap which was becoming available towards the end of the nineteenth century. The main advantage however, was that the complete refining process in the open-hearth took some eight to ten hours so that there was ample time for control and adjustment of the composition of the product.

Until 1878 only those pig irons low in sulphur and phosphorus were suitable for steel making since, unlike silicon, manganese and carbon, these impurities were not oxidised and removed in the slag. Then came the research of Thomas and Gilchrist which enabled large quantities of high-phosphorus pig iron available in Britain to be converted to steel. To achieve this they added lime to the furnace charge thus producing a *basic slag* which would combine with phosphorus after the latter had been oxidised:

---

* The process was similar in principle to modern methods of carburising for case-hardening (19.20).

$$3CaO + P_2O_5 \rightarrow Ca_3(PO_4)_2$$

Calcium phosphate

This calcium phosphate joined the basic slag.

Thus both the Bessemer and open-hearth processes flourished in Britain and, subsequently, elsewhere for almost a century. In either case the process was said to be 'acid' or 'basic'. The acid processes were so called because they utilised low-phosphorus pig irons and therefore did not require the addition of lime to the charge. The slag formed was acid since it contained an excess of silica (derived from oxidised silicon) and to match this the furnace was lined with silica bricks. Those pig irons rich in phosphorus required the charge to be treated with lime and this produces a basic slag. This *basic* slag would quickly attack ordinary *acid* silica brick furnace linings and so furnaces used in basic steel making had to be lined with a *basic* refractory such as 'burnt' magnesite (MgO) or 'burnt' dolomite (MgO.CaO).

7.34   As low-phosphorus ore became scarce in Britain more and more steel was produced by basic processes. Gradually the basic open-hearth became the dominant steel-making process because of its capability of producing high-quality steel from high-phosphorus raw materials. Nevertheless vast quantities of *mild* steel continued to be made by the Bessemer process; though one of its chief disadvantages was that since only the impurities present in the initial pig iron were available as fuel, to keep the charge molten during the 'blow' no scrap could be added and the pig iron composition had to be between close limits.

Of the air blown into the Bessemer converter only 20% by volume had a useful function in oxidising the impurities. This of course was the oxygen. The remaining 80% (mainly nitrogen) entered the converter cold and emerged as hot gas, thus carrying heat away from the converter and reducing the thermal efficiency of the process. Moreover small amounts of nitrogen dissolved in the steel during the 'blow'. This increased the hardness of the product and frustrated the demand for mild steel of increasing ductility by the motor-car manufacturers and others.

7.35   In 1952 a new approach to these problems was made in steel plants at Linz and Donnawitz in Austria. Here, instead of blowing air through molten pig iron as in the Bessemer process, pure oxygen was injected into the surface of molten pig iron via a water-cooled 'lance'. This process—called the *L-D process*—was made possible by the introduction of cheap 'tonnage' oxygen and though this was the first major steel-making process not to be developed in Britain, it is only fair to say that Bessemer had been aware of the advantages of using oxygen rather than air in his original process. Unfortunately in the nineteenth century oxygen was far too expensive to produce on a large scale.

Basic Oxygen Steelmaking (BOS)

7.36   Following the introduction of L-D steelmaking in 1952 a spate of modifications of the process followed. Thus both the Kaldo process

(Sweden) and the Rotor process (West Germany) were popular for a time and it is inevitable that variations of the general oxygen method will continue to be developed. Up to the time of publication all such modifications have had the following features in common:

(i) an oxygen blast is used to oxidise impurities in the original raw material, these oxidised impurities being drawn off in the slag;

(ii) the processes are chemically basic so that phosphorus removal is effective.

The BOF (basic oxygen furnace) is a pear-shaped vessel of up to 400 tonnes capacity, lined with magnesite bricks covered with a layer of dolomite. Scrap is first loaded into the converter followed by the charge of molten pig iron. Oxygen is then blown at the surface of the molten charge through a water-cooled lance which is lowered through the mouth of the converter (Fig. 7.2).

As soon as the oxidising reaction commences lime, fluorspar and millscale are admitted to the converter to produce a slag on the surface which will collect the impurities oxidised from the charge. At the end of the 'blow' the slag is run off first and the charge of steel then transferred to a ladle for casting as ingots.

BOS has the following major advantages over competing processes:

(i) It is rapid—the cycling time is about forty-five minutes;

(ii) Nitrogen contamination is very low so that deep-drawing quality mild steel is produced;

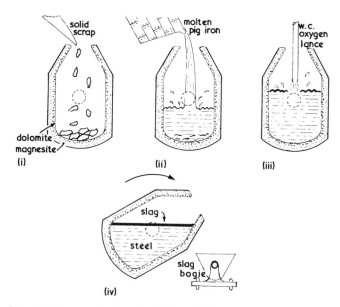

Fig. 7.2 Stages in the manufacture of steel in a BOF.
The water-cooled oxygen lance may be up to 0.5m in diameter and its tip between 1 and 3m above the surface of the charge—depending upon the composition of the latter.

(iii) Thermal efficiency is high because heat is not carried away by nitrogen as in the former Bessemer process. Hence the charge may include 40%—and in some circumstances 50%—scrap;

(iv) A wide variety of both scrap and pig iron can be used.

The development of BOS has rendered the Bessemer process completely obsolete whilst the open-hearth process is now used only in Eastern Europe, India and a few Latin American plants.

Electric Arc Steelmaking

7.37   This is the only alternative steelmaking process which is significant at present and its operation is complementary to BOS rather than competitive. Originally electric-arc furnaces were used for the manufacture of small amounts of high-grade tool steels and alloy steels. In a modern integrated steel plant it is widely used to melt process scrap and other medium-grade material which can be bought cheaply and then up-graded to produce very high-quality steel. By this means the high cost of electrical energy is largely offset. As electricity offers a chemically neutral method of providing heat, the chemical conditions in the furnace can be altered at will to produce either oxidising or *reducing* slags. The latter favour the removal of sulphur from the charge, making the process one in which sulphur removal is definite.

The furnace (Fig. 7.3) employs carbon rods which strike an arc on to the charge. The lining is basic allowing the addition of lime and millscale in order to produce a basic oxidising slag for the effective removal of phosphorus from the charge as well as any remnant silicon or manganese. Often the slag is then removed to be replaced by a basic *reducing* slag composed of lime, anthracite and fluorspar. This removes sulphur from the charge:

$$FeS + C + CaO \rightarrow Fe + CO + CaS \ (joins\ slag)$$

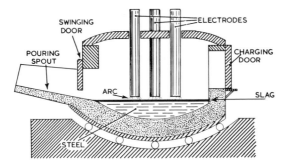

Fig. 7.3   The principles of the electric-arc furnace for steelmaking.

Hence the main advantages of the arc process are:
    (i) Removal of sulphur is reliable;
    (ii) Conditions are chemically 'clean' and contamination of the charge is impossible;
    (iii) Temperature can be accurately controlled;
    (iv) Carbon content can be adjusted between fine limits;
    (v) The addition of alloying elements can be made with precision.
Currently about a quarter of Britain's steel production comes from electric processes. The remainder is from BOS.

The Microstructural Nature of Carbon Steels

7.40 Despite the development of many sophisticated alloys in recent years ordinary steel seems likely to remain the most important engineering alloy available. Hence it has been considered desirable to make a preliminary study of the structures and properties of carbon steels at this stage in preparation for a more detailed study later in the book.

It is impossible adequately to study the structure of a steel, or any other alloy, without reference to what are called 'thermal equilibrium diagrams' —or 'phase diagrams'. Probably some readers will have been introduced to the iron–carbon thermal equilibrium diagram during preliminary studies of materials science. The purpose of this chapter is to clarify such ideas as those readers may have formulated on the subject and also to introduce other readers to this important field of physical metallurgy. Both phase diagrams in general, and that for iron and carbon in particular, will be discussed in succeeding chapters. We will begin by studying the method of construction and also the interpretation of a simple thermal equilibrium diagram by reference to some tin–lead alloys.

7.41 Most readers will be aware that there are two main varieties of tin–lead solder. Best-quality tinman's solder contains 62% tin and 38% lead* and its solidification begins and ends at the same temperature— 183°C (Fig. 7.4(iii)). Plumber's solder, however, contains 33% tin and 67% lead, and whilst it begins to solidify at about 265°C, solidification is not complete until 183°C (Fig. 7.4(i)). Between 265 and 183°C, then, plumber's solder is in a pasty, partly solid state which enables the plumber to 'wipe' a joint with the aid of his 'cloth' (20.21).

From observations such as these it can be concluded that the temperature range over which a tin–lead alloy solidifies depends upon its composition. On further investigation it will be found that an alloy containing 50% tin and 50% lead will begin to solidify at 220°C, and be completely solid at 183°C; whilst one containing 80% tin and 20% lead will begin to solidify at 200°C and finish solidifying at 183°C.

From the data accumulated above we can draw a diagram which will indicate the state in which any given tin–lead alloy (within the range of

---

* Whilst for reasons of economy tinman's solder often contains less than 62% tin (20.21), the latter composition is ideal, since the solder will melt and freeze quickly at a fixed temperature.

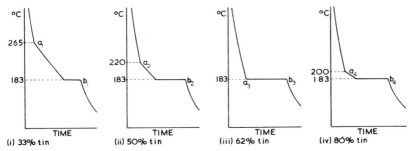

Fig. 7.4 Temperature/time cooling curves for various tin/lead alloys. Points (*a*) indicate the temperature at which solidification begins, and points (*b*) the temperature at which it ends.

compositions investigated) will exist at any given temperature (Fig. 7.5). This diagram has been obtained by plotting the temperatures at which the alloys mentioned above begin and finish solidifying, on a temperature-composition diagram. All points—a_1, a_2, a_3, a_4—at which the various alloys begin to solidify, are joined, as are the points—b_1, b_2, b_3, b_4—where solidification is complete.

Any alloy represented in composition and temperature by a point above *AEB* will be in a completely molten state, whilst any alloy similarly represented by a point below *CED* will be completely solid. Likewise, any alloy whose temperature and composition are represented by a point between *AE* and *CE* or between *EB* and *ED* will be in a part liquid–part solid state.

7.42 Such a diagram is of great use to the metallurgist, and is called a *thermal-equilibrium diagram*—or *phase diagram*. The meaning of the term

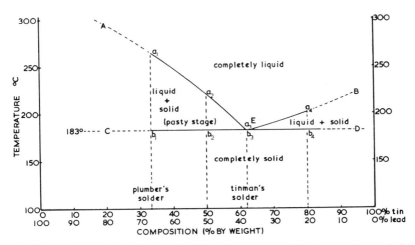

Fig. 7.5 A diagram showing the relationship between composition, temperature and physical state for the range of tin/lead alloys studied. This is part of the tin/lead thermal equilibrium diagram.

'equilibrium' in this thermodynamical context will become apparent as a result of later studies in this book, but for the moment we will consider an everyday example which goes some way to illustrate its meaning.

On a hot summer's day we can produce a delightfully refreshing drink by putting a cube of ice into a glass of lager. The contents of the glass, however, are not in thermal equilibrium with the surroundings, and as heat-transfer takes place into the lager the ice ultimately melts and the liquid warms up, so that the whole becomes more homogeneous if less palatable. Rapid cooling, as we shall see later, often produces an alloy structure which, like the ice and lager, is not in thermal equilibrium at room temperature. The basic difference between the ice–lager mixture and the non-equilibrium metallic structure is that the former is able to reach 'structural' equilibrium with ease, due to the great mobility of the constituent particles, but in the case of the metallic structure rearrangement of the atoms is more difficult, since they are retained by considerable forces of attraction in an orderly pattern in a crystal lattice. A non-equilibrium metallic structure produced by rapid cooling may therefore be retained permanently at room temperature.

7.43   If we assume that a series of alloys has been cooled slowly enough for structural equilibrium to obtain, then the thermal-equilibrium diagram will indicate the relationship which exists between composition, temperature and microstructure of the alloys concerned. By reference to the diagram, we can, for an alloy of any composition in the series, find exactly what its structure or physical condition will be at any given temperature. We can also in many cases forecast with a fair degree of accuracy the effect of a particular heat-treatment on the alloy; for in modern metallurgy heat-treatment is not a process confined to steels, but is applied also to many non-ferrous alloys. These are two of the more important uses of the thermal-equilibrium diagram as a metallurgical tool. Let us now proceed with our preliminary study of the iron–carbon alloys, with particular reference to their equilibrium diagram.

7.50   Plain carbon steels are generally defined as being those alloys of iron and carbon which contain up to 2.0% carbon. In practice most ordinary steels also contain appreciable amounts of manganese residual from a deoxidation process carried out prior to casting. For the present, however, we shall neglect the effects of this manganese and regard steels as being simple iron–carbon alloys.

7.51   As we have seen (3.14), the pure metal iron, at temperatures below 910°C, has a body-centred cubic structure, and if we heat it to above this temperature the structure will change to one which is face-centred cubic. On cooling, the change is reversed and a body-centred cubic structure is once more formed. The importance of this reversible transformation lies in the fact that up to 2.0% carbon can dissolve in face-centred cubic iron, forming what is known as a 'solid solution',* whilst

---

* We shall deal more fully with the nature of solid solutions in the next chapter, and for the present it will be sufficient to regard a solid solution as being very much like a liquid solution in that particles of the added metal are absorbed without visible trace, even under a high-power microscope, into the structure of the parent metal.

in body-centred cubic iron no more than 0.02% carbon can dissolve in this way.

7.52 As a piece of steel in its face-centred cubic form cools slowly and changes to its body-centred cubic form, any dissolved carbon present in excess of 0.02% will be precipitated, whilst if it is cooled rapidly enough such precipitation is prevented. Upon this fact depends our ability to heat-treat steels—and, in turn, the present advanced state of our twentieth-century technology.

7.53 The solid solution formed when carbon atoms are absorbed into the face-centred cubic structure of iron is called *Austenite* and the extremely low level of solid solution formed when carbon dissolves in body-centred cubic iron is called *Ferrite*. For many practical purposes we can regard ferrite as having the same properties as pure iron. In most text-books on metallurgy the reader will find that the symbol γ ('gamma') is used to denote both the face-centred cubic form of iron and the solid-solution austenite, whilst the symbol α ('alpha') is used to denote both the body-centred cubic form of iron existing below 910°C and the solid-solution ferrite. The same nomenclature will be used in this book.

When carbon is precipitated from austenite it is not in the form of elemental carbon (graphite), but as the compound iron carbide, Fe_3C, usually called *Cementite*. This substance, like most other metallic carbides, is very hard, so that, as the amount of carbon (and hence, of cementite) increases, the hardness of the slowly cooled steel will also increase.

7.54 Fig. 7.5 indicates the temperatures at which *solidification* begins and ends for any homogeneous *liquid* solution of tin and lead. In the same way Fig. 7.6 shows us the temperatures at which *transformation* begins and ends for any *solid* solution (austenite) of carbon and face-centred cubic iron. Just as the melting point of either tin or lead is lowered by adding each to the other, so is the allotropic transformation temperature of face-centred cubic iron altered by adding carbon. Fig. 7.6 includes only a part of the whole iron–carbon equilibrium diagram, but it is the section which we make use of in the heat-treatment of carbon steels. On the extreme left of this diagram is an area labelled 'ferrite'. This indicates the range of temperatures and compositions over which carbon can dissolve in body-centred cubic (α) iron. On the left of the sloping line *AB* all carbon present is dissolved in the body-centred cubic iron, forming the solid-solution fer-rite, whilst any point representing a composition and temperature to the right of *AB* indicates that the solid-solution α is saturated, so that some of the carbon contained in the steel will be present as cementite. The signifi-cance of the slope of *AB* is that the solubility of carbon in body-centred cubic iron increases from 0.006% at room temperature to 0.02% at 723°C. Temperature governs the degree of solubility of solids in liquids in exactly the same way.

7.55 We will now study the transformations which take place in the structures of three representative steels which have been heated to a tem-perature high enough to make them austenitic and then allowed to cool slowly. If a steel containing 0.40% carbon is heated to some temperature above U_1 it will become completely austenitic (Fig. 7.6(i)). On cooling again to just below U_1 (which is called the 'upper critical temperature' of the steel), the structure begins to change from one which is face-centred cubic to one which is body-centred cubic. Consequently, small crystals of body-centred cubic iron begin to separate out from the austenite. These body-centred cubic crystals (Fig. 7.6(ii)) retain a small amount of carbon (less than 0.02%), so we shall refer to them as crystals of ferrite. As the temperature continues to fall the crystals of ferrite grow in size at the expense of the austenite (Fig. 7.6(iii)), and since ferrite is almost pure iron, it follows that most of the carbon present accumulates in the shrinking crystals of austenite. Thus, by the time our piece of steel has reached L_1 (which is called its 'lower critical temperature') it is composed of approximately half ferrite (containing only 0.02% carbon) and half austenite, which now contains 0.8% carbon. The composition of the austenite at this stage is represented by E. Austenite can hold no more than 0.8% carbon in solid solution at this temperature (723°C), therefore, as the temperature falls still farther, the carbon begins to precipitate as cementite. At the same time ferrite is still separating out and we find that these two substances, ferrite and cementite, form as alternate layers until all the remaining austenite is used up (Fig. 7.6(iv)). This laminated structure of ferrite and cementite, then, will contain exactly 0.8% carbon, so that it will account for approximately half the volume of our 0.4% carbon steel. It is an example of what, in metallurgy, we call a *eutectoid* (8.43). This particular eutectoid is known as *Pearlite* because when present on the etched surface of steel it acts as a 'diffraction grating', splitting up white light into its component spectrum colours and giving the surface a 'mother of pearl' sheen. In order to be able to see these alternate layers of ferrite and cementite of which pearlite is composed, a metallurgical microscope capable of a magnification in the region of 500 diameters is necessary.

Any steel containing less than 0.8% carbon will transform from austenite to a mixture of ferrite and pearlite in a similar way when cooled from its austenitic state. Transformation will begin at the appropriate upper critical temperature (given by a point on CE which corresponds with the composition of the steel) and end at the lower critical temperature of 723°C. The relative amounts of ferrite and pearlite will depend upon the carbon content of the steel (Fig 7.7), but in every case the ferrite will be almost pure iron and the pearlite will contain exactly 0.8% carbon.

7.56 A steel containing 0.8% carbon will not begin to transform from austenite on cooling until the point E is reached. Then *transformation* will begin and end at the same temperature (723°C), just as tinman's solder *solidifies* at a single temperature (183°C). Since the steel under consideration contained 0.8% carbon initially, it follows that the final structure will be entirely pearlite (Fig 7.6(vi)).

7.57 A steel which contains, say, 1.2% carbon will begin to transform from austenite when the temperature falls to its upper critical at U_2. Since the carbon is this time in excess of the eutectoid composition, it will begin to precipitate first; not as pure carbon but as needle-shaped crystals of cementite round the austenite grain boundaries (Fig 7.6(viii)). This will cause the austenite to become progressively less rich in carbon, and by the

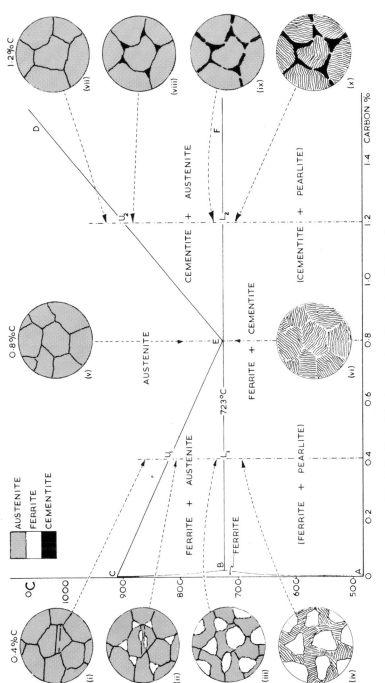

Fig. 7.6 Part of the iron–carbon thermal equilibrium diagram.

time a temperature of 723°C has been reached the remaining austenite will contain only 0.8% carbon. This remaining austenite will then transform to pearlite (Fig 7.6(x)), as in the two cases already dealt with.

Any steel containing more than 0.8% carbon will have a structure consisting of cementite and pearlite if it is allowed to cool slowly from its austenitic state. Since the pearlite part of the structure always contains alternate layers of ferrite and cementite in the correct proportions to give an overall carbon content of 0.8% for the pearlite, it follows that any variation in the total carbon content of the steel above 0.8% will cause a corresponding variation in the amount of primary cementite present. (The terms 'primary cementite' and 'primary ferrite' are used to denote that cementite or ferrite which forms first, before the residual austenite transforms to pearlite.)

A plain carbon steel which contains less than 0.8% carbon is generally referred to as a *hypo-eutectoid* steel, whilst one containing more than 0.8% carbon is known as a *hyper-eutectoid* steel. Naturally enough, a plain carbon steel containing exactly 0.8% carbon is called a *eutectoid* steel.

7.60   So far we have been dealing only with the types of structure produced when plain carbon steels are cooled slowly from the austenitic condition. Such conditions prevail during industrial processes such as normalising and annealing. By very rapid cooling from the austenitic condition, such as would be obtained by water-quenching, another structure, called *Martensite*, is formed. This does not appear on the equilibrium diagram simply because it is not an equilibrium structure. Rapid cooling has prevented equilibrium from being reached.

7.61   As most readers will already know, martensite is very hard indeed. Unfortunately it is also rather brittle, and the steel is used in this condition only when extreme hardness is required. To increase the steel's toughness after quenching (at the expense of a fall in hardness) the steel can be *tempered*. A modification in the structure will take place depending upon the tempering temperature. This temperature will vary between 250 and 650°C according to the combination of mechanical properties required in the finished component. Whatever temperature is used, tempering assists the microstructure to proceed in some measure back towards equilibrium, with the precipitation of microscopical particles of cementite in varying amounts from the original martensitic structure. The type of structure formed by tempering at about 400°C was formerly known as *Troostite*, whilst that produced in the region of 600°C used to be called *Sorbite*. Metallurgists now discourage the use of these terms for reasons which will be explained later when the heat-treatment of steel is discussed more fully in Chapters 11 and 12.

The Uses of Plain Carbon Steels

7.70   By varying the amount of carbon in a steel, and by selecting a heat-treatment programme suited to that carbon content, we are able to produce a vast range of different mechanical properties such as are available in no other metallic alloy. Moreover, carbon steel is a relatively inexpensive alloy when compared with non-ferrous alloys generally. Small wonder then that steel is by far our most important engineering alloy. Possibly its most serious fault is that it rusts and we must often spend considerable amounts of money on protecting its surface from atmospheric corrosion (21.10).

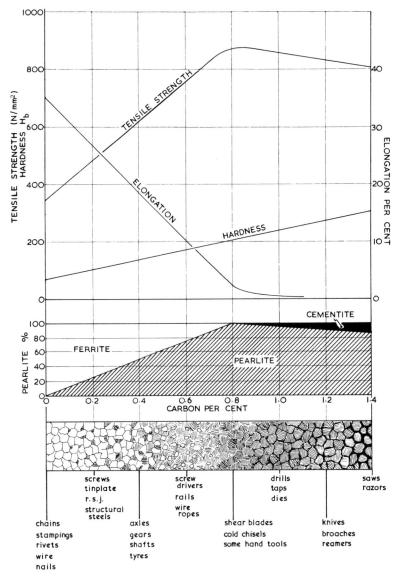

Fig. 7.7 A diagram showing the relationship between carbon content, microstructure and mechanical properties of plain carbon steels in the *normalised* condition. Typical uses of these steels are also indicated.

As shown in Fig. 7.7, the hardness of a plain carbon steel increases progressively with increase in carbon content, so that generally the low- and medium-carbon steels are used for structural and constructional work, whilst the high-carbon steels are used for the manufacture of tools and other components where hardness and wear-resistance are necessary.

Commercial plain carbon steels may be classified into five groups as indicated in Table 7.1 which, along with Fig. 7.7, indicates some of the common uses of these alloys.

Table 7.1

Type of steel	Percentage carbon	Uses
Dead mild	0.05–0.15	Chain, stampings, rivets, wire, nails, seam-welded pipes, mattresses, hot- and cold-rolled strip for many purposes
Mild	0.10–0.20	Structural steels, RSJ, screws, machine parts, tin-plate, case-hardening, drop-forgings, stampings
	0.20–0.30	Machine and structural work, gears, free-cutting steels, shafting, levers, forgings
Medium carbon	0.30–0.40	Connecting-rods, shafting, wire, axles, fish-plates, crane hooks, high-tensile tubes, forgings
	0.40–0.50	Crankshafts, axles, gears, shafts, die-blocks, rotors, tyres, heat-treated machine parts
	0.50–0.60	Loco tyres, rails, laminated springs, wire ropes
High carbon	0.60–0.70	Drop-hammer dies, set-screws, screw-drivers, saws, mandrels, caulking tools, hollow drills
	0.70–0.80	Band saws, anvil faces, hammers, wrenches, laminated springs, car bumpers, small forgings, cable wire, dies, large dies for cold presses
	0.80–0.90	Cold chisels, shear blades, cold setts, punches, rock drills, some hand tools
Tool steels	0.90–1.00	Springs, high-tensile wire, axes, knives, dies, picks
	1.00–1.10	Drills, taps, milling cutters, knives, screwing dies
	1.10–1.20	Ball bearings, dies, drills, lathe tools, woodworking tools
	1.20–1.30	Files, reamers, knives, broaches, lathe and wood-working tools
	1.30–1.40	Saws, razors, boring and finishing tools, machine parts where resistance to wear is essential

Exercises

1. Show how the exploitation of new iron ore fields has helped to change the balance of world power during the present century. (7.10)
2. What advantages has 'direct reduction' over the blast-furnace process for pig iron production? (7.25)
3. Outline the essential chemistry common to all modern steel-making processes. (7.33–7.36)
4. Discuss both the economic and technical advantages of modern BOS processes as compared with the obsolete Bessemer process. (7.36)
5. Why has electric-arc steelmaking survived despite the high cost of electric power? (7.37)

6. Relate the changes in microstructure to the properties of normalised steels as the carbon content increases from 0.1 to 1.2%. Comment on the choice of carbon content for some typical engineering applications. (7.50)
7. Show how the mechanical properties of tensile strength, % elongation and hardness for normalised plain-carbon steels vary with carbon content up to 1.2%.

   Sketch and label the microstructures of the following normalised steels (i) 0.2%C; (ii) 0.8%C; (iii) 1.2%C. (Fig 7.7)

Bibliography

Peters, A. J., *Ferrous Production Metallurgy*, John Wiley, 1982.
United States Steel Corporation, *The Making, Shaping and Treatment of Steel*.
BS 970: 1973 and 1988 *Wrought Steels in the Form of Blooms, Billets, Bars and Forgings* (Part 1—carbon steels).
BS 4659: 1971 *Tool Steels*.

8

The Formation of Alloys

8.10 Pure metals are rarely used for engineering purposes except where high electrical conductivity, high ductility or good corrosion resistance are required. These properties are generally at a maximum value in pure metals, but such mechanical properties as tensile strength, yield point and hardness are improved by alloying. In most cases the manufacture of an alloy involves melting two or more metals together and allowing the mixture to solidify in a suitably-shaped mould. The main object of this chapter is to study the type of microstructure which is produced when the alloy solidifies.

8.11 Broadly speaking, we are only likely to obtain a metallurgically useful alloy when the two metals in question dissolve in each other in the liquid state to form a completely homogeneous liquid solution. There are exceptions to this rule, as, for example, the suspension of undissolved lead particles in a free-cutting alloy (6.64). The globules of lead are undissolved in the molten, as well as in the solidified alloy. In general however, a useful alloy is formed only when the metals in question are mutually soluble in the liquid state forming a single liquid which is uniform in composition throughout.

8.12 When such a homogeneous liquid solution solidifies one of the following conditions is possible:

(*a*) The solubility ends and the two metals which were mutually soluble as liquids become totally insoluble as solids. They therefore separate out as particles of the two pure metals.

(*b*) The solubility prevailing in the liquid state may be retained either completely or partially in the solid state. In the former case a single *solid solution* will be formed, whilst in the latter case a mixture of two different solid solutions will result.

(*c*) As solidification proceeds the two metals give rise to what is loosely termed an *intermetallic compound* but which in many cases might more accurately be described as an *intermediate phase*.

8.13 Any one of the separate substances referred to in 8.12 is called a *phase*. Thus a solid phase may be either a pure metal or a solid solution or some intermediate phase of two or more metals so long as it exists in the microstructure as a separate entity.

A law, based on thermodynamical considerations, and known as *the Phase Rule*, was formulated originally in 1876 by Josiah W. Gibbs, Professor of Mathematical Physics at Yale. Certain principles, derived from this Phase Rule (9.24) are of use to us in a study of physical metallurgy. For example, it can be demonstrated that in a solid binary* alloy *not more than two* phases can co-exist. Therefore a binary alloy may contain:

(*a*) Two pure metals existing as separate entities in the structure. This is a very rare occurrence, since in most cases there is a slight solubility of one metal in the other.

(*b*) A single complete solid solution of one metal dissolved in the other.

(*c*) A mixture of two solid solutions if the metals are only partially soluble in each other.

(*d*) A single intermediate phase of some type.

(*e*) A mixture of two different intermediate phases.

(*f*) A solid solution and an intermediate phase mixed together—and so on.

8.14 The general procedure in making an alloy is first to melt the metal having the higher melting point and then to dissolve in it the metal with the lower melting point, stirring the mixture so that a homogeneous liquid solution is formed. When this liquid solidifies the type of structure which is produced depends largely upon the relative physical and chemical properties of the two metals and in part on their relative positions in the Periodic Classification (1.69). If they are very near to each other in the Periodic Table it is *probable*—though not inevitable—that they will form mixed crystals (ie a solid solution) in which the atoms of each metal are randomly placed in a common crystal pattern. This is rather like the effect which would be produced if a bricklayer used both red and blue bricks *indiscriminately* in building a wall. If the red and blue bricks were of approximately equal size then a strong and reasonably regular wall would result even though it looked a bit untidy due to the lack of pattern in placing the different coloured bricks.

If the two metals are widely separated in the Periodic Table so that their chemical properties, and probably their atomic sizes, differ considerably then it is *possible*—but again, not certain—that they will combine to form some sort of chemical compound (an 'intermetallic compound'). Between these two extremes the atoms of the two metals may associate with each other in different patterns to form transitional phases known collectively as *intermediate phases*.

We will now consider in some detail the nature of the phases and structures mentioned above.

---

* Containing *two* metals.

The Solid Solution

8.20 A solid solution is formed when two metals which are mutually soluble in the liquid state remain dissolved in each other after crystallisation. If the resulting structure is examined, even with the aid of a high-power electron microscope, no trace can be seen of the parent metals as separate entities. Instead one sees crystals consisting of a homogeneous solid solution of one metal in the other, and if the metals are completely soluble in each other there will be crystals of one type only.

8.21 It will be clear that in order that one metal may dissolve in another to form a solid solution, its atoms must fit in some way into the crystal lattice of the other metal. This may be achieved by the formation of either a 'substitutional' or an 'interstitial' solid solution (Fig. 8.1).

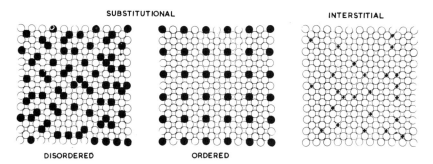

Fig. 8.1 Ways in which solid solution can occur.

8.22 Interstitial solid solutions can be formed only when the atoms of the added element are very small compared with those of the parent metal, thus enabling them to fit into the interstices or spaces in the crystal lattice of the parent metal. Not only can this occur during solidification but, also in many cases, when the parent metal is *already solid*. Thus, carbon can form an interstitial solid solution with FCC iron during the solidification of steel, but it can also be absorbed by solid iron provided the latter is heated to a temperature at which the structure is face-centred cubic. This is the basis of carburising steel (19.20). Nitrogen is also able to dissolve interstitially in solid steel, making the nitriding process (19.40) possible; whilst hydrogen, the atoms of which are very small indeed, is able to dissolve in this way in a number of solid metals, usually producing mechanical brittleness as a result.

8.23 In a substitutional solid solution the atoms of the *solvent* or parent metal are replaced in the crystal lattice by atoms of the *solute* metal. This substitution can be either 'disordered' or 'ordered' (Fig. 8.1). Solubility of one metal in another is governed in part by the relative sizes of their atoms. Thus disordered solid solutions are formed over a wide range of alloy compositions when the atoms—or, properly, their positive *ions*—are of similar size but when the atomic diameters differ by more than 14% of that of the atom of the solvent metal then solubility will be slight because the *atomic size factor* is unfavourable.

The size factor effect arises from the strain which is induced in the crystal lattice by a badly misfitting solute atom. The atoms of the solvent metal are displaced either outwards or inwards according as the solute atom is either larger or smaller than the atom of the solvent metal. This distortion of the crystal lattice from the ideal shape increases the energy of the crystal and tends to restrict solubility in it.

Electrochemical properties of the two metals also influence the extent to which they will form a solid solution. Thus if the two metals have very similar electrochemical properties, generally indicated by a close proximity to each other in the Periodic Table, then they may form solid solutions over a wide range of compositions but if one metal is very strongly electro-positive and the other only weakly electropositive as shown by a wide separation in the Table, then they are very likely to form some type of compound.

As might be expected, the lattice structures of the two metals will influence the extent to which they are likely to form substitutional solid solutions. If the structures are similar then it is reasonable to expect that one metal can replace the other satisfactorily in its lattice. Thus two FCC metals like copper and nickel will form a continuous series of solid solutions—assisted by the fact that atom sizes of these two metals are almost the same.

The valency of one metal relative to the other will also affect the tendency to form a solid solution. A metal of lower valency is more likely to dissolve one of higher valency than vice versa. Thus copper will dissolve considerable amounts of many other metals giving useful ranges of solid solution alloys.

To summarize, a high degree of substitutional solid solubility between two metals is likely to be obtained when:

1. The difference in atomic radii is less than 14%.
2. Electrochemical properties are similar.
3. The two metals have similar crystal structures.

Even though the size factor of the two metallic atoms may favour their forming a solid solution they are more likely to form compounds if one is much more strongly electropositive than the other.

When a disordered solid solution crystallises from the liquid state there is a natural tendency for the core of the dendrite to contain rather more atoms of the metal with the higher melting point, whilst the outer fringes of the crystal will contain correspondingly more atoms of the metal of the lower melting point. Nevertheless, within such a 'cored' crystal, once formed, diffusion does take place; and this diffusion tends to produce uniform distribution of each kind of atom. Diffusion is only appreciable at relatively high temperatures, for which reason a rapidly cooled alloy will be appreciably cored and a slowly cooled alloy only slightly cored. Prolonged annealing will remove coring completely by allowing diffusion to take place and produce a uniform solid solution. The driving force behind diffusion is the reduction in strain energy produced by the fact that as solute atoms

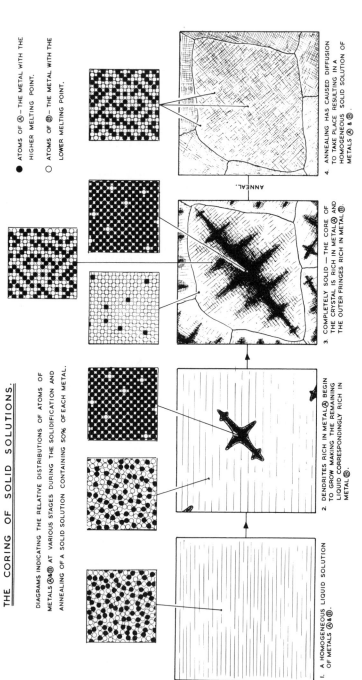

THE CORING OF SOLID SOLUTIONS.

DIAGRAMS INDICATING THE RELATIVE DISTRIBUTIONS OF ATOMS OF METALS Ⓐ&Ⓑ AT VARIOUS STAGES DURING THE SOLIDIFICATION AND ANNEALING OF A SOLID SOLUTION CONTAINING 50% OF EACH METAL.

● ATOMS OF Ⓐ – THE METAL WITH THE HIGHER MELTING POINT.

○ ATOMS OF Ⓑ – THE METAL WITH THE LOWER MELTING POINT.

ANNEAL.

1. A HOMOGENEOUS LIQUID SOLUTION OF METALS Ⓐ&Ⓑ.

2. DENDRITES RICH IN METAL Ⓐ BEGIN TO GROW MAKING THE REMAINING LIQUID CORRESPONDINGLY RICH IN METAL Ⓑ.

3. COMPLETELY SOLID — THE CORE OF THE CRYSTAL IS RICH IN METAL Ⓐ AND THE OUTER FRINGES RICH IN METAL Ⓑ.

4. ANNEALING HAS CAUSED DIFFUSION TO TAKE PLACE RESULTING IN A HOMOGENEOUS SOLID SOLUTION OF METALS Ⓐ & Ⓑ.

Fig. 8.2 The variations in composition which are possible in a solid solution.

move away from regions of high concentration the overall distortion of the crystal is reduced. The effects of the phenomena of coring and diffusion are illustrated in Fig. 8.2.

8.24 In an interstitial solid solution diffusion can take place with relative ease, since the solute atoms, being small, can move readily through the crystal lattice of the parent metal, but in substitutional solid solutions the mechanism of diffusion is still open to speculation. It was stated earlier (4.15) that faults can occur in crystals, giving rise to dislocations which facilitate slip. Similarly, it is suggested that what are called 'vacant sites' can exist in a metal crystal. These are positions in the crystal lattice which are not occupied by an atom (Fig. 8.3).

Since there is always some difference in size between the solute atoms and the atoms of the parent metal, the presence of a solute atom will cause some distortion in the lattice of the parent metal. This distortion will be minimised by the association of a solute atom with a vacant site as shown in Fig. 8.4A. The associated pair of vacant site with solute atom can migrate easily through the crystal in stages as indicated in Fig. 8.4 B–F. The rate of diffusion will depend to some extent upon the number of vacant sites, but also upon the concentration gradient of the solute atoms within the structure of the solvent metal (Fig. 8.6); and upon the temperature at which diffusion is taking place. The general result will be for solute atoms to diffuse away from regions of high concentration towards regions of low

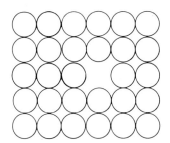

Fig. 8.3 A 'vacant site' in a crystal lattice.

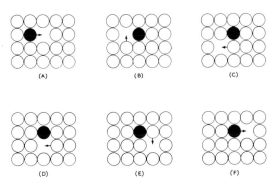

Fig. 8.4 The diffusion of a solute atom which is associated with a vacant site.

concentration and diffusion will ultimately cease when a uniform distribution of solute atoms has been achieved.

The Schottky defect (3.62) is formed by a similar progressive movement of atoms to that described above, whereby a vacancy is left behind in the lattice and a corresponding atom is deposited at the surface of the metal. Such a process takes place due to internal strains within the lattice.

8.25   As long ago as 1855 Adolph Fick proposed rules governing the diffusion of substances in general. Here we are concerned with their application to diffusion in metallic alloys. *Fick's Law* states that the flow or movement (J) of atoms across unit area of a plane (Fig. 8.5) at any given instant is proportional to the concentration gradient $\left(\dfrac{\delta c}{\delta x}\right)_t$ at the same instant but of opposite sign, ie,

$$J \propto -\left(\frac{\delta c}{\delta x}\right)_t \qquad J = -D\left(\frac{\delta c}{\delta x}\right)_t \tag{1}$$

Here D, the *diffusion coefficient*, is a constant for the system but will vary for other compositions of alloy. It is in units $\dfrac{(\text{length})^2}{\text{time}}$ and is usually expressed as $10^{-4} \text{m}^2/\text{s}$. The significance of the negative sign preceding D is that the movement of solute atoms is taking place 'down' the concentration gradient.

When $\dfrac{c}{x} = 0$, then $J = 0$ and this satisfies the requirement that diffusion ceases when the concentration gradient reaches zero as the solid solution becomes homogeneous (line CD in Fig. 8.6). In practice coring is never completely eliminated. It is reduced quite rapidly in the early stages of an annealing process but traces still remain even after prolonged treatment —mathematically of course CD (Fig. 8.6) only becomes horizontal after an infinitely long time interval. Traces of coring are however generally acceptable in most industrial processing.

Here we have considered diffusion in only one direction—parallel to the x-axis in which the concentration gradient is operating. In practice diffusion will also take place along y and z axes.

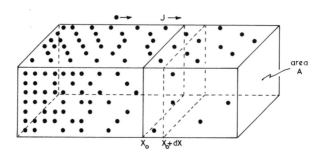

Fig. 8.5.

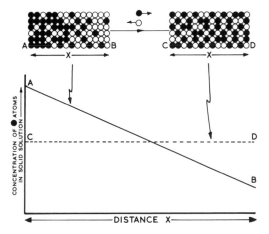

Fig. 8.6 *AB* is the initial concentration gradient and *CD* the concentration gradient after a long time during which diffusion has been taking place.

There are similarities in principle between Fick's Law and other Laws such as Ohm's Law. The concentration gradient operates in a similar way to potential difference as the driving force and D is independent of the value of $\frac{\delta c}{\delta x}$ in the same way as electrical resistance is independent of potential difference. Similar principles also govern heat flow through solids where thermal conductivity is independent of the magnitude of the temperature gradient. All such properties, ie electrical and thermal conductivities, viscosity and diffusion are known as *transport properties* of a substance.

Fick's Law assumes that diffusion is proceeding under steady-state conditions and that the concentration gradient at any given point will not change with time. That is, we are as it were 'feeding in' solute atoms at one end of a piece of alloy and extracting an equal number at the other end. In practice this will not be the case and in a given piece of alloy matter will be conserved: its overall or 'average' composition remains constant and only the positions of its atoms vary in terms of x and t.

Consider the situation where, due to coring in an alloy, a certain distribution of concentration is set up and that subsequently, due to heat activation (eg annealing), the atoms diffuse so as to attempt to reach a uniform concentration (eliminate coring). Concentrations are therefore changing with time and atoms must be accumulating in the region between plane x_0 and plane $(x_0 + dx)$ or moving from it (Fig. 8.5). Hence the number crossing area A of the plane, x_0, is not equal to that crossing the same area at $(x_0 + dx)$. The flux entering this section is:

$$J_{x_0} = -D \left(\frac{\delta c}{\delta x} \right)_{x=x_0}$$

The flux leaving the section can be written: $J_{x_0 + dx}$ where

$$J_{x_0 + dx} = J_{x_0} + \left(\frac{\delta J}{\delta x}\right)_{x=x_0} . dx + \ldots + \text{(The higher terms can be neglected)}$$

The rate of movement of atoms from the section x_0 $x_0 + dx$ is equal to the difference between the two values of AJ and also equal to the volume of the section, $A.dx$ times the rate of decrease of c:

$$-\frac{\delta J}{\delta x} . A dx = \frac{\delta c}{\delta t} . A dx$$

or

$$\frac{\delta J}{\delta x} = -\frac{\delta c}{\delta t} \tag{2}$$

Combining (2) with Fick's Law (1) and eliminating J:

$$\frac{\delta c}{\delta t} = -\frac{\delta}{\delta x}\left(-D\frac{\delta c}{\delta x}\right)$$

or

$$\frac{\delta c}{\delta t} = D\frac{\delta^2 c}{\delta x^2} \tag{3}$$

assuming that D is a constant independent of the concentration, c. This used to be known as Fick's Second Law of Diffusion but is now more generally referred to as *The diffusion equation*. Since c depends upon x and t it could be written as $c_{(x, t)}$.

In this system of three equations (1) is an experimental law linking the flow of atoms at any point with the concentration gradient there, (2) is the continuity equation expressing the fact that atoms cannot disappear whilst (3) combines these two equations. Equation (1) applies only to steady-state conditions where these conditions do not vary with time. Equation (3) applies to the general case where the concentration gradient at a certain plane changes with time. Values of $\delta c/\delta t$ and $\delta^2 c/\delta x^2$ can be determined experimentally in order to determine the value of D for a given set of conditions. Differential coefficients vary with the nature of the solute atoms, the nature of the solvent structure and with temperature.

8.26 In Fig. 8.1, and the illustrations which follow, both solvent and solute atoms in substitutional solid solutions have been shown of equal size, whilst in interstitial solid solutions solute atoms have been shown as very small and fitting easily into the interstices of the lattice. In the real world such is not the case; both solvent and solute atoms have been represented in this way in order not to complicate the illustration unduly.

Atoms of different elements forming substitutional solid solutions will never be of exactly the same size and for this reason some distortion of the lattice structure of the parent metal will be caused by the presence of atoms—whether larger or smaller—of the solute metal (Fig. 8.7(i)). Similarly solute atoms of a size similar to those of the parent metal will sometimes dissolve interstitially (Fig. 8.7(ii)). However, Fig. 8.7 does not represent the situation completely since we are only describing the coplanar distribution of atoms, ie along the x and y axes. In practice the distribution of solute atoms and the distortion they produce will also occur along the z axis since crystals are three-dimensional in nature.

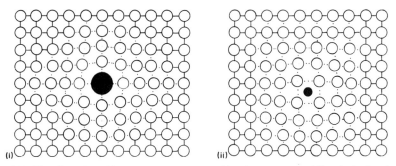

Fig. 8.7   The distortion of lattice structures caused by the presence of (i) substitutional, (ii) interstitial solute atoms.
Only a coplanar representation can be given here. In practice of course crystals are three-dimensional.

Intermediate Phases

8.30   In the foregoing section we have been dealing with disordered substitutional solid solutions in which there is no 'ordering' of the atoms of the solute metal within the lattice of the solvent metal. Such solutions are formed when the electrochemical properties of the metals are similar. However, when two metals have widely divergent electrochemical properties they are likely to associate to form a *chemical compound*. Thus strongly electropositive magnesium will combine with weakly electropositive tin to form the substance Mg_2Sn. This is generally described as an *intermetallic compound*. Between these two extremes of substitutional solid solution on the one hand and intermetallic compound on the other, phases are formed which exhibit a gradation of properties according to the degree of association which occurs between the atoms taking part. These phases are collectively termed *intermediate phases*. At one extreme we have true intermetallic compounds whilst at the other ordered structures which can be more accurately classed as secondary solid solutions.

8.31   These intermediate phases can be classified into three main groups:

1 Intermetallic compounds   in which the laws of chemical valence are apparently obeyed as in Mg_2Sn, Mg_2Pb, Mg_3Sb_2 and Mg_3Bi_2. These valence compounds are generally formed when one metal (such as magnesium) has chemical properties which are strongly metallic, and the other metal (such as antimony, tin or bismuth) chemical properties which are only weakly metallic and, in fact, bordering on those of non-metals. Frequently such a compound has a melting point which is higher than that of either of the parent metals. For example, the intermetallic compound Mg_2Sn melts at 780°C, whereas the parent metals magnesium and tin melt at 650 and 232°C respectively. This is an indication of the high strength of the chemical bond in Mg_2Sn.

2 Electron Compounds   As was shown earlier (1.70) the chemical valence of a metal is a function of the number of electrons in the outer 'shell' of the atom, whilst the nature of the metallic bond is such that wholesale sharing of numbers of electrons takes place in the crystal structure of a pure metal.

In these 'electron compounds' the normal valence laws are not obeyed, but in many instances there is a fixed ratio between the total number of valence bonds of all the atoms involved and the total number of atoms in the empirical formula of the compound in question.

There are three such ratios, commonly referred to as Hume-Rothery ratios:

(i) Ratio 3/2 (21/14)—β structures, such as CuZn, Cu_3Al, Cu_5Sn, Ag_3Al, etc.

(ii) Ratio 21/13—γ structures, such as Cu_5Zn_8, Cu_9Al_4, $Cu_{31}Sn_8$, Ag_5Zn_8, $Na_{31}Pb_8$, etc.

(iii) Ratio 7/4 (21/12)—ε structures, such as $CuZn_3$, Cu_3Sn, $AgCd_3$, Ag_5Al_3, etc.

Thus, in the compound CuZn, copper has a valence of 1 and zinc a valence of 2, giving a total of 3 valences and hence a ratio of 3 valences to 2 atoms. In the compound $Cu_{31}Sn_8$ copper has a valence of 1 and tin a valence of 4. Therefore 31 valences are donated by the copper atoms and 32 (ie, 4×8) by the tin atoms, making a total of 63 valences. In all, 39 atoms are present in the empirical formula of $Cu_{31}Sn_8$ therefore the ratio

$$\frac{\text{Total number of valences}}{\text{Total number of atoms}} = \frac{63}{39} \text{ or } \frac{21}{13}$$

These Hume-Rothery ratios have been valuable in relating structures which were apparently unrelated. There are, however, many electron compounds which do not fall into any of the three groups mentioned above, nor are they valence compounds. Electron compounds are formed as a result of the nature of the metallic bond and the stability of such a substance depends upon electron concentration and the lattice structure involved.

3 Size-factor Compounds   These are intermediate phases in which compositions and crystal structures arrange themselves in such a way as to allow the constituent atoms to pack themselves closely together.

In the *Laves phases* compositions are based upon the general formula AB_2, eg $MgNi_2$, $MgCu_2$, $TiCr_2$ and $MnBe_2$. Their formation depends upon the fact that the constituent atoms vary in size by about 22.5% but that they can none the less pack closely together in crystal structures.

A very important group of size-factor compounds are the *interstitial compounds* formed between some transition metals and certain small non-metallic atoms. When the solid solubility of an *interstitially* dissolved element is exceeded a compound is precipitated from the solid solution. In this type of compound the small non-metal atoms still occupy interstitial positions but the overall crystal structure of the compound is different from that of the original interstitial solid solution. Compounds of this type have metallic properties and comprise hydrides, nitrides, borides and carbides of which TiH_2, TiN, Mn_2N, TiB_2, TaC, W_2C, WC, Mo_2C and Fe_3C are typical. All of these compounds are extremely hard and the carbides find application in tool steels and cemented-carbide cutting materials. Fe_3C is of course the phase cementite of ordinary carbon steels. Many of these carbides are extremely refractory, having melting points well in excess of 3000°C.

8.32 Many intermediate phases are extremely hard and brittle so that a small ingot of such a substance may often be crushed to a powder by gentle pressure in the jaws of a vice. This is particularly true of valence compounds where covalent or ionic bonds tend to replace the metallic bond, thus making the substance essentially non-metallic in its properties as indicated by brittleness and a low electrical conductivity. Consequently such phases are of limited use in engineering alloys and then in relatively small proportions of the total microstructure.

Small quantities of intermediate phases—usually electron compounds—are generally present in bearing alloys. Here the hard particles of compound are embedded in a matrix of tough solid solution. The compound resists wear and has a low coefficient of friction, whilst the solid solution provides the necessary tough matrix capable of withstanding mechanical shock and high compressive forces.

Intermediate phases are often easy to identify as such under the microscope. Frequently they are of an unexpected colour but this is explained by the fact that crystal structures and lattice distances are different from those of the parent metals. (Lattice distances control the wavelength of light reflected from a surface.) Thus $Cu_{31}Sn_8$ is pale blue whilst Cu_2Sb is bright purple. Since intermediate phases are of ordered structure, coring is not present and the microstructure appears uniform.

8.33 The good high-temperature properties of many of the stable intermetallic compounds suggests that they could be used as engineering materials, particularly in the aerospace industries, but for their low ductility and extremely poor impact toughness. Nevertheless research is now in progress with a view to improving ductility and hence 'fabricability' of some of the more promising compounds.

Thus the compound Ni_3Al has a melting point of 1390°C and an ordered FCC structure. Single crystals of the compound are very ductile, as are most FCC structures (4.12), but in the normal polycrystalline state it suffers from extreme grain-boundary brittleness. The addition of as little as 0.05%

boron raises the percentage elongation from zero to 50 as the fracture mode is changed from being intercrystalline to a mixture of intercrystalline and transcrystalline. Formability is therefore dramatically improved. This effect is probably due to the segregation of boron at the crystal boundaries causing some disorder in the boundary region. Stresses due to dislocation 'pile-ups' there are more evenly distributed and can be relieved by slip *across* the boundary rather than by fracture *along* the boundary only.

Eutectics and Eutectoids

8.40 When two metals which are completely soluble in the liquid state but completely insoluble in the solid state solidify, they do so by crystallising out as alternate layers of the two pure metals. Thus a laminated type of structure is obtained and is termed a *eutectic*. This eutectic mixture is always of a fixed composition for the two metals concerned and it melts at a temperature *below* the melting points of either of the metals.

Consider the two metals cadmium and bismuth which are completely soluble in each other as liquids but completely insoluble as solids. If increasing amounts of cadmium are added to bismuth then the freezing point of bismuth (271°C) is progressively depressed (along *AE* in Fig. 8.8). Similarly if increasing amounts of bismuth are added to cadmium then the freezing point of cadmium (321°C) decreases progressively (along *BE*). These depression-of-freezing point lines meet at a minimum in *E* which is called the *eutectic point*. This indicates the composition of the eutectic mixture (60% wt. Bi/ 40% wt. Cd) and also the temperature (140°C) at which it freezes. The relative thicknesses of the layers of bismuth and cadmium will be so adjusted as to give an overall *eutectic* composition as indicated by *E*. Suppose the *overall* composition of a bismuth–cadmium alloy were given by *X*. This indicates that bismuth is present in the molten alloy in excess of the eutectic composition. Therefore some bismuth will crystallise out first until by the time that the temperature has fallen to

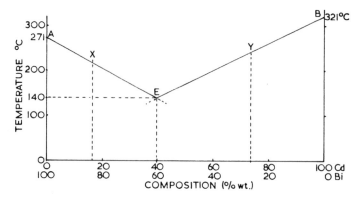

Fig. 8.8 The eutectic point (*E*) for cadmium/bismuth alloys.

140°C the remaining liquid will be of composition E. This liquid will then crystallise out as eutectic. Similarly if the overall composition of the initial liquid mixture were given by Y then some cadmium would solidify first until the remaining liquid were of composition E. Eutectic would then form as before, ie as alternate layers of pure bismuth and pure cadmium. The mechanism of this type of crystallisation will be discussed more fully in the next chapter (9.40).

8.41 In a eutectic of two *pure* metals there is no question whatsoever of solution, since the layers of the pure metals forming the eutectic can be seen quite clearly under the microscope, at magnifications usually between 100 and 500, as definite separate entities. The formation of the eutectic is therefore a result of insolubility being introduced when the alloy solidifies.

At this point it must be admitted that it is very doubtful whether complete insolubility exists in the solid state in any alloy system. There is nearly always some solid solubility however slight it may be. However, if the two metals in question are partially soluble in each other in the solid state, we may still obtain a eutectic on solidification, but it will be a eutectic composed of alternate layers of two saturated solid solutions. One layer will consist of metal A saturated with metal B, and the other a layer of metal B saturated with metal A. This statement may at first cause some confusion in the mind of the reader, but it must be understood that the two layers are not one and the same as far as composition is concerned, since one layer is rich in metal A and the other is rich in metal B. Consider as a parallel case in liquid solutions the substances water and ether. If a few spots of ether are added to a test-tube nearly filled with water and the tube shaken, a single solution of ether in water will result. If, on the other hand, a few spots of water are added to a test-tube nearly filled with ether, and the tube shaken, a single solution of water in ether will result. When, however, we put into the test-tube equal volumes of ether and water and again shake, we find that we are left with two layers. The upper layer will be ether saturated with water, and the lower layer will be water saturated with ether. The upper layer is a solution containing most of the ether, and the lower layer a different solution containing most of the water. We have a similar situation in the case of metallic solid solutions, in that one phase of a eutectic may be a solid solution rich in metal A and the other phase a solid solution rich in metal B.

8.42 Not all eutectics, whether of pure metals or solid solutions, appear under the microscope in the well-defined laminated form mentioned above. Usually layers of one of the phases are embedded in a matrix of the other phase, and quite often these layers are broken or disjointed. Fig. 17.2 illustrates the aluminium–silicon eutectic structure in which the thinner silicon layers (the eutectic mixture contains only 11.7% silicon) are broken into plate form and are embedded in an aluminium matrix. This is indeed fortunate since silicon is brittle and if present as continuous layers would seriously impair the properties of this very useful alloy.

Again, surface tension may influence crystallisation and the final eutectic may consist of globules of one phase embedded in a matrix of the other. The effect of surface tension will be accentuated by slow rates of cooling which allow layers of one phase to break up; first, into smaller, thicker plates, and ultimately, into rounded globules (Fig. 11.8).

8.43 Sometimes a solid solution which has already formed in an alloy transforms at a lower temperature to a eutectic type of structure. When this happens, the eutectic type of structure produced is called a *eutectoid*, since it was not formed from a liquid solution like a eutectic, but from a solid solution. The transformation of the solid solution, austenite, to the eutectoid, pearlite (7.55), is an example of this type of change. Cementite nuclei form at random at the crystal boundaries of the austenite, and these nuclei initiate the growth of cementite plates in the direction in which the concentration of carbon in the austenite is highest. As a result of the extraction of the carbon from the surrounding austenite to form cementite, ferrite will nucleate alongside the cementite. In this way cementite and ferrite plates will develop alongside each other (Fig. 8.9)

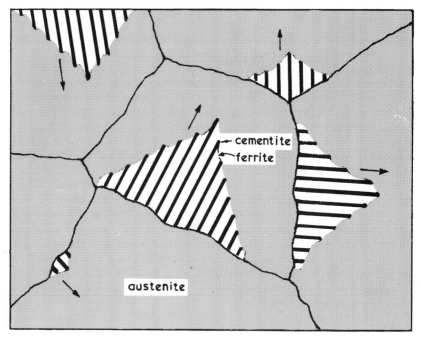

Fig. 8.9 The mechanism of the austenite → pearlite transformation.

If an alloy is cooled very rapidly from a temperature above that at which the eutectoid begins to form, the solid solution (or possibly a modified form of it such as martensite in steels) is often retained. This demonstrates that the eutectoid has not formed from the liquid but had been produced at a later stage.

8.50 Many useful alloys consist structurally of a single uniform solid solution in which increased strength is developed as suggested in the following section (8.60). Such alloys are generally also ductile and corrosion resistant (21.71). Yet other useful alloys are of duplex structure, that is, they contain two different phases in the microstructure. These two different phases may be solid solutions in a eutectic type of structure or they may be a solid solution along with some form of intermediate phase. Since many intermediate phases are weak and brittle it is usually only those which are hard and fairly strong which find use in metallurgical alloys, and then only in relatively small quantities. In addition to their use in bearing alloys as already mentioned, such phases may be employed to increase strength by some form of precipitation hardening process (9.92) or dispersion hardening mechanism (8.62).

It is the function of the metallurgist, by control of his alloy compositions along with suitable mechanical and thermal treatments, to produce alloys which will provide a set of physical and mechanical properties which fulfil the requirements of the engineer.

Strengthening Mechanisms in Alloys

8.60   As mentioned earlier in this chapter, the main reason for alloying is to increase the yield strength of a metal. This involves impeding the movement of dislocations by making alterations to the structure on approximately the atomic scale.

8.61 Solid-solution Hardening   In a cold-worked metal the presence of a dislocation causes distortion of that part of the lattice structure near to it. This distortion, and the energy associated with it, can be reduced by the presence of solute atoms. *Large* substitutional solute atoms will reduce distortion if they take up positions where the lattice structure is being *stretched* (Fig. 8.10(i)) due to the presence of a dislocation; whilst *small* substitutional solute atoms will have a similar effect if they replace solvent atoms in regions where the lattice is being *compressed* (Fig. 8.10(ii)). When these so-called 'atmospheres' of solute atoms are produced, the movement of dislocations will be impeded and a greater stress must be applied to move them. That is to say, the yield point has been raised.

In interstitial solid solutions the relatively small solute atoms will tend to occupy positions where the lattice is being extended (Fig. 8.10(iii)) since

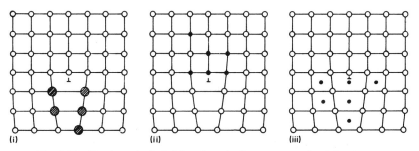

(i)   (ii)   (iii)

Fig. 8.10   The 'pegging' of dislocations by the presence of solute atoms.

interstitial gaps will be larger in these regions. This hypothesis affords an explanation of the yield-point effect in mild steel (2.31), and was a significant triumph in support of the dislocation theory in its earlier days.

In Fig. 8.11 the yield point, A, corresponds to the stress necessary to move dislocations away from their atmosphere of interstitial carbon atoms by which they have become anchored, whilst the point B indicates the stress necessary to keep the dislocations moving once movement has begun.

This yield-point effect in mild steel causes non-uniform localised deformation, which leads to the formation of 'stretcher strains' (or 'Lüder's lines') on the surface of such products as deep-drawn motorcar bodies. Such markings disfigure the surface, and their formation is avoided by applying slight initial deformation to the stock material, for example, by means of a light rolling pass. As this deformation is taking place in compression instead of in tension, stretcher strains will not form.

Assume that, during this treatment, the stock material was stressed to the point X (Fig. 8.11). Since the material has now yielded, no further yield point will be encountered during the subsequent deep-drawing process. Thus, the stress-strain curve will follow XZ. If, however, the pre-stressed steel is allowed to remain at room temperature for a few months, or if it is heated to 300°C for a few hours, carbon atoms are able to move back to their initial positions relative to the dislocations and form their original 'atmosphere', so that the yield-point effect returns. Since there are now more dislocations than were present in the original unstressed material, the new yield point will be at Y. This effect, which is known as 'strain-ageing', can be caused by the presence of as little as 0.003% carbon or nitrogen. It is sometimes used to strengthen springs after coiling, but its effect in material destined for deep drawing must be avoided. Consequently, such material is usually deep-drawn within twenty-four hours of the pre-stressing operation mentioned above.

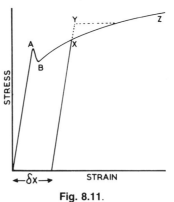

Fig. 8.11.

8.62 Dispersion Hardening   The presence of small separate particles in the microstructure can impede the movement of dislocations provided that these particles are stronger than the matrix in which they are embedded. The degree of strengthening produced also depends on the size of particles, their distance apart and the tenacity of the bond between particles and matrix. The passage of a dislocation through an alloy containing isolated particles is represented in Fig. 8.12. Assuming that the particles are stronger than the matrix, the dislocation cannot pass through them, but if the stress is high enough, the dislocation can by-pass them leaving a 'dislocation loop' around each particle. This will make the passage of a second dislocation much more difficult, particularly since dislocations have greater difficulty in passing between particles which are near to each other.

Particles of a separate phase are often precipitated from a solid solution during cooling. The cementite platelets in pearlite are an example, and their presence effectively strengthens steel. Such particles which have separated to form their own individual crystal structures are referred to as non-coherent precipitates. Prior to the actual precipitation of separate particles, however, an intermediate state often exists where atoms associated with the *solute* material begin to form groups *within the crystal lattice of the solvent metal*. These are small regions of less than one hundred atoms on an edge, and are called 'coherent precipitates'. Although still part of the continuous crystal structure of the matrix, they are attempting to form their own separate crystals, which will have a different crystal pattern. The presence of these coherent precipitates produces severe elastic distortions which impede the movement of dislocations. They are generally more effective than non-coherent precipitates. The nature of coherent precipitates, which give rise to the 'precipitation hardening' of some aluminium alloys, beryllium bronzes and a number of other alloys, will be dealt with more fully later (9.92).

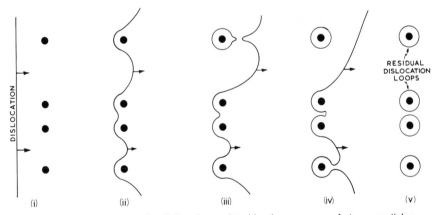

Fig. 8.12   The movement of a dislocation resisted by the presence of strong particles.

Exercises

1. Show how the positions of two metals relative to each other in the Periodic Table are likely to affect the manner in which they associate in a metallic alloy. (8.14)

2. Discuss the factors which may limit or even prevent the formation of a substitutional solid solution in a binary alloy. (8.23)
3. Describe, with the aid of sketches, the various types of solid solution which may be formed in metallic alloys. Show, in each case, how the structure of the solid solution affects its behaviour during heat-treatment, giving typical examples among important engineering alloys. (8.22)
4. Describe, with the aid of sketches, how a cored structure develops in a substitutional solid solution. Outline a theory which seeks to explain how the coring can be dispersed when such a solid solution is annealed at a sufficiently high temperature. (8.23–8.24)
5. State Fick's First Law of Diffusion and show how it applies to the diffusion which takes place in a metallic solid solution. What factors influence the rate of such diffusion? (8.25)
6. (a) Into which Hume-Rothery ratios do the following electron compounds fit: (i) $AuZn_3$ (ii) AgCd; (iii) Cu_5Cd_8; (iv) Cu_5Si? (b) What are interstitial *compounds* and why are many of them important in engineering alloys? (8.31)
7. Explain the following terms: (i) substitutional solid solution; (ii) interstitial solid solution; (iii) electron compound; (iv) valence compound. What factors affect the formation of these phases? (8.23; 8.24 and 8.31)
8. What are the main characteristics which define a eutectic?
   Distinguish between *eutectic* and *eutectoid*. (8.40 and 8.43)
9. (*a*) Explain, with the aid of sketches, why binary solid solutions are usually stronger than the pure metals they contain. (*b*) Show how the Dislocation Theory helps to explain the 'yield-point effect' peculiar to normalised low-carbon steels. (8.61)

Bibliography

Cahn, R. W., *Physical Metallurgy*, North Holland, 1980.
Goldschmidt, H. J., *Interstitial Alloys*, Butterworths, 1967.
Hume-Rothery, W., Smallman, R. E. and Howarth, C. W., *The Structure of Metals and Alloys*, Institute of Metals, 1988.
Hume-Rothery, W., *The Structure of Alloys of Iron*, Pergamon Press, 1966.
Martin, J. W. and Doherty, R. D., *Stability of Microstructures in Metallic Systems*, Cambridge University Press, 1976.
Porter, D. A. and Easterling, K. E., *Phase Transformations in Metals and Alloys*, Van Nostrand Reinhold Co., 1981.

9

Thermal Equilibrium Diagrams

9.10 The subject of phase equilibrium was briefly discussed in Chapter 7 to introduce the reader to this very important field of study in as painless a manner as possible. The reader will have acquired a general impression of the significance of equilibrium diagrams, since no useful examination of the microstructures of alloys is possible without reference to the appropriate diagram, which, for this reason, becomes one of the metallurgists most useful 'tools'. With its aid he can predict with precision the structure of a given alloy at any given temperature provided that the alloy's thermal-treatment history is known. In the thermodynamical sense, 'equilibrium' refers to the state of balance which exists, or which tends to be attained, between the phases in the structure of an alloy after a chemical or physical change has taken place. In some cases equilibrium may not be reached for long periods after the change has begun. In fact, rapid cooling (or 'quenching') to room temperature may suppress the chemical or physical change involved to such an extent that equilibrium will never be reached so long as the alloy remains at room temperature. Reheating (or 'tempering') followed by slow cooling to room temperature will allow the physical or chemical change to proceed in some degree towards completion; the extent to which equilibrium is thus attained being dependent upon the temperature reached, the time the alloy remains at that temperature and the rate of cooling to room temperature or at whatever other temperature equilibrium is being studied.

9.11 We have already seen that physical and chemical changes which take place after solidification in carbon steel do so with relative ease because the small carbon atoms can move quickly through the crystal lattice of face-centred cubic iron. Thus, during the process of normalising, a steel will attain structural equilibrium. In some alloy systems, however, physical changes which take place after solidification are so sluggish that it is doubtful if equilibrium is ever attained. Often we have no means of telling whether or not equilibrium has been reached, even after extremely slow rates of cooling involving periods of days or even weeks. There is therefore a tendency these days to refer to these charts as 'constitutional diagrams' instead of 'thermal-equilibrium diagrams'. The latter term, however, is still in general use, and we shall employ it throughout this book.

9.12   The thermal-equilibrium diagram is in reality a chart which shows the relationship between the composition, temperature and structure of any alloy in a series. Let us consider briefly how these diagrams are constructed.

A pure metal will complete its solidification without change of temperature (3.12), whilst an alloy will solidify over a range of temperature which will depend upon the composition of the alloy (7.41). Consider, for example, a number of alloys of different composition containing the two metals A and B which form a series of solid solutions (Fig. 9.1). For successive compositions containing diminishing amounts of the metal A, freezing commences at a_1, a_2, a_3, etc., and ends at b_1, b_2, b_3, etc. Thus, if we join all points a_1, a_2, a_3, etc., we shall obtain a line called the *liquidus*, indicating the temperature at which any given alloy in the series will commence to solidify. Similarly, if we join the points b_1, b_2, b_3, etc., we have a line, called the *solidus*, showing the temperature at which any alloy in the series will become completely solid. Hence the liquidus can be defined as the line above which all indicated alloy compositions represent completely homogeneous liquids, whilst the solidus can be defined as the line below which all represented alloy compositions of A and B are completely solid. For temperatures and compositions corresponding to the co-ordinates of points between the two lines both liquid solutions and solid solutions can co-exist in equilibrium.

9.13   Such a system as this occurs when the two metals are soluble in

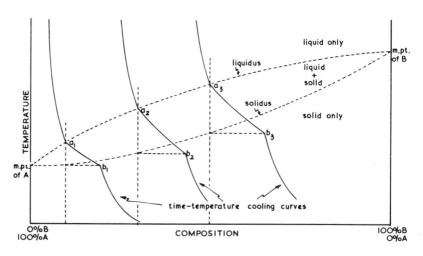

Fig. 9.1 The construction of a simple equilibrium diagram of the solid solution type using cooling curves of a series of alloys.

each other in all proportions in both the liquid and solid states, as in the case of copper and nickel or gold and silver. When, on the other hand, the metals are completely insoluble in the solid state a eutectic type of equilibrium diagram represents the system (Fig. 9.2). Since the eutectic part of the structure of any alloy in the series solidifies at the same temperature, and is the portion of the alloy with the lowest melting point, the line *ACEDB* will be the solidus whilst *AEB* is the liquidus.

9.14 Diagrams of this type can be constructed by using the appropriate points (obtained from time–temperature cooling curves) which indicate where freezing began and where it was complete. Thus the liquidus and solidus temperatures for a number of alloys in the series between 100% A and 100% B can be obtained and then plotted on a temperature-composition diagram as indicated in Figs. 9.1 and 9.2. Such cooling curves will determine the positions of the liquidus and solidus; and if the alloy system is one in which further structural changes occur after the alloy has solidified, then the metallurgist must resort to other methods of investigation to determine the phase boundary lines. These methods include the use of X-rays, electrical-conductivity measurements, dilatometric methods (which measure changes in volume) and exhaustive microscopical examination following the quenching of representative specimens from different temperatures.

9.15 The construction of equilibrium diagrams, then, is the result of much tedious experimental work, and, since the conditions under which the work is carried out can be so variable, it is not surprising that the values assigned to compositions and temperatures at which phase changes occur are under constant review by research metallurgists. The reader will, therefore, find that the equilibrium diagrams printed in books on metallurgy often differ in small detail. In general, however, the variations are so small as not to affect the treatment of most of the industrially

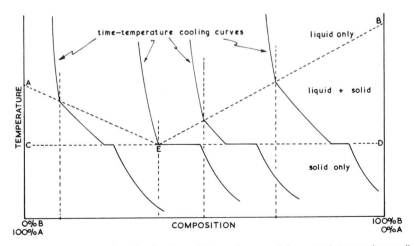

Fig. 9.2 The construction of a simple equilibrium diagram of the eutectic type using cooling curves of a series of alloys.

important alloys. As an example, the accepted value for the carbon content of a completely eutectoid carbon steel (7.55) has varied between 0.80 and 0.89% during recent years, whilst the eutectoid temperature (the lower critical temperature of plain carbon steels) has varied between 698 and 732°C.

The Phase Rule

9.20   In the previous chapter a 'phase' was described as either a pure metal, a solid solution or some form of intermediate compound so long as it was a *single homogeneous substance*. Similarly a single element or compound can exist in different physical forms, each one a separate phase. Thus the chemical substance 'H_2O' can exist as ice, water or water vapour, each of these conditions constituting a 'phase'. The 'phase structure' of H_2O is governed by the prevailing conditions of temperature and pressure as indicated in the phase diagram in Fig. 9.3. This diagram shows the effects which variations of temperature and pressure will have on the stability of a given phase. The 'fusion curve' is an almost vertical line and this indicates that variation in pressure has very little effect on the melting point of ice. The same is true of the effects of pressure on the melting points of alloys for which reason we can neglect pressure when studying the phase equilibrium of liquid or solid alloys.

The 'vaporisation curve' however indicates that the boiling point of water is very dependent on pressure. Water boils at 100°C when the pressure is equivalent to 760 mm of mercury. If the pressure is reduced to, say, 300 mm of mercury then water will boil at about 50°C which is the reason why tea cannot be brewed successfully near the summit of Everest.

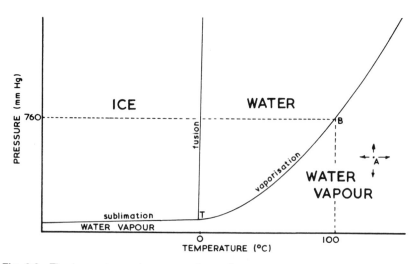

Fig. 9.3   The ice–water–water-vapour phase diagram.

The 'sublimation curve' shows that at low pressures solid ice can vaporise —or 'sublime'—without passing through the liquid phase. Some substances, notably carbon dioxide ('dry ice') and iodine, do this at normal atmospheric pressures when heated to the 'sublimation temperature'.

9.21 In 1876 Gibbs (8.13) introduced his Phase Rule which predicts the relationship between the number of phases, the number of components (or pure chemical substances) and the conditions of temperature and pressure which prevail. The general form of the rule can be written:

$$P + F = C + E$$

where P = the number of different phases present;
C = the number of components (stable chemical substances);
F = the degrees of 'freedom', or 'variance';
E = the environmental factors (temperature and pressure).

9.22 The implications of this rule can be explained more easily by reference to examples. Consider point A in Fig. 9.3. Here there is only *one* phase in the system (water vapour) and *one* component (the chemical compound, H_2O) whilst there are *two* environmental factors (temperature and pressure) to be considered. Hence, since—

$$P + F = C + E$$
$$1 + F = 1 + 2$$
$$\text{ie} \quad F = 2$$

This means that *two* degrees of freedom are involved so that both temperature and pressure can be altered *independently* without altering the number of phases in the system, which is consequently said to be *bivariant*.

Now consider the point B. This indicates that water is boiling at 100°C and 760 mm pressure so we have *two* phases (water and water vapour) present at the same time, but still *one* component (H_2O). Applying the Phase Rule:

$$2 + F = 1 + 2$$
$$\text{or} \quad F = 1$$

This means that only *one* degree of freedom is available and if temperature is increased pressure must also be increased in sympathy if the water is to continue to boil steadily. Conversely if, say, pressure were reduced then the boiling point of water would decrease in sympathy so that conditions were still represented by a point on the 'vaporisation curve'. Failure to observe these conditions would mean that *one* phase would disappear —either the water would cease to boil or it would turn very quickly to steam. The system represented by B is said to be *univariant*.

At the triple point T all three phases, ice, water and water vapour can exist together. Hence:

$$3 + F = 1 + 2$$
$$\text{or} \quad F = 0$$

In this case there is no degree of freedom and if either temperature or pressure are altered then at least one of the phases will disappear. This system is *invariant*.

9.23 We will now consider the Phase Rule as it applies to a study of metallic alloys. Here the *components* in the system are *pure metals* or *elements* from which the alloy is made, whilst the *phases* produced will be either *liquids, solid solutions, intermediate compounds* or, rarely, *pure metals*.

We are usually dealing with alloy systems in the liquid and solid states at atmospheric pressure. Since small variations in atmospheric pressure will have virtually no effect on the stability of phases we can neglect the effect of pressure and standardise at 'one atmosphere'. As we are now dealing with only one variable environmental factor, temperature, we can adapt Gibbs' Phase Rule so:

$$P + F = C + 1$$

Moreover since one environmental factor has disappeared we can still express a two-component system in a two-dimensional diagram, variation in composition replacing variation in pressure. We will now examine such a system (Fig. 9.4).

Point *A* represents a completely homogeneous molten alloy containing a single liquid phase composed of the two components bismuth and cadmium. Hence:

$$1 + F = 2 + 1$$
$$\text{or} \quad F = 2$$

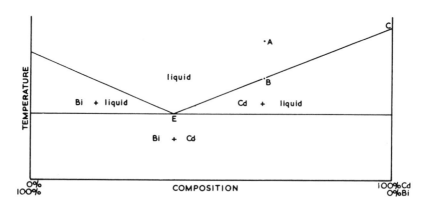

Fig. 9.4 The bismuth–cadmium phase (or thermal equilibrium) diagram.

There are two degrees of freedom so that both temperature and composition may be varied *independently* without altering the single-phase structure.

On cooling the alloy to B the liquid becomes saturated with cadmium so that solid cadmium begins to form. There are now two phases present and:

$$2 + F = 2 + 1$$
$$\text{and } F = 1$$

Thus there is one degree of freedom and if the temperature is altered the composition of the remaining liquid must alter so that B remains on CE.

At E solid cadmium and solid bismuth co-exist with remaining liquid so that there are three phases and:

$$3 + F = 2 + 1$$
$$\text{and } F = 0$$

Therefore the system at E is invariant and any alteration of temperature or composition will lead to the disappearance of at least one of the phases. Conversely if the alloy of composition indicated by E is cooling, then the temperature must remain constant until the liquid has crystallised completely. That is, a eutectic crystallises at a constant temperature.

9.24 In the interpretation of thermal equilibrium diagrams it is often more convenient to make use of a set of definitions and rules which are derived from Gibbs' Phase Rule and also from some observations made by the French metallurgist A. Portevin:

(*a*) The areas of the diagram are called 'phase fields', and on crossing any sloping boundary line from one field to the next, the number of phases will always change by one, ie two single-phase fields will always be separated by a double-phase field containing both of the phases. In a binary system three phases can only exist together at a point, such as the eutectic point E in Fig. 9.4.

(*b*) In Fig. 9.5 the point P represents some particular alloy mixture in both composition and temperature. The diagram indicates that at P both liquid and solid will exist together. If a 'temperature horizontal' (usually known as a *tie-line*) is drawn through P the composition of the solid is given by X, and the composition of the liquid solution in equilibrium with it by Y. P itself represents the overall composition of the mixture.

(*c*) The relative amounts of both liquid and solid at P are given by the relative lengths PX and PY—

$$\begin{array}{ccc} \text{Weight of solid solution} & & \text{Weight of liquid solution} \\ \text{(composition } X) \times PX & = & \text{(composition } Y) \times PY \end{array}$$

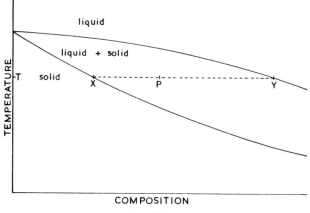

Fig. 9.5.

This is generally referred to as the *Lever Rule* since it is as though we were taking moments about *P* as the fulcrum. The relationship can be expressed in more convenient form:

$$\frac{\text{Weight of solid solution (composition } X)}{\text{Weight of liquid solution (composition } Y)} = \frac{PY}{PX}$$

(d) A phase which does not occupy a field by itself, but appears only in a two-phase field, is either a pure metal or an intermediate phase

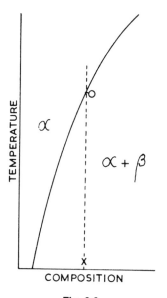

Fig. 9.6.

(usually an intermetallic compound) of invariable composition. In fact the phase 'field' will be infinitely narrow such as is represented by the 100% bismuth ordinate in Fig. 9.4 or the Mg_2Sn ordinate in Fig. 9.15.

(e) If we draw a vertical line on a diagram this will represent the composition of a given alloy (X in Fig. 9.6). If, on following along this line, a phase boundary is crossed (at O) the number of phases will change by one at this point, ie a phase (β) will be either precipitated or absorbed. (In this instance β will be absorbed when the alloy X is heated above O and precipitated again when the alloy X cools below O.)

We will now consider the main types of thermal equilibrium diagram which are of use in studying metallic alloy systems. Generally a useful alloy will only be formed when the two metals are *completely* soluble in each other in the liquid state; but in some instances the two metals are only *partially* soluble as liquids. We will begin with a brief study of one such case.

Case I—Two metals which are only partially soluble in each other in the liquid state

9.30   Fig. 9.7 represents the phase equilibrium conditions for the zinc–lead system above 500°C. As in all of these equilibrium diagrams the base line shows the composition varying between 100% of one component at one side of the diagram to 100% of the other component at the opposite side. In most diagrams the percentages are by weight but in some cases it is more relevant to indicate percentages in terms of numbers of atoms. The ordinate represents temperature. Thus point P represents a mixture consisting of 70% by weight of zinc and 30% by weight of lead at a temperature of 700°C.

9.31   The diagram (Fig. 9.7) shows that at 500°C molten zinc will dissolve no more than 2% lead (point A) giving a *single* solution of lead in zinc. At the same temperature molten lead will dissolve about 5% zinc (point B) thus giving a *single* solution of zinc in lead. However, any alloy at 500°C which has a composition given by some point between A and B will consist of *two* separate layers of two different liquid solutions—the upper layer will be zinc containing 2% dissolved lead (composition A) and the lower layer will be lead containing 5% dissolved zinc (composition B).

9.32   The slope of the line $ADJC$ indicates that as the temperature rises lead becomes increasingly soluble in molten zinc. Similarly the slope of BEC shows that, under similar conditions, the solubility of zinc in molten lead also increases. These two solubility curves meet at 798°C indicating that at temperatures above this, the two metals become completely soluble in each other so that a single liquid phase will be present in the mixture whatever its overall composition.

9.33   Consider a molten mixture consisting of 70% zinc and 30% lead (composition H). At 500°C this will consist of a saturated layer of lead in

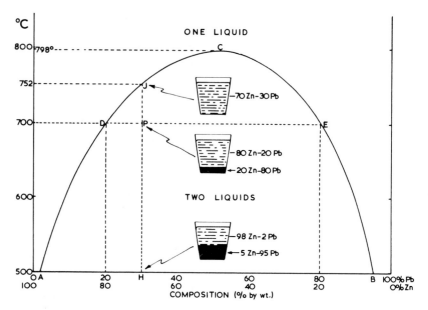

Fig. 9.7   The zinc–lead thermal equilibrium diagram above 500°C.

zinc (composition *A*) floating on top of a saturated layer of zinc in lead (composition *B*). That is, it is a two-phase mixture.

If the temperature is now raised the amount of lead which dissolves in zinc increases as does the amount of zinc which dissolves in lead. Thus at 700°C zinc will now dissolve 20% lead (composition *D*) whilst lead will dissolve 20% zinc (composition *E*). Using the Lever Rule we can calculate the relative proportions by weight of each layer:

$$\frac{\text{Weight of zinc-rich layer}}{\text{(composition } D) \times DP} = \frac{\text{Weight of lead-rich layer}}{\text{(composition } E) \times PE}$$

or

$$\frac{\text{Weight of zinc-rich layer (composition } D)}{\text{Weight of lead-rich layer (composition } E)} = \frac{PE}{DP}$$

Substituting values from the diagram:

$$\frac{\text{Weight of zinc-rich layer } (80\% \text{Zn–}20\% \text{Pb})}{\text{Weight of lead-rich layer } (80\% \text{Pb–}20\% \text{Zn})} = \frac{80 - 30}{30 - 20}$$

$$= \frac{5}{1}$$

As the temperature of this 70–30 mixture is raised still further to 752°C the two liquid phases merge into one, as the two metals become completely soluble in each other at this composition.

Case II—Two metals mutually soluble in all proportions in the liquid state becoming completely insoluble in the solid state

9.40 It is extremely doubtful whether such a situation as this really exists in practice since most solid metals appear to dissolve quantities, however small, of other metals. However, in the case of bismuth and cadmium the mutual solid solubility is so small that we can neglect it for the sake of our argument and assume that the metals are completely insoluble in the solid state. Bismuth, a heavy, brittle, slightly pink metal, is one of the least 'metallic' of metals as indicated by its position near to the non-metals in the Periodic Table. It forms a complex rhombic-type crystal structure in which the metallic bond is partly replaced by a mixture of covalent bonds and van der Waals forces. Cadmium on the other hand crystallises as a normal CPH structure so that it is not surprising that these two metals with such unlike lattice structures do not form mixed crystals but separate out almost completely on solidification. The bismuth–cadmium thermal-equilibrium diagram is shown in Fig. 9.8.

9.41 Let us consider the solidification of an alloy of composition x, ie containing about 80% cadmium and 20% bismuth. When the temperature

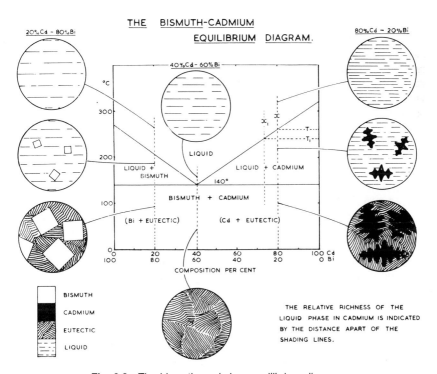

Fig. 9.8 The bismuth–cadmium equilibrium diagram.

falls to T, crystal nuclei of pure cadmium begin to form in accordance with rule (b) stated in 9.24. (The temperature horizontal or tie-line, T, cuts the liquidus at the chosen composition, x, and the other phase boundary is the 100% cadmium ordinate.) Since pure cadmium is deposited, it follows that the liquid which remains becomes correspondingly richer in bismuth. Therefore the composition of the liquid moves to the left—say, to x_1— and, as indicated by the diagram, no further deposition of cadmium takes place until the temperature has fallen to T_1. When this happens more cadmium is deposited, and dendrites begin to develop from the nuclei which have already formed. The growth of the cadmium dendrites, on the one hand, and the consequent enrichment of the remaining liquid in bismuth, on the other, continues until the temperature has fallen to 140°C. The remaining liquid then contains 40% cadmium and 60% bismuth, ie the eutectic point E has been reached.

At this point the two metals are in equilibrium in the liquid, but, due to the momentum of crystallisation, the composition swings a little too far past the point E, resulting in the deposition of a little too much cadmium. In order that equilibrium shall be maintained, a swing back in composition across the eutectic point takes place by the deposition of a layer of bismuth. In this way the composition of the liquid oscillates about E by depositing alternate layers of cadmium and bismuth, whilst the temperature remains at 140°C until the remaining liquid has solidified. Thus the final structure will consist of primary crystals of cadmium which formed between the temperature T and 140°C, and a eutectic consisting of alternate layers of cadmium and bismuth which formed at 140°C.

9.42 Had the original liquid contained less than 40% cadmium, then crystals of pure bismuth would have formed first, causing the composition of the remaining liquid to move to the right until ultimately the point E was reached as before, and the final liquid contained 40% cadmium and 60% bismuth. This remaining liquid would solidify as eutectic in the manner already described.

9.43 If the original liquid contained exactly 40% cadmium and 60% bismuth at the outset, then no solidification whatever would occur until the temperature had fallen to 140°C. Then a structure composed entirely of eutectic would be formed as outlined above.

9.44 In all three cases mentioned, the eutectic part of the structure will be of constant composition and will always contain 40% cadmium and 60% bismuth. Any variation either side of this in the *overall* composition of the alloy will be compensated for by first depositing appropriate amounts of either primary cadmium or primary bismuth, whichever is in excess of the eutectic composition. It is important to realise that there is no question of solid solubility existing in any way in the final structure, whatever its composition. With the aid of a microscope, we can see the two pure metals cadmium and bismuth as separate constituents in the microstructure. In other words, this is a case of complete insolubility in the solid state.

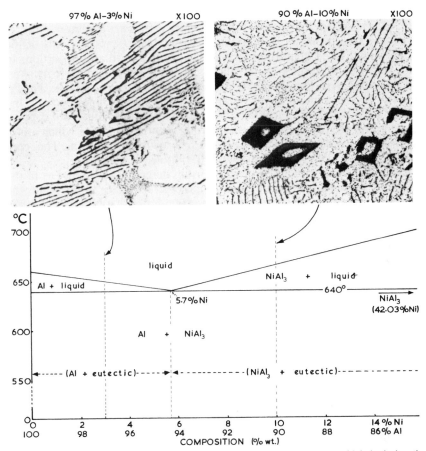

Plate 9.1 Part of the aluminium–nickel thermal equilibrium diagram which includes the eutectic system between aluminium and the intermetallic compound NiAl₃.

The photomicrographs of two representative alloys in the cast condition are shown. The 97–3 alloy contains approximately equal amounts of primary aluminium crystals (the light areas) and eutectic. The structure of the 90–10 alloy consists of primary NiAl₃ crystals (dark) and eutectic. Note the regular 'geometric' shape of the primary NiAl₃ crystals—this is often a feature of intermetallic compounds.

The light aluminium layers in the lamellar eutectic are much broader than those of the dark NiAl₃ because the eutectic composition contains only 5.7% nickel.

Case III—Two metals, mutually soluble in all proportions in the liquid state, remain mutually soluble in all proportions in the solid state

9.50 A number of pairs of metals fulfil these conditions and are generally those in which:

(i) the two atoms concerned are compatible in terms of the atomic size factor (8.23);

(ii) the electrochemical properties of the two metals are very similar;
(iii) the crystal structures of the two pure metals are similar in pattern.

If these conditions are fulfilled it is reasonable to expect atoms of the second metal to be able to replace atoms of the first in all proportions to form a disordered solid solution. Such is the case in the alloy systems: Ag–Au; Ag–Pd; Au–Pt; Bi–Sb; Co–Ni; Cu–Ni; Cu–Pt; Fe–Pt; Ni–Pt; and Ta–Ti. (In some of these systems further phase changes occur in the solid state due to polymorphic transformations (3.14)). Since the copper–nickel alloys are the only ones of the group mentioned above which are of use in engineering it is proposed to deal with this system. The copper–nickel thermal-equilibrium diagram is shown in Fig. 9.9.

Again this is a simple type of equilibrium diagram, and since, as in the cadmium–bismuth system, no transformations take place in the solid, the diagram consists of two lines only—the liquidus and solidus. Above the liquidus we have a uniform liquid solution for any alloy in the series, whilst below the solidus we have a single solid solution for any alloy, though in the cast condition, as we shall see, the solid solution may vary in concentration due to coring. Between the liquidus and solidus both liquid and solid solutions co-exist.

9.51   Applying the Phase Rule to that portion of the diagram above the liquidus:

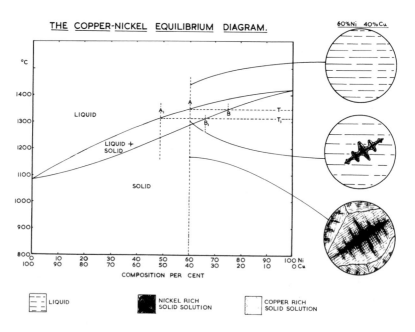

Fig. 9.9 The copper–nickel equilibrium diagram.
The microstructures indicated are those obtained under *non-equilibrium* conditions of solidification.

$$P + F = C + 1$$
$$1 + F = 2 + 1$$
$$\text{and } F = 2$$

since there are two degrees of freedom both temperature and composition can be altered over fairly wide limits without upsetting the stability of the single phase. However, for a point between the liquidus and solidus, ie in 'liquid + solid' field:

$$2 + F = 2 + 1$$
$$\text{and } F = 1$$

There is thus one degree of freedom, and if either temperature or composition are altered independently of the other, then the system becomes unstable so that some solid will melt or some liquid will solidify. Thus any alteration in temperature must be accompanied by a change in overall composition if the *proportions* of liquid and solid are to remain the same.

9.52   Consider the freezing of an alloy of composition A (Fig. 9.9), ie containing 60% nickel and 40% copper. Assume that cooling is taking place fairly rapidly, such as would be the case during the industrial casting of slab ingots into cast-iron moulds. Above $T°C$ the alloy will exist as a uniform liquid solution, but as the temperature falls to $T°C$, dendrites of solid solution will begin to form. They will not however be dendrites of composition A but dendrites of composition B. This composition B is indicated by drawing a tie-line from the liquidus at A (which represents the composition and temperature of the original liquid) to the solidus,

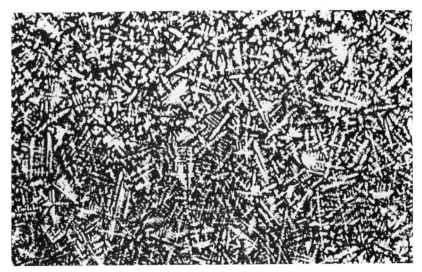

Plate 9.2 70–30 cupro-nickel in the chill-cast condition. × 80.
This shows a mass of dendrites of heavily cored solid solution. The lighter cores are nickel-rich whilst the copper-rich infilling has etched a darker colour.

which it cuts at B. Thus the dendrites which form will contain approximately 75% nickel (composition B) and, since the original liquid contained only 60% nickel (composition A), it follows that the remaining liquid will contain an even lower proportion of nickel. Hence its composition will move to the left, say to A_1.

Solidification will continue when the temperature falls further to T_1, and this time a layer of solid of composition B_1 will be deposited. This is less rich in nickel than the original seed crystals and, as crystallisation proceeds, successive layers will contain less and less nickel, and consequently more and more copper, until ultimately the liquid is used up.

Clearly, then, a non-uniform solid solution is formed, and whilst its *overall* composition will be 60% nickel and 40% copper, due to the coring effect the initial skeleton of a crystal will contain about 75% nickel and its outside fringes only about 50% nickel.

9.53 The situation is further influenced, however, by diffusion (8.23), which is taking place simultaneously with crystallisation. Due to the fact that successive layers of the alloy which deposit are richer in nickel than the remaining liquid, a concentration gradient is set up which tends to make copper atoms diffuse inwards towards the centre of the dendrite, whilst nickel atoms move outwards towards the copper-rich liquid. In the above case we have assumed very rapid cooling so that little or no diffusion has been possible, and that under these conditions the structure has been unable to reach equilibrium so that a heavily cored structure forms. Since we are dealing with non-equilibrium conditions, only a limited amount of information can be obtained from the equilibrium diagram—in this case a guide as to why the cored structure develops. If cooling were slow, diffusion could take place extensively so that coring would be slight in the final crystals. Annealing at a sufficiently high temperature will also permit diffusion to proceed (8.24), resulting in the formation of almost completely uniform solid-solution crystals.

9.54 We will now consider the cooling of the same alloy under conditions which are ideally slow so that complete equilibrium is attained at each stage of solidification. The liquid (composition A, Fig. 9.10) will begin to solidify at temperature T by depositing nuclei of composition B. This causes the composition of the remaining liquid to move to the left, but, due to the prevailing slow cooling, *diffusion is able to keep pace with*

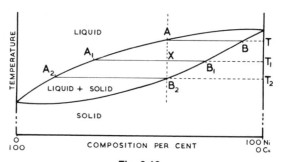

Fig. 9.10.

solidification so that, as the composition of the liquid follows the liquidus from A to A_2, the composition of the solid follows the solidus from B to B_2.

Thus, at some temperature T_1, the composition of the *uniform* solid solution is given by B_1, whilst that of the remaining homogeneous liquid in equilibrium with it is given by A_1. Since the *overall* composition of the alloy is indicated by X (which is the same as A) then, by the Lever Rule:

$$\frac{\text{Weight of solid solution (composition } B_1)}{\text{Weight of remaining liquid solution (composition } A_1)} = \frac{XA_1}{XB_1}$$

At the temperature T_2 the last trace of liquid (composition A_2) has just solidified at the crystal boundaries and, by the process of diffusion, its composition has changed to that of the remainder of the structure which has adjusted itself to uniform B_2 throughout. A and B_2 of course represent the same composition, which is obvious, since a *single uniform* liquid solution has been replaced by a *single uniform* solid solution.

9.55   When two metals form a substitutional solid solution some distortion of the lattice is inevitable due to influences such as different atom size. In some cases the two metals are *only just* compatible and although they are still soluble in all proportions in the solid state a phase diagram of the type shown in Fig. 9.11 results. A minimum melting point (32.8°C) is produced where liquidus and solidus are *congruent*. The significance of this minimum point is that the metals are so incompatible in solid solution that there is a strong tendency for them to return to the disordered liquid state

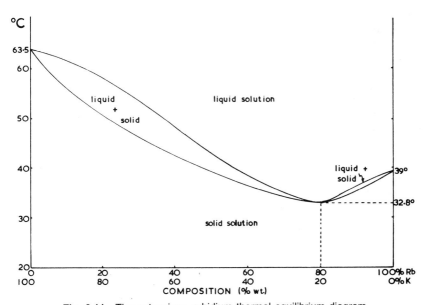

Fig. 9.11 The potassium–rubidium thermal equilibrium diagram.

at as low a temperature as possible. In fact a little more incompatibility would lead to a complete 'divorce' to the extent that the metals would separate forming a eutectic, probably of two solid solutions. We shall in fact deal with this type of system in the next case.

Case IV—Two metals mutually soluble in all proportions in the liquid state but only partially soluble in the solid state

9.60 This case is in effect intermediate between the two dealt with in Cases II and III and the thermal equilibrium diagram is, in consequence, a sort of hybrid of these two. It represents a stage where incompatibility has exceeded that suggested in Fig. 9.11. As in the cadmium–bismuth system (where the metals are completely incompatible in the solid state) a eutectic is formed, but in this case it is a eutectic of two solid solutions instead of two pure metals. It is in fact the 'general' case and purists may argue that, as such, it should have been dealt with before either of the preceding particular cases. The systems have, however, been dealt with in order of complexity rather than logical classification in order to assist the student who may be encountering the subject for the first time.

The only new feature of this system, as compared with those foregoing, is that we have phase boundaries occurring below the solidus, indicating that phase changes can take place in the solid. On the tin–lead thermal-equilibrium diagram (Fig. 9.12), which we shall use as an example of this type of system, such phase boundaries as these are indicated by the lines *BC* and *FG*. A line such as *BC* or *FG* is termed a *solvus* and indicates a change in solubility of one metal in another in the solid state. Part of the tin–lead system was used in Chapter 7 as an introduction to equilibrium diagrams. We will now consider it in its entirety.

9.61 As the diagram shows, we have two solid solutions in the system, since:

(a) Tin will dissolve up to a maximum of 2.6% lead at the eutectic temperature, forming the solid solution α.
(b) Lead will dissolve up to a maximum of 19.5% tin at the eutectic temperature, giving the solid solution β.

(The different phases present in an alloy series are generally lettered from left to right across the diagram, using letters of the Greek alphabet for convenient reference.)

The slope of the phase boundaries *BC* and *FG* indicates that both the solubility of lead in tin (α) and the solubility of tin in lead (β) decrease as the temperature falls. This is a normal phenomenon with solid solutions as it is with liquid solutions. In Fig. 9.13 the solubility of a salt in water at different temperatures is compared with the *solid* solubility of one metal in another at different temperatures. In each case solubility increases as

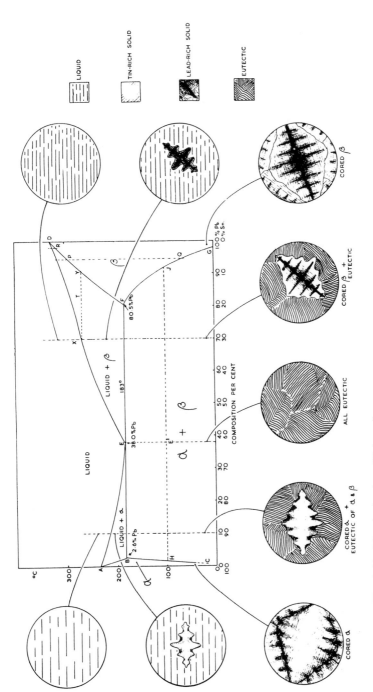

Fig. 9.12 The tin—lead equilibrium diagram
The microstructures indicated are those obtained under *non-equilibrium* conditions of solidification.

LIQUID

TIN-RICH SOLID

LEAD-RICH SOLID

EUTECTIC

CORED β

CORED β + EUTECTIC

ALL EUTECTIC

CORED α + EUTECTIC OF α & β

CORED α

100% Pb 0% Sn

°C

300

200

100

LIQUID

LIQUID + β

183°

38 % Pb

80.5%Pb

LIQUID + α

2.6% Pb

α + β

α

β

COMPOSITION PER CENT

100 90 80 70 60 50 40 30 20 10 00
0   10 20 30 40 50 60 70 80 90 100

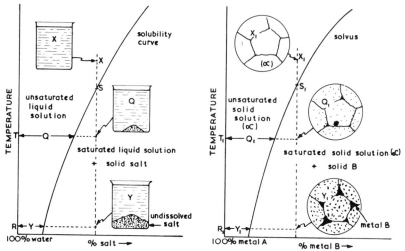

Fig. 9.13 An analogy between the liquid solubility of salt in water and the partial solid solubility of one metal in another.

temperature rises and saturated solutions are formed in a similar manner. X and X_1 each represent a mixture which at that temperature is an *unsaturated* solution, ie it could dissolve more solute. As the temperature falls to S (S_1) the solution (liquid or solid) reaches a saturation point so that solid salt (or metal B) begin to precipitate. At T (T_1) a considerable quantity Q (Q_1) of solute (either salt or metal B) is still in solution but as the temperature falls to R (R_1) this is reduced to a very small amount in each case— Y (Y_1) as salt (or metal B) is progressively precipitated from solution.

9.62   In the case of the tin–lead alloys the sloping boundaries BC and FG indicate that changes take place in the solid. Thus as the temperature falls:

   (i) lead becomes less soluble in solid tin (α) along BC;
   (ii) tin becomes less soluble in solid lead (β) along FG.
Several different structures may be formed, depending upon the alloy composition. For example, let us consider:

   (*a*) An alloy of composition X (70% lead–30% tin). This will begin to solidify when the temperature falls to T and dendrites of composition Y will deposit. The alloy continues to solidify in the manner of solid solution until at 183°C the last layer of solid to form will be composition F (80.5% lead–19.5% tin) and the remaining liquid will be of composition E (the eutectic composition with 38% lead and 62% tin).

   The remaining liquid now solidifies by depositing, in the form of a eutectic, alternate layers of α and β, of the compositions B and F respectively. If this structure now cools *slowly* to room temperature the compositions of the solid solutions α and β will follow the lines BC and FG, ie the solid solution α will become progressively poorer in lead and the solid solution β will become poorer in tin, until at, say, 100°C α will contain less than 1% lead and β will contain less than 10% tin. The

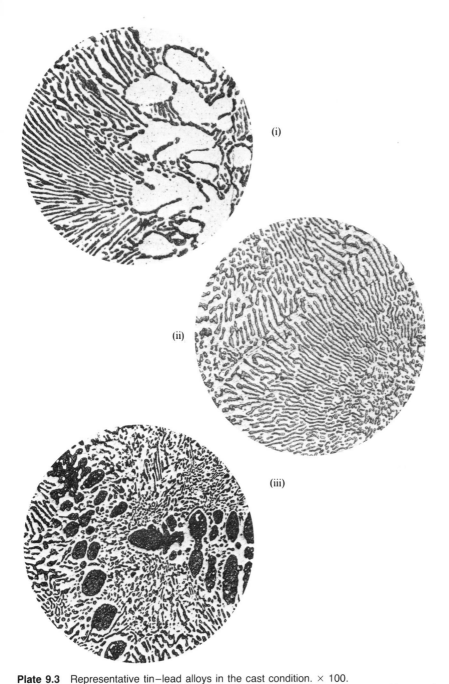

Plate 9.3 Representative tin–lead alloys in the cast condition. × 100.
(i) 75 tin–25 lead. Primary crystals of α (light) in a matrix of eutectic consisting of layers of α (light) and β (dark).
(ii) 62 tin–38 lead. Completely eutectic in structure—layers of α (light) and β (dark).
(iii) 50 tin–50 lead. Primary crystals of β (dark) in a matrix α + β eutectic as in (i) and (ii). Note the small 'islands' of α *within* the primary β crystals in (iii). These were precipitated as the alloy cooled slowly from 183°C to ambient temperature and the β changed in composition along the steeply sloping solvus *FG* (Fig. 9.12).

proportions of α and β will also vary from *EF/EB* at 183°C to *E'J/E'H* at 100°C. In the same way, provided we assume that the structure cooled slowly enough to allow the primary dendrites of β to reach a uniform composition *F*, these dendrites will now alter in composition to *J* at 100°C, and in so doing precipitate some α of composition *H* in order to adjust their own composition. The α precipitated will join that already present in the eutectic.

(*b*) If the alloy contained, say, 95% lead, and was cooled slowly enough to prevent coring, solidification would be complete at *P* and a uniform solid solution β would result. On continuing to cool slowly, further solid changes would begin at *Q* with the precipitation of small amounts of α at the grain boundaries of the β. This α would increase in amount as the temperature fell, and the β became progressively poorer in tin. Hence the final structure would consist of crystals of uniform β containing about 98% lead with small amounts of α precipitated at the crystal boundaries.

(*c*) If the original alloy contained more than 98% lead and was cooled slowly, the structure would remain entirely β throughout, after its solidification had been completed at *R*.

9.63   In actual practice it is unlikely that cooling would be slow enough to prevent some coring from taking place. Since the initial β dendrites would then be relatively rich in lead, this could lead to the formation of small amounts of α at the β grain boundaries in (*c*) and more than the expected amount of α in (*b*). In case (*c*) this α would be absorbed on annealing. 'Solution annealing' is an important process used in the treatment of many alloys to absorb some constituent which has been precipitated due to coring. After treatment the alloy will be soft and ductile and able to receive cold-work. Solution annealing is also an integral part of most strengthening processes based on precipitation hardening, the principles of which will be discussed later (9.92).

Case V—A system in which a peritectic transformation is involved

9.70   Sometimes in an alloy system two phases which are already present interact at a fixed temperature to produce an entirely new phase. This is known as a peritectic transformation. The term is derived from the Greek 'peri', which means 'around', since, during the transformation process, the original solid phase becomes surrounded or coated by the transformation product. Frequently one of the interacting phases is a liquid though this is not a precondition and sometimes two solids will take part—albeit very slowly—in a peritect*oid* transformation. A peritectic transformation occurs between the phase δ and the remaining liquid (producing austenite) in the iron–carbon system (11.20) but here we will consider the peritectic transformation which takes place in the platinum–silver system, part of which is shown in Fig. 9.14.

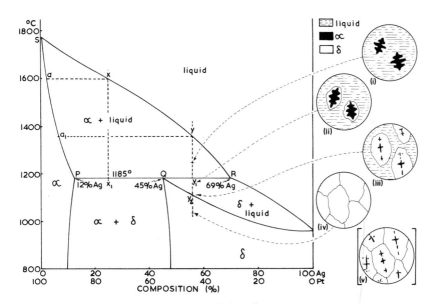

Fig. 9.14 The platinum–silver thermal equilibrium diagram.
(i) A slowly cooled alloy containing about 75% silver will first begin to solidify as uniform α
(ii) Below 1185°C these α crystals will become coated with δ due to peritectic interaction
with the remaining liquid. (iii) The original α will be transformed slowly to δ as the temperature
falls. (iv) If cooling is *extremely slow* all α will be absorbed and the remaining liquid will
solidify as uniform δ. (v) In practice cooling will generally be so rapid as to 'trap' some α
'cores' in the final structure, producing a non-equilibrium structure.

9.71 If we ignore coring effects, that is, assume a very slow rate of
cooling, the peritectic transformation will occur only in those platinum–silver mixtures containing between 12 and 69% silver. Let us first consider a
liquid of composition x, ie containing about 25% silver. This will begin to
solidify by depositing dendrites of the solid solution α of composition a. If
the rate of cooling is slow enough for diffusion to remove the effects of
coring, by the time the temperature has fallen to 1185°C the structure will
consist of α dendrites containing 12% silver (composition P) and a remaining liquid containing 69% silver (composition R), since the composition of
α will change as it follows the solidus SP. At this stage:

$$\frac{\text{Weight of } \alpha \text{ (composition } P)}{\text{Weight of remaining liquid (composition } R)} = \frac{x_1 R}{x_1 P}$$

9.72 At 1185°C, the peritectic temperature, the α dendrites begin to
interact with the remaining liquid and form a new solid solution, δ, ie

α + liquid ⇌ δ

This solid solution δ contains 55% platinum–45% silver (composition Q).
However the overall composition of the alloy is x_1 (75% platinum–25%

silver) and it follows that not all of the α will be used up, since it is in excess of that required to produce a structure entirely of δ. Therefore, all of the liquid is used up and some α remains, so that the final structure will consist of α containing 12% silver (composition P), and δ containing 45% silver (composition Q). These phases will be in a ratio given by:

$$\frac{\text{Weight of } \alpha \text{ (composition } P)}{\text{Weight of } \delta \text{ (composition } Q)} = \frac{x_1 Q}{x_1 P}$$

9.73 Let us now consider a liquid alloy of initial composition y. This will begin to solidify by depositing crystals of α of composition a_1, and if we again neglect coring, the crystals will gradually change in composition to P when the temperature has fallen to 1185°C, and again the composition of the remaining liquid will be R. At this stage—

$$\frac{\text{Weight of } \alpha \text{ (composition } P)}{\text{Weight of remaining liquid (composition } R)} = \frac{y_1 R}{y_1 P}$$

When the peritectic transformation takes place, we shall find that this time there is an excess of liquid because y_1 lies between Q and R. The α will therefore be completely used up, and just below 1185°C we shall be left with the new solid solution, δ, and some remaining liquid in the ratio—

$$\frac{\text{Weight of } \delta \text{ (composition } Q)}{\text{Weight of remaining liquid (composition } R)} = \frac{y_1 R}{y_1 Q}$$

As the temperature continues to fall slowly, this remaining liquid will solidify as δ, which will change in composition along Qy_2. At y_2 solidification will be complete and the structure will consist of crystals of uniform δ of the same composition as that of the original liquid. In practice it is unlikely that the rate of cooling would be slow enough to permit equilibrium being reached. Consequently, though the equilibrium structure is indicated as being completely δ (in the case of alloy y), remnants of the original α dendrites may be present in the rapidly-cooled structure. The original α dendrites, produced at temperatures above 1185°C, became coated with a protective envelope of δ as the interaction with the liquid began and subsequent diffusion was far too slow to promote dispersal of the remaining α.

Case VI—Systems containing one or more intermediate phase

9.80 Frequently two metals will not only form limited solid solutions with each other but, in other suitable proportions, will also interact to form one or more intermediate phases which may vary in nature from secondary solid solutions to true intermetallic compounds. The formation of any intermediate phase further complicates the thermal equilibrium diagram by introducing extra boundary lines and since some alloy systems include a number of intermediate phases, formed at different compositions of the two metals, the resulting diagram becomes increasingly complex.

9.81 We will first examine the tin–magnesium system (Fig. 9.15) as an example containing a single intermediate phase—in this case the intermetallic compound Mg_2Sn (containing 29.1% magnesium). This system is represented by two simple equilibrium diagrams joined at B. The section AB represents a eutectiferous series of the substances pure tin and Mg_2Sn. Here no solid solubility is involved and section AB can therefore be interpreted in the manner of Case II (the bismuth–cadmium system) dealt with in 9.40. The section BC is a eutectiferous series of the Case IV type (9.60) involving Mg_2Sn and the solid solution α. If we consider the complete system extending from pure tin to pure magnesium then, clearly, these two sections become joined as shown, forming a single diagram. Since Mg_2Sn is an intermetallic compound of single fixed composition its 'phase field' is represented on the diagram by a vertical line at 29.1% magnesium. A particularly interesting feature of this diagram is that pure Mg_2Sn has a melting point (778°C) which is higher than those of either of the constituent pure metals. This is an indication of the stability and strong chemical bonding in this valence-type compound.

9.82 We will now consider the solidification of a representative alloy in the series, ie alloy 'x' (60% tin–40% magnesium). This will begin to solidify when the temperature falls to T and the tie-line from x cuts the next phase boundary at y which represents the pure intermetallic com-

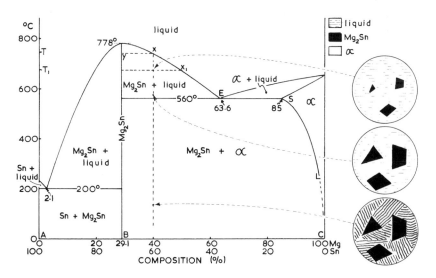

Fig. 9.15 The tin–magnesium thermal equilibrium diagram, consisting essentially of two eutectiferous series—tin–Mg_2Sn and Mg_2Sn–magnesium.

pound Mg_2Sn. Therefore crystals of Mg_2Sn begin to form. Since these crystals are relatively tin-rich (70.9% tin) the composition of the remaining liquid moves to the right to 'x_1' and as the temperature falls to T_1 more Mg_2Sn solidifies. Mg_2Sn continues to solidify in this way until at 560°C the remaining liquid contains 63.6% magnesium–36.4% tin—the composition, E, of the $Mg_2Sn + \alpha$ eutectic.

The remaining liquid then solidifies as eutectic at 560°C by depositing alternate layers of Mg_2Sn and the solid solution α, containing 85% magnesium–15% tin (composition S). When completely solid the structure will consist of primary crystals of the intermetallic compound Mg_2Sn surrounded by a eutectic of Mg_2Sn and α. If the solid alloy is then allowed to cool slowly below 560°C, α will change in composition along the solvus line SL by depositing particles of Mg_2Sn which will join the Mg_2Sn constituent already present in the eutectic.

9.83 The large size of lead atoms compared with those of many other metals contributes to the fact that lead forms relatively few solid solutions. Thus lead and gold are to all intents and purposes insoluble in each other in the solid state but, due to differences in the electrochemical properties of the metals two intermediate phases, Au_2Pb and $AuPb_2$ are formed at the appropriate compositions. These two intermediate phases are produced as a result of peritectic interactions and the resultant phase diagram is shown in Fig. 9.16.

9.84 Consider an alloy containing 40% lead. This will begin to solidify

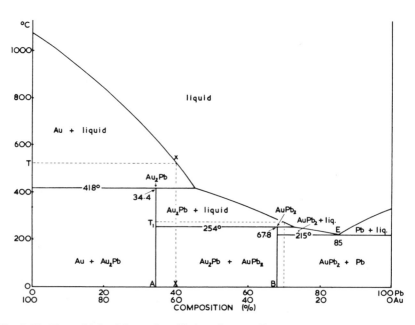

Fig. 9.16 The gold–lead thermal equilibrium diagram. Two intermediate phases are produced as a result of peritectic transformations.

at about 500°C (T) by forming dendrites of gold, but on cooling to 418°C, the first peritectic temperature, these gold dendrites undergo a peritectic interaction with the remaining liquid to produce crystals of the intermediate phase, Au_2Pb:

$$Au + liquid \rightleftharpoons Au_2 Pb$$

These crystals of Au_2Pb continue to grow as the temperature falls, but at 254°C a further peritectic interaction occurs resulting in the formation of another intermediate phase $AuPb_2$:

$$Au_2Pb + liquid \rightleftharpoons AuPb_2$$

During this second transformation all of the liquid is used up first and some Au_2Pb remains along with the new phase $AuPb_2$. Assuming that the system has cooled slowly to room temperature these two phases will be present in the ratio:

$$\frac{\text{Weight of } Au_2Pb}{\text{Weight of } AuPb_2} = \frac{BX}{AX}$$
$$= \frac{67.8 - 40}{40 - 34.4}$$
$$= \frac{27.8}{5.6}$$
$$\simeq \frac{5}{1}$$

9.85 An alloy containing 70% lead will begin to solidify at about 280°C (T_1) by depositing crystals of Au_2Pb but at 254°C a peritectic interaction results in the formation of $AuPb_2$. Here all of the Au_2Pb is used up and the remaining liquid continues to solidify as the new phase $AuPb_2$ until at 215°C the remaining liquid is of composition E (85% lead–15% gold). A eutectic consisting of alternate layers of $AuPb_2$ and pure lead is then formed so that the final structure consists of primary $AuPb_2$ in a eutectic of $AuPb_2$ and lead.

In both of these cases it has been assumed that cooling has been slow enough to permit equilibrium to be attained throughout so that peritectic transformations proceed to completion. In practice this may not be so. Thus in the 40% lead alloy more than the indicated quantity of Au_2Pb may be present in the microstructure whilst in the 70% lead alloy some Au_2Pb cores may be present even though the diagram indicates a $AuPb_2$–lead structure.

9.86 Sometimes an intermediate phase is of variable composition. Thus the phase β in the tantalum–nickel phase diagram (Fig. 9.17) is based on $TaNi_3$ but it may be thought of as being able to dissolve either excess tantalum or excess nickel so that the phase field occupies width on the diagram.

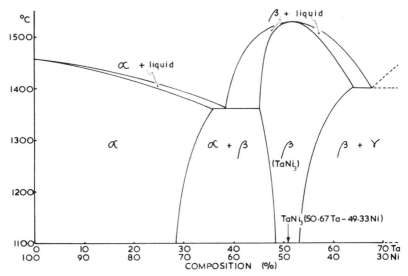

Fig. 9.17 Part of tantalum–nickel thermal equilibrium diagram. β is an intermediate phase which varies in composition about TaNi₃.

Precipitation from a Solid Solution

9.90 When the temperature of a solid solution falls such that it reaches a state of saturation, any further fall in temperature will lead to the precipitation of some second phase. This phenomenon was mentioned in 9.61 and is similar in principle to the precipitation which takes place from saturated liquid solutions when these continue to cool.

Precipitation from a solid solution, however, takes place much more sluggishly than from a liquid solution because of the greater difficulty of movement of the solute atoms in a solid solution, particularly if it is of the substitutional type. We must, therefore, consider carefully the relationship between the rate of cooling of a saturated solid solution and the extent to which precipitation can take place—and consequently equilibrium be attained—either during the cooling process or subsequently.

9.91 Precipitation under Equilibrium Conditions Fig. 9.18 represents part of a system such as was described in 9.60. Here metal B is partially soluble in metal A in the solid state forming the solid solution α. Any B in excess of solubility at any given temperature is precipitated, under equilibrium conditions, as the phase β, which in this case may be either a solid solution or an intermediate phase*.

Consider the slow cooling of some alloy of composition X from the temperature T_u. At T_u the solid solution α is unsaturated and this state of affairs prevails until the temperature falls to T_s. Here the solid solution

---

* Since the remainder of the diagram is not given we cannot know whether β is a solid solution as indicated in 9.60 or some intermediate phase as indicated in the various examples discussed in 9.80.

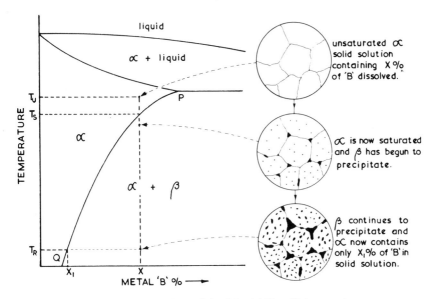

Fig. 9.18 Changes in partial solid solubility with temperature.

reaches saturation and when the temperature falls below T_s, precipitation may begin. Random nucleation takes place at the grain boundaries as well as on certain crystallographic planes in the α matrix, and nuclei of β begin to form. As the alloy cools, precipitation of β continues, and since β is rich in B, the composition of α changes progressively along the solvus line PQ. Because β is richer in B than is α, B must diffuse through α in order to reach the growing nuclei of β. This tends to reduce the concentration of B rather more rapidly in that α near to the β nuclei than in those regions of α which are far away from any nuclei. The concentration gradient is, however, too small to cause rapid diffusion of the B atoms, and, since the temperature is falling, regions of α far from any β nuclei ultimately become supersaturated with excess B. Consequently more β nuclei will form and begin to grow.

Precipitates which form under equilibrium conditions in this manner are generally non-coherent. That is, the new phase which has formed has a crystal structure which is entirely its own and is completely separate from the surrounding matrix from which it was precipitated. Its strengthening effect on the alloy as a whole is somewhat limited (8.62) and generally much less than that resulting from the presence of a coherent precipitate, the formation of which will be dealt with in the next section.

Assuming that the temperature has fallen to T_R and that the precipitation of β has taken place under equilibrium conditions, the structure will consist of non-coherent particles of β (usually of such a size as can be seen easily using an ordinary optical microscope) in a matrix of α which is now of composition X_1, ie it contains much less B than did the original α (composition X).

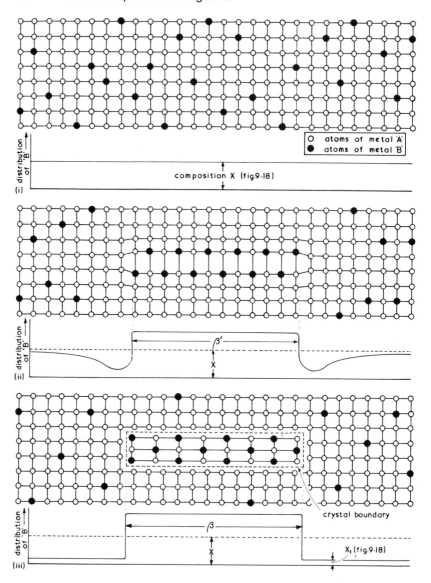

Fig. 9.19   The formation of coherent and non-coherent precipitates.
In (i) the alloy has been heated to T_u (Fig. 9.18) and then cooled rapidly (water quenched)
The structure consists of homogeneous solid solution α, supersaturated with B. (ii) Heating
to some selected temperature causes B atoms to migrate and form clusters within the α
lattice. These clusters produce β', which although preliminary to the formation of β, is still
continuous—or *coherent*—with the original α lattice. (iii) Here equilibrium has been attained
—the alloy has been heated to a sufficiently high temperature (9.91) so that β has been
rejected from α as a *non-coherent* precipitate. The lattice parameters of β no longer 'match
up' sufficiently well to those of α as do those of the intermediate β'. Hence a crystal boundary
now separates α and β.

9.92 Precipitation under Non-equilibrium Conditions   We will now assume that the alloy X, which has been retained at temperature T_U, long enough for its structure to be completely homogeneous α, is cooled very rapidly to room temperature (T_R). This could be achieved by quenching it in cold water. In most cases treatment such as this will prevent any precipitation from taking place and we are left with α solid solution which, at T_R, is now super-saturated with B. The structure will not be in equilibrium and is said to be in a *metastable* state. It has an urge to return to a state of equilibrium by precipitating some β. At room temperature such precipitation will be unlikely to occur due to the extreme sluggishness of movement of B atoms, but if the temperature is increased, diffusion will begin and then accelerate as the temperature continues to rise.

The alloy is held at some selected temperature so that diffusion occurs at a low but definite rate. The temperature chosen is well below T_S in order to ensure that non-coherent precipitation does not occur. During this heat-treatment clusters of atoms of both A and B, but with the overall composition of the phase β, slowly form groups at many points in the α lattice. *These clusters have the important property that their lattice structures are continuous with that of the α matrix*, and there is no discontinuous interface as exists between α and β, which has precipitated during cooling under equilibrium conditions. This unbroken continuity of the two lattices is known as *coherency*. Since the cluster size is extremely small and the rate of diffusion very slow, a large number of these coherent nuclei will form and the chosen temperature will not be high enough to allow the formation of a separate β structure. Instead, this intermediate structure— we will call it β'—is produced, and the mismatching between β' and the α matrix leads to distortion in the α lattice in the neighbourhood of these nuclei. Such distortions will hinder the movement of dislocations and so the strength and hardness increase. The greater the number of nuclei and the larger they are, provided that coherency is retained, the greater the strength and hardness of the structure as a whole. Time and temperature will influence this 'precipitation' procedure and, hence, also the ultimate mechanical properties which result. If the temperature is high (T_1, Fig. 9.20) a high rate of diffusion prevails, and this in turn leads to the formation

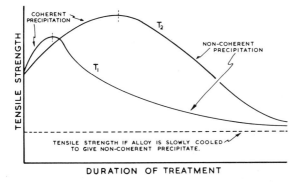

Fig. 9.20

of relatively few nuclei, which, however, will grow to a large size. At a lower temperature (T_2) a larger number of nuclei will form and grow slowly so that strength increases slowly.

Alloys Containing More Than Two Metals

9.100   In this chapter we have dealt only with binary alloys, that is, alloys containing two different metals. In the case of ternary alloys, namely, those containing three different metals, a third variable is obviously introduced, so that the system can no longer be represented by a two-dimensional diagram. Instead we must work in *three* dimensions in order adequately to represent the alloy system. The base of the three-dimensional diagram will be an equilateral triangle, each pure metal being represented by an apex of the triangle. Temperature will be represented by an ordinate perpendicular to the triangular base. Fig. 9.21 illustrates a ternary-alloy system of the metals A, B and C in which the binary systems A–B, B–C and A–C are all of the simple eutectic type. The three binary liquidus *lines* now become liquidus *surfaces* in the three-dimensional system, and these surfaces intersect in three 'valleys' which drain down to a point of minimum temperature, vertically above E_{ABC}. This is the ternary eutectic point of the system, its temperature being lower than that of any of the binary eutectic points E_{AB}, E_{BC} or E_{AC}.

Equilibrium diagrams for alloys of four or more metals cannot be represented in a single diagram since more than three dimensions would be required. In practice such a system can often usefully be represented in the form of a pseudo-binary diagram in which the concentration of one component is varied whilst the others are kept constant—this is equivalent to taking a vertical 'slice' through a ternary system, parallel to one of the end faces. One of these pseudo-binary systems is shown in Fig. 14.1. Such systems can be interpreted in the same way as an ordinary binary system.

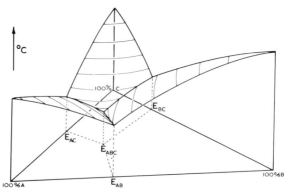

Fig. 9.21   Diagram representing a three-dimensional ternary system, of three metals A, B and C.
The curved isothermal contour lines meet binary eutectics which converge to a ternary eutectic point, E_{ABC}.

We have dealt here only with equilibrium diagrams of the simple funda-mental types. Many of those which the reader will encounter are far more complicated and often contain such a multitude of different phases as almost to exhaust the Greek alphabet. This is particularly true of equilib-rium diagrams which represent those copper-base alloy systems in which a number of intermetallic compounds are formed and in which peritectic reactions are also common. In general, we are interested only in those parts of the diagram near to one end of the system, where we usually find a solid solution with, possibly, small amounts of an intermetallic com-pound. The interpretation then becomes much simpler, and the reader has been provided with sufficient information in this chapter to deal with most of the alloy systems likely to be encountered.

Rapid Solidification Processes (RSP)

9.110   It was explained earlier (3.18) that the crystal size present in a casting is dependent upon the rate at which the metal is cooling when it reaches the solidification temperature. Large castings cool at a slow rate of the order of 1°C/s and will contain crystals of average grain diameter in the region of 5 mm. However with very rapid cooling rates of approxi-mately 10^4°C/s the crystal size will be no more than 10^{-3} mm, whilst if the cooling rate is increased further to 10^6°C/s a completely amorphous struc-ture can be retained with some alloys. That is, the structure is non-crystalline and similar to that of glass.

9.111   The refinement—or elimination—of grain attendant upon RSP may also produce the following structural changes:

1 The effects of minor segregation (3.41) are reduced since the same amount of impurity is spread over a vastly increased area of grain boun-dary so that its effect is 'diluted'. In amorphous structures segregation will be virtually eliminated since grain boundaries are absent. In either case greater homogeneity is achieved.
2 Rapid cooling can greatly extend the limits of solid solubility by introduc-ing increased supersaturation. This can enhance precipitation hardening.
3 Some phases may be retained at ambient temperatures which could not be obtained by orthodox quenching methods.

Reduction in crystal size and the consequent reduction in minor segre-gation result in much higher strength, Young's modulus and in resistance to corrosion (21.70); whilst magnetic properties can be improved in suitable alloys. With some aluminium alloys the amounts of iron, manganese, cobalt and nickel can be increased because more of these elements will be retained in solid solution, due to rapid solidification, and so increase strength. Improved properties of rapidly solidified superalloys of both nickel and chromium as well as of titanium alloys are reported; whilst amorphous alloys of iron, silicon and boron are very suitable for use in transformer cores since they show an extremely low remanence (14.35) and hysteresis loss.

9.112 The practical difficulties involved in attaining very high rates of cooling limit rapid solidification products to either very fine powder or thin foil. Nevertheless methods of achieving the high rates necessary to produce extremely small crystals or, ultimately, an amorphous structure, are ingenious and varied.

Smooth spherical powders in a wide range of alloys can be made by pouring the molten alloy through a small-diameter refractory nozzle. Below the nozzle jets of a suitable inert gas impinge on the metal stream 'atomising' it and scattering it as small droplets. These are rapidly cooled by the gas and collected as a fine powder at the base of the 'atomisation tower'.

In spray-deposition processes a 'gas-atomised' stream of metal droplets is directed on to a cooled metal surface. Fig. 9.22(i) indicates the basis of a method by which strip may be formed; whilst in the *Osprey* process the sprayed droplets build up on a cooled rotating former to produce discs, rings or billets for subsequent hot working. Spray-deposition has been used to produce high-speed steel forms up to 1.5 tonnes in mass.

Ribbon or foil can be manufactured direct from the melt by a variety of 'melt-spinning' processes the principles of which are indicated in Fig. 9.22(ii). Here molten metal is fed on to the surface of a continuously cooled rotating wheel, to produce thin foil which subsequently parts from the wheel as contraction occurs. Thin foil in iron–neodymium–boron alloys is produced by this method and then consolidated for the manufacture of permanent magnets.

Pulsed laser techniques giving cooling rates as high as 10^{12}°C/s have been developed recently and will increase the scope of this work.

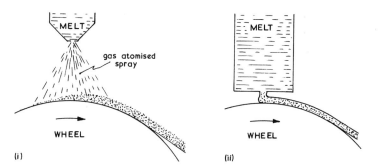

Fig. 9.22 Typical rapid solidification processes: (i) spray forming; (ii) a melt-spinning process.

Exercises

1. Fig. 9.23 represents the magnesium–silicon thermal equilibrium diagram.
    (i) Given that X is a chemical compound derive its formula. (Atomic masses: Mg-24.3; Si-28.1);
    (ii) Annotate the diagram completely;

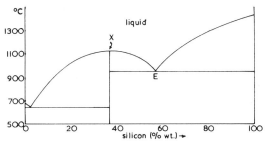

Fig. 9.23.

   (iii) What phases can co-exist at E?
   (iv) What will happen if the temperature of the alloy represented by E is raised by a small amount?
   (v) What are the compositions of the *phases* present in an alloy containing 80% Si–20% Mg overall, at a temperature of 1100°C?
   (vi) In what proportions by mass will the phases in (v) be present?
   (vii) Describe, step by step, what happens as an alloy containing 50% of each element cools slowly from 1200°C to 700°C;
   (viii) What will be the compositions of the phases present at 700°C?
   (ix) In what proportions will the phases in (viii) be present?
2. Beryllium (m.pt. 1282°C) and silicon (m.pt. 1414°C) are completely soluble as liquids but completely insoluble as solids. They form a eutectic at 1090°C containing 61% silicon.
   Draw the thermal equilibrium diagram and explain, with the aid of sketches, what happens when alloys containing (i) 10% silicon, (ii) 70% silicon, solidify completely. (9.40)
3. Fig. 9.24 represents the bismuth–antimony thermal equilibrium diagram.
   (a) Would you expect the diagram to be of this type? Give reasons.
   (b) Consider an alloy containing 70% Sb–30% Bi which is cooling *slowly*:
      (i) At what temperature will solidification begin?
      (ii) What will be the composition of the first solid to form?
      (iii) What will be the compositions of the phases present at 500°C?
      (iv) In what proportions by mass will the phases in (iii) be present?
      (v) What will be the composition of the last trace of liquid which solidifies?
      (vi) At what temperature will solidification be complete?
   (c) Explain what would happen if the alloy containing 70% Sb–30% Bi cooled

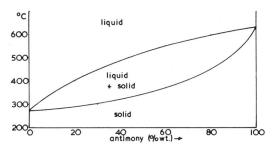

Fig. 9.24.

rapidly from 600°C to 300°C. Sketch the type of microstructure you would expect in the solid alloy. (9.50)

4. Again referring to Fig. 9.24, at what temperature will a 60% Sb–40% Bi alloy contain 25% by mass of liquid and 75% by mass of solid? What will be the compositions of liquid and solid at this temperature?

5. Two metals A and B have melting points 600°C and 450°C respectively. The following results indicate the temperatures associated with discontinuities in the cooling curves of the alloys indicated:

%B	10	20	35	50	55	60	75	90	95
°C	545	495	410	330	300	315	365	415	430
°C	450	300	300	300	—	300	300	300	370

If the maximum and minimum percentage solubilities of the two metals are 20% B in A, 10% B in A, and 10% A in B, 5% A in B, sketch and label the equilibrium diagram. (Assume solubility lines are straight.)
   (i) At what temperature will an alloy containing 30% B begin to solidify?
   (ii) Describe the cooling of an alloy containing 30% B and sketch typical microstructures.
   (iii) What proportions of α and β would you expect in the eutectic alloy at the eutectic temperature and at 0°C? (9.60)

6. Draw a thermal equilibrium diagram representing the system between two metals, X and Y, given the following data:
   (i) X melts at 1000°C and Y at 800°C;
   (ii) X is soluble in Y in the solid state to the extent of 10.0% at 700°C and 2.0% at 0°C;
   (iii) Y is soluble in X in the solid state to the extent of 20.0% at 700°C and 8.0% at 0°C;
   (iv) a eutectic is formed at 700°C containing 40.0% X and 60.0% Y.
   Describe what happens when an alloy containing 70.0% X solidifies and cools slowly to 0°C.
   Sketch the microstructures of an alloy containing 15.0% Y (*a*) after it has cooled slowly to 0°C; (*b*) after it has been heated for some time at 700°C and then water quenched. (9.60 and 9.92)

7. Fig. 9.25 shows part of the aluminium–magnesium thermal equilibrium diagram.
   (i) What is the maximum solid solubility of magnesium in aluminium?
   (ii) Over what temperature range will an alloy containing 7% Mg exist as a single solid phase?
   (iii) At what temperature does an alloy containing 5% Mg begin to melt on heating?
   (iv) An alloy containing 16% Mg is at 520°C. What are the compositions of the phases present?
   (v) In what proportions will the phases in (iv) be present?
   (vi) What is the percentage increase in solubility of magnesium in aluminium in an alloy containing 12% Mg as the temperature rises slowly from 60°C to 350°C?
   (vii) Sketch the structure of an alloy containing 10% Mg (*a*) slowly cooled from 450°C; (*b*) water quenched from 450°C
   (viii) Which will be the stronger in (vii)—(*a*) or (*b*)? (9.90)

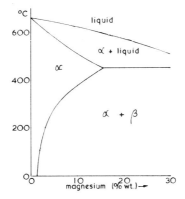

Fig. 9.25.

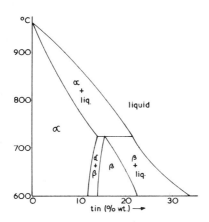

Fig. 9.26.

8. Fig. 9.26 shows part of the silver–tin thermal equilibrium diagram. Consider an alloy which contains 18% by mass of tin, cooling under equilibrium conditions:
   (i) At what temperature does solidification begin?
   (ii) What is the composition of the initial solid which forms?
   (iii) At what temperature is solidification complete?
   (iv) What is the composition of the last trace of liquid?
   (v) What are the natures and compositions of the phases present at 710°C?
   (vi) In what proportions by mass are these phases present at 710°C?
   (vii) Outline, step by step, the structural changes which occur as the alloy cools between 850°C and 600°C. Illustrate your answer with sketches of the phase structures at each stage. (9.70)
9. The tin–indium phase diagram is shown in Fig. 9.27.
   (i) Annotate it completely starting from the *right-hand* side of the diagram;
   (ii) Indicate on the diagram (*a*) any eutectic points (*b*) any peritectic points;
   (iii) What is the maximum solubility of tin in indium?

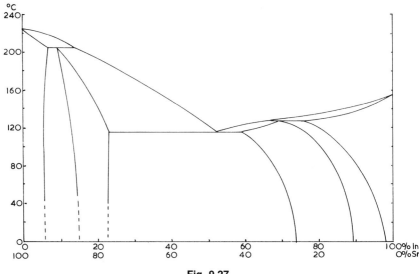

Fig. 9.27.

10. Tungsten dissolves 5.0% platinum at 2460°C to form a solid solution α. The composition of the liquid in equilibrium with α at this temperature is 61.0% tungsten–39.0% platinum. At 2460°C a peritectic reaction takes place between α and the remaining liquid forming a solid solution β, which contains 63.0% tungsten–37.0% platinum. At 1700°C, α contains 4.8% platinum and β contains 37.1% platinum.

   Draw the thermal equilibrium diagram between 3500°C and 1700°C and explain what happens when liquids containing (i) 80.0% tungsten; (ii) 62.0% tungsten solidify slowly. (M.pts.: Tungsten—3400°C; Platinum—1773°C) (9.70)

11. Fig. 9.28 shows part of the gold–tin thermal equilibrium diagram.
   (i) Derive formulae for the three compounds X, Y and Z (Atomic masses: Au—197.0; Sn—118.7);

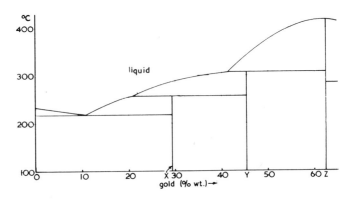

Fig. 9.28.

(ii) Annotate the diagram;
(iii) Describe, step by step, the phase changes which occur in an alloy containing 75% Sn-25% Au as it cools slowly from 350°C to 0°C. (9.70)

Bibliography

Cahn, R. W., *Physical Metallurgy*, North Holland, 1980.
Copper Development Association, *Equilibrium Diagrams for Binary Copper Alloys*.
Elliot, R. P., *Constitution of Binary Alloys (First Supplement)*, McGraw-Hill, 1965.
Ferguson, F. D. and Jones, T. K., *The Phase Rule*, Butterworths, 1958.
*Hansen, M., *Constitution of Binary Alloys*, McGraw-Hill, 1958.
Martin, J. W., *Precipitation Hardening*, Pergamon, 1968.
Martin, J. W., *Micromechanisms in Particle-hardening Alloys*, Cambridge University Press, 1979.
Massalski, T. B., *Binary Alloy Phase Diagrams (Vols. I and II)*, American Society of Metals, 1989.
Rhines, R. N., *Phase Diagrams in Metallurgy*, McGraw-Hill, 1956.
Shunk, F. A., *Constitution of Binary Alloys (Second Supplement)*, McGraw-Hill, 1969.
Smithells, C. J., *Metals Reference Book*, Butterworths, 1983.

* This work was first published in Berlin in 1936 as *Der Aufbau der Zweistofflegierungen* but is still a major reference book for phase diagrams when used in conjunction with the Supplements by Elliot and Shunk.

10
Practical Metallography

10.10 Aloys Beck von Widmanstätten lived to the venerable age of ninety-five, and when he died in 1849, had been, successively, owner of a printing works; an editor in Graz; manager of a spinning mill near Vienna; and, between 1806 and 1816, director of the State Technical Museum in Vienna. In 1808 he discovered that some meteorites, when cut and polished, developed a characteristic structure when subsequently oxidised by heating in air. Later, he found that etching with nitric acid gave better results and revealed the type of metallurgical structure which still bears his name. It may therefore be argued that von Widmanstätten originated metallographic examination. The microscope, however, was not used in this direction until 1841, when Paul Annosow used the instrument to examine the etched surfaces of Oriental steel blades.

In the early 1860s Professor Henry C. Sorby of Sheffield developed a technique for the systematic examination of metals under the microscope and can therefore lay claim to be the founder of that branch of metallurgy known as microscopical metallography. Some of the photomicrographs he produced are excellent even by modern standards.

It would not be possible in a book of this type to deal exhaustively with the subject of the microscopical examination of metals. However, it is hoped that what follows may help the reader to prepare representative microstructures for himself, and so equip him to be able, as a practising engineer, to trace many of the more common causes of failure which are attributed to microstructural defects.

Much useful work can be done with a minimum of equipment. Whilst a simple but good quality metallurgical microscope is an essential purchase, much of the other apparatus required to prepare specimens for the microscope can be home-made quite cheaply. Such simple 'tools' will enable a trained eye to evaluate most of the common microstructural defects encountered in commercial metals and alloys. Metallurgical knowledge of this type can only be accumulated as a result of practice and experience. Ability in this branch of metallurgical technique is, like beauty, in the eye of the beholder.

The Preparation of Specimens for Microscopical Examination

10.20   In preparing a specimen for microscopical examination it is first necessary to produce in it a surface which appears perfectly flat and scratch-free when viewed with the aid of a microscope. This involves first grinding the surface flat, and then polishing it to remove the marks left by grinding. The polishing process causes a very thin layer of amorphous metal to be burnished over the surface of the specimen, thus hiding the crystal structure. In order to reveal its crystal structure the specimen is 'etched' in a suitable reagent. This etching reagent dissolves the 'flowed' or amorphous layer of metal.

That, briefly, is the basis of the process employed in preparing a specimen for examination; but first it will be necessary to select a representative sample of the material under investigation.

10.21 Selecting the Specimen   The selection of a specimen for microscopical examination calls for a little thought, since a large body of metal may not be homogeneous either in composition or crystal structure. Sometimes more than one specimen will be necessary in order adequately to represent the material. In some alloys the structure may also exhibit 'directionality', as, for example, in wrought iron. In a specimen of the latter, cut parallel to the direction of rolling, the slag will appear as fibres elongated in the direction of rolling, whilst a section cut at right angles to the direction of rolling will show the slag as apparent spherical inclusions, and give no hint to the fact that what is being observed is a cross-section *through* slag fibres.

For the examination of surface defects a specimen must be chosen so that a section through the surface layer is included in the face to be polished. Surface cracks and the like should be investigated by cutting a piece of metal containing the crack and mounting it in bakelite or a similar compound. The surface to be polished is then ground sufficiently so that a section through the crack is obtained.

A specimen approximately 20 mm diameter or 20 mm square is a convenient size to handle. It is difficult to grind a perfectly flat surface on smaller specimens, and these are best mounted as described below. The specimen should not be more than 12 mm thick, or it may rock during polishing, producing a bevelled surface.

10.22   When it is necessary to preserve an edge, or when a specimen is so small that it is difficult to hold it flat on the emery paper, the specimen may be mounted in a suitable compound. This can be done most satisfactorily by using a proprietary plastics mounting material. These are generally either acrylic, polyester or epoxide resins. The kit supplied often consists of a powder and two liquids. When these are mixed polymerisation takes place and a hard plastics substance is produced which will retain a metal specimen during and after the polishing operation. Only simple apparatus of the type shown in Fig. 10.1 is required. The specimen is placed on a suitable flat surface and the two L-shaped retaining pieces

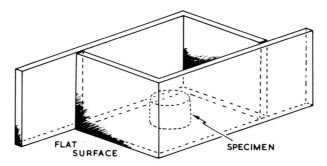

Fig. 10.1 Method of mounting specimen in plastic material where no pressure is required.

arranged around it. (If possible, these retaining members should be held in position with a small clamp.) The specimen is covered with the powder, and this is then moistened with the first liquid. The whole is then saturated with the liquid 'hardener'. In about twenty minutes the mass will have set hard and the L-shaped members can be detached.

Specimens can be mounted more quickly by using some thermosetting substance, such as bakelite or, alternatively, a transparent thermoplastics material. These substances mould at about 150°C, which is usually too low a temperature to cause any structural change in the specimen. They can be ground and polished easily and do not promote any electrolytic action during etching. A small mould (Fig. 10.2) is required in conjuction with a press capable of giving pressures up to about 25 N/mm². Indeed some of the thermoplastics materials mould satisfactorily at such low pressures that a sturdy bench vice can be used to apply the necessary force to this small moulding unit.

After placing the specimen, the powder and the plunger in the mould the latter is heated by means of a special electric heater which encircles it. If this is not available a bunsen burner will suffice. In either case a thermometer should be inserted in a hole provided in the plunger so that overheating of the mould is avoided. Some mounting powders decompose at high temperatures, with the formation of dangerously high pressures.

10.23 Grinding and Polishing the Specimen It is first necessary to obtain a reasonably flat surface on the specimen. This can be done either by using a fairly coarse file or, preferably, by using a motor-driven emery belt. If a file is used it will be found easier to obtain a flat surface by rubbing the specimen on the file than by filing the vice-held specimen in the orthodox way. Skilled workshop technologists may wince at the thought of such a procedure, but it is guaranteed to produce a flat surface for those readers who, like the author, possess negligible skill in the use of a file. Whatever method is used, care must be taken to avoid overheating the specimen by rapid grinding methods, since this may lead to alterations in the microstructure. When the original hack-saw marks have been ground out, the specimen (and the operator's hands) should be thoroughly washed in order to prevent carry-over of filings and dirt to the polishing papers.

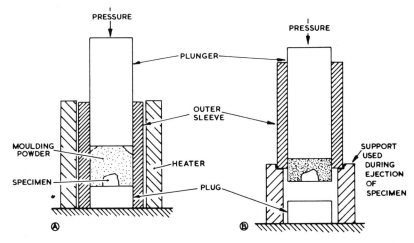

Fig. 10.2   Mould for mounting specimens in plastic materials when pressure is necessary. (A) Moulding the mount; (B) ejecting the finished mount.

PLATE 10.1A

PLATE 10.1B

Fig. Plate 10.1   10.1A A modern hand grinding 'deck'.
Four strips of emery paper (waterproof base) are clamped on the sloping flat base plate. A stream of water flows down the papers into the sump and thence to a drain, flushing away grit and swarf as well as lubricating the surfaces. *(Courtesy of Metallurgical Services Laboratories Ltd. Betchworth, Surrey)*
10.1B A typical rotary polishing machine for finishing metallographic specimens.
Suitable cloths are fixed to the rotating discs, which are normally covered as shown when not in use to prevent contamination of the cloth. Similar machines are also used for rapid grinding with emery discs *(Courtesy of Metallurgical Services Laboratories Ltd. Betchworth, Surrey)*.

Intermediate and fine grinding is then carried out on emery papers of progressively finer grade. These must be of the very best quality, particularly in respect of uniformity of particle size. With modern materials not more than four grades are necessary (220, 320, 400 and 600 from coarse to fine), since by using a paper with a waterproof base wet grinding can be employed. Indeed dry grinding processes are now rarely used particularly since it has been recognised that the dust of many heavy metals is dangerously toxic. Rotary grinding 'decks' are available on to which discs of grinding paper are clamped. These are driven by two-speed motors and are fitted with water drip-feeds and suitable drains. Nevertheless equally good results can be obtained (in a slightly longer time) by hand-grinding the specimen on a home-made table of the type illustrated in Fig. 10.3. Strips of water-proof emery paper, approximately 300 mm × 50 mm, are obtainable for this purpose. The current of water which is flushed across the surface not only acts as a lubricant but also carries away particles of grit and swarf which might damage the surface being ground. The specimen is drawn back and forth along the entire length of the No. 220 paper, so that scratches produced are roughly at right angles to those produced by the preliminary grinding operation. In this way it can easily be seen when the original scratches produced by the primary grinding operation have been completely removed. If the specimen were ground in the same direction so that the new scratches were parallel to the original ones this would be virtually impossible. Having removed the primary grinding marks, the specimen is washed free of No. 220 grit. Grinding is then continued on the

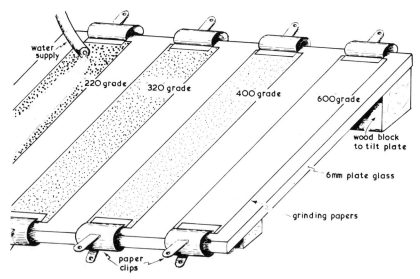

Fig. 10.3 A home-made grinding table for metallographic specimens.
The moveable water supply is transferred from one paper to the next as grinding proceeds along the table. It would be easy to replace the wood blocks with a 'perspex' frame stuck together with Araldite. The complete unit stands in an old photographic dish which is fitted with a suitable drain and operated over a sink.

No. 320 paper, again turning the specimen through 90° and polishing until the previous scratch marks have been removed. This process is repeated with the No. 400 and No. 600 papers.* Light pressure should be used at all stages otherwise particles of coarse grit, which may be trapped in the surface of the paper, will cause deep score marks in the surface of the specimen and these will take longer to remove and may necessitate returning to a coarser paper.

In using rotary tables the same technique is used in that the specimen is turned at right angles to the previous direction of grinding on passing from one paper to the next. Here it is even more necessary to employ light pressures since a torn paper will be the inevitable result of the edge of the specimen digging into the surface when high-speed wheels are used.

So far the operation has been purely one of grinding, and if our efforts have been successful we shall have finished with a specimen whose surface is covered by a series of parallel grooves cut by the particles on the last emery paper to be employed. The final polishing operation is somewhat different in character and really removes the ridged surface layer by means of a burnishing operation. When polishing is complete the ridges have been completely removed, but the mechanism of polishing is such that it leaves a 'flowed' or amorphous layer of metal on the surface (Fig 10.4). This hides the crystal structure and must be dissolved by a suitable etching reagent.

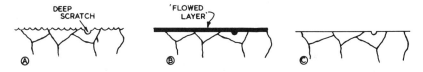

Fig. 10.4   (A) Grooves produced in the metal surface by the final grinding operation. A deep scratch produced by a particle of coarse grit is shown. (B) Final polishing has produced a 'flowed layer'. This may cover the deep scratch as shown, rendering it invisible. (C) Etching has removed the flowed layer thus revealing the crystal structure beneath. Unfortunately the deep scratch is also visible again.

Irons and steels are polished by means of a rotating cloth pad impregnated with a suitable polishing medium. 'Selvyt' cloth was widely used to cover the polishing disc but special hard-wearing cloths are now available and are generally more suitable for this purpose. Numerous polishing powders were popular in the past. These included alumina, magnesia, chromium oxide and iron(III) oxide ('jeweller's rouge'), most of which were messy in use. 'Gamma alumina' (aluminium oxide), prepared by calcining pure ammonium alum, still has its devotees. Finer grades of this material are supplied as a thick suspension in water. A little of this is applied to the pad and worked into the surface with *clean* finger tips. A constant drip of water is fed to the rotating pad which should be run at

* *Fine* grinding is now possible using modern high-grade papers, down to grit 2400 or 4000.

low speeds until the operator has acquired the necessary manipulative technique. Light pressures should be used, since heavy pressures are more likely to result in scratches being formed by grit particles embedded deep in the cloth. Moreover, the use of light pressure is less likely to result in the specimen being suddenly projected across the laboratory.

Most metallographic specimens are now polished using one of the proprietary diamond-dust polishing compounds. In these materials the graded diamond particles are carried in a 'cream' base which is soluble in both water and the special polishing fluid, a few spots of which are applied to the polishing pad in order to lubricate the work and promote even spreading of the compound. These compounds are graded and colour-coded according to particle size (in micrometres). For polishing irons and steels it is generally convenient to use a two-stage technique necessitating two polishing wheels. Preliminary polishing is carried out using a 6 μm particle size. The specimen is then washed and finished on the second wheel using a 1 μm material. Since these compounds are expensive, it is desirable that the operator should have some manipulative skill in order that frequent changing of the polishing pad is not made necessary due either to tearing of the cloth or lack of cleanliness in working. At the same time the polishing cloths have a much longer life than is the case with those polishing media which tend to dry on the cloth and render it unfit for further use. The oil-based lubricant used with diamond pastes keeps the cloth in good condition so that it can be used intermittently over long periods, provided that the pad is kept covered to exclude dust and grit. In this way the higher cost of the diamond paste can be offset.

Non-ferrous specimens are best 'finished' by hand on a small piece of 'Selvyt' cloth wetted with 'Silvo'. Polishing should be accomplished with a circular sweep of the hand, instead of the back-and-forth motion used in grinding. As at every other stage, absolute cleanliness should be observed if a reasonably scratch-free surface is to be obtained. Copper alloys can be polished quickly by passing from the No. 400 grade of emery paper to a piece of good-quality chamois leather wetted with 'Brasso'. The grinding marks are very quickly removed in this way, and final polishing is then accomplished with 'Silvo' and 'Selvyt' cloth. These polishing agents may be found unsuitable in a few cases because of an etching action on the alloy. It is then better to use magnesium oxide (magnesia).

When the specimen appears to be free from scratches it is thoroughly cleaned and examined under the microscope using a magnification of about 50 or 100. If satisfactorily free from scratches the specimen can be examined for inclusions, such as manganese sulphide (in steel) or slag fibres (in wrought iron), before being etched.

To summarise, the most important factors affecting a successful finish are:

(a) Care should be taken not to overheat the specimen during grinding. In steel this may have a tempering effect.

(b) Absolute cleanliness is essential at every stage.

(c) If a specimen has picked up deep scratches in the later stages of grinding it is useless to attempt to remove them on the polishing pad. If a specimen is polished for too long on the pad its surface may become rippled.

(*d*) Apply light pressure at all times during grinding and polishing.

10.24 Etching the Specimen   Before being etched the specimen must be absolutely clean, otherwise it will undoubtedly stain during etching. Nearly every case of failure in etching can be traced to inadequate cleaning of the specimen so that a film of grease still remains.

The specimen should first be washed free of any adhering polishing compound. The latter can be rubbed from the *sides* of the specimen with the fingers, but care must be exercised in touching the polished face. The best way to clean this is by very gently smearing the surface with a finger-tip dipped in grit-free soap solution, and washing under the tap. Even now the specimen may be slightly greasy, and the final film of grease is best removed by immersing the specimen in boiling ethanol ('white' industrial methylated spirit) for about two minutes. The ethanol should not be heated over a naked flame, but preferably by an electrically heated water-bath.

Proprietary degreasing solutions are now obtainable. The fact that some of these are perfumed brought forth the predictable ribald comments from some of the author's students. However, it is suspected that the function of the perfume is to mask the identity of the simple organic solvents used. They seem to have no advantage over white methylated spirit or trichloromethane (carbon tetrachloride) as de-greasants.

From this point onwards the specimen must not be touched by the fingers but handled with a pair of nickel crucible tongs. It is removed from the ethanol and cooled in running water before being etched. With specimens mounted in thermoplastic materials it may be found that the mount is dissolved by hot ethanol. In such cases swabbing with a piece of cotton wool soaked in caustic soda solution may be found effective for degreasing.

When thoroughly clean, the specimen is etched by being plunged into the etching reagent and agitated vigorously for several seconds. The specimen is then *quickly* transferred to running water to wash away the etching reagent, and then examined to see the extent to which etching has taken place. Such inspection is carried out with the naked eye. If successfully etched the surface will appear slightly dull, and in cast materials the individual crystals may actually be seen without the aid of the microscope. If the surface is still bright further etching will be necessary. The time required for etching varies with different alloys and etching reagents. Some alloys can be etched sufficiently in a few seconds, whilst some stainless steels, being resistant to attack by most reagents, require as much as thirty minutes.

After being etched the specimen is washed in running water and then dried by immersion for a minute or so in boiling ethanol. If it is withdrawn from the ethanol and shaken with a flick of the wrist to remove the surplus, it will dry almost instantaneously. For mounted specimens, the mounts of which are affected by boiling ethanol, it is better to spot a few drops of ethanol on the surface of the specimen. The surplus is then shaken off and the specimen held in a current of warm air from a hair drier. The specimen must be dried *evenly and quickly*, or it will stain. If an efficient hair drier is available the boiling ethanol bath can be dispensed with and the spotting method used for all specimens.

A summary of the most useful etching reagents is given in Tables 10.1, 10.2, 10.3 and 10.4.

10.25   Both polishing and etching can be carried out electrolytically. This involves setting up an electrolytic cell in which the surface of the specimen acts as the anode. By choosing a suitable electrolyte and appropriate current conditions, the surface of the specimen can be selectively dissolved to the required finish. Not only can 'difficult' metals and alloys be attacked in this way, but a high-quality, scratch-free finish can be produced. The description of detailed techniques is, however, beyond the scope of this book. The traditional methods of polishing and etching already described are adequate for the successful preparation of most metallurgical materials.

Table 10.1   *Etching Reagents for Iron, Steels and Cast Irons*

Type of etchant	Composition	Characteristics and uses
Nital	2 cm^3 nitric acid; 98 cm^3 ethanol (industrial methylated spirit)	The best general etching reagent for irons and steels. Etches pearlite, martensite, tempered martensite and bainite, and attacks the grain boundaries of ferrite. For pure iron and wrought iron, the concentration of nitric acid may be raised to 5 cm^3. To resolve pearlite, etching must be very light. Also suitable for ferritic grey cast irons and blackheart malleable irons.
Picral	4 g picric acid; 96 cm^3 ethanol	Very good for etching pearlite and spheroidised structures, but does not attack the ferrite grain boundaries. It is the most suitable reagent for all cast irons, with the exception of alloy and completely ferritic cast irons.
Alkaline sodium picrate	2 g picric acid; 25 g sodium hydroxide; 100 cm^3 water	The sodium hydroxide is dissolved in the water and the picric acid then added. The whole is heated on a boiling water-bath for 30 minutes and the clear liquid poured off. The specimen is etched for 5–15 minutes in the boiling solution. Its main use is to distinguish between ferrite and cementite. The latter is stained black, but ferrite is not attacked.
Mixed acids and glycerol	10 cm^3 nitric acid; 20 cm^3 hydrochloric acid; 20 cm^3 glycerol; 10 cm^3 hydrogen peroxide	Suitable for nickel–chromium alloys and iron–chromium-base austenitic steels. Also for other austenitic steels, high chromium–carbon steels and high-speed steel. Warm the specimen in boiling water before immersion
Acid ammonium peroxydisulphate	10 cm^3 hydrochloric acid; 10 g ammonium peroxydisulphate; 80 cm^3 water	Particularly suitable for stainless steels. Must be freshly prepared for use.

Table 10.2 *Etching Reagents for Copper and its Alloys*

Type of etchant	Composition	Characteristics and uses
Ammonical ammonium peroxydisulphate	20 cm^3 ammonium hydroxide (0.880); 10 g ammonium peroxydisulphate; 80 cm^3 water	A good etchant to reveal the grain boundaries of pure copper, brasses and bronzes. Should be freshly made to give the best results.
Ammonia–hydrogen peroxide	50 cm^3 ammonium hydroxide (0.880); 20–50 cm^3 hydrogen peroxide (3% solution); 50 cm^3 water	The best general etchant for copper, brasses and bronzes. Etches grain boundaries and gives moderate contrast. The hydrogen peroxide content can be varied to suit a particular alloy. Used for swabbing or immersion, and should be freshly made as the hydrogen peroxide deteriorates.
Acid iron(III) chloride	10 g iron(III) chloride; 30 cm^3 hydrochloric acid; 120 cm^3 water	Produces a very contrasty etch on brasses and bronzes. Darkens the β in brasses. Can be used following a grain-boundary etch with the ammonium peroxydisulphate etchant. Use at full strength for nickel-rich copper alloys. Dilute 1 part with 2 parts of water for copper-rich solid solutions in brass, bronze and aluminium bronzes.
Acid dichromate solution	2 g potassium dichromate; 8 cm^3 sulphuric acid; 4 cm^3 saturated sodium chloride solution; 100 cm^3 water	Useful for aluminium bronze and complex brasses and bronzes. Also for copper alloys of beryllium, manganese and silicon, and for nickel silvers.

Table 10.3 *Etching Reagents for Aluminium and Alloys*

Type of etchant	Composition	Characteristics and uses
Dilute hydrofluoric acid	0.5 cm^3 hydrofluoric acid; 99.5 cm^3 water	The specimen is best swabbed with cotton wool soaked in the etchant. A good general etchant.
Caustic soda solution	1 g sodium hydroxide; 99 cm^3 water	A good general etchant for swabbing.
Keller's reagent	1 cm^3 hydrofluoric acid; 1.5 cm^3 hydrochloric acid; 2.5 cm^3 nitric acid; 95 cm^3 water	Particularly useful for duralumintype alloys. Etch by immersion for 10–20 seconds.

NB On no account should hydrofluoric acid or its fumes be allowed to come into contact with the skin or eyes. Care must be exercised with all strong acids.

Table 10.4 *Etching Reagents for Miscellaneous Alloys*

Type of etchant	Composition	Characteristics and uses
Ethanoic and nitric acids	3 cm^3 glacial ethanoic (acetic) acid; 4 cm^3 nitric acid; 16 cm^3 water	Useful for *lead and its alloys* (use freshly prepared and etch for 4–30 minutes). 5% Nital is also useful for lead and its alloys.
Ethanoic acid and hydrogen peroxide	30 cm^3 glacial ethanoic (acetic) acid; 10 cm^3 hydrogen peroxide (30% solution)	Suitable for *lead–antimony* alloys. Etch for 5–20 seconds.
Acid iron(III) chloride	10 g iron(III) chloride; 2 cm^3 hydrochloric acid; 95 cm^3 water	Suitable for *tin-rich bearing metals*. Other tin-rich alloys can be etched in 5% Nital.
Dilute hydrochloric acid in alcohol	1 cm^3 hydrochloric acid; 99 cm^3 alcohol (industrial methylated spirit)	For *zinc and its alloys*.1% Nital is also useful.
Iodine solution	10 g iodine crystals; 30 g potassium iodide; 100 cm^3 water	The best etchant for *cadmium–bismuth alloys*.
Mixed nitric acid ethanoic acid	50 cm^3 nitric acid; 50 cm^3 glacial ethanoic (acetic) acid	Suitable for *nickel and monel metal*.Should be freshly prepared.

The Metallurgical Microscope

10.30   The metallurgical microscope is similar in optical principles to any other microscope, but differs from some of them in the method by which the specimen is illuminated. Most biological specimens can be prepared as thin, transparent slices mounted between sheets of thin glass, so that illumination can be arranged simply, by having a source of light *behind* the specimen. Metals, however, are opaque substances, and since they must be illuminated by frontal lighting, it follows that the source of light must be *inside* the microscope tube itself. This is usually accomplished, as in Fig. 10.5, by means of a small plain-glass reflector, R, placed inside the tube. With this system of illumination much of the light is lost both by transmission when it first strikes the plate and, by reflection, when the returning ray from the specimen strikes the inclined plate again. Nevertheless, a small 6-volt bulb is usually sufficient as a source of illumination. The width of the beam is controlled by the iris diaphragm, *D*. Generally speaking, this should be partly closed so that the beam of light is just sufficient to cover the back component of the objective lens. An excess of light, reflected from the sides of the microscope tube, will cause light-scatter and, consequently, 'glare' in the field of view.

The optical system of the microscope consists of two lenses, the objective, *O*, and the eyepiece, *E*. The former is the more important and expensive of the two lenses, since it has to resolve the fine detail of the object being examined. Good-quality objectives are corrected for chromatic and

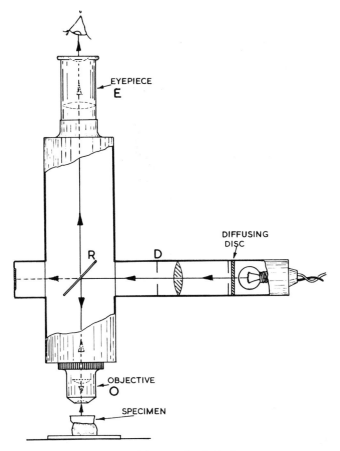

Fig. 10.5 The basic optical system of the metallurgical microscope.
This illustrates the system used in an early instrument. However, the optical system is fundamentally similar in more sophisticated modern instruments.

spherical aberrations, and hence, like camera lenses, are of compound construction. The magnification given by the objective depends upon its focal length—the shorter the focal length, the higher the magnification. In addition to magnification, resolving power is also important. This is defined as the ability of a lens to show clearly separated two lines which are very close together. In this way resolution can be expressed as a number of lines per mm. Thus resolving power depends upon the quality of the lens, and it is useless to increase the size of the image, either by extending the tube length of the microscope or by using a higher-power eyepiece, beyond a point where there is a falling off of resolution. A parallel example in photography is where a small 90 mm × 60 mm snapshot, on enlarging, fails to show any more *detail* than it did in its original size and in fact shows blurred outlines as a result.

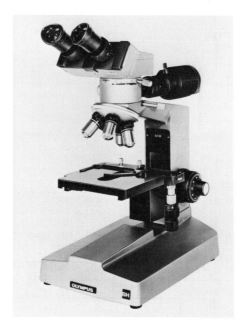

Plate 10.2 An 'Olympus' metallurgical microscope equipped for binocular vision.
Four objectives of different focal lengths are carried on a rotatable turret. Illumination is by
either tungsten or quartz halogen light source. *(Courtesy of Metallurgical Services Laboratories Ltd, Betchworth, Surrey).*

However, there is a definite limit to the extent to which it is worthwhile
developing *microscope* objectives since at high magnifications we are dealing with dimensions of the same order as the wavelength of light itself so
that definition then begins to suffer considerably. Consequently microscope lenses are cheap compared with high-quality camera lenses where
the main restriction is the 'resolving power' of the photographic emulsion
which accepts the image formed by the lens.

The eyepiece is so called because it is the lens nearest the eye. It is a
relatively simple lens and its purpose is to magnify the image produced by
the objective. Eyepieces are made in a number of 'powers', usually $\times$ 6,
$\times$ 8, $\times$ 10 and $\times$ 15.

The overall approximate magnification of the complete microscope
system can be calculated from:

$$\text{Magnification} = \frac{T.N}{F}$$

where T = the tube length of the microscope measured from the back
component of the objective to the lower end of the eyepiece;
F = the focal length of the objective; and
N = the power of the eyepiece.

Thus for a microscope having a tube length of 200 mm and using a 16-mm
focal-length objective and a $\times$ 10 eyepiece, the magnification would be

$\dfrac{200 \times 10}{16}$, ie 125. Most metallurgical microscopes have a standard tube length of 200 mm. Assuming a tube length of this value, approximate magnifications with various objectives of standard focal lengths will be found in Table 10.5.

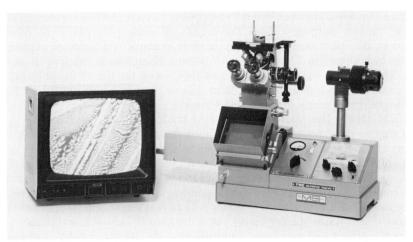

Plate 10.3 The ultimate in optical microscopy!
Here the 'Olympus' PME microscope is of the 'inverted' type. The apparatus incorporates a reflex screen for direct viewing and also photography on sheet film, as well as for TV viewing in colour or monochrome. *(Courtesy of Metallurgical Services Laboratories Ltd, Betchworth, Surrey).*

Table 10.5 *Approximate magnifications with different combinations of objective and eyepiece (assuming a tube length of 200 mm)*

Focal length of objective (mm)	Power of eyepiece	Approximate magnification
32	× 6	37.5
	× 8	50
	× 10	62.5
16	× 6	75
	× 8	100
	× 10	125
8	× 6	150
	× 8	200
	× 10	250
4	× 6	300
	× 8	400
	× 10	500
2 (oil-immersion objective)	× 6	600
	× 8	800
	× 10	1000

Most modern microscope objectives are engraved with a 'multiplying power' which is relative to the tube length of the microscope for which they are designed. The overall magnification is then obtained simply by multiplying the 'power' of the objective by that of the eyepiece, eg a × 40 objective used in conjunction with a × 10 eyepiece will give an overall magnification of × 400.

Many sophisticated modern microscopes contain built-in refinements fulfilling various purposes. Thus the facility of using polarised light often helps in the identification of a phase. For example, particles of cuprous oxide in copper (16.22) which, with normal illumination, appear as sky-blue globules can be made to glow like rubies with the use of the polariser. Microscopes equipped for dark-field illumination are very useful in the photography of many low-contrast subjects.

10.31 Using the Microscope   Most readers will be familiar with the fact that in ordinary camera photography the depth of field reasonably in focus at the focal plane depends upon the distance of the object from the camera lens. Thus with a landscape view everything will be fairly sharply in focus between about ten metres and infinity. However, if the subject is only one metre from the lens then, assuming it is sharply in focus itself, a zone of only a few centimetres either side of one metre will be reasonably in focus. A microscope lens works at a distance of only a few millimetres from the object and it follows therefore that the 'depth of field' is negligible. Consequently, as mentioned earlier (10.20) it is essential to provide the specimen with an *absolutely flat* surface and to mount the specimen so that its surface is normal to the axis of the instrument. This is most easily achieved by fixing the specimen to a microscope slide by means of a piece of plasticine. Normality is assured by using a mounting ring, as shown in Fig. 10.6. Obviously, the mounting ring must have perfectly parallel end faces. Mounting may not be necessary for specimens which have been set in bakelite, since the top and bottom faces of the bakelite mount are usually parallel, so that it can be placed directly on to the table of the microscope.

The specimen is brought into focus by using first the coarse adjustment and then the fine adjustment. It should be noted that the lenses supplied are generally designed to work at a fixed tube length (usually 200 mm), under which conditions they give optimum results. Therefore, the tube carrying the eyepiece should be drawn out the appropriate amount (a scale is usually engraved on the side of the tube). Slight adjustments in tube length may then be made to suit the individual eye.

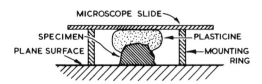

Fig. 10.6   Mounting a specimen for examination under the microscope.

Finally, the iris in the illumination system should be closed to a point where illumination just begins to decrease. This will limit glare due to internal reflections in the tube.

It is a mistake to assume that high magnifications in the region of 500 or 1000 are always the most useful. In fact, they will often give a completely meaningless impression of the structure, since the field under observation will be so small. Directional properties in wrought structures or dendritic formation in cast structures are best seen using low powers of × 40 to × 100. Even at × 40 a single crystal of, say, cast 70–30 brass may completely fill the field of view. The dendritic pattern, however, will be clearly apparent, whereas at × 500 only a small area between two dendrite arms would fill the field of view. As a matter of routine, a low-power objective should always be used first to gain a general impression of the structure before it is examined at a high magnification.

10.32 The Care of the Microscope   Care should be taken never to touch the surface of optical glass with the fingers, since even the most careful cleaning may damage the surface, particularly if it has been 'bloomed' (coated with magnesium fluoride to increase light transmission). In normal use dust may settle on a lens, and this is best removed by sweeping gently with a high-quality camel-hair brush.

If a lens becomes accidentally finger-marked, this is best dealt with by wiping gently with a good-quality lens-cleaning tissue (such as Green's No. 105) moistened with xylol. Note that the operative word is *wipe* and not *rub*. Excess xylol must be avoided, as it may penetrate into the mount of the lens and soften the cement holding the components together.

High-power objectives of the oil-immersion type should always be wiped clean of cedar-wood oil before the latter has a chance to harden. If hardening takes place due to the lens being left standing for some time, then the oil will need to be removed by xylol, but the use of the latter should be avoided when possible.

If special lens tissues are not available soft, well-washed linen may be used to clean lenses. It is far superior to chamois leather, which is likely to retain particles of grit, and to silk which has a tendency to scratch the surface of soft optical glass.

10.33 The Electron Microscope   Whilst much of the routine micro-examination of metals is carried out at low magnifications in the region of × 100, it is often necessary in metallurgical research to be able to examine structures at very high magnifications. Unfortunately the highest magnification possible with an ordinary optical microscope is in the region of × 2000. Above this magnification the dimensions being dealt with are comparable with the wavelength of light itself. Indeed, since blue light is of shorter wavelength than red light, it is advantageous to view specimens by blue light when examining very fine detail at high magnifications.

For very high-power microscopy (between × 2000 and several millions) light rays can be replaced by a beam of electrons. The bending or *refracting* of light rays in an optical microscope is achieved by using a suitable glass-lens system. A similar effect is produced in the electron microscope by using an electro-magnetic 'lens' to refract the electron beam. This 'lens' consists of coil systems which produce the necessary electro-magnetic field to focus the electron beam.

Whereas light rays are reflected to a very high degree from the surface of a metal, electrons are transmitted. They are able to pass through a thin metallic foil, but will be absorbed by greater thicknesses of metal. We must therefore view the specimen by transmission instead of by reflection, and in order to examine the actual surface of a metal the replica method is generally used. This involves first preparing the surface of the specimen and then coating it with some suitable plastic material, a very thin film of which will follow the contours of the metallic surface. The thin plastic mould, or replica of the surface, is then separated from the specimen and photographed by transmitted electrons using the electron microscope.

Following the development of the orthodox electron microscope outlined very briefly above, other instruments of allied design have become available. Thus, the high-voltage electron microscope permits the use of thicker replica specimens and at the same time offers greater magnifications. The scanning electron microscope differs in principle from the orthodox instrument in that electrons generated and *reflected* from the actual surface of the specimen are used to produce the image. The image is built up by scanning the surface in a manner similar to that employed in a synchronously-scanned cathode ray tube. An image of very high resolution is produced but an even more important feature is the *great depth of field* available as compared with an optical instrument. This enables *unprepared* surfaces, eg fractures, to be scanned at magnifications up to 50 000.

In a somewhat different category is the field-ion microscope in which the image is formed by gas ions rather than by electrons. With this instrument magnifications can be measured in millions enabling details on the atomic scale such as dislocations, vacancies, grain boundary defects and very small precipitates to be examined. Single atoms, particularly the larger ones, come just within the range of this instrument.

A recent development is the scanning *acoustic* microscope (SAM) which enables the materials scientist to examine both surface and subsurface structures. The essential feature of this rather complex instrument is a ruby 'lens' which focuses high-frequency vibrations (of up to 1.5 GHz) on to the region being examined. The reflected vibrations, following the same path as the incident waves, are received by a transducer and converted into electrical signals. By scanning in raster pattern an image can be built up which is displayed on a 'TV monitor'. The main feature of SAM is that it can examine both surfaces and subsurfaces non-destructively and with a minimum of specimen preparation.

Macro-Examination

10.40   Useful information about the structure of a piece of metal can often be obtained without the aid of a microscope. Such investigation is usually referred to as 'macro-examination' and may be carried out with the naked eye or by using a small hand magnifying lens.

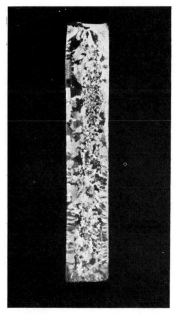

Plate 10.4   10.4A A vertical section through a small ingot of aluminium.
Normal size. Etched by swabbing with a 20% solution of hydrofluoric acid.
10.4B Section through a 'close-form' forging.
Deep-etched in boiling 50% hydrochloric acid for 30 minutes to reveal the flow lines. The strength of the teeth is increased by the continuous 'fibre' produced by forging, × ½. *(Courtesy of Forgings and Presswork Ltd, Birmingham).*

The method of manufacture of a component can often be revealed by an examination of this type (Plate 10.4в and Fig. 6.5), whilst defects, such as the segregation of antimony–tin cuboids in a bearing metal (Plate 18.1), can also be effectively demonstrated. Macro-examination is also a means of assessing crystal size, particularly in cast structures (Plate 10.4в), whilst various forms of heterogeneity, such as is shown in the distribution of sulphide globules in cast steel, can be similarly examined.

Usually, medium grinding is sufficient to produce a satisfactory surface for examination, and for large components an emery belt will be found almost indispensable. Polishing is not necessary, and for most specimens grinding can be finished at Grade 320 paper. Bearing in mind the rougher nature of the work, it will be realised that grinding should not be carried out on papers which are used for the preparation of specimens for examination under the microscope. Discarded papers from micro-polishing processes can be used for the preparation of macro-sections.

After being ground, the specimen should be washed to remove grit, but it will not generally be necessary to degrease it in view of the fact that macroscopic etching is often prolonged and removes more than the flowed layer. The fibrous structure of a forged-steel component, for example, is revealed by deep-etching the component in boiling 50% hydrochloric acid for up to fifteen minutes. After being etched, the specimen is washed and dried in the usual way, though fine detail is often more clearly seen when the component is wet. Suitable etchants for the macroscopic examination of various alloys are shown in Table 10.6.

If a macro-section such as that shown in Plate 10.4B is etched very deeply by immersion in boiling 50% hydrochloric acid for thirty minutes or more, a tolerable 'ink print' can be taken from its surface. To do this a blob of printer's ink is squeeged thoroughly on to a flat piece of plate glass using a photographic roller-type squeegee. The object is to obtain a very thin but uniform film of ink on the *roller*. This is then carefully rolled over the deep-etched surface using very light pressure so that only the raised ridges receive ink. The inked surface is then gently pressed on to a sheet of white smooth paper when an ink print should be obtained. Better contact between paper and inked section is achieved if the paper (which must be on an absolutely flat surface) is backed by a couple of thicknesses of blotting paper.

Table 10.6 *Etching Reagents for Macroscopic Examination*

Material to be etched	Composition of etchant	Working details
Steel	50 cm^3 hydrochloric acid; up to 50 cm^3 water	Use boiling for 5–15 minutes. Reveals flow lines; the structure of fusion welds; cracks; and porosity; also the depth of hardening in tool steels.
	25 cm^3 nitric acid; 75 cm^3 water	Similar uses to above, but can be used as a cold swabbing reagent for large components.
	1 g copper(II) chloride; 4 g magnesium chloride; 2 cm^3 hydrochloric acid; 100 cm^3 ethanol.	*Stead's Reagent.* The salts are dissolved in the smallest possible amount of hot water along with the acid. Shows dendritic structure in cast steel. Also phosphorus segregation—copper deposits on areas *low* in phosphorus.
Copper and its alloys	25 g iron(III) chloride; 25 cm^3 hydrochloric acid; 100 cm^3 water	Useful for showing the dendritic structure of α solid solutions.
	50 cm^3 ammonium hydroxide (0.880); 50 cm^3 ammonium peroxydisulphate (5% solution); 50 cm^3 water.	Useful for alloys containing the β-phase.
Aluminium and its alloys	20 cm^3 hydrofluoric acid* 80 cm^3 water	Degrease the specimen first in tetrachloromethane and then wash in *hot* water. Swab with etchant.
	45 cm^3 hydrochloric acid; 15 cm^3 hydrofluoric acid; 15 cm^3 nitric acid; 25 cm^3 water	A much more active reagent—care should be taken to avoid contact with the skin.

On no account should hydrofluoric acid or its fumes be allowed to come into contact with the skin or eyes. Wear rubber gloves when handling it. It can lead very quickly to the loss of fingers.

Sulphur Printing

10.50 This affords a useful means of determining the distribution of sulphides in steel. The specimen should first be ground to Grade 320 emery paper and then thoroughly degreased and washed. Meanwhile a sheet of single-weight matt photographic bromide paper is soaked in a 2% solution of sulphuric acid for about five minutes. It is then removed from the solution and any surplus drops are wiped from the surface.

The emulsion side of the paper is then placed on the surface of the specimen and gently rolled with a squeegee to expel any air bubbles and surplus acid from between the surfaces. Care must be taken that the paper does not slide over the surface of the specimen, for which reason matt paper is preferable. For small specimens the paper can be laid emulsion side upwards on a flat surface and the specimen then pressed firmly into contact with it; care again being taken to prevent slipping between the paper and the specimen.

After about five minutes* paper and specimen can be separated, and it will be found that the paper has been stained brown where it was in contact with particles of sulphide. The sulphuric acid reacts with the sulphides to produce the gas hydrogen sulphide, H_2S:

$$MnS + H_2SO_4 = MnSO_4 + H_2S$$
$$FeS + H_2SO_4 = FeSO_4 + H_2S$$

The liberated hydrogen sulphide then reacts with the silver bromide, $AgBr$, in the photographic emulsion to form a dark-brown deposit of silver sulphide, Ag_2S:

$$2AgBr + H_2S = 2HBr + Ag_2S$$

The print is rinsed in water and 'fixed' for ten minutes in a solution containing 100 g of 'hypo' (sodium thiosulphate) in 1 litre of water. The function of this treatment is to dissolve any surplus silver bromide, which would otherwise darken on exposure to light. Finally, the print is washed for thirty minutes in running water, and dried.

Exercises

1. By reference to Figs. 15.1 and 15.2 show how it is possible for an inexperienced operator to make a misleading interpretation of a microstructure as it appears under the microscope. (10.21)
2. Why is it necessary to wash specimens thoroughly between each stage of the process during grinding and polishing? (10.23)
3. Why is it not possible for *optical* microscopes to be used at magnifications greater than about × 2000? (10.33)
4. Describe, with the aid of sketches, the optical system used in the metallurgical

---

* A corner of the paper may be lifted from time to time to check the progress of printing, taking care not to allow the paper to slip.

microscope. What is meant by 'resolving power' as applied to the objective lens used in such a microscope?.

A metallurgical microscope employs a tube length of 200 mm. What ocular magnification will be obtained when this microscope is used in conjunction with a 4 mm objective and a × 10 eyepiece? (10.30)

5. What are the general objectives of the *macro*-examination of a metallic component as compared with the *micro*-examination of a metal? (10.40)

Bibliography

Grundy, P. T. and Jones, G. A., *Electron Microscopy in the Study of Materials*, Edward Arnold, 1976.

Pickering, F. B., *The Basis of Quantitative Metallography*, Metals and Metallurgy Trust, 1976.

Modin, H. and Modin, S. (Trans Kinnane, G. G.) *Metallurgical Microscopy*, Butterworths, 1973.

Venables, J. A. (Ed.), *Developments in Electron Microscopy and Analysis*, Academic Press, 1978.

Vander Voort, G. F., *Metallography Principles and Practice*, McGraw-Hill, 1984.

BS 5166:1974 *Method for Metallographic Replica Techniques of Surface Examination*.

11

The Heat-Treatment of Plain Carbon Steels—(I)

11.10 Most modern schoolboys appear to be aware of the fact that a piece of carbon steel can be hardened by plunging it into cold water from a condition of bright red heat. Unfortunately many of them assume that similar treatment will harden *any* metallic material. Which illustrates the danger of feeding unrelated and unexplained facts to schoolboys! Be that as it may there are numerous examples where metallurgical technology has predated its scientific understanding. Steel has been hardened by quenching for many centuries, yet it was only during the present century that a reasonable scientific explanation of the phenomenon was forthcoming.

In the first part of this chapter we shall consider the development of equilibrium structures in steels in greater detail than was possible in Chapter 7. We shall follow this with a study of those heat-treatment processes which depend upon equilibrium being reached in the structure of the steel under treatment.

11.11 What is generally called the 'iron–carbon thermal equilibrium diagram' is illustrated in Fig. 11.1. Strictly speaking it should be named the 'iron–iron carbide metastable system' since, theoretically at least, iron carbide is not a completely stable phase. Nevertheless iron carbide precipitates from austenite, during ordinary conditions of cooling, in preference to the theoretically more stable graphite. Once formed iron carbide—or cementite—is quite stable and for our purposes it will be satisfactory to regard it as an equilibrium phase.

The iron–carbon diagram is of the type dealt with in 9.60, that is, where two substances are completely soluble in each other in the liquid state but are only partially soluble in the solid state. The diagram is modified in shape as a result of the polymorphic changes occurring in iron at 910°C and 1400°C. However, despite the apparent complexity of the diagram, we have only three important phases to consider, namely:

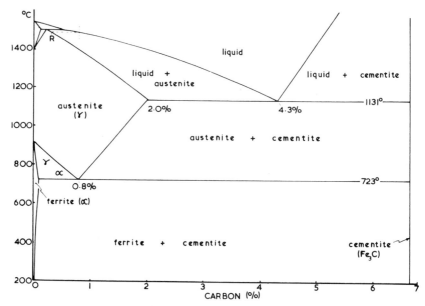

Fig. 11.1   The iron–carbon thermal equilibrium diagram.

(i) *austenite* (γ), the solid solution formed when carbon dissolves in face-centred cubic iron in amounts up to 2.0%;

(ii) *ferrite* (α), a very dilute solid solution of carbon in body-centred cubic iron and containing at the most only 0.02% carbon;

(iii) *cementite*, or iron carbide, Fe₃C, an interstitial compound (8.31) of iron and carbon containing 6.69% carbon.

For the sake of clarity the important areas of the iron–carbon diagram are shown in greater detail in Figs. 11.2, 11.3 and 11.4.

The reader may well have seen different values ascribed to the salient points of the iron–carbon diagram, depending upon the age of the publication, its author and even the country of origin. In fact during the professional lifetime of the present author the carbon content of austenite at the eutectoid point has been accepted as 0.89; 0.85; 0.83; 0.80 and 0.77% —and not necessarily in that order! At the same time the eutectoid, or lower critical, temperature has been quoted between 698 and 732°C; whilst the maximum solubility of carbon in austenite (at 1131°C) has been given different values between 1.7 and 2.08%. This lack of precision leading to a variation of no less than 15% in the value ascribed to the carbon content of the eutectoid composition, is due to the large number of variable influences prevailing during the *experimental* determination of these salient points combined with the rather imprecise methods which are available, mainly microexamination, to make such determinations.

Having redrawn the iron–carbon diagrams at each new edition of this book a number of times in the past, this author has decided to settle on

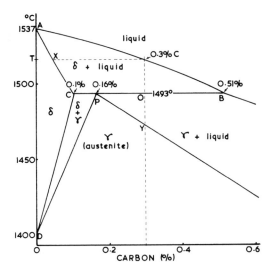

Fig. 11.2   The upper section of the iron–carbon diagram which includes a peritectic trans-formation.

the 'round' figures of 0.8% and 2.0% as the eutectoid and maximum carbon contents of austenite respectively. However the student should perhaps be warned—take note of the particular values used by your lecturer or you may lose marks in examinations.

11.20   Let us now consider the type of structure likely to be produced in a large steel sand-casting, containing 0.3% carbon, as it solidifies and cools slowly to room temperature. Such an alloy will begin to solidify at temperature T (Fig. 11.2) by forming dendrites of the solid solution δ of composition X. These dendrites will develop and change in composition along XC due to diffusion promoted by the slow rate of cooling, until at 1493°C they will be of composition C (0.1% carbon). The remaining liquid will have become correspondingly enriched in carbon and will be of composition B (0.51% carbon).

$$\frac{\text{Weight of }\delta\ (0.1\%\ \text{carbon})}{\text{Weight of liquid }(0.51\%\ \text{carbon})} = \frac{OB}{OC}$$
$$= \frac{(0.51-0.3)}{(0.3-0.1)}$$
$$= \frac{0.21}{0.2}$$
$$= 1.05/1$$

Application of the Lever Rule (above) indicates that at this stage there will be approximately equal amounts of δ and liquid. At 1493°C a peritectic interaction takes place between the remaining liquid and the dendrites of δ, resulting in the disappearance of the latter and the formation of austenite (γ) of composition P (0.16% carbon).

$$\frac{\text{Weight of austenite (0.16\% carbon)}}{\text{Weight of liquid (0.51\% carbon)}} = \frac{OB}{OP}$$
$$= \frac{(0.51 - 0.3)}{(0.3 - 0.16)}$$
$$= \frac{0.21}{0.14}$$
$$= 1.5/1$$

Thus there is now one-and-a-half times as much austenite as there is remaining liquid, and, as the temperature falls, the remaining liquid solidifies as austenite which will change in composition along PY. Solidification will be complete at Y and the austenite crystals will be of uniform composition containing 0.3% carbon. Since carbon is dissolved interstitially it can diffuse rapidly through the face-centred cubic structure of the austenite and because our large steel sand-casting will be cooling slowly there will be virtually no coring remaining in the structure.

11.21   Apart from considerable grain growth of the austenite no further change will take place in the microstructure until the line FE (Fig. 11.3) is reached at F'. The temperature represented by F' is called the *upper critical* (or A₃) *temperature* of this 0.3% carbon steel. Thus the upper critical temperature of a steel varies with its carbon content and will be

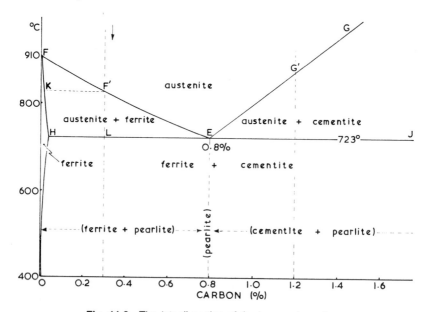

Fig. 11.3 The 'steel' portion of the iron–carbon diagram.

represented by the appropriate point on *FEG*. As the temperature of our 0.3% carbon steel falls below F' the face-centred cubic austenite becomes unstable and the polymorphic transformation (3.14) to body-centred cubic ferrite (α) begins. Thus, crystals of ferrite nucleate within the austenite crystals and grow progressively by absorbing the austenite structure. Since the ferrite which forms first contains very little carbon (K) it follows that the shrinking crystals of austenite will become increasingly rich in carbon. Since transformation from austenite to ferrite is accompanied by diffusion the composition of the ferrite will change slightly along KH whilst the austenite will change in composition along $F'E$. At 723°C (L) the ferrite will contain 0.02% carbon and the remaining austenite 0.8% carbon, and:

$$\frac{\text{Weight of ferrite } (0.02\,\%\,\text{C})}{\text{Weight of austenite } (0.8\,\%\,\text{C})} = \frac{LE}{LH}$$
$$= \frac{(0.8-0.3)}{(0.3-0.02)}$$
$$= \frac{0.5}{0.28}$$
$$= 1.79/1$$

Thus there is almost twice as much ferrite present as there is austenite.

11.22 At 723°C the remaining austenite transforms to the eutectoid pearlite by forming alternate layers of ferrite and cementite as previously described (7.55 and 8.43). The temperature, 723°C, at which pearlite is formed is called the *lower critical* (or A_1) *temperature*, and is the same for carbon steels of all compositions since the eutectoid temperature is constant, ie *HEJ* is horizontal. Since the whole of the austenite remaining at 723°C has transformed to pearlite it follows that the proportions ferrite/pearlite will be 1.79/1 as calculated above. That is, the microstructure would show roughly twice as much ferrite as pearlite.

11.23 A 0.8% carbon steel will begin to solidify at approximately 1470°C by depositing dendrites of austenite of composition R (Fig. 11.1) and, when solidification is complete, the structure will consist of crystals of austenite of overall composition 0.8% carbon. As the steel cools slowly, the structure becomes uniform by rapid diffusion and no further structural change will take place until the point E (Fig. 11.3) is reached. For a steel of this composition the upper and lower critical temperatures coincide and the austenitic structure transforms at this temperature to one which is totally pearlitic.

11.24 A 1.2% carbon steel will solidify in a similar way to the 0.8% carbon steel by forming austenite crystals of an overall carbon content of 1.2%. As the temperature falls to the upper critical for this alloy at G' (Fig. 11.3) needles of primary cementite begin to precipitate at the crystal boundaries of the austenite (at least the cementite appears to be needle-like in form in a *two-dimensional* microscope view but in fact we will be seeing cross-sections through flat cementite plates). Since cementite is being deposited the remaining austenite will be rendered less rich in carbon, so

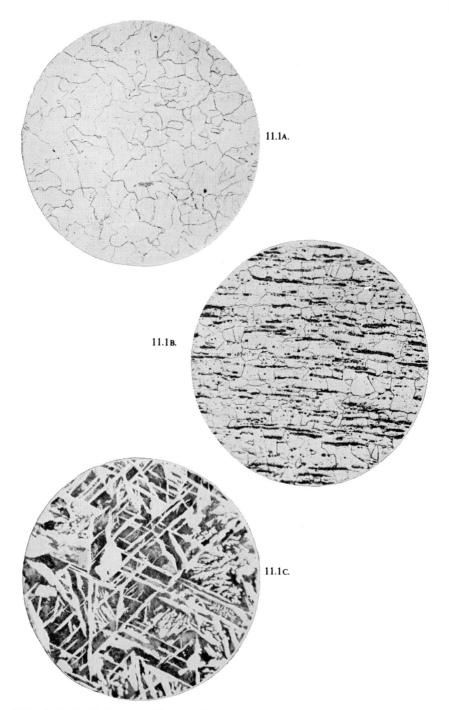

Plate 11.1 11.1A Commercially pure ('Armco') iron, showing crystals of ferrite. × 100. Etched in 2% nital.
11.1B Wrought iron (longitudinal section), showing slag fibres in a background of ferrite. × 100. Etched in 2% nital.
11.1C 0.5% carbon steel in the cast condition showing Widmanstätten structure of ferrite (light) and pearlite (dark), × 50. Etched in 2% nital.

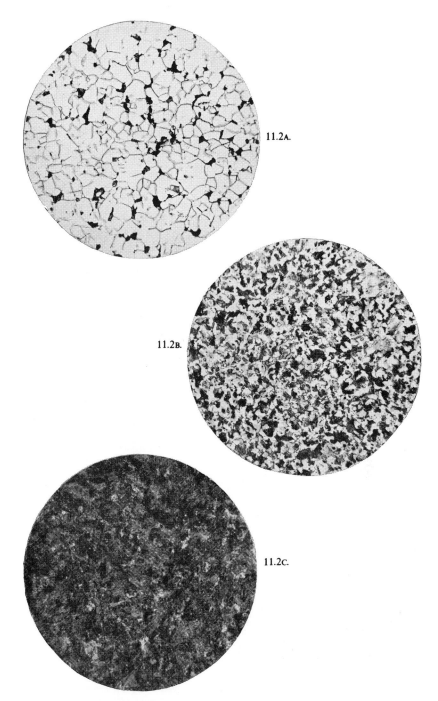

Plate 11.2 11.2A 0.15% carbon steel in the normalised condition.
Ferrite and a small amount of pearlite (dark). × 100. Etched in 2% nital.
11.2B 0.5% carbon steel, normalised.
Roughly equal amounts of ferrite and pearlite. × 100. Etched in 2% nital.
11.2C 0.8% carbon steel, normalised.
All pearlite. × 100. Etched in 2% nital.

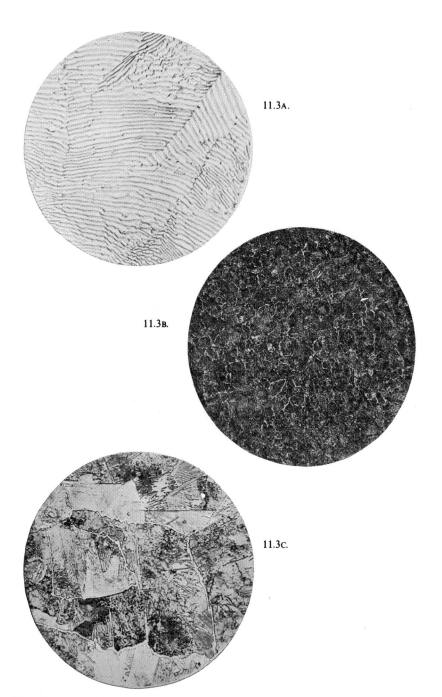

11.3A.

11.3B.

11.3C.

Plate 11.3 11.3A The structure of lamellar pearlite revealed by a micrograph taken at × 1000. Etched picral−nital. *(Courtesy of United Steel Companies Ltd, Rotherham)*
11.3B 1.3% carbon steel, normalised.
Network of free cementite around the patches of pearlite, × 100. Etched in 2% Nital. *(Courtesy of Hadfields Ltd, Sheffield).*
11.3C Similar to 11.3B, but higher magnification reveals the lamellar nature of the pearlite as well as the network of free cementite. × 750. Etched picral−nital. *(Courtesy of United Steel Companies Ltd, Rotherham).*

that its composition will move to the left and when the temperature has fallen to 723°C the remaining austenite will contain 0.8% carbon. As before, pearlite will now form, giving a final structure of primary cementite in a matrix of pearlite.

11.25   Thus, in a steel which has been permitted to cool slowly enough to enable it to reach structural equilibrium, we shall find one of the following structures:

(a) With less than 0.006% carbon it will be entirely ferritic. In practice, such an alloy would be classed as commercially pure iron.

(b) With between 0.006% and 0.8% carbon the structure will contain ferrite and pearlite. The relative proportions of ferrite and pearlite appearing in the microstructure will vary according to the carbon content, as shown in Fig. 7.7.

(c) With exactly 0.8% carbon the structure will be entirely pearlitic.

(d) With between 0.8% and 2.0% carbon the structure will consist of cementite and pearlite, in relative amounts which depend upon the carbon content.

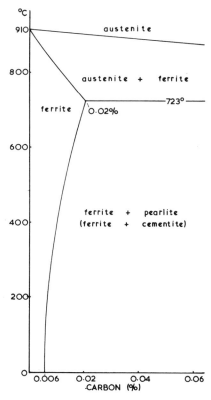

Fig. 11.4   The 'ferrite area' of the iron–carbon diagram, showing the very low solubility of carbon in body-centred cubic iron.

11.26 The composition of the pearlite area in the microstructure of any plain carbon steel is always the same, namely 0.8% carbon, and if the overall carbon content is either greater or smaller than this, then it will be compensated for by variation in the amount of either primary ferrite or primary cementite. The hardness of a slowly cooled steel increases directly as the carbon content, whilst the tensile strength reaches a maximum at the eutectoid composition (Fig. 7.7). These properties can be modified by heat-treatment, as we shall see in this chapter and the next.

Impurities in Steel

11.30 Most ordinary steels contain appreciable amounts of manganese, residual from the deoxidation process (3.21). Impurities such as silicon, sulphur and phosphorus (7.21) are also liable to be present in the finished steel. The effect of such impurities on mechanical properties will depend largely upon the way in which these impurities are distributed throughout the structure of the steel. If a troublesome impurity is *heavily cored* in the structure it can be expected to have a far more deleterious effect than if the same quantity of impurity were *evenly distributed* throughout the structure. Excessive coring concentrates the impurity in the grain-boundary regions often producing the effect of very brittle intergranular films. The extent to which coring of a particular element is likely to occur will be indicated by the distance apart of the solidus and liquidus lines at any temperature on the appropriate equilibrium diagram. Thus in Fig. 11.5A the relative compositions of solid (S) and liquid (L) are very far apart at any temperature and this may lead to excessive coring. Since relatively pure metal is solidifying it follows that the bulk of the impurity element becomes concentrated in the metal which solidifies last—in the grain boundary regions. In Fig. 11.5B, however, the compositions of the solid (S) and the liquid (L) remain close to each other throughout solidification and this will lead to a relatively even distribution of the impurity element throughout the microstructure and a consequent lack of dangerous crystal-boundary concentrations of brittle impurity.

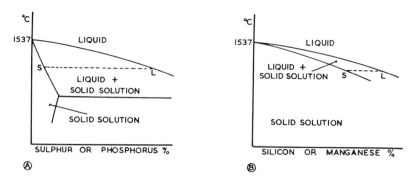

Fig. 11.5.

The crystals in solid steel are never extensively cored with respect to silicon and manganese, and since these elements have a high solid solubility in steel they are unlikely ever to appear as separate constituents in the microstructure. In solid solution in amounts up to 0.3% therefore their *direct* effect is minimal. Sulphur and phosphorus, on the other hand, segregate appreciably and if present in sufficient amounts could precipitate during solidification, as their respective iron compounds, at the austenite grain boundaries. The effect would be aggravated by the relatively low solubilities of these elements in steel.

11.31 Manganese   is not only soluble in austenite and ferrite but also forms a stable carbide, Mn_3C. In the nomenclature of the heat-treatment shop, manganese 'increases the depth of hardening' of a steel, for reasons which will be discussed in Chapter 13. It also improves strength and toughness. Manganese should not exceed 0.3% in high-carbon steels because of a tendency to induce quench cracks particularly during water-quenching.

11.32 Silicon   imparts fluidity to steels intended for the manufacture of castings, and is present in such steels in amounts up to 0.3%. In high-carbon steels silicon must be kept low, because of its tendency to render cementite unstable (as it does in cast iron (15.22)) and liable to decompose into graphite (which precipitates) and ferrite.

11.33 Phosphorus   is soluble in solid steel to the extent of almost 1%. In excess of this amount the brittle phosphide Fe_3P is precipitated. In solution phosphorus has a considerable hardening effect on steel but it must be rigidly controlled to amounts in the region of 0.05% or less because of the brittleness it imparts, particularly if Fe_3P should appear as a separate constituent in the microstructure.

In rolled or forged steel the presence of phosphorus is indicated by what are usually termed 'ghost bands' (Fig. 11.6c). These are areas (which

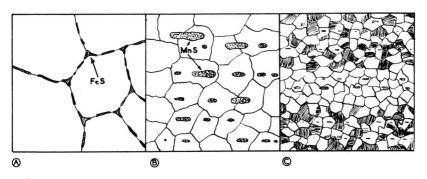

Fig. 11.6   (A) The segregation of iron(II) sulphide (FeS), at the crystal boundaries in steel. × 750.
(B) The formation of isolated manganese sulphide (MnS) globules when manganese is present in a steel. × 200.
(C) 'Ghost bands' or areas lacking in pearlite, which indicate the presence of phosphorus. × 75.

naturally become elongated during rolling) containing no pearlite, but instead, a high concentration of phosphorus. The presence of phosphorus and absence of pearlite will naturally make these ghost bands planes of weakness, particularly since, being areas of segregation, other impurities may be present in the ghosts.

11.34 Sulphur is the most deleterious impurity commonly present in steel. If precautions were not taken to render it harmless it would tend to form the brittle sulphide, FeS. Sulphur is completely soluble in molten steel but on solidification the solubility falls to 0.03% sulphur. If the effects of extensive coring, referred to above, are also taken into account it will be clear that amounts as low as 0.01% sulphur may cause precipitation of the sulphide at the crystal boundaries. In this way the austenite crystals would become virtually coated with brittle films of iron(II) sulphide. Since this sulphide has a fairly low melting point, the steel would tend to crumble during hot-working. Being brittle at ordinary temperatures, iron(II) sulphide would also render steel unsuitable for cold-working processes, or, indeed, for subsequent service of any type.

It would be very difficult, and certainly very expensive, to reduce the sulphur content to an amount less than 0.05% in the majority of steels. To nullify the effects of the sulphur present an excess of manganese is therefore added during deoxidation. Provided that about five times the theoretical manganese requirement is added, the sulphur then forms manganese sulphide, MnS, in preference to iron(II) sulphide. The manganese sulphide so formed is *insoluble in the molten steel*, and some is lost in the slag. The remainder is present as fairly large globules, distributed throughout the steel, but since they are insoluble, they will not be associated with the structure when solidification takes place. Moreover, manganese sulphide is plastic at the forging temperature, so that the tendency of the steel to crumble is removed. The manganese sulphide globules become elongated into threads by the subsequent rolling operations (Fig. 11.6A and B).

11.35 Nitrogen   Atmospheric nitrogen is absorbed by molten steel during the manufacturing process. Whether this nitrogen combines with iron to form nitrides or remains dissolved interstitially after solidification (Fig. 11.7), it causes serious embrittlement and renders the steel unsuitable for severe cold-work. For this reason mild steel used for deep-drawing operations must have a low nitrogen content.

Due to the method of manufacture, Thomas steel was particularly suspect and had nitrogen contents as high as 0.02% probably leading to the presence of brittle Fe_4N in the structure (Fig. 11.7). This was more than four times the average nitrogen content of open-hearth steel adequate for deep-drawing operations. Naturally the modern 'oxygen' processes (7.36) can produce mild steel with a very low nitrogen content (below 0.002%), since little or no nitrogen is present in the blast to the molten charge. Such steel is obviously ideal for deep-drawing. It is difficult, however, to prevent some atmospheric nitrogen from being absorbed, since the molten steel is in contact with the atmosphere during teeming.

11.36 Hydrogen   ions dissolve interstitially in solid steel and are thus able to migrate within the metal, resulting in embrittlement as shown by

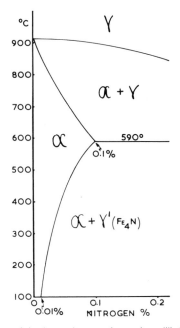

Fig. 11.7   Part of the iron–nitrogen thermal equilibrium diagram.

a loss in ductility. This hydrogen may be dissolved during the steel-making process but is more likely to be introduced from moisture in the flux coating of electrodes during welding, or released at the surface during an electroplating or acid-pickling operation. Hydrogen ions released during surface corrosion may also be absorbed.

The presence of hydrogen in steels can result in so-called 'delayed fracture', that is fracture under a static load during the passage of time. Such failure may occur after several hours at a stress of no more than 50% of the 0.2% proof stress. The effect is very dependent on the strain rate so that whilst ductility is considerably impaired during slow tensile tests, impact values are little affected.

In steels the mechanism of hydrogen embrittlement seems to be associated with the interstitial movement of hydrogen ions to positions at or near lattice faults, and also to regions of high tri-axial stress; in each case causing the nucleation of cracks and consequent premature failure. This would explain why failure is more likely with the passage of time during which hydrogen ions are able to migrate. Much of this dissolved hydrogen can be dispersed during a low-temperature (200°C) annealing process in a hydrogen-free atmosphere.

The Heat-treatment of Steel

11.40   Because of the solid-state structural changes which take place in suitable alloys, steels are among the relatively few engineering alloys which can be usefully heat-treated in order to vary their mechanical properties. This statement refers, of course, to heat-treatments other than simple stress-relief annealing processes.

Heat-treatments can be applied to steel not only to harden it but also to improve its strength, toughness or ductility. The type of heat-treatment used will be governed by the carbon content of the steel and its subsequent application.

11.41   The various heat-treatment processes can be classified as follows:
 (*a*) annealing;
 (*b*) normalising;
 (*c*) hardening;
 (*d*) tempering;
 (*e*) treatments which depend upon transformations taking place at a single predetermined temperature during a given period of time (isothermal transformations).

In all of these processes the steel is heated fairly slowly to some predetermined temperature, and then cooled, and it is the *rate of cooling* which determines the resultant structure of the steel and, hence, the mechanical properties associated with it. The final structure will be independent of the rate of heating, provided this has been slow enough for the steel to reach structural equilibrium at its maximum temperature. The subsequent rate of cooling, which determines the nature of the final structure, may vary between a drastic water-quench and slow cooling in the furnace.

Annealing

11.50   The term 'annealing' describes a number of different thermal treatments which are applied to metals and alloys. Annealing processes for steels can be classified as follows:

11.51 Stress-relief Annealing   The recrystallisation temperature of mild steel is about 500°C, so that, during a hot-rolling process recrystallisation proceeds simultaneously with rolling. Thus, working stresses are relieved as they are set up.

Frequently, however, we must apply a considerable amount of cold-work to mild steels, as, for example, in the drawing of wire. Stress-relief annealing then becomes necessary to soften the metal so that further drawing operations can be carried out. Such annealing is often referred to as 'process' annealing, and is carried out at about 650°C. Since this temperature is well above the recrystallisation temperature of 500°C, recrystallisation will be accelerated so that it will be complete in a matter of minutes on attaining the maximum temperature. Prolonged annealing may in fact cause a deterioration in properties, since although ductility may increase further, there will be a loss in strength. A stage will be reached where grain growth becomes excessive, and where the layers of cementite in the patches of pearlite begin to coalesce and assume a globular form so that the identity of the eutectoid is lost (Fig. 11.8). In fact, the end-product

Fig. 11.8   The spheroidisation of pearlitic cementite.

would be isolated globular masses of cementite in a ferrite matrix. The result of this 'balling-up' of the pearlitic cementite is usually called 'deteriorated' pearlite.

It should be noted that process annealing is a *sub-critical* operation, that is, it takes place below the lower critical temperature (A_1). Consequently, reference to the iron–carbon diagram is not involved. Although recrystallisation is promoted by the presence of internal energy remaining from the previous cold-working process, there is *no phase change* and the constituents ferrite and cementite remain in the structure throughout the process. The balling-up of the pearlitic cementite is purely a result of surface-tension effects which operate at the temperatures used.

Process annealing is generally carried out in either batch-type or continuous furnaces, usually with some form of inert atmosphere derived from burnt 'town gas' or other hydrocarbons. The carbon dioxide present in such mixtures does not react with the surface of iron at the relatively low temperature of 650°C. At higher temperatures, however, it may behave as an oxidising agent.

11.52 Spheroidising Anneals 'The spheroidisation of pearlitic cementite' may sound a somewhat ponderous phrase. In fact, it refers to the balling-up of the cementite part of pearlite mentioned above. This phenomenon is utilised in the softening of tool steels and some of the air-hardening alloy steels. When in this condition such steels can be drawn and will also machine relatively freely.

The spheroidised condition is produced by annealing the steel at a temperature between 650 and 700°C, that is, just *below* the lower critical temperature (A_1), so that, again, the iron–carbon diagram is not involved in our study. Whilst no basic phase change takes place, surface tension causes the cementite to assume a globular form (Fig. 11.8) in a similar way to which droplets of mercury behave when mercury is spilled. If the layers of cementite are relatively coarse they take rather a long time to break up, and this would result in the formation of very large globules of cementite. This in turn would lead to tearing of the surface during machining. To obviate these effects it is better to give the steel some form of quenching treatment prior to annealing in order to refine the distribution of the cementite. It will then be spheroidised more quickly during annealing and will produce much smaller globules of cementite. These small globules will not only improve the surface finish during machining but will also be dissolved more quickly when the tool is ultimately heated for hardening.

11.53 Annealing of Castings   As stated earlier (11.20), the cast structure of a large body of steel is extremely coarse. This is due mainly to the slow rates of solidification and subsequent cooling through the austenitic range. Thus, a 0.35% carbon steel will be completely solid in the region of 1450°C, but, if the casting is large, cooling, due to the lagging effect of the sand mould, will proceed very slowly down to the point (approximately 820°C) where transformation to ferrite and pearlite begins. By the time 820°C has been reached, therefore, the austenite crystals will be extremely large. Ferrite, which then begins to precipitate in accordance with the equilibrium diagram, deposits first at the grain boundaries of the austenite, thus revealing, in the final structure, the size of the original austenite grains. The remainder of the ferrite is then precipitated along certain crystallographic planes within the lattice of the austenite. This gives rise to a directional precipitation of the ferrite, as shown in Fig. 11.9 and Plate 11.1c, representing typically what is known as a Widmanstätten structure. This type of structure was first encountered by Widmanstätten in meteorites (10.10), which may be expected to exhibit a coarse structure in view of the extent to which they are overheated during their passage through the upper atmosphere. The mesh-like arrangement of ferrite in the Widmanstätten structure tends to isolate the stronger pearlite into separate patches, so that strength, and more particularly toughness, are impaired. The main characteristics of such a structure are, therefore, weakness and brittleness, and steps must be taken to remove it either by heat-treatment or by mechanical working. Hot-working will effectively break up this coarse as-cast structure and replace it by a fine-grained material, but in this instance we are concerned with retaining the actual shape of the casting. Heat-treatment must therefore be used to effect what limited refinement of grain is possible, but it should be noted that the crystal size after heat treatment will be greater than that achieved by hot-working.

11.54   The most suitable treatment for a large casting involves heating it slowly up to a temperature about 40°C above its upper critical (thus the annealing temperature depends upon the carbon content of the steel, as shown in Fig. 11.10), holding it at that temperature only just long enough for a uniform temperature to be attained throughout the casting and then allowing it to cool slowly in the furnace. This treatment not only introduces the improvements in mechanical properties associated with fine grain but also removes mechanical strains set up during solidification.

As the lower critical temperature (723°C) is reached on heating, the patches of pearlite transform to austenite but these new crystals of austenite are *very small since each patch of pearlite gives rise to many new austenite crystals*. It is upon this fact that the complete success of this type of annealing process depends. As the temperature rises, the Widmanstätten-type plates of ferrite are dissolved by the austenite until, when the upper critical temperature is reached, the structure consists entirely of fine-grained austenite. Cooling causes reprecipitation of the ferrite, but, since the new austenite crystals are small, the precipitated ferrite will also be distributed as small particles. Finally, as the lower critical temperature is

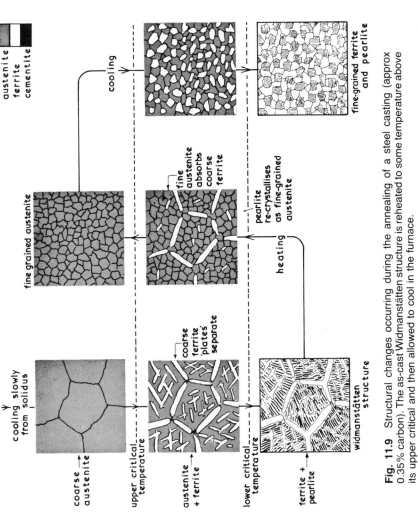

Fig. 11.9 Structural changes occurring during the annealing of a steel casting (approx 0.35% carbon). The as-cast Widmanstätten structure is reheated to some temperature above its upper critical and then allowed to cool in the furnace.

austenite
ferrite
cementite

cooling

fine grained austenite

fine-grained ferrite and pearlite

fine austenite absorbs coarse ferrite

pearlite re-crystallises as fine-grained austenite

heating

coarse ferrite 'plates' separate

cooling slowly from solidus

coarse austenite

upper critical temperature

austenite + ferrite

lower critical temperature

ferrite + pearlite

widmanstätten structure

reached, the remaining small patches of austenite will transform to pearlite. The structural changes taking place during annealing are illustrated diagrammatically in Fig. 11.9.

11.55 Whilst the tensile strength is not greatly affected by this treatment, both toughness and ductility are improved as shown by the following values for a cast carbon steel:

Condition	Tensile strength	Percentage elongation	Bend test
Specimen 'as cast'.	470 N/mm²	18	40°
Specimen annealed.	476 N/mm²	34	180° (without fracture)

11.56 Overheating during annealing, or heating for too long a period in the austenitic range, will obviously cause grain growth of the newly formed austenite crystals, leading to a structure almost as bad as the original Widmanstätten structure. For this reason the requisite annealing temperature should not be exceeded, and the casting should remain in the austenitic range only for as long as is necessary to make it completely austenitic. In fact, castings are sometimes air-cooled to about 650°C and then cooled more slowly to room temperature, by returning to a furnace to prevent stresses due to rapid cooling from being set up.

11.57 Excessive overheating will probably cause oxidation, or 'burning', of the surface, and the penetration by oxide films of the crystal boundaries following decarburisation of the surface. Such damage cannot be

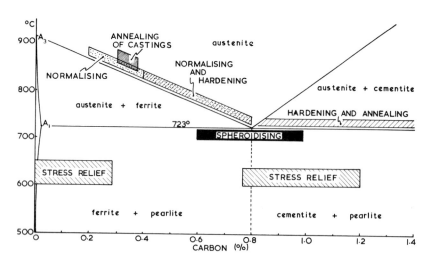

Fig. 11.10 The heat-treatment temperature ranges of classes of carbon steels in relation to the equilibrium diagram.

repaired by heat-treatment, and the castings can only be scrapped. To prevent 'burning' some form of inert atmosphere must be used in the annealing furnace in order to limit contact between the castings and atmospheric oxygen.

11.58   If annealing is carried out at too low a temperature, remnants of the as-cast structure will be apparent in the form of undissolved skeletons of Widmanstätten ferrite. As the temperature falls on cooling, the ferrite which did dissolve tends to reprecipitate on the existing ferrite skeleton so that the final structure resembles the unannealed.

Normalising

11.60   Normalising resembles the 'full' annealing of castings described in 11.53 in that the maximum temperature attained is similar. It is in the method of cooling that the processes differ. Whilst, in annealing, cooling is retarded, in normalising the steel is removed from the furnace and allowed to cool in still air. This relatively rapid method of cooling limits grain growth in normalising, whilst the ferrite/cementite lamellae in pearlite will also be much finer. For both reasons the mechanical properties are somewhat better than in an annealed component. Moreover, the surface finish of a normalised article is often superior to that of an annealed one when machined, since the high ductility of the latter often gives rise to local tearing of the surface.

11.61   The type of structure obtained by normalising will depend largely upon the thickness of cross-section, as this will affect the rate of cooling. Thin sections will give a much finer grain than thick sections, the latter differing little in structure from an annealed section.

Exercises

1. Calculate the relative proportions by mass of ferrite and cementite in pearlite (Fig. 11.1 and Fig. 11.4).
2. By reference to Fig. 11.2 explain what happens as a steel containing 0.12% C cools slowly between 1550°C and 1450°C (11.20).
3. One particular plain carbon steel may be described as a 'binary alloy' of exact 'peritectic' composition and as being 'hypo-eutectoid' in nature.
   (i) Define the three terms in inverted commas.
   (ii) Sketch qualitatively the relevant portion of the equilibrium diagram and on it indicate clearly the composition of the steel.
   (iii) Discuss the changes of structure that would occur if an alloy of this composition were cooled under equilibrium conditions from its molten state down to room temperature. (8.13, 7.57 and 11.20)
4. Fig. 11.11 shows part of the iron–carbon thermal equilibrium diagram.
   (i) Make an accurate assessment of the upper critical temperature of a steel containing 0.45% C;
   (ii) What proportions by mass of primary ferrite and pearlite will be present in a 0.45% C steel which has been normalised?

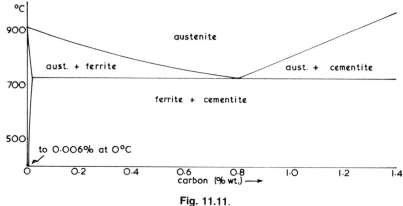

Fig. 11.11.

(iii) Between what temperatures would primary cementite be deposited when a steel containing 1.3% C were cooled slowly from 1200°C to ambient temperature?

(iv) What effects would such treatment be likely to have on the resultant mechanical properties of the steel in (iii)? (11.20 and 11.24)

5. By using Fig. 11.11 determine what phases are present, the compositions of each of these phases and the proportions in which they exist for a 0.15% C steel at (i) 900°C; (ii) 800°C and (iii) 700°C. Assume equilibrium in each case. (11.20)

6. Draw and describe a Widmanstätten structure and explain how such a structure arises in a hypo-eutectoid steel. (11.53)

7. Compare and contrast the objectives in the following heat-treatments:
   (i) stress-relieving of cold-worked mild steel;
   (ii) annealing of steel and castings;
   (iii) spheroidising annealing of tool steel.
   Show how the thermal treatment differs in each case. (11.50)

8. How does *normalising* differ from *annealing* as applied to steels? What are the advantages of the normalising process in respect of final properties? (11.60)

Bibliography

Higgins, R. A., *Engineering Metallurgy (Part II)*, Edward Arnold, 1986.

Honeycombe, R. W. K., *Steels: Microstructure and Properties*, Arnold, 1981.

Hume-Rothery, W., *The Structures of Alloys of Iron*, Pergamon, 1966.

Leslie, W.C ., *The Physical Metallurgy of Steels*, McGraw-Hill, 1982.

Pickering, F.B., *Physical Metallurgy and Design of Steels*, Applied Science, 1978.

Samuels, L. E., *Optical Microscopy of Carbon Steels*, American Society for Metals, 1980.

Thelning, K-E., *Steel and its Heat-treatment: Bofors Handbook*, Butterworths, 1984.

BS 970:1983 *Wrought Steels in the Form of Blooms, Billets, Bars and Forgings (Part I)*

BS 3100:1984 *Specifications for Steel Castings for General Engineering Purposes*

12

The Heat-Treatment of Plain Carbon Steels—(II)

12.10 In the previous chapter those heat-treatment processes were discussed in which the steel component was permitted to reach a state of thermal equilibrium at ambient temperature. That is, cooling took place sufficiently slowly to allow a pearlitic type of microstructure to form. Such treatments are normally only useful for improving the toughness and ductility of a steel component, and when increased hardness is required it is necessary to quench, or cool, the component sufficiently rapidly, in order to prevent the normal pearlitic structure from being formed. If a combination of strength and toughness is necessary then a further 'tempering' process may follow quenching. Alternatively one of the isothermal treatments may be used to replace the dual treatments of quenching and tempering.

12.11 Prior to the development of metallurgy as a science many of the processes associated with the hardening of steel were clothed in mystery. For example, it was thought that the water of Sheffield possessed certain magical properties, and it is said that an astute Yorkshire business man once exported it in barrels to Japan at considerable profit. In point of fact the high quality of Sheffield steel was a measure of the craftsmanship used in its production. Similarly, it is reported that Damascus steel swords were hardened by plunging the blade, whilst hot, into the newly decapitated body of a slave and stirring vigorously. Some metallurgists have suggested, possibly more out of cynicism than scientific accuracy, that hardening would be assisted by nitrogen absorption from the blood of the slave during this somewhat gruesome procedure. James Bowie, originator of the Bowie knife in the days of the 'Wild West', is said to have quenched his knives *nine times* in succession in panther oil.

In this chapter, then, we shall deal with the production of structures, other than pearlite, in plain carbon steels, and seek to explain the relationship which exists between the mechanical properties and the crystal structure produced by the treatments employed.

Hardening

12.20 When a piece of steel, containing sufficient carbon, is cooled rapidly from above its upper critical temperature it becomes considerably harder than it would be if allowed to cool slowly. The degree of hardness produced can vary, and is dependent upon such factors as the initial quenching temperature; the size of the work; the constitution, properties and temperature of the quenching medium; and the degree of agitation and final temperature of the quenching medium.

12.21 Whenever a metallic alloy is quenched there is a tendency to suppress structural change or transformation. Frequently, therefore, it is possible to 'trap' a metallic structure as it existed at a higher temperature and so preserve it at room temperature. This is usually an easy matter with alloys in which transformation is sluggish, but in iron–carbon alloys the reverse tends to be the case. Here, transformation, particularly that of austenite to pearlite, is rapid and is easily accomplished during ordinary air-cooling to ambient temperature. This is due largely to the polymorphic transformation which takes place but also to rapid diffusion of carbon atoms in the face-centred cubic lattice of iron. The rapid diffusion of carbon atoms is a result of their smaller size and the fact that they dissolve interstitially (This also leads to the absence of coring with respect to carbon in cast steels.)

When a plain carbon steel is quenched from its austenitic range it is not possible to trap austenite and so preserve it at room temperature. Instead, one or other phases is obtained intermediate between austenite on the one hand and pearlite on the other. These phases vary in degree of hardness, but all are harder than either pearlite or austenite.

12.22 Water quenching of a steel containing sufficient carbon produces an extremely hard structure called *martensite* which appears under the microscope as a mass of uniform needle-shaped crystals (Plate 12.1A). These 'needles' are in fact cross-sections through lens- or discus-shaped crystals—another instance of the misleading impression sometimes given by the two-dimensional image offered by the metallurgical microscope. Since martensite is of uniform appearance even at very high magnifications it follows that the carbon is still *in solution* in the iron and has not been precipitated as iron carbide as it would have been if the steel had been cooled under equilibrium conditions. However, X-ray crystallographic examination of martensite shows that despite very rapid cooling which has prevented the precipitation of iron carbide, the lattice structure has nevertheless changed from FCC (face-centred cubic) to something approaching the BCC (body-centred cubic) structure which is normally present in a steel cooled slowly to ambient temperature. This BCC type structure is considerably supersaturated with carbon since at ambient temperatures only 0.006% carbon is retained in solution under equilibrium conditions. Consequently the presence of dissolved carbon in amounts of, say, 0.5% can be expected to cause considerable distortion of the structure and in fact produces one which is body-centred tetragonal.

The transformation of a single crystal of martensite from austenite appears to be achieved in about 10^{-7} seconds. How can this change in structure take place so rapidly? It is suggested that a process of *diffusionless phase transformation* is involved, that is, there is an extremely limited movement of iron and carbon atoms into positions more nearly approaching equilibrium at the lower temperature. The lattice structure of austenite is represented in Fig. 12.1. This is the FCC structure with carbon atoms able to occupy interstitial positions as indicated*. If we regard the superimposed figure, indicated in heavier line with a base ABCD, this shows how FCC austenite can be regarded as a body-centred tetragonal structure and it is thought that martensite transformation involves a change from this structure to a true body-centred tetragonal structure with very little consequent movement of the iron atoms. In Fig. 12.2 the alteration in dimensions of a unit cell from the original austenite to the new BCT (body-centred tetragonal) is indicated. Thus the unit becomes more 'squat' in shape as the 'a_0' axis shrinks and the $a_0/\sqrt{2}$ axis expands. But for the presence of the carbon atoms inherited from the austenite the structure would transform to simple BCC ferrite. The interstitially dissolved carbon atoms have little chance to move and must fit into the new structure where they will cause considerable distortion since in a body-centred structure there are far fewer interstitial sites available. Since not all of the interstitial sites will be occupied by carbon atoms a structure something like that shown in Fig. 12.3 will form.

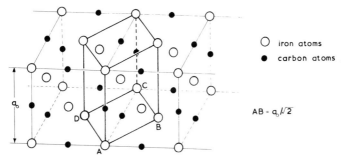

○ iron atoms
● carbon atoms

$AB = a_0/\sqrt{2}$

Fig. 12.1   The face-centred cubic structure of austenite showing its relationship to a body-centred tetragonal cell based on *ABCD*.

The actual change from FCC to BCT involves a very small movement of atoms and probably proceeds in a manner similar to that in mechanical twinning. Movement of dislocations due to the shear involved can have an effect akin to severe work-hardening. This, in conjunction with the great distortion produced by the interstitial carbon atoms helps to explain the great hardness and negligible ductility of martensite. The presence of any carbon in excess of 0.02% will frustrate the formation of a simple BCC structure when such a steel is quenched from the austenitic range. The

* In steel a maximum of less than one in six of these positions are ever occupied by a carbon atoms.

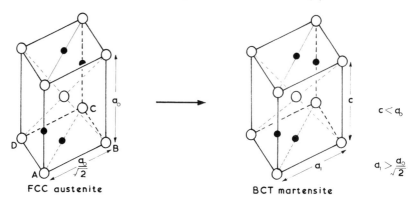

Fig. 12.2　The transformation of FCC austenite to BCT martensite.
The austenite 'tetragonal unit' shown above is the one outlined in Fig. 12.1. The sides, a_o, have contracted to 'c', and the sides $a_0/\sqrt{2}$ have expanded to a_1.

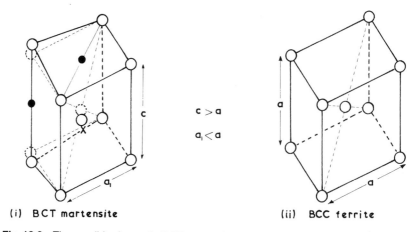

Fig. 12.3　The possible shape of a BCT martensite cell containing only one interstitial carbon atom (i). In any one martensite crystal all carbon atoms occupy the same interstitial position on the 'c' axis. (Iron atom X is displaced by the carbon atom of the cell 'above' it). In (ii) a normal BCC ferrite cell is shown for comparison.

degree of distortion existing in the resulting tetragonal martensite will be proportional to the overall carbon content. Consequently as the carbon content increases so does hardness.

Less severe quenching gives rise to a structure known as *Bainite*. This phase appears under the microscope, at magnifications in the region of × 100, as black patches (Plates 12.1 B and C), but a higher magnification of × 1000 shows that it is of a laminated nature something like pearlite. The growth of bainite (Fig. 12.4) differs from that of pearlite in that ferrite nucleates first followed by carbide, whereas in pearlite it is the carbide

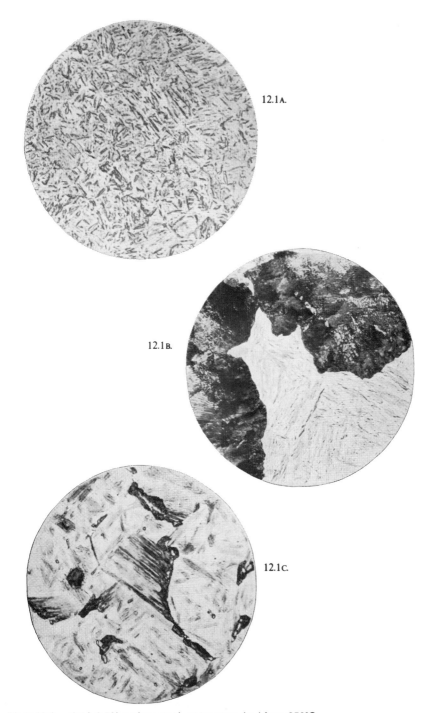

12.1A.

12.1B.

12.1C.

Plate 12.1 12.1A 0.5% carbon steel, water quenched from 850°C.
Entirely martensite. × 100. Etched in 2% nital.
12.1B 0.5% carbon steel, oil quenched from 780°C.
Bainite (dark) and martensite. × 750. Etched in picral–nital. *(Courtesy of United Steel Companies Ltd. Rotherham).*
12.1C 0.2% carbon steel, water quenched from 870°C on a falling gradient.
Acicular bainite (dark) and martensite. × 1000. Etched in picral–nital. *(Courtesy of United Steel Companies Ltd. Rotherham).*

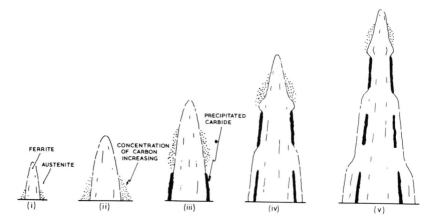

FERRITE

AUSTENITE

CONCENTRATION
OF CARBON
INCREASING

PRECIPITATED
CARBIDE

(i)   (ii)   (iii)   (iv)   (v)

Fig. 12.4 The growth of bainite.
As the ferrite crystals grow, so the concentration of carbon in the surrounding austenite increases until a point is reached where carbide is rejected.

which nucleates first (8.43). Bainite growth takes place quickly because the driving force is increased by a greater degree of non-equilibrium at the lower temperatures at which it is formed. Consequently particle size is too small to be seen by low-power microscopy.

Still slower rates of cooling produce normal pearlite, the coarseness of the ferrite and cementite laminations depending upon the rate of cooling. Thus, normalising leads to the formation of a fairly fine-grained structure whilst annealing produces coarse-grained structures.

12.23   In practice, factors such as composition, size and shape of the component to be hardened dictate the rate at which it shall be cooled. Generally no attempt is made to harden plain carbon steels which contain less than 0.25% carbon since the increase in hardness produced would be so small and non-uniform for reasons which will become apparent later in this chapter (12.45). Large masses of steel of heavy section will obviously cool more slowly than small articles of thin section when quenched, so that whilst the surface skin may be martensitic, the core of a large section may be bainitic because it has cooled more slowly. If, however, small amounts of such elements as nickel, chromium or manganese are added to the steel, it will be found that the martensitic layer is much thicker than with a plain carbon steel of similar carbon content and dimensions which has been cooled at the same rate. Alloying elements therefore 'increase the depth of hardening', and they do so by slowing down the transformation rates. This is a most important feature, since it enables an alloy steel to be hardened by much less drastic quenching methods than are necessary for a plain carbon steel. The liability to produce quench-cracks, which are often the result of water-quenching, is reduced in this way. Design also affects the susceptibility to quench-cracking. Sharp variations in cross-section and the presence of sharp angles, grooves, notches and rectangular holes are all likely to cause the formation of quench-cracks. Consequently when mass production is involved it is often more satisfactory to use a low-alloy steel containing small amounts of the cheaper elements like manganese which can then be oil quenched on a conveyor-belt system. This not only cuts labour costs but eliminates the human element from quenching, as well as minimising distortion and cracking and providing a more uniform product.

12.24   The quenching medium is chosen according to the rate at which it is desired to cool the steel. The following list of media is arranged in order of quenching speeds:

    5% Caustic soda
    5–20% Brine
    Cold water
    Warm water
    Mineral oil
    Animal oil
    Vegetable oil

The very drastic quench resulting from the use of caustic soda solution is used only when extreme hardness is required in components of simple shape. For more complicated shapes an oil-quenched alloy steel would give better results. Originally animal oils obtained from the blubber of seal and whale were used for this purpose, but the near extinction of the whale has brought to an end the extremely barbaric practice of whaling by all civilised nations. (We must be vigilant nevertheless that whaling does not begin again as apparently there are considerable numbers of so-called gourmets in some countries who wish to eat these noble mammals.) Most quenching oils are now of mineral origin and are obtained during the refining of crude petroleum.

In addition to the rate of heat abstraction such factors as flash point, viscosity and chemical stability are important. A high flash point is necessary to reduce fire risks, whilst high viscosity will lead to loss of oil by 'drag-out', ie oil clinging to the work piece as it is withdrawn from the quenching bath. Atmospheric oxidation and other chemical changes generally led to a thickening of whale oil and the formation of thick scum. On the other hand some mineral oils 'crack' or break down to simpler compounds of lower boiling point which will volatilise in use leaving a thicker, more viscous mixture behind.

Water solutions of synthetic polymers such as polyalkalene glycol are now replacing oils for many quenching operations. Not only do they eliminate fire risks, smoke and unpleasant fume but are generally less expensive. Moreover, having lower viscosities, loss by drag-out is reduced. Less contamination of the work results and degreasing prior to subsequent operations is unnecessary.

12.25.   To harden a piece of steel, then, it must be heated to between 30 and 50°C above its upper critical temperature and then quenched in some medium which will produce in it the desired rate of cooling. The medium used will depend upon the composition of the steel and the ultimate properties required. Symmetrically shaped components are best quenched 'end-on', and all components should be agitated in the medium during quenching.

Tempering

12.30 A fully hardened carbon tool steel is relatively brittle, and the presence of stresses set up by quenching make its use, in this condition, inadvisable except in cases where extreme hardness is required. Hence it is customary to re-heat—or 'temper'—the quenched component so that internal stresses will be relieved and brittleness reduced. Medium-carbon constructional steels are also tempered but here the temperatures are somewhat higher so that strength and hardness are sacrificed to some extent in favour of greater toughness and ductility.

12.31 During tempering, which is always carried out *below* the lower critical temperature, martensite tends to transform to the equilibrium structure of ferrite and cementite. The higher the tempering temperature the more closely will the original martensitic structure revert to this ferrite-cementite mixture and so strength and hardness fall progressively, whilst toughness and ductility increase (Fig. 12.5). Thus by choosing the appropri-

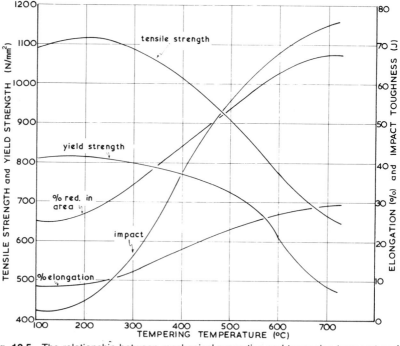

Fig. 12.5 The relationship between mechanical properties and tempering temperature for a steel containing 0.5% carbon and 0.7% manganese in the form of a bar 25mm diameter, previously water quenched from 830°C.

ate tempering temperature a wide range of mechanical properties can be achieved in carbon steels.

12.32   The structural changes which occur during the tempering of martensite containing more than 0.3% carbon, take place in three stages:

1st stage   At about 100°C, or possibly even lower, the existing martensite begins to transform to another form of martensite, containing only 0.25% carbon, together with very fine particles of a carbide. However, this carbide is not ordinary cementite but one containing rather more carbon and of a formula approximately Fe_5C_2. It is designated ε-carbide. No alteration in the microstructure is apparent under an ordinary optical microscope because the ε-carbide particles are so small, but the electron microscope reveals them as films about 2×10^{-8}m thick. At this stage a slight increase in hardness may occur because of the presence of the finely-dispersed but hard ε-carbide. Brittleness is significantly reduced as quenching stresses disappear in consequence of the transformation. At 100°C the transformation proceeds very slowly but increases in speed up to 200°C.

2nd stage   This begins at about 250°C when any 'retained austenite' (12.43) begins to transform to bainite. This will cause the martensite 'needles' to etch a darker colour and formerly this type of structure was known as *troostite*. A further slight increase in hardness may result from the replacement of austenite by much harder bainite.

3rd stage   At about 350°C the ε-carbide begins to transform to ordinary cementite and this continues as the temperature rises. In the meantime the remainder of the carbon begins to precipitate from the martensite—also as cementite—and in consequence the martensite structure gradually reverts to one of ordinary BCC ferrite. Above 500°C the cementite particles coalesce into larger rounded globules in the ferrite matrix. This structure was formerly called *sorbite* but both this term and that of troostite are now no longer used by metallurgists who prefer to describe these structures as 'tempered martensite'.

Due to the increased carbide precipitation which occurs as the temperature rises the structure becomes weaker but more ductile, though above 550°C strength falls fairly rapidly with little rise in ductility (Fig. 12.5).

12.33   Tempering can be carried out in a number of ways, but, in all, the temperature needs to be fairly accurately controlled. As the steel is heated, the oxide film which begins to form on a bright, clean surface first assumes a pale-yellow colour and gradually thickens with increase in temperature until it is dark blue. This is a useful guide to the tempering of tools in small workshops where pyrometer-controlled tempering furnaces are not available and is the time-honoured method of heat-treating high-quality hand-made wood-working tools. Table 12.1 shows typical colours obtained on clean surfaces when a variety of components are tempered to suitable temperatures. Such a colour–temperature relationship is only applicable to plain carbon steels. Stainless steels, for example, oxidise less easily, so that the colours obtained will bear no relationship to the temperatures indicated in the table. Moreover, the oxide film colour is only a reliable guide when the component has been progressively raised in temperature. It does not apply to one which has been maintained at a fixed temperature for some time, since here the oxide film will be thicker and darker in any case. In addition, the human element must also be taken into account, so that, in general, tempering in a pyrometer-controlled furnace is more successful.

12.34   Furnaces used for tempering are usually of the batch type (13.20 —Part II). They employ either a circulating atmosphere or are of the liquid-bath type. Liquids transfer heat more uniformly and have a greater heat capacity, and this ensures an even temperature throughout the furnace. For low temperatures oils are often used, but higher temperatures demand the use of salt baths containing various mixtures of sodium nitrite and potassium nitrate. These baths can be used at about 500°C, but above that temperature either mixtures of chlorides or lead baths are necessary. Another popular furnace, in which the temperature can be varied easily and controlled thermostatically, is the circulating-air type. Here, uniform temperatures up to 650°C can be obtained by using fans to circulate the atmosphere, first over electric heaters, and then through a wire basket holding the charge.

Failing a pyrometer-controlled furnace, temperature-indicating paints and crayons are useful in determining the tempering temperature of small components, provided some method of uniform heating is available. Such indicators do indeed record the *actual temperature reached by the component*, which is more than can be said for a pyrometer controlling a furnace which is in the hands of an unskilled operative.

Table 12.1   *Tempering Colours for Plain-carbon-steel Tools*

Temperature (°C)	Colour	Type of component
220	Pale yellow	Scrapers; hack saws; light turning and parting tools
230	Straw	Screwing dies for brass; hammer faces; planing and slotting tools
240	Dark straw	Shear blades; milling cutters; paper cutters; drills; boring cutters and reamers; rock drills
250	Light brown	Penknife blades; taps; metal shears; punches; dies; wood-working tools for hard wood
260	Purplish-brown	Plane blades; stone-cutting tools; punches; reamers; twist drills for wood
270	Purple	Axes; gimlets; augers; surgical tools; press tools
280	Deeper purple	Cold chisels (for steel and cast iron); chisels for wood; plane cutters for soft woods
290	Bright blue	Cold chisels (for wrought iron); screw-drivers
300	Dark blue	Wood saws; springs

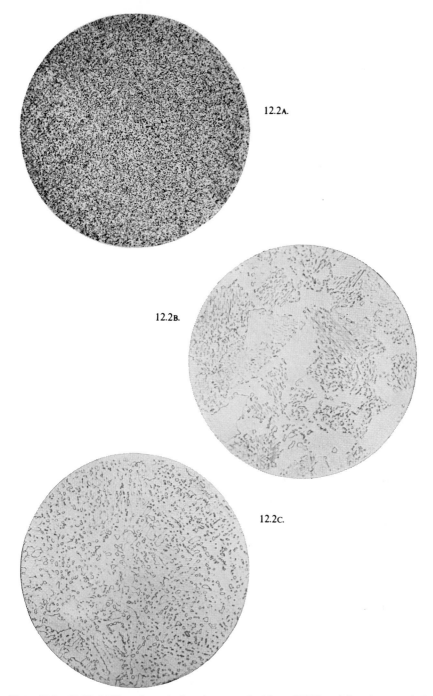

12.2A.

12.2B.

12.2C.

Plate 12.2 12.2A 0.5% carbon steel, water quenched from 850°C and then tempered at 600°C.
Spheroid carbide in ferrite. × 250. Etched in 2% nital.
12.2B 0.5% carbon steel, normalised and then annealed for 48 hours at 670°C
The pearlite cementite has become spheroidised. × 750. Etched in picral–nital. *(Courtesy of United Steel Companies Ltd., Rotherham).*
12.2C 0.5% carbon steel, water quenched and then tempered for 48 hours at 670°C.
Spheroidised carbide has in this case been precipitated from martensite; making the distribution more even than in 12.2B. × 750. Etched in picral–nital. *(Courtesy of United Steel Companies Ltd. Rotherham).*

Isothermal Transformations

12.40   As pointed out earlier in this chapter (12.22), the microstructure and properties of a quenched steel are dependent upon the rate of cooling which prevails during quenching. This relationship, between structure and rate of cooling, can be studied for a given steel with the help of a set of isothermal transformation curves which are known as TTT (Time–Temperature–Transformation) curves. The TTT curves for a steel of eutectoid composition are shown in Fig. 12.6. They indicate the time necessary for transformation to take place and the structure which will be produced when austenite is *supercooled* to any predetermined transformation temperature.

12.41   Such curves are constructed by taking a large number of similar specimens of the steel in question and heating them to just inside the austenitic range. These specimens are divided into groups each of which

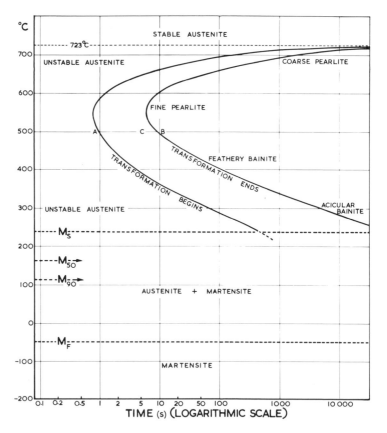

Fig. 12.6   Time–temperature–transformation (TTT). Curves for a plain carbon steel of eutectoid composition.
Martensitic transformation is not complete until approx −50°C. Consequently a trace of 'retained austenite' may be expected in a steel quenched to room temperature.

is quickly transferred to an 'incubation' bath at a different temperature. At predetermined time intervals individual specimens are removed from their baths and quenched in water. The microstructure is then examined to see the extent to which transformation had taken place at the holding temperature. Let us assume, for example, that we have heated a number of specimens of eutectoid steel to just above 723°C and have then quenched them into molten lead at 500°C (Figs. 12.6 and 12.7). Until one second has elapsed transformation has not begun, and if we remove a specimen from the bath in less than a second, and then quench it in water, we shall obtain a completely martensitic structure, proving that at 500°C after one second ('A' on Figs. 12.6 and 12.7) the steel was still completely austenitic. The production of martensite in the viewed structure is entirely due to the final water-quench. If we allow the specimen to remain at 500°C for ten seconds ('B' on Figs. 12.6 and 12.7) and then water-quench it, we shall find that the structure is composed entirely of bainite in feather-shaped patches, showing that after ten seconds at 500°C transformation to bainite was complete. If we quenched a specimen after it had been held at 500°C for five seconds ('C' on Figs. 12.6 and 12.7) we would obtain a mixture of bainite and martensite, showing that, at the holding temperature (500°C), the structure had contained a mixture of bainite and austenite due to the incomplete transformation of the latter. By repeating such treatments at different holding temperatures we are able, by interpreting the resulting microstructures, to construct TTT curves of the type shown in Fig. 12.6.

Because of the very rapid transformations, austenite → martensite (and even austenite → pearlite) which are involved, it is obvious that considerable practical difficulties arise during laboratory investigations of this type. Since it is impossible to change the temperature from 730°C to 500°C (in the example described above) *in zero time* and again from 500°C to 0°C *in zero time*, we must do the best we can by using very small specimens which are *thin* enough to reach quenching bath temperatures as quickly as possible. For this reason small specimens about the size of a 1p piece are appropriate (Fig. 12.8). These can be attached to suitable 'handles' to facilitate their very rapid transfer between baths. If the incubation bath is of molten metal this will also provide the maximum quenching rate on transfer from the austenitising bath.

12.42   The horizontal line (Fig. 12.6) representing the temperature of 723°C is, of course, the lower critical temperature above which the structure of the eutectoid steel in question consists entirely of stable austenite. Below this line austenite is unstable, and the two approximately C-shaped curves indicate the time necessary for the austenite → ferrite + cementite transformation to begin and to be completed following rapid quenching to any predetermined temperature. Transformation is sluggish at temperatures just below the lower critical, but the delay in starting, and the time required for completion, decrease as the temperature falls towards 550°C. In this range the greater the degree of undercooling, the greater is the urge for the austenite to transform, and the rate of transformation reaches a maximum at 550°C. At temperatures just below 723°C, where transformation takes place slowly, the structure formed will be coarse pearlite, since

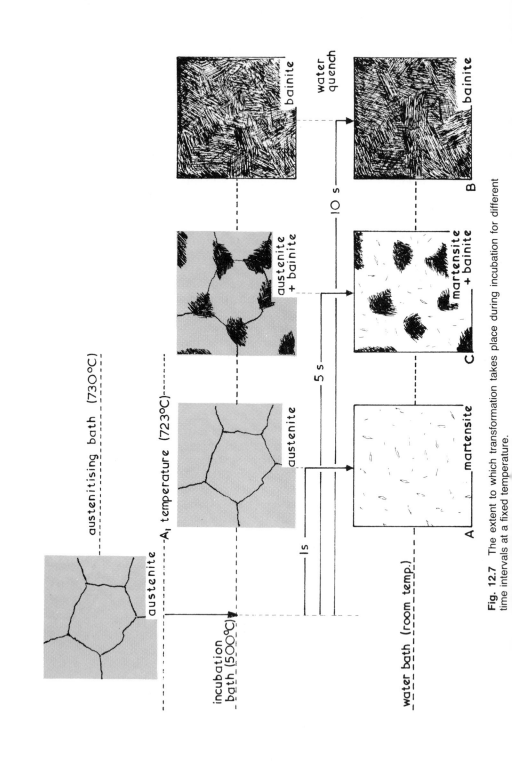

Fig. 12.7 The extent to which transformation takes place during incubation for different time intervals at a fixed temperature.

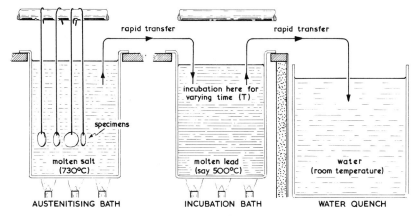

Fig. 12.8 The thermal treatment sequence used in the derivation of a set of TTT curves. The thin specimens used are about the diameter of a 1p coin.

there is plenty of time for diffusion to take place. In the region just above 550°C, however, rapid transformation results in the formation of very fine pearlite.

12.43 At temperatures between 550 and 220°C transformation becomes more sluggish as the temperature falls, for, although austenite becomes increasingly unstable, the slower rate of diffusion of carbon atoms in austenite at lower temperatures outstrips the increased urge of the austenite to transform. In this temperature range the transformation product is bainite. The appearance of this phase may vary between a feathery mass of fine cementite and ferrite for bainite formed around 450°C; and dark acicular (needle-shaped) crystals for bainite formed in the region of 250°C.

The horizontal lines at the foot of the diagram are, strictly speaking, not part of the TTT curves, but represent the temperatures at which the formation of martensite will begin (M_s) and end (M_f) during cooling of austenite through this range. It will be noted that the M_f line corresponds approximately to −50°C. Consequently if the steel is quenched in water at room temperature, some 'retained austenite' can be expected in the structure since at room temperature transformation is incomplete. This retained austenite, however, will amount to less than 5% of the austenite which was present at the M_s temperature. In fact, at 110°C (Fig. 12.6) 90% of the austenite will have transformed to martensite.

12.44 These TTT curves indicate structures which are produced by transformations which take place *isothermally*, that is, at a fixed single temperature and specify a given 'incubation' period which must elapse before transformation begins. There is no direct connection between such isothermal transformations and transformations which take place under continuous cooling at a constant rate from 723°C to room temperature. Thus it is not possible to superimpose curves which represent continuous cooling on to a TTT diagram. Modified TTT curves which are related to continuous rates of cooling can, however, be produced. These are similar in shape to the true TTT curves, but are displaced to the right, as shown in Fig. 12.9. On this diagram are superimposed four curves, *A*, *B*, *C* and *D*, which represent different rates of cooling.

Curve *A* represents a rate of cooling of approximately 5°C per second such as might be encountered during normalising. Here transformation will begin at *X* and can be completed at *Y*, the final structure being one of fine pearlite. Curve *B*, on the other hand, represents very rapid cooling at a rate of approximately 400°C per second. This is typical of conditions prevailing during a water-quench, and transformation will not begin until 220°C, when martensite begins to form. The structure will consist of 90% martensite at 110°C and so contain a little retained austenite at room temperature. The lowest rate at which this steel (of eutectoid composition) can be quenched, in order to obtain a structure which is almost wholly martensitic, is represented by curve *C* (140°C per second). This is called the critical cooling rate for the steel, and if a rate lower than this is used some fine pearlite will be formed. For example, in the case of the curve *D*, which represents a cooling rate of about 50°C per second, transformation would begin at *P* with the formation of some fine pearlite. Transformation, however, is interrupted in the region of *Q* and does not begin again

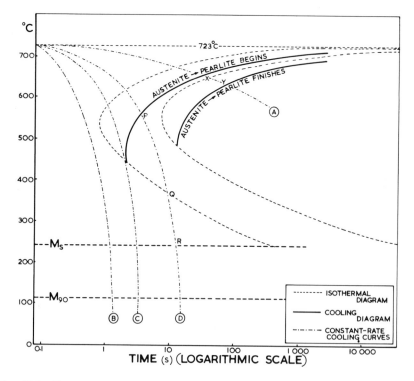

Fig. 12.9 The relationship between TTT curves and curves representing continuous cooling.

until the M_s line is reached at R, when the remaining austenite begins to transform to martensite. Thus the final structure at room temperature is a mixture of pearlite, martensite and traces of retained austenite.

12.45   The TTT curves illustrated in Fig. 12.6 are those for a steel of eutectoid composition. If the carbon content is either above or below this, the curves will be displaced to the left so that the critical cooling rate necessary to produce a completely martensitic structure will be greater. In order to obtain a structure which is entirely martensitic the steel must be cooled at such a rate that the curve representing its rate of cooling does not cut into the 'nose' of the modified 'transformation begins' curve in the region of 550°C. Obviously, if the steel remains in this temperature range for more than one second, then transformation to pearlite will begin. Hence the need for drastic water-quenches to produce wholly martensitic structures in plain carbon steels.

For a steel containing less than 0.3% carbon the transformation-begins curve has moved so far to the left (Fig. 12.10(i)) that it has become impossible to obtain a wholly martensitic structure however rapidly it is cooled. Large quantities of ferrite will inevitably precipitate when the transformation-begins curve is unavoidably cut in the upper temperature ranges. The resulting structure will be most unsatisfactory since hard martensite will be interspersed with soft ferrite.

Fortunately, the addition of alloying elements has the effect of slowing down transformation rates so that the TTT curves are displaced to the right of the diagram. This means that much slower rates of cooling can be used, in the form of oil- or even air-quenches, and a martensitic structure still obtained. Small amounts of elements, such as nickel, chromium and manganese, are effective in this way and this is one of the most important effects of alloying. In Fig. 12.10(ii) representing the TTT curves for a low alloy steel (covered by BS970:945M38) the 'nose' of the transformation-

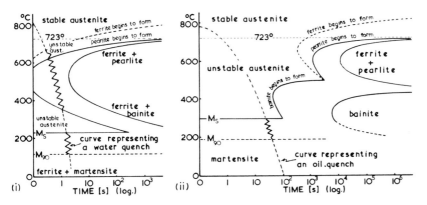

Fig. 12.10   (i) TTT curves for a 0.35% carbon steel, indicating that it is impossible to produce a wholly martensitic structure even by drastic water quenching. (ii) TTT curves for a low-alloy steel. Since the curves are displaced significantly to the right it is now possible to obtain a completely martensitic structure by oil quenching.

begins curve is displaced well to the right and in fact a 'double nose' is formed. Even when a continuous-cooling curve (representing an oil-quench) is superimposed on this isothermal diagram, it will be seen that there is no transformation until the M_s line is reached and the structures will be wholly martensitic. Since this diagram represents the TTT curves for a hypo-eutectoid steel, ferrite precipitation will begin before pearlite formation as indicated by the broken line.

12.46   We will now consider one or two practical applications arising from this study of modified isothermal transformation curves. Let us first examine the conditions under which a fairly large body of steel will cool, when quenched. The core will cool less quickly than the outside skin, and since its cooling curve B (Fig. 12.11A) cuts into the nose of the 'transformation-begins' curve, we can expect to find some fine pearlite in the core, whilst the surface layer is entirely martensitic. This feature is usually referred to as the 'mass effect of heat-treatment' (12.50). Even if we are able to cool the component quickly enough to obtain a completely martensitic structure, as indicated in Fig. 12.11B there will be such a considerable time interval CD between both core and surface reaching a martensitic condition that this will lead to quench-cracks being formed. These cracks will be caused by stresses set up as the volume change, associated with the austenite $\rightarrow$ martensite transformation, progresses from the skin to the core of the section. Supposing, however, we cool the steel under conditions of the kind indicated in Fig. 12.11c. Here the steel is quenched into a bath at temperature E and left there long enough to permit it to reach a uniform temperature throughout. It is then removed from the bath and allowed to cool so that martensite will begin to form at F. The net result is that, by allowing the core to attain the same temperature as the surface whilst in the bath at temperature E, we have prevented a big temperature gradient from being set up between the surface and the core of the specimen at the moment when martensite begins to form. The final air-cooling will not be rapid enough to allow a large temperature gradient to be set up, and both core and surface will become martensitic at approximately the same time, thus minimising the tendency towards quench-cracking. The success of this treatment, which is known either as *martempering* or *marquenching* lies in cooling the steel quickly enough past the 'nose' of the modified transformation-begins curve. Once safely past that point, relatively slow cooling will precipitate martensite. If we should cut into the 'nose' fine pearlite will begin to form. It is obvious that for plain carbon steels this type of treatment will have its limitations and will be applicable to components of *thin* section only, otherwise the temperature gradient set up within the section would be too great to prevent pearlite formation in the core.

With suitable steels an *ausforming* operation can be combined with martempering. Whilst the steel is at the holding temperature the austenitic structure is heavily worked—up to 90% sectional reduction in area being applied. The density of dislocations is increased by working and these 'jammed' dislocations are retained because the relatively low temperature does not permit recrystallisation. The steel is then allowed to cool and so transform to martensite which will *retain the high density of* dislocations,

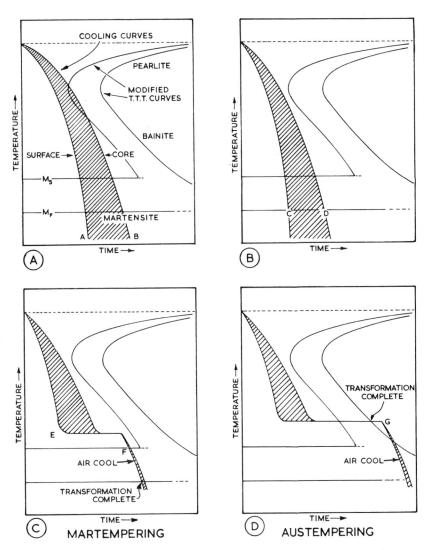

Fig. 12.11 (A) and (B) illustrate the effects of mass during normal quenching. (C) and (D) show how these effects may be largely overcome in martempering and austempering.

so that its strength is further increased. Moreover, the grain is fine due to the low forming temperature, whilst a fine dispersion of carbides, presumably deposited during forming, contributes some degree of particle hardening. Some low-alloy steels, having a deep 'bay' between the pearlite and bainite 'noses' of the TTT curve (Fig. 12.10(ii)) can be treated in this way and will develop tensile strengths in the region of 3000 N/mm^2.

12.47 Isothermal transformation offers a method by which we can obtain a tempered type of structure without the preliminary drastic water-quench. Such a treatment, known as *austempering*, is illustrated in Fig. 12.11D. Here the steel is quenched into a bath at a temperature above that at which martensite can be formed and allowed to remain there long enough for transformation to be complete at *G*. Since transformation to bainite is complete at *G*, the steel can be cooled to room temperature at any desired rate, but air-cooling is preferable. We have thus succeeded in obtaining a structure which is similar in properties, though not in microstructure, to that of tempered martensite which is obtained by quenching and tempering. The drastic water-quench from above the upper critical temperature, however, has been avoided. Austempering is therefore a process of considerable importance when heat-treating components of intricate section. Such components might distort or crack if they were heat-treated by the more conventional methods of quenching and tempering.

Although most of our modern knowledge of the basic mechanisms of isothermal transformation phenomena stem from the research carried out by Bain and Davenport from the late nineteen-thirties onwards, industrial processes based on what we now understand as isothermal transformation in steel, predated the work of Bain and his associates by many years. Such a process was *patenting*, a treatment employed in the manufacture of high-tensile steel wire. This wire, containing between 0.35 and 0.95% carbon, is first austenitised at temperatures up to 970°C (depending of course on the carbon content) by passing it through heated tubes in order to minimise decarburisation. From the austenitising furnace it passes into a bath of molten lead at 400–500°C where it remains for a period long enough for a structure of feathery bainite to develop. This structure is sufficiently ductile to permit cold-working of up to 90% sectional reduction in area, without annealing being necessary. Tensile strengths of over 2000 N/mm^2 are attainable as a result of the combination of heat-treatment and cold-work. Wire for piano, guitar and other musical instrument strings can be made in this way as well as high-tensile wire ropes.

The heat-treatment of spades, forks and other thin-section garden tools by a method akin to austempering also pre-dated Bain by many years. There are many such instances where industrial processes were developed by trial and error over many years, long before the underlying scientific explanation was forthcoming. Nevertheless modern austempering processes are finding increasing use. For example, even the steel toecaps of industrial boots are heat-treated in this manner.

12.48   Finally, mention must be made of *isothermal annealing*. In this process the steel is heated into the austenitic range and then allowed to transform as completely as possible in the pearlitic range. The object of such treatment is generally to soften the steel sufficiently for subsequent cold-forming or machining operations.

The nature of the pearlite formed during transformation is influenced by the initial austenitising temperature. An austenitising temperature which is little above the upper critical for the steel promotes the formation of spheroidal pearlitic cementite during isothermal annealing, whilst a higher austenitising temperature favours the formation of lamellar pearlitic cementite. The pearlite structure is also influenced by the temperature at which isothermal transformation takes place, as would be expected. Transformation just below the lower critical temperature leads to the formation of spheroidal pearlitic cementite since precipitation is slow, whilst at lower temperatures transformation rates are higher and lamellar cementite tends to form. A structure containing spheroidal cementite is generally preferred for lathe work and cold-forming operations, whilst one with lamellar cementite is often used where milling or drilling are involved. It is claimed that isothermal annealing gives more uniform properties than does an ordinary annealing process.

Hardenability and Ruling Section

12.50   Brief mention of the workshop term 'mass effect' in connection with the heat treatment of steel has already been made (12.46). If a piece of carbon steel is of heavy cross-section it will probably be impossible to cool it quickly enough to produce a uniformly martensitic structure throughout, even by the most severe quenching. Such a section would be likely to have a soft un-hardened core due to its relatively slow cooling rate (Fig. 12.11A), whilst a piece of steel of thin section quenched in a similar way would be martensitic throughout (Fig. 12.12). This difficulty may be remedied to some extent by adding alloying elements to the steel. These reduce the critical rates of austenite transformation and make it possible to get a martensitic structure throughout quite thick sections even when the less drastic oil- or air-quenching processes are used.

12.51   This is one of the most important functions of alloying, but to avoid the misuse of steels due to lack of understanding of their properties it became necessary for manufacturers to specify the maximum diameter or *ruling section* of a bar, up to which the stated mechanical properties would apply following heat-treatment. That is, some users failed to appreciate that these low-alloy steels still had a critical cooling rate even though it was much lower than that of a plain carbon steel of similar carbon

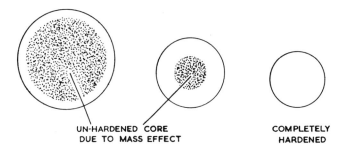

UN-HARDENED CORE
DUE TO MASS EFFECT

COMPLETELY
HARDENED

Fig. 12.12   The effects of mass produced in the structure on quenching. The heavier sections will cool too slowly to be entirely martensitic.

content. If the ruling section is exceeded then the properties across the section will not be uniform since hardening of the core will not be complete.

12.52   The Jominy end-quench test is of great practical use in determining the hardenability of steel. Here a standard test piece is made (Fig. 12.13A) and heated up to its austenitic state. It is then dropped into position in a frame, as shown in Fig. 12.13B, and quenched at its end only, by means of a pre-set standard jet of water at 25°C. Thus different rates of cooling are obtained along the length of the bar. After the cooling, a 'flat', 0.4 mm deep, is ground along the side of the bar and hardness determinations made every millimetre along the length from the quenched end. The results are then plotted as in Fig. 12.14.

These curves show that the depth of hardening of a nickel–chromium steel is greater than that of a plain carbon steel of similar carbon content, whilst the depth of hardening of a chromium–molybdenum steel is greater than that of the nickel–chromium steel.

With modifications, the results of the Jominy test can be used as a basis in estimating the 'ruling section' of a particular steel. There is no simple mathematical relationship between the two, however, and it is often more satisfactory to find by trial and error how a particular section will harden, after a preliminary Jominy test has been conducted.

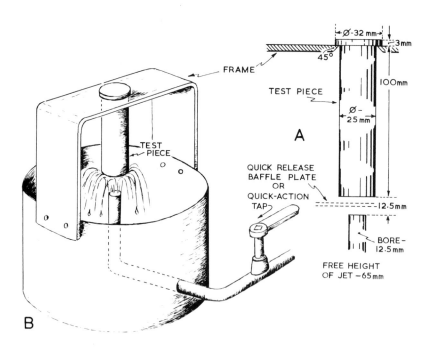

Fig. 12.13   The Jominy end-quench test.
(A) The standard form of test piece used. (B) A simple type of apparatus for use in the test.

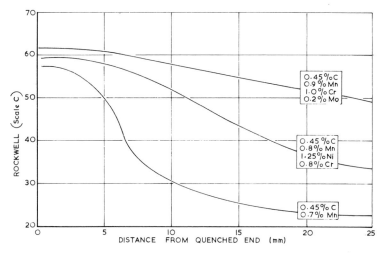

Fig. 12.14 The relative depth of hardening of three different steels as indicated by the Jominy test.

British Standard Specifications for Carbon Steels

12.60 British Standard Specifications for wrought steels containing carbon and manganese (including some free-cutting steels) are dealt with in Part 1 of BS 970. The original 'En' designation which was used to identify individual steels within BS 970 was abandoned as long ago as 1972 but 'old habits die hard' and one still finds these En numbers being quoted. Some claim that the new numbers are 'too cumbersome' but in fact they are very useful in that the six digit designation used indicates both carbon and manganese contents as well as other important information relating to supply of the steel.

The series 000 to 199 has been allocated for the first three digits in respect of steels containing carbon and manganese. These digits represent one hundred times the mean manganese content. The letter 'A' or 'M' has been introduced as the fourth digit to indicate whether the steel is to be supplied to analysis (A) or mechanical property (M) requirements. When hardenability requirements are included in the specification the letter 'H' is used as the fourth digit. Finally the fifth and sixth digits represent one hundred times the mean carbon content. (Since only one steel in this Part of the standard contains 1.00% carbon, the two digits '99' are used to describe it so as to avoid the use of a three-digit suffix throughout the standard.)

As an example BS 970:060A52 describes a steel containing 0.50–0.55% carbon; 0.50–0.70% manganese supplied according to analysis (A) requirements.

The free-cutting carbon-manganese steels (6.63) are prefixed by the

Table 12.2 Representative Compositions and Properties for Some Steels Covered in BS970: Part I

| Type | Composition (%) | | | Heat-treatment | Typical Mechanical properties | | | | | |
	BS970:	C	Mn		Ruling Section (mm)	Tensile strength (N/mm²)	Yield strength (N/mm²)	Elongation (%)	Izod (J)	Hardness (Brinell)
Low carbon	015A03	0.03	0.15	Not heat-treatable—used where ductility and formability are required	—	—	—	—	—	—
	050A12	0.12	0.50		—	—	—	—	—	—
	060A17	0.17	0.60		—	—	—	—	—	—
'20' carbon	070M20	0.20	0.70	Normalised at 880–910°C. Water-quenched 880–910°C; tempered 550–660°C.	150	430	215	21	—	150
					20	620	355	20	41	180
'30' carbon	080M30	0.30	0.80	Normalised at 860–890°C. Water quenched 860–890°C; tempered 550–660°C.	150	495	250	20	—	165
					20	695	415	16	34	205
'40' carbon	080M40	0.40	0.80	Normalised at 830–860°C. Water quenched 830–860°C; tempered 550–660°C.	150	540	280	16	20	180
					20	770	465	16	34	230
'50' carbon	080M50	0.50	0.80	Normalised at 810–840°C. Oil quenched 810–840°C; tempered 550–660°C.	150	620	310	14	—	205
					12.7	925	570	12	—	275
'55' carbon	070M55	0.55	0.70	Normalised at 810–840°C. Oil quenched 810–840°C; tempered 550–660°C.	63.5	695	355	12	—	230
					20	930	575	12	—	275
High carbon	080A62	0.62	0.80	Heat-treated to give required combination of hardness and toughness—generally used as tool steels			—	—	—	—
	080A72	0.72	0.80				—	—	—	—
	080A83	0.83	0.80				—	—	—	—
	060A86	0.86	0.60				—	—	—	—
	060A99	1.00	0.60				—	—	—	—

(Medium carbon—mainly constructional steels.)

(Compositions for some of the carbon-manganese free-cutting steels will be found in Table 6.2) For typical uses of steels in the above Table refer to Table 7.1 and Fig. 7.7.

series 200 to 240 where the second and third digits are roughly one hundred times the sulphur content. Thus BS 970:216M28 describes a free-cutting steel containing 0.24–0.32% carbon and 0.12–0.20% sulphur supplied according to mechanical property (M) requirements.

A few representative carbon steels along with their BS 970 designations, mean compositions, heat-treatments and typical mechanical properties are shown in Table 12.2.

Exercises

1. Examine Table 12.2 and explain why the following steels (covered in BS 970): 070M20; 080M30; 080M40 and 080M50 have progressively lower normalising and quenching temperatures. (12.20)
2. Outline, using sketches, the theory which seeks to explain the development of the martensite structure when a carbon steel is water quenched. (12.22)
3. An annealed 0.4% C steel bar is cold-worked and placed with one end in a furnace at 900°C whilst the other end is maintained at room temperature. After a few hours the bar is quenched in cold water. Describe the structures you would expect to find along the length of the bar. (11.51 and 12.21)
4. Sketch and label the 'steel part' of the iron–carbon thermal equilibrium diagram. With reference to the diagram describe the structural changes which occur when a cast 0.5%C steel is:
   (i) slowly heated to 900°C;
   (ii) slowly cooled from 900°C;
   (iii) quenched from 900°C;
   (iv) quenched from 900°C and reheated to 650°C.
   Sketch each microstructure, including that of the steel in the cast condition, and comment *qualitatively* on the mechanical properties you would expect as a result of each treatment. (11.53, 12.21 and 12.32)
5. Both annealing and tempering are processes used to soften steel. Outline the conditions when these treatments would be used, and indicate any difficulties that may be encountered in practice. (11.50 and 12.30)
6. Four thin pieces of the same 0.8% C rolled-steel rod are heat-treated differently as follows:
   (i) heated to 680°C for twenty-four hours and cooled in still air;
   (ii) water-quenched from 750°C;
   (iii) heated to 730°C, quenched into molten lead at 400°C, allowed to remain there for five minutes and then cooled;
   (iv) heated to 1200°C and cooled in still air.
   Sketch the type of microstructure produced in each specimen and explain the mechanism of its formation. (11.52; 12.20; 12.46 and 11.21)
7. Using diagrammatic TTT curves, explain the reasons for the addition of alloying elements to steels to overcome the limitations of carbon steels in heat-treatment. (12.50)
8. Explain fully what is meant by the 'ruling section' of a steel and discuss its significance in the choice of steels for engineering design.
   Outline *one* experimental procedure which is helpful in assessing the ruling section of a steel. (12.50 and 12.52)
9. The following results were obtained during Jominy end-quench tests carried out under similar conditions on two steels of similar carbon content;

Dist. from quenched end (mm)	2	5	7	10	15	20	25	30	40	50
Hardness (H_V), steel. 1.	604	512	321	290	254	250	246	241	236	235
Hardness (H_V), steel. 2.	605	590	585	580	577	556	549	535	500	437

Draw, on squared paper, the Hardness/Distance-from-quenched-end curve for each steel and comment on them.

What would you deduce regarding the possible compositions of each steel? (12.52)

Bibliography

Brooks, C. R., *Heat-treatment of Ferrous Alloys*, Hemisphere Publishing Corp., 1979

Higgins, R. A., *Engineering Metallurgy (Part II)*, Edward Arnold, 1986.

Honeycombe, R. W. K., *Steels: Microstructure and Properties*, Edward Arnold, 1981

Hume-Rothery, W., *The Structures of Alloys of Iron*, Pergamon, 1966.

Petty, E. R., *Martensite: Fundamentals and Technology*, Longmans, 1970.

Pickering, F. R., *Physical Metallurgy and Design of Steels*, Applied Science, 1978.

Roberts, G. A. and Carey, R. A., *Tool Steels*, American Society for Metals, 1980.

Thelning, K-E., *Steel and its Heat-treatment: Bofors Handbook*, Butterworths, 1984

United States Steel, *Isothermal Transformation Diagrams*, 1963.

Wilson, R., *Metallurgy and Heat-treatment of Tool Steels*, McGraw-Hill, 1975.

BS 970: 1983 *Wrought Steels in the Form of Blooms, Billets, Bars and Forgings*.

BS 4659: 1971 *Tool Steels*.

BS 4437: 1987 *Method for End Quench Hardenability Test for Steel (Jominy Test)*.

13

Alloy Steels

13.10 The earliest recorded attempt to produce an alloy steel was made in 1822 at the instigation of Michael Faraday in his searches for better cutting tools and non-corrodible metals for reflectors. To develop the latter he attempted to alloy iron with a number of rare elements including silver, gold, platinum and rhodium, and, incidentally chromium; but it was not until ninety years later that in 1922 Brearley discovered the stainless properties of high-chromium steel.

The first *successful* attempt at producing an alloy steel was the result of the researches of the British metallurgist Robert Mushet, who had made Henry Bessemer's process a viable proposition by introducing deoxidation with manganese. Mushet's 'self-hardening' tungsten–manganese steel produced in 1868 was indeed the forerunner of modern high-speed steel.

Systematic research into the properties of alloy steels dates from Sir Robert Hadfield's discovery of high-manganese steel in 1882. Some time afterward Dr. J. E. Stead said of this steel: 'Hadfield had surprised the whole metallurgical world with the results obtained. The material produced was one of the most marvellous ever brought before the public.' Large quantities of this alloy are produced to-day by the Sheffield firm which still carries the inventor's name. Indeed the prowess of the Sheffield steelmakers of the nineteenth century confirms the maxim that 'the hand which wields the ladle rules the World'.

So-called plain-carbon steels contain up to 1.0% manganese, residual from deoxidation and desulphurisation processes, but it was generally accepted that an *alloy* steel was one containing more than 0.1% molybdenum or vanadium; or 0.3% tungsten, cobalt or chromium; or 0.4% nickel; or 2.0% manganese. The modern generation of *micro-alloyed* steels (13.140) has rendered this definition obsolete—in so far as one was necessary—since additions of as little as 0.0005% of some elements are now made to influence properties effectively.

The principal objectives in adding alloying elements to steel are:

(i) to improve and extend the existing properties of plain carbon steels;
(ii) to introduce new properties not available in plain carbon steels.

Thus the addition of small quantities of nickel and chromium will produce a general improvement in the basic mechanical properties of strength and toughness, whilst larger amounts of these elements will introduce new phenomena such as the stabilisation of austenite at ambient temperatures, accompanied by the loss of ferromagnetism and, of course, a very high resistance to corrosion. The alloying elements added may either simply dissolve in the ferrite or they may combine with some of the carbon, forming carbides, which associate with the iron carbide already present.

The decision to use alloy steels will not be taken lightly since they are expensive materials as compared with plain carbon steels. This is to be expected when it is realised that the unit costs of metals like nickel and chromium are many times that of ordinary medium carbon steel. Nevertheless this extra cost may often be partly offset by the greater ease with which most alloy steels can be heat-treated, making possible automatic programming of the heat-treatment cycle with the use of relatively unskilled labour.

The principal effects which alloying elements have on the microstructure and properties of a steel can be classified as follows:

13.11 The Effect on the Polymorphic* Transformation Temperatures
The polymorphic transformation temperatures which concern us here are those at 910°C where the $\alpha \rightleftharpoons \gamma$ transformation occurs; and at 1400°C where the $\gamma \rightleftharpoons \delta$ change takes place. That is, when BCC (α) iron is heated above 910°C it transforms to FCC (γ) iron and if heated further to 1400°C it changes again to BCC (δ) iron. These transformations are reversible on cooling. The temperatures 910°C and 1400°C are designated A_3 and A_4 respectively (Fig. 13.1). (The A_1 temperature is at 723°C—the 'lower critical temperature'—where the austenite $\rightleftharpoons$ pearlite transformation occurs in plain-carbon steels; whilst the A_2 temperature is at 769°C, the Curie point, above which pure iron ceases to be ferromagnetic. The Curie point has no metallographic significance.)

Some elements, notably nickel, manganese, cobalt and copper, raise the A_4 temperature and lower A_3 as shown in Fig. 13.1A. Therefore these elements, when added to a carbon steel tend to stabilise austenite (γ) still further and increase the range of temperature over which austenite can exist as a stable phase. Other elements, the most important of which include chromium, tungsten, vanadium, molybdenum, aluminium and silicon, have the reverse effect, in that they tend to stabilise ferrite (α) by raising the A_3 temperature and lowering the A_4, as indicated in Fig. 13.1B. Such elements restrict the field over which austenite may exist, and thus form what is commonly called a 'gamma (γ) loop'.

Many of the elements of the austenite-stabilising group have a FCC crystal structure like that of austenite. They therefore dissolve substitutionally with ease in austenite and consequently resist and retard the transformation of austenite to ferrite. Carbon itself has the same effect on the $\gamma \rightarrow \alpha$ transformation in iron, as indicated in the iron–carbon diagram,

---

* The term 'allotropic' is often used to describe transformations of this type. However, 'polymorphic' is more precise since we refer to changes in *crystal* structures only, in this instance.

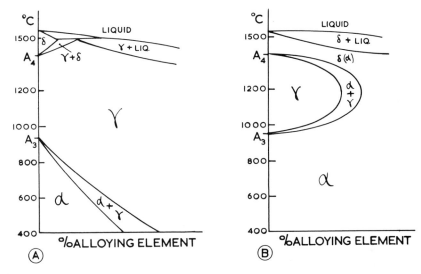

Fig. 13.1   Relative effects of the addition of an alloying element on the polymorphic transformation temperatures at A_3 and A_4
(A) Tending to stablise γ, and (B) tending to stabilise α.

because it dissolves interstitially in FCC iron but not significantly in BCC iron. This group of elements retards the precipitation of carbides and this also has the effect of stabilising austenite over a wider range of temperature.

The ferrite-stabilising elements are principally those which, like α-iron have a BCC crystal structure. They will therefore dissolve substitutionally more readily in α-iron than in γ-iron, thus stabilising ferrite (α) over a wider temperature range. As shown in Fig. 13.1B, progressive increase in one or more of the stabilising elements will cause a point to be reached, beyond the confines of the γ-loop, where the γ-phase cannot exist at any temperature. Thus the addition of more than 30% chromium to a steel containing 0.4% carbon would lead to the complete suppression of the polymorphic transformations, and such a steel would no longer be amenable to normal heat-treatment (Fig. 13.10).

13.12 The Effect on the Formation and Stability of Carbides   Some alloying elements form very stable carbides when added to steel (Fig. 13.2). This generally has a hardening effect on the steel, particularly when the carbides formed are harder than iron carbide itself. Such elements include chromium, tungsten, vanadium, molybdenum, titanium and niobium, often forming interstitial compounds (8.31). Some of these elements form separate hard carbides in the microstructure, eg Cr_7C_3, W_2C, Mo_2C and VC, whilst double or complex carbides containing iron and one or more other metals are also formed. In highly-alloyed tool steels these single and complex carbides are often of the general formulae M_6C; $M_{23}C_6$ and MC

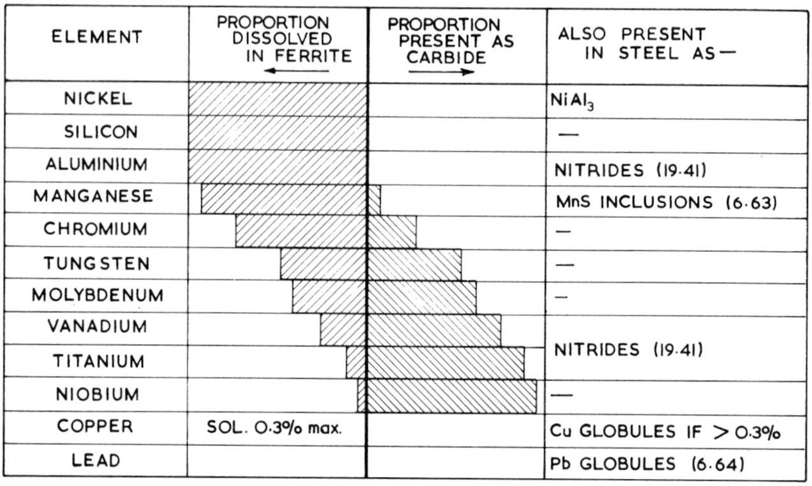

ELEMENT	PROPORTION DISSOLVED IN FERRITE	PROPORTION PRESENT AS CARBIDE	ALSO PRESENT IN STEEL AS—
NICKEL			$NiAl_3$
SILICON			—
ALUMINIUM			NITRIDES (19·41)
MANGANESE			MnS INCLUSIONS (6·63)
CHROMIUM			—
TUNGSTEN			—
MOLYBDENUM			—
VANADIUM			
TITANIUM			NITRIDES (19·41)
NIOBIUM			—
COPPER	SOL. 0·3% max.		Cu GLOBULES IF > 0·3%
LEAD			Pb GLOBULES (6·64)

Fig. 13.2 The physical states in which the principle alloying elements exist when in steel.

where 'M' represents the *total* metal atoms. Thus M_6C is represented by Fe_4W_2C and Fe_4Mo_2C, whilst MC is represented by WC and VC.

Other elements, whilst not being particularly strong carbide formers, nevertheless contribute towards the stability of carbides generally. Manganese is such an element, its weak carbide-forming tendency being indicated by its position in Fig. 13.2. However, its general effect is to increase the stability of other carbides present.

Yet another group of elements, notably nickel, cobalt, silicon and aluminium, have little or no chemical affinity for carbon and in fact have a graphitising effect on the iron carbide; that is, they tend to make it unstable so that it breaks up, releasing free graphitic carbon. Therefore, if it is necessary to add appreciable amounts of these elements to a steel it can be done only when the carbon content is very low. Alternatively, if a carbon content in the intermediate range is necessary, one or more of the elements of the first group, namely the carbide stabilisers, must be added to counteract the effects of the graphitising element. As an example very few steels containing nickel as the sole alloying element are available—most low-alloy steels contain both nickel and chromium.

13.13 The Effect on Grain Growth   The rate of crystal growth is accelerated, particularly at high temperatures, by the presence of some elements, notably chromium. Care must therefore be taken that steels containing elements in this category are not overheated or, indeed, kept for too long at a high temperature, or brittleness, which is usually associated with coarse grain, may be increased.

Fortunately, grain growth is retarded by other elements notably vanadium, titanium, niobium, aluminium and to a small extent, nickel. Steels containing these elements are less sensitive to the temperature conditions of heat-treatment. Vanadium is possibly the most potent grain-refining element. As little as 0.1% will inhibit grain-growth by forming finely-dispersed carbides and nitrides which, being relatively insoluble at high temperatures, act as barriers to grain-growth. Titanium and niobium behave in a similar manner, whilst in high-alloy tool steels the carbides of tungsten and molybdenum reduce grain growth at the necessarily high heat-treatment temperatures.

High-grade steels are initially deoxidised, or 'killed', with ferromanganese but receive a final deoxidation, before being cast, with controlled quantities of aluminium. The final product contains sufficient 'trace' aluminium to make it inherently fine-grained.

13.14 The Displacement of the Eutectoid Point   The addition of any alloying element to carbon steel diminishes the solubility of carbon in austenite and so results in a displacement of the eutectoid point towards the left of the equilibrium diagram. That is, an alloy steel will be completely pearlitic even though it contains less than 0.8% carbon (Fig. 13.3). This explains why low-alloy steels contain less carbon than plain carbon steels of similar characteristics and uses.

At the same time the A_1 (or eutectoid) temperature is altered by alloying. The ferrite-stabilisers (chromium, tungsten, molybdenum, titanium, etc.) raise the eutectoid temperature in the same way that they raise A_3; whilst the austenite-stabilisers (nickel and manganese) lower the eutectoid temperature (Fig. 13.4).

Thus the addition of 2.5% manganese to a steel containing 0.65% carbon will give it a completely pearlitic structure in the normalised condition, along with a reduction in the eutectoid temperature to about 700°C (Fig. 13.5). Similarly, although a high-speed steel may contain only 0.7% carbon, its microstructure exhibits masses of free carbide due to the dis-

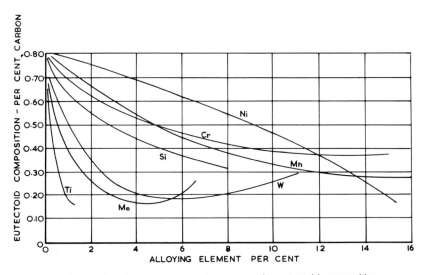

Fig. 13.3   The effect of alloying elements on the eutectoid composition.

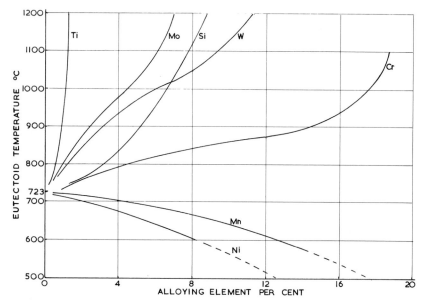

Fig. 13.4   The effect of alloying elements on the eutectoid temperature.

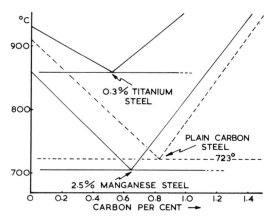

Fig. 13.5   The effects of manganese and titanium on the displacement of the eutectoid point in steel.

placement of the eutectoid point far to the left by the effects of the alloying elements (totalling about 25%) which are present. At the same time the eutectoid temperature in high-speed steel is raised to about 850°C.

13.15 The Retardation of the Transformation Rates   We have already seen that the TTT curves for a plain-carbon steel are displaced to the right due to the effects produced by the addition of alloying elements (12.45). Thus the addition of alloying elements renders the austenite → pearlite transformation increasingly sluggish at temperatures between

500°C and 700°C and so reduces the critical cooling rate necessary to obtain martensite. This feature of alloying in steels has obvious advantages and all alloying elements, with the exception of cobalt, will reduce transformation rates.

In order to obtain a completely martensitic structure in the case of a plain 0.8% carbon steel, we must cool it from above 723°C to room temperature in approximately one second. This treatment involves a very drastic quench, generally leading to distortion or cracking of the component. By adding small amounts of suitable alloying elements such as nickel and chromium, we reduce this critical cooling rate to such an extent that a less drastic oil-quench is rapid enough to produce a fully martensitic structure. Further increases in the amounts of alloying elements will so reduce the rate of transformation that such a steel can be hardened by cooling in air. 'Air-hardening' steels have the particular advantage that comparatively little distortion is produced during hardening. Alternatively, such a steel, containing 4¼% nickel and 1¼% chromium and which will air-harden in thin sections, can be hardened completely through sections up to 0.15 m diameter by oil-quenching. This aspect of alloying is one of the greatest value since it makes possible uniform hardening of mass-produced components by conveyor-belt methods operated by unskilled labour.

A minor disadvantage is that all alloying elements, except cobalt, also lower the M_s and M_f temperatures to the extent that for many alloy steels, M_f is well below ambient temperatures resulting in the retention of some austenite in the quenched structure.

13.16 The Improvement in Corrosion-resistance The corrosion resistance of steels is substantially improved by the addition of elements such as aluminium, silicon and chromium. These elements form thin but dense and adherent oxide films which protect the surface of the steel from further attack. Of the elements mentioned, chromium is the most useful when mechanical properties have to be considered. When nickel also is added in sufficient quantities, the austenitic structure is maintained at room temperature, so that, provided the carbon content is kept low, a completely solid-solution type of structure prevails. This also helps in maintaining a high corrosion-resistance limiting the possibility of electrolytic attack (21.30).

13.17 Effects on the Mechanical Properties One of the main reasons for alloying is to effect improvements in the mechanical properties of a steel. These improvements are generally the results of physical changes already referred to. For example, hardness is increased by elements which stabilise the carbides; strength is increased by all alloying elements since they dissolve in ferrite; and toughness is improved by elements which refine the grain. Many formulae have been devised, by the use of which the approximate tensile strength of an alloy steel may be estimated from its known composition. For example, Walters takes a basic tensile strength for pure iron of 250 N/mm² and multiplies this by factors for each alloying element present. The factor for each element changes as its quantity changes. Such factors for most of the principal alloying elements are shown in Fig. 13.6 and are true for pearlitic steels in the normalised condition.

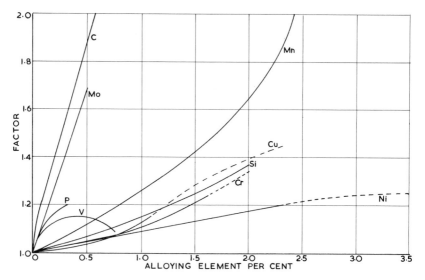

Fig. 13.6 Walters' factors for estimating the tensile strength of pearlitic steels in the normalised condition.

They give best results for steels with carbon contents below 0.25% and within the intermediate alloy range.

13.18 The Combined Effects of Alloying on the Microstructure   In the simplest alloy steel three components are present, ie iron, carbon and the alloying element. Consequently this system can only be represented completely as a ternary equilibrium diagram (9.100). The use of these three-dimensional diagrams is inconvenient to say the least and it is generally more useful to fix one of the variables in the system. Thus we can take a 'vertical' section from a ternary diagram at some pre-determined quantity of the alloying element and in this way produce a 'pseudo-binary' diagram in which % carbon and temperature are the variables (Fig. 14.1). Conversely we could take a 'horizontal' section from the ternary diagram at some temperature and thus be in a position to read off the microstructural effects of varying % carbon and % alloying element independently at that fixed temperature.

However, we have seen that in addition to microstructural changes associated with stabilising either austenite or ferrite and also alterations in the position of the eutectoid point, alloying affects the rate at which the austenite → pearlite transformation occurs. We are generally more interested in the structure which will be produced either by normalising or by quenching, rather than the structure formed by slow cooling under equilibrium conditions. Consequently attempts have been made from time to time to produce simple diagrams which will indicate the type of structure likely to be obtained, given the composition of the steel and some specific cooling rate to ambient temperature.

Many years ago Guillet produced some simple diagrams (Figs. 13.8B and

13.10b) which showed the general relationship between composition and microstructure for alloy steels cooled slowly to ambient temperatures. These diagrams were later developed to give rather fuller information of a practical nature (Fig. 13.12). Guillet diagrams allowed for the effects of only one alloying element to be considered and modern alloy steels contain as many as three or more elements in addition to iron and carbon. To represent the structures of such steels graphically the most satisfactory method is to place the alloying elements into two groups according to whether they stabilise austenite or ferrite and to ascribe factors to each element according to the potency of its stabilising effect. Some years ago Schaeffler diagrams were introduced to assist in assessing the microstructural effects of welding on the region around the weld in stainless steels. These diagrams have been modified by a number of workers to include steels containing elements in addition to nickel and chromium (Fig. 13.7). It must be appreciated that these diagrams deal specifically with microstructures which are likely to be formed due to heating and cooling occurring in the region of a weld. Nevertheless the use of such a diagram can give an indication of how the structure of a steel is likely to be affected by similar thermal treatment.

Example 1. A steel to BS 970:835M30 contains 4.25% Ni; 1.25% Cr; 0.25% Mo; 0.45% Mn; 0.25% Si and 0.3% C.

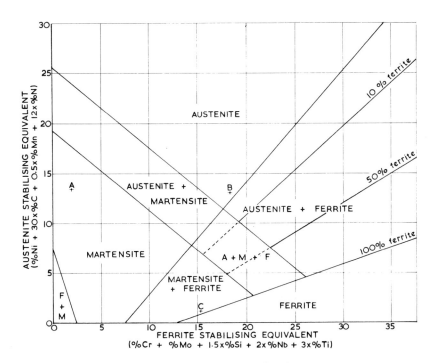

Fig. 13.7 A modified Schaeffler diagram.

Austenite stabilising
equivalent (from Fig. 13.7) = 4.25(Ni) + (0.45 × 0.5(Mn)) + (0.3 ×
$\qquad$ 30(C))
$\qquad$ = 4.25 + 0.225 + 9.0
$\qquad$ = 13.475

Ferrite stabilising
$\qquad$ equivalent = 1.25(Cr) + 0.25(Mo) + (0.25 × 1.5(Si))
$\qquad$ = 1.25 + 0.25 + 0.375
$\qquad$ = 1.875

These coordinates are represented by point A in Fig. 13.7 indicating that
the steel will have a martensitic structure after air-cooling from its austenitic condition. It is in fact an air-hardening steel.

Example 2. A stainless steel (304S15) contains 17.5% Cr; 11.0% Ni;
1.2% Mn; 0.6% Si and 0.05% C.

Austenite stabilising equiv. = 11.0(Ni) + (1.2 × 0.5(Mn)) + (0.05 ×
$\qquad$ 30(C))
$\qquad$ = 11.0 + 0.6 + 1.5
$\qquad$ = 13.1

Ferrite stabilising equiv. $\qquad$ = 17.5(Cr) + (0.6 × 1.5(Si))
$\qquad$ = 17.5 + 0.9
$\qquad$ = 18.4

which is represented by B in Fig. 13.7 showing that the steel will have an
austenitic structure following air-cooling to ambient temperatures.

Example 3. A stainless iron (403S17) contains 14.0% Cr; 0.03% C; 0.8%
Si and 0.5% Mn.

Austenite stabilising equiv. = (0.03 x 30(C)) + (0.5 × 0.5(Mn))
$\qquad$ = 1.15

Ferrite stabilising equiv. $\qquad$ = 14.0(Cr) + (0.8 × 1.5(Si))
$\qquad$ = 15.2

These coordinates are represented by point C suggesting that the structure will be mainly ferritic but that it may also contain a little martensite
when air-cooled from a high temperature.

Nickel Steels

13.20   Nickel is extensively used in alloy steels for engineering purposes,
generally in quantities up to about 5.0%. When so used, its purpose is to
increase strength and toughness. It is also used in larger amounts in stainless steels (13.44) and in maraging steels (13.101). Nickel is mined extensively in CIS; the Pacific island of New Caledonia; Australia; Cuba and
the Philippines, but by far the major producing country is Canada from
whence Britain imports her supplies.

13.21   The addition of nickel to a plain-carbon steel tends to stabilise
the austenite phase over an increasing temperature range, by raising the
A_4 temperature and depressing the A_3. Thus, the addition of 25% nickel
to pure iron renders it austenitic, and so non-magnetic, even after slow

cooling to ambient temperatures. The structure obtained as a result of this treatment can be estimated by reference to the simple Guillet diagram (Fig. 13.8B).

13.22   Nickel does not form carbides and in fact its presence in steel makes iron carbide less stable so that it tends to have a graphitising influence. For this reason nickel is never added to a high-carbon steel unless it is accompanied by elements which have a strong carbide-stabilising influence. Alloy steels containing nickel as the sole alloying element are low-carbon, low-nickel steels. If a higher carbon content is desired, then the manganese content is usually increased, since manganese is a carbide stabiliser.

Nickel increases the tensile strength and toughness of ferrite in low alloy steels by simple substitutional solid solution. It also has a grain-refining effect which makes the low-nickel, low-carbon steels very suitable for case-hardening (19.31), since grain growth will be limited during the prolonged period of heating in the region of 900°C.

13.23   In high-nickel steels containing small amounts of carbon, nickel introduces considerable thermal hysteresis in the polymorphic transformation. Thus a steel containing 15% nickel will not begin to transform on cooling until a temperature of about 250°C has been reached, when martensite begins to form. On re-heating the structure, however, martensite does not begin to change to austenite until a temperature approaching 600°C. This hysteresis, or lag, in transformation is fundamental to the use of nickel in maraging steels (13.101).

13.24   The popularity of steels containing nickel as sole alloying element has decreased in recent years mainly because of the graphitising effect which nickel may have on cementite. Those which are still available normally carry a manganese content of more than 1.0% to provide carbide stabilisation. 3% and 5% nickel steels are still available for case-hardening

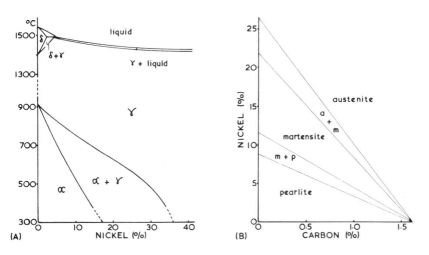

Fig. 13.8   The effect of nickel as an alloying element in steel. (A) Its influence on the stability of γ. (B) A Guillet-type diagram.

Table 13.1 *Nickel Steels and Alloys*

Type of steel	BS 970 designation	Composition (%)			Condition	Typical mechanical properties						Heat treatment	Uses
		C	Mn	Ni		Ruling Section: (mm)	Yield point (N/mm²)	Tensile strength (N/mm²)	Elong-ation (%)	Redn. in area (%)	Izod (J)		
1% nickel	503M40	0.4	1.50	1.0	Quenched & tempered at 600°C	—	494	695	25	55	91	Oil-quench from 850°C; temper between 550 and 660°C, and cool in oil or air.	Crankshafts, axles, connecting rods, other parts in the automobile industry, general engineering purposes.
1% nickel cast steel	—	0.15	1.2	1.1	—	—	—	—	—	—	—	—	Offshore oil rigs—as an alternative to fabricated weldments (superior in fatigue).
3% nickel case-hardening	—	0.12	0.45	3.0	Carburised, refined and water-quenched	38	510	772	20	60	83	After carburising, refine by oil-quenching from 860°C. Then harden by water-quenching from 770°C.	Case-hardening; crown wheels, gudgeon pins, differential pinions, camshafts.
5% nickel case-hardening	—	0.12	0.40	5.0	Carburised, refined and oil-quenched	28.5	602	849	22	47	68	After carburising, refine by oil-quenching from 850°C. Then harden by oil quenching from 760°C.	Heavy-duty case-hardened parts; bevel pinions, gudgeon-pins, gear-box gears, worm shafts.
Thermal expansion alloy	—	0.05	—	36.0	—	—	—	—	—	—	—	Non-hardenable (except by cold-work)	Constant-coeficient, low-expansion nickel/iron alloy used for temperature-control equipment (thermostats, etc.).
Thermal expansion alloy	—	0.05	—	48.0	—	—	—	—	—	—	—	Non-hardenable (except by cold-work).	Higher-temperature thermostats and glass/metal sealing applications.

but these have been largely replaced by nickel–chromium; nickel–molybdenum or nickel–chromium–molybdenum steels.

13.25 Nickel reduces the coefficient of thermal expansion of iron–nickel alloys progressively, until with 36% nickel expansion is extremely small. 'Invar' (36Ni; C–less than 0.1) was developed originally for accurate measuring tapes used in land surveys. In 1920 C. E. Guillaume received the Nobel Prize for Physics in recognition for its invention. It was also used for pendulum rods in master clocks and similar alloys are now employed for such diverse applications as the lining of tanks in vessels carrying 'liquid natural gas' and in the many types of thermostat in modern heating equipment.

Another valuable property of the high-nickel alloys is high magnetic permeability (14.30). An alloy containing 51 Ni–49Fe has a similar coefficient of expansion to that of glass making it useful in the production of 'reed switches'. In this simple device (Fig. 13.9) two 'reeds' are aligned and then sealed in a closed glass envelope containing an inert gas. To close the switch a magnetic field is applied from outside the tube, the reeds operating as simple cantilever springs. The switch is used extensively in semi-electronic telephone exchanges.

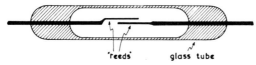

Fig. 13.9 A miniature 'reed' switch using a 51 Ni–49 Fe alloy.

Chromium Steels

13.30 The bulk of metallic chromium produced is used in the manufacture of alloy steels and in the electro-plating industry. The main producers of chromium are South Africa, CIS, Albania, Turkey, Zimbabwe, the Philippines and India. Britain's chief imports of the metal are from South Africa and the Philippines.

13.31 It is often assumed that the addition of chromium to a steel will automatically increase its hardness, but this can only take place when sufficient carbon is present. The increase in hardness is due mainly to the fact that chromium is a carbide stabiliser and forms the hard carbides Cr_7C_3 or $Cr_{23}C_6$ or, alternatively, double carbides with iron. All of these carbides are harder than ordinary cementite. In low-carbon steels the addition of chromium increases strength, with some loss in ductility, due to its forming a solid solution in ferrite.

13.32 Chromium lowers the A_4 temperature and raises the A_3 temperature, forming the closed γ-loop already mentioned. In this way it stabilises the α-phase at the expense of the γ-phase. The latter is eliminated entirely, as shown in Fig. 13.10 if more than 11% chromium is added to pure iron, though with carbon steels a greater amount of chromium would be

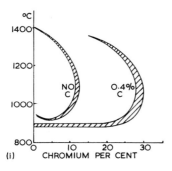 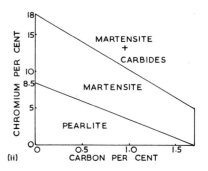

Fig. 13.10 The effects of chromium as an alloying element. (i) The effect of carbon and chromium additions on the 'γ-loop' in chromium steels—the γ field lies inside the relevant 'loop'; α + γ in the hatched band and α outside the loop. Thus a steel represented by a composition to the right of the loop will not harden on quenching since it is already ferritic in structure; (ii) the Guillet-type diagram.

necessary to have this effect. Provided that the composition of the steel falls to the left of the γ-loop, chromium will give rise to a much greater depth of hardening, due to the retardation of the transformation rates which it also produces.

13.33 The main disadvantage in the use of chromium as an alloying element is its tendency to promote grain growth, with the attendant brittleness that this involves. Care must therefore be taken to avoid overheating or holding for too long at the normal heat-treatment temperature.

13.34 Low-chromium, low-carbon steels are popular for case-hardening, whilst low-chromium, medium-carbon steels are used for axles, connecting rods and gears. Low-chromium steels containing 1.0% carbon are extremely hard and are useful for the manufacture of ball-bearings, drawing dies and parts for grinding machines.

13.35 Chromium is also added in larger amounts—up to 25%—and has a pronounced effect in improving corrosion-resistance, due to the protective layer of oxide which forms on the surface. This oxide layer is extremely thin, and these steels take a high polish. These alloys are ferritic and non-hardening (except by cold-work) provided that both carbon and nitrogen are kept to very low limits. The Schaeffler diagram (Fig. 13.7) indicates that the presence of either carbon or nitrogen will tend to produce increasing amounts of martensite in a 13% chromium steel and so make it brittle if cooled at a rate such as would be encountered in welding. Consequently in modern ferritic stainless irons containing 13% chromium the total amount of carbon and nitrogen is kept very low. Nevertheless there is still sufficient of these 'interstitial elements' present to cause intergranular corrosion in the presence of strong electrolytes and the alloys are generally 'stabilised' (20.93 and 21.71) by the addition of small quantities of titanium and/or niobium. Best mechanical properties and formability are obtained when these additions are at a minimum necessary to impart corrosion resistance. The combined amount of these 'stabilisers' can be calculated from:

$$Ti + Nb = 0.2 + 4 (C + N)$$

When carefully refined stainless irons contain no more than 0.01% each of carbon and nitrogen the above formula indicates a total requirement of titanium and niobium of no more than 0.28%.

As the Schaeffler diagram indicates, it is far easier to stabilise a ferritic structure in those stainless irons containing 17–26% chromium and less than 0.1% carbon.

Ferritic stainless irons are used widely in the chemical engineering industries. Lower-grade alloys are used for domestic purposes such as stainless-steel sinks; and in food containers, refrigerator parts, beer barrels, cutlery and table-ware. The best-known alloy in this group is 'stainless iron', containing 13% chromium and usually less than 0.05% carbon. Recently British Steel have developed a similar alloy ('Hyform 409') for the manufacture of corrosion-resistant motor-car exhaust systems. This contains 12% chromium which gives optimum formability whilst still retaining adequate corrosion resistance. Nitrogen and carbon are kept to a minimum and titanium is added as a stabiliser. Such an exhaust system would outlast many made in the usual mild steel. It will be up to 'consumer pressure' to back British Steel in this venture—otherwise we will continue to get the rotten cars we deserve.

13.36 If the carbon content exceeds 0.1% the alloy is a true stainless steel and is amenable to hardening by heat-treatment. The most common alloy in this group contains 13% Cr–0.3% C. This is a cutlery-blade steel which is not radically different from the composition proposed by Harry Brearley when he introduced the first stainless steel in 1913, and improvements in recent years have been entirely in the field of heat-treatment

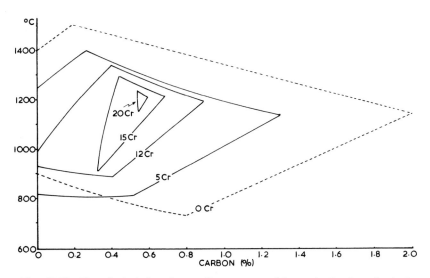

Fig. 13.11   The effect of chromium on the presence of the austenite phase in steel.

Table 13.2a *Low-chromium Steels*

Type of Steel	BS 970 designation	Composition (%) C	Mn	Cr	Condition	Ruling section (mm)	Yield point (N/mm²)	Tensile strength (N/mm²)	Elongation (%)	Reduction area (%)	Izod (J)	Hardness (Brinell)	Heat-treatment	Uses
'60' C–Cr steel	526M60	0.60	0.65	0.65	Oil-quenched and tempered at 200°C.	63.5	—	—	—	—	—	700	Oil-quench from 800–850°C. Temper: (1) for cold-working tools at 200–300°C; (2) for hot-working tools at 400–600°C.	General blacksmith's and boilermaker's tools and chisels. Hot and cold sates. Swages. Hot-stamping and forging dies. Builder's, mason's and miner's tools. Spring collets, chuck and vice jaws. Turning mandrels and lathe centres.
					Oil-quenched and tempered at 400°C.		620	850	15		34	250		
1% Cr steel	530M40	0.45	0.90	1.00	Oil-quenched and tempered at 650°C.	28.5	680	850	18	54	95	250	Oil-quench from 860°C; temper from 550–700°C.	Agricultural machine parts, machine-tool components. Lining plates, paddles and drums for concrete and tar mixers. Excavator cutting blades and teeth. Automobile axles, connecting-rods and steering arms. Spanners and small tools.
						100	500	700	22	61	122	200		
1% Cr rail steel	—	0.75	1.0	1.1	—	—	—	1140	8	—	—	340	—	Side-wear situations (tightly-curved railway track).
1% C–Cr steel	534A99	1.00	0.45	1.40	Hardened	28.5	—	—	—	—	—	850	Oil-quench from 810°C; temper at 150°C	Ball- and roller-bearings. Roller- and ball-races. Instrument pivots and spindles. Cams. Small rolls.

Typical mechanical properties

Table 13.2b *Stainless High-chromium Steels and Irons*

Type of Steel	BS 970 (Pt 4) designation	Composition (%)			Condition	Ruling section (mm)	Typical mechanical properties				Heat-treatment	Uses
		C	Mn	Cr			Yield point (N/mm^2)	Tensile strength (N/mm^2)	Elongation (%)	Hardness (Brinell)		
Stainless iron	403S17	0.04	0.45	14.0	Soft	—	340	510	31	—	Non-hardening except by cold-work.	Wide range of domestic articles—particularly forks and spoons. Can be pressed, drawn and spun.
Stainless iron	410S21	0.10	0.50	13.0	Oil-quenched from 1000°C, and tempered at 750°C	28.5	371	571	33	170	Oil-quench, water-quench or air-cool from 950–1000°C; temper at 650–750°C.	Turbine-blade shrouding rivets, split pins, golf-club heads, solid-drawn tubes; structural and ornamental work.
Stainless steel	420S37	0.22	0.50	13.0	Oil-quenched from 960°C, and tempered at 700°C	50	633	757	26	220	Oil-quench, water-quench or air-cool from 950–1000°C; temper at 500–750°C.	A general-purpose alloy—not in contact with non-ferrous metal parts or graphite packing. Valve and pump parts.
Stainless steel	420S45	0.3	0.50	13.0	Cutlery temper / Spring temper	6 / —	— / —	1670 / 1470	— / —	534 / 450	Oil- or water-quench or air-cool from 975–1040°C; temper (for cutlery) at 150–180°C; temper (for springs) at 400–450°C.	Specially for cutlery and sharp-edged tools. Approximately pearlitic in structure when normalised.
Stainless iron	430S15	0.05	0.50	17.0	Soft	—	370	525	27	—	Non-hardening—except by cold-work.	Can be deep-drawn or spun.
High-carbon chromium tool steel	BS 4659: BD3	2.10	0.30	12.5	Oil-quenched and tempered at 200°C	—	—	—	—	850	Heat slowly to 750–800°C and then raise to 960–990°C, and oil-quench (small sections can be air-cooled); temper at between 150 and 400°C for 30–60 minutes.	Blanking punches, dies and shear-blades for hard, thin materials. Dies for moulding abrasive powders (ceramics); Master gauges, hobbing dies, threadrolling dies.

enabling much better cutting edges to be obtained. Consequently in post-war years these steels have been developed for razor blades, hand tools, garden tools and food-processing equipment as well as for knife blades.

Due to displacement of the eutectoid point to the left, this 13Cr–0.3 C steel is of approximately eutectoid composition when annealed and slowly cooled. It has a martensitic structure when air-hardened but it is essential that the hardening temperature is high enough (975–1040°C) or undissolved carbides may be present after hardening, giving rise to electrolytic corrosion. With higher carbon contents the formation of brittle carbide networks is more likely particularly in structures which have been annealed and then inadequately solution treated. This type of carbide precipitation leaves the matrix locally depleted in chromium so that it becomes anodic (21.71) to the remainder of the chromium-rich structure and electrolytic corrosion follows.

Compositions and properties of the more important chromium steels are given in Tables 13.2a and b.

Nickel–Chromium Steels

13.40   The addition of either nickel or chromium *singly* to a steel can have some adverse effects. Whilst nickel tends to inhibit grain growth during heat-treatment, chromium accelerates it, thus producing brittleness under shock. Meanwhile, chromium tends to form stable carbides, making it possible to produce high-chromium, high-carbon steels, whilst nickel has the reverse effect in promoting graphitisation. The deleterious effects of each element can be overcome, therefore, if we add them in conjunction with each other. Then, the tendency of chromium to cause grain growth is nullified by the grain-refining effect of the nickel, whilst the tendency of nickel to favour graphitisation of the carbides is counteracted by the strong carbide-forming tendency of the chromium.

13.41   At the same time other physical effects of each element are additive, as for example, increase in strength due to the formation of substitutional solid solutions in the ferrite; increase in corrosion resistance and also the retardation of transformation rates during heat-treatment. This latter effect is particularly useful in rendering drastic water-quenches avoidable.

13.42   In general, the low-nickel, low-chromium steels contain rather more than two parts of nickel to one part of chromium. Those with up to 3.5Ni and 1.5Cr are oil-hardened from 810–850°C, followed by tempering at 150–650°C according to the properties required. Some of these steels suffer from 'temper brittleness' when tempered in the range 250–400°C. This is shown by low resistance to shock in impact tests. If tempered above 400°C, therefore, it is necessary to cool the steel quickly in oil through the range in which temper-brittleness develops. The effect is minimised but not completely eliminated by adding small amounts of molybdenum (13.50), thus producing the range of well-known 'nickel-chrome-moly' steels. In fact most of the steels in this group now contain molybdenum.

13.43   The nickel–chromium steels are very adaptable and useful alloys. They forge well and also machine well in the softened condition whilst their mechanical properties can be varied considerably by the treatment given. When the nickel content is increased to about 4% and chromium to about 1.5% an air-hardening steel is obtained. Such a steel is very useful for the manufacture of complex shapes which have to be hardened, and which would be likely to distort if water- or oil-quenching were attempted.

13.44   The high-nickel, high-chromium steels are all stainless alloys containing less than 0.1% carbon which is virtually a troublesome impurity, expensive to reduce below 0.04%. The most popular is the '18–8' stainless steel with 18Cr–8Ni. The introduction of nickel to a ferritic 18% Cr alloy considerably enlarges the γ-loop of the phase diagram. It also decreases the M_s temperature so that with 8% Ni the M_s temperature is below ambient temperatures and austenite is therefore retained after slow cooling. 18–8 stainless steel takes a good polish and resists corrosion by many relatively corrosive organic and inorganic reagents. When cold-worked these austenitic steels strain harden quickly and it was more than twenty years after their introduction before they became widely used for domestic purposes. Improved tool design and shaping methods solved many of the production problems but even so a 12Cr–12Ni alloy is far more ductile and still sufficiently corrosion resistant for use as table ware.

13.45   Chromium has a relatively high affinity for oxygen and consequently oxidises very easily. Nevertheless the oxide film which forms rapidly on the surface of chromium, though extremely thin, is also very stable and strongly adherent to the surface which it therefore protects from further attack. When in solid solution, either in ferrite or in austenite, it bestows these corrosion-resisting properties upon iron particularly when more than 12% chromium is present. Although the film is extremely thin it builds up immediately the surface is polished.

In the presence of concentrated nitric acid, a powerful oxidising agent, it has been shown that stainless steel begins to dissolve almost as quickly as would mild steel but that, immediately, a thin oxide film is formed providing a protective passive layer. As long as oxidising conditions are present in the environment the film repairs itself should abrasion take place, but in the presence of *non-oxidising* corrosive media such as concentrated hydrochloric acid or strong chloride solutions, corrosion may occur, particularly in impingement attack where the oxide film is broken and is unable to repair itself because of the absence of oxidising conditions. Fortunately 8–10% Ni renders stainless steel more resistant to attack by hydrochloric acid, chloride solutions and other non-oxidising media—yet another instance where the action of nickel and chromium is complementary and indicating why 18–8 stainless steels are so widely used both in the chemical industries and elsewhere.

13.46   The carbon content is kept as low as is economically possible since the presence of precipitated carbides in the microstructure reduces corrosion-resistance. Even with a carbon content below 0.1% slow cooling of the steel to ambient temperature will cause carbide precipitation to take place, and this considerably reduces corrosion-resistance, because the

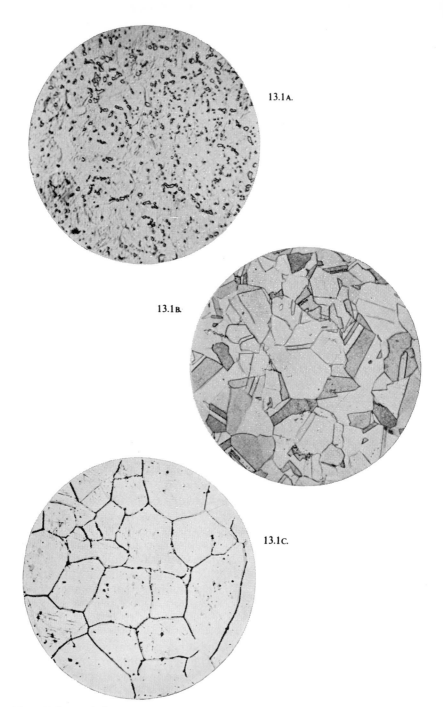

13.1A.

13.1B.

13.1C.

Plate 13.1   13.1A Stainless cutlery steel (13% chromium; 0.3% carbon) oil-hardened from 950°C.
Particles of carbide in a martensite matrix. × 1500, *(Courtesy of Messrs Edgar Allen & Co Ltd, Sheffield)*
13.1B Stainless steel (18% chromium; 8% nickel; 0.1% carbon) oil quenched from 1100°C. Twinned crystals of austenite. × 250. Etched in acidified iron (III) chloride. *(Courtesy of Messrs Hadfields Ltd. Sheffield)*
13.1C 18/8 Stainless steel oil quenched from 1100°C and then reheated at 650°C for 1 hour. Carbides precipitated at the grain boundaries of the austenite. × 250. Etched in acidified iron(III) chloride. *(Courtesy of Messrs Hadfields Ltd, Sheffield).*

Table 13.3a *Low chromium–nickel steels*
(See also low-nickel, low-chromium case-hardening steels (Table 19.1).)

BS 970 desig-nation	Composition (%)					Condition	Typical mechanical properties						Heat-treatment	Uses
	C	Mn	Ni	Cr	Other elements		Ruling section (mm.)	Yield point (N/mm²)	Tensile strength (N/mm²)	Elong-ation (%)	Reduc-tion area (%)	Izod (J)		
655M13	0.12	0.45	3.3	1.0	—	Carburised and double-quenched	28.5	850	925	13	40	39	After carburising, refine by heating to 850°C and then cool in oil or air. Harden by oil-quench from 770°C. Temper at 150°C.	A case-hardening steel of high core strength and hard-wearing surface.
653M31	0.3	0.6	3.0	0.8	Mo 0.65 (optional)	Oil-quenched and tempered at 600°C	63.5	800	1000	16	—	47	Oil-quench from 820–840°C; temper between 550 and 650°C. Cool in oil to avoid temper-brittleness if molybdenum is absent.	Highly-stressed parts in aero-, auto- and general engineering, eg differential shafts, stub axles, connecting-rods, high-tensile studs, pinion shafts. (For heavy sections the addition of molybdenum is advisable.)
							150	585	770	20	—	54		
640M40	0.35	0.8	1.25	0.6	—	Oil-quenched and tempered at 600°C	28.5	800	925	22	59	88	Oil-quench from 830–850°C, temper between 180 and 220°C. Cool in oil. Suffers from temper-brittleness if tempered between 250 and 400°C.	Small and medium-sized sections for gears, differential pinions, etc. Automobile connecting-rods, crankshafts, axles, bolts and screwed parts.
							150	525	695	22	79	114		

Table 13.3b *High nickel-chromium stainless steels*

BS 970 (Pt 4) designation	Composition (%)					Condition	Typical mechanical properties						Heat-treatment	Uses
	C	Mn	Ni	Cr	Other elements		Yield point (N/mm²)	Tensile strength (N/mm²)	Elonga-tion (%)	Reduc-tion area (%)	Izod (J)	Brinell		
—	<0.03	0.75	0.75	11.5	Ti 0.25 Si 0.5	Soft	335	550	27	—	90	160		Low-cost ferritic stainless iron—beet sugar industries, coal mining, etc
302S25	0.05	0.8	8.5	18.0	—	Softened Cold-rolled sheet	278 803	618 896	50 30	50 —	143 —	170 —		Particularly suitable for domestic and decorative purposes.
321S20	0.05	0.8	8.5	18.0	Ti 0.6	Softened Cold-rolled	278 402	649 803	45 30	50 42	104 —	180 225	quickly from 1050°C to keep carbides in solution.	A 'weld-decay' proof steel (fabrications may be used in the as-welded condition). Used extensively in plant for the manufacture of nitric acid. An austenitic steel.
—	0.05	0.8	12.5	12.5	—	Softened	232	579	50	50	—	160		A deep-drawing austenitic alloy—suitable for table-ware and kitchen equipment.
304S15	0.03	0.8	10.0	18.5	—	Softened	232	618	60	65	104	160		A low-carbon alloy suitable for deep-drawing. Used mainly for domestic purposes, etc.

Designation	C	Ni	Cr	Other	Condition							Remarks
347S17	0.04	0.8	10.0	18.0	Nb 1.0	Softened	263	618	58	65	104	175	Immune from weld-decay, due to the stabilising effect of niobium. Used in welded plant where corrosive conditions are very severe, eg in nitric acid plant
320S17	0.04	0.8	13.0	17.5	Mo 2.75 Ti 0.6	Softened	278	618	50	50	104	180	Molybdenum improves the corrosion-resistance to such reagents as sulphuric and phosphoric acids, chloride solutions and organic acids, such as acetic acid.
325S21	0.05	1.5	8.5	18.0	Mo 0.3 Ti 0.6 S 0.25	Softened	232	587	53	54	65	175	A free-cutting stainless steel in which the machinability is increased by the presence of manganese and sulphur.
—	0.04	5.0	12.5	21.2	Mo 2.2 Nb 0.2 V 0.2 N 0.3	Cold-worked	—	1690	—	—	—	—	Fittings in marine environments. Chemical process plant-pumps, valves; etc. (Nitronic 50)
—	<0.025	—	1.2	27.5	Mo 3.5	Soft	515	620	30	—	—	—	A ferritic steel
—	<0.025	—	24	20	Mo 6	Soft	275	620	45	—	—	—	An austenitic steel

Very hostile conditions, eg. power-station condenser tubes using sea water

Non-hardening except by cold-work. (Cool)

precipitation of chromium carbide particles leads to impoverishment of the surrounding austenite in chromium, so that electrolytic action (21.30) is induced between the chromium-depleted region and the adjoining chromium-rich regions. It is therefore necessary to heat the alloy to 1050°C and then quench it, in order to retain the carbon in solid solution. Carbide precipitation leads to a defect usually known as 'weld-decay' in fabricated articles (20.93). This defect may be reduced by adding small amounts of titanium, molybdenum or niobium; elements which have a strong carbide-forming tendency and thus 'tie up' most of the carbon as carbides at a high temperature, so that this carbon is no longer available to form chromium carbide during slow-cooling following a welding process. In this way any variation in chromium content across the microstructure is avoided and post-welding treatment is unnecessary.

13.47 High-strength Stainless Steels   The development of the aerospace industries created a demand for high-strength stainless steels and these alloys are now finding use in many other fields. In some austenitic stainless steels part of the nickel content has been replaced by manganese to reduce cost, manganese like nickel stabilising austenite. These steels are sometimes strengthened by the addition of nitrogen which, unlike carbon, is less likely to introduce corrosion problems. Steels of this type are further strengthened by cold-work. In the USA the 'Nitronic' austenitic steels are in this class. 'Nitronic 33' contains 18Cr; 12Mn; 3.2Ni; 0.32N and 0.05C and has roughly double the yield strength of a conventional 18–8 stainless steel in the annealed form. Examination of this composition in relation to the Schaeffler diagram (Fig. 13.7) indicates that it will remain completely austenitic after welding and that the formation of martensite will be unlikely. 'Nitronic 50' contains 21.2Cr; 12.5Ni; 5Mn; 2.2Mo; 0.2Nb; 0.2V; 0.3N and 0.04C. It has a high resistance to corrosion and after cold-work attains a tensile strength of 1690 N/mm^2.

Some stainless steels are precipitation hardened and may be either martensitic or austenitic in structure. Martensitic steels of this type usually contain 17Cr; 4Ni; 4Cu and small amounts of molybdenum, aluminium or niobium. They are solution treated at about 1000°C and then air-cooled to give a martensitic structure. Subsequent precipitation treatment at 500–600°C results in the formation of coherent precipitates based on compounds of nickel with molybdenum, aluminium or niobium which further strengthen the martensitic structure. Austenitic steels of this type contain 17Cr; 10Ni and up to 0.25P. These are precipitation hardened at 700°C to produce coherent precipitates based on carbides and phosphorus compounds which will introduce lattice strain and inhibit the movement of dislocations.

The compositions, properties and uses of some of the more important nickel–chromium steels are given in Tables 13.3a and b.

Steels Containing Molybdenum

13.50   The USA is the dominant producer of molybdenum, much of her output being a by-product of several copper mines; Canada, Chile, CIS and China are other significant producers. At the end of last century practically the whole of the world production of molybdenum was used in making the chemical reagent ammonium molybdate for the analytical determination of phosphorus in iron, steel, and fertilisers. Today the principal use of molybdenum is in the manufacture of alloy steels.

As mentioned earlier in this chapter (13.42), one of the main uses of molybdenum is to reduce the tendency to 'temper-brittleness' in low-nickel, low-chromium steels. Additions of about 0.3% molybdenum are usually sufficient in this respect, and the resultant steels retain a high impact value, irrespective of the rate of cooling after tempering.

Among alloy steels the 'nickel-chrome-moly' steels possess the best all-round combination of properties, particularly where high tensile strength combined with good ductility is required in large components. These steels are much less influenced by the mass effects of heat-treatment, the transformation rates of the nickel–chromium steels being still further reduced by the presence of molybdenum, which therefore contributes considerably to depth of hardening.

13.51   Molybdenum is also added to chromium steels to produce a general improvement in machinability and mechanical properties, whilst nickel–molybdenum steels are very suitable for case-hardening. Molybdenum dissolves in ferrite which it strengthens considerably and also forms a hard, stable carbide Mo_2C as well as double carbides such as Fe_4Mo_2C and $Fe_{21}Mo_2C_6$. Small amounts of molybdenum are very effective in reducing the transformation rates, particularly austenite $\rightarrow$ pearlite, ie the nose of the TTT curve is displaced considerably to the right. Molybdenum raises the high-temperature strength and creep resistance of high-temperature alloys and also enhances the corrosion resistance of stainless alloys particularly to chloride solutions. In modern high-speed steels (14.10) much of the tungsten has been replaced by molybdenum. Molybdenum-bearing steels are listed in Table 13.4.

Steels Containing Vanadium

13.60   In 1801 Del Rio found a new metal in a Mexican ore. He called it 'erythronium'—from the Greek *erythros* meaning 'red '—because it formed red compounds. Later, Sefström discovered what he thought was a new metal in some Swedish iron ore and named it 'vanadium' (after the Scandinavian goddess, Vanadis) but almost immediately the eminent chemist Wöhler identified Sefström's vanadium as being Del Rio's erythronium. To-day South Africa is the leading producer of the metal, followed by the USA, CIS, Finland and Norway.

13.61   Plain vanadium steels are used only to a limited extent, but

Table 13.4 *Steels Containing Molybdenum*

Type of Steel	BS 970 desig- nation	Composition (%)					Condition	Ruling Section (mm)	Typical mechanical properties						Heat-treatment	Uses
		C	Mn	Ni	Cr	Mo			Yield point (N/mm^2)	Tensile strength (N/mm^2)	Elong- ation (%)	Reducn. area (%)	Izod (J)	Brinell		
Mn–Mo steel	605M56	0.35	1.6	—	—	0.27	Oil-quenched and tempered at 600°C	22	800	1010	16	57	93	293	Oil-quench from 830–850°C; temper at 550–650°C and cool in oil or air.	A substitute for more highly alloyed nickel–chromium steels.
								150	525	700	22	62	54	201		
Ni–Mn– Mo steel	785M19	0.2	1.6	0.55	—	0.25	Oil-quenched and tempered	150	463	617	22	—	54	179	Oil-quench from 860–900°C; temper at 600–650°C; cool in air.	*Railway engineering:* connecting-rods, spring hanger and slide-bar bolts, draw gears, etc. *General engineering:* parts of harvesters, dredging machines, crushers, rods for well-boring. Has very good fatigue- and shock-resisting properties.
1% Cr– Mo steel	709M40	0.4	0.65	—	1.1	0.3	Oil-quenched and tempered at 600°C	28.5	895	1080	15	54	41	311	Oil-quench from 840–860°C; temper at 500–700°C, and cool in oil or air.	Suitable for crank-shafts and connecting-rods, stub axles, etc. As a substitute for 3% nickel steel (it machines more readily and is free from temper brittleness). Zinc die-casting dies.
								150	525	695	22	67	54	201		

Steel	Designation	C	Mn	Ni	Cr	Mo	Condition	Section (mm)			%				Heat treatment	Applications
1¼% Mn–Ni–Cr–Mo steel	945M38	0.4	1.35	0.75	0.45	0.2	Oil-quenched from 850°C and tempered at 600°C.	28.5	895	1005	21	62	47	291	Oil-quench from 830–850°C; temper at 550–660°C; cool in air.	Automobile and general-engineering components requiring a tensile strength of 700–1000 N/mm²
								150	525	690	22	—	55	200		
1½% Ni–Cr–Mo steel	817M40	0.4	0.55	1.5	1.1	0.3	Oil-quenched and tempered at 200°C	28.5	1470	1544	12	47	27	440	Oil-quench from 830–850°C; 'light' temper 180–200°C; 'full' temper 550–650°C; cool in oil or air.	Differential shafts, crankshafts and other highly-stressed parts where fatigue and shock-resistance are important. In the lightly tempered condition it is suitable for automobile and machine-tool gears. (It can be surface-hardened effectively in cyanide bath.)
							Oil-quenched and tempered at 600°C	150	590	770	20	—	54	223		
3% Ni–Cr–Mo steel	830M31	0.3	0.6	3.25	0.8	0.3	Oil-quenched and tempered at 600°C	100	895	1080	15	—	40	310	Oil-quench from 820–840°C; temper at 550–650°C; cool in oil or air.	Parts of thin section where maximum ductility and shock-resistance are required, eg connecting-rods, inlet valves, cylinder studs, valverockers.
								150	680	850	18	—	55	250		
4¼% Ni–Cr–Mo steel	835M30	0.3	0.45	4.25	1.25	0.25	Air-hardened and tempered at 200°C	63.5	1470	1700	14	45	35	440	Air-harden from 820–840°C; temper at 150–200°C, and cool in air.	An air-hardening steel for aero-engine connecting-rods, valve mechanisms, gears differential shafts and other highly-stressed parts. Can be further surface-hardened by cyanide or carburising.
½% Mo–boron steel	—	0.10	0.6	—	B 0.03	0.5	Normalised at 950°C, then stress-relieved for 3 hours at 600°C	—	432	571	16	—	—	—	Normalise at 930–980°C; stress-relieve at 590–610°C for 3 hours.	High creep stress up to 400°C; pressure vessels; heat exchanger; reactor vessels; gas-turbine and power-plant components.

chromium–vanadium steels containing up to 0.2% vanadium are widely used for small and medium sections. The mechanical properties resemble those of low nickel–chromium steels but usually show a higher yield stress and percentage reduction in area. Chromium–vanadium steels are also easier to forge, stamp and machine, but are more susceptible to mass effects of heat-treatment than corresponding nickel–chromium steels of similar section.

13.62   Like nickel, vanadium is a very important grain refiner in steels and as little as 0.1% is effective in restricting grain growth during normal hardening processes. Vanadium is present in the microstructure as finely dispersed carbides and nitrides and since these do not dissolve at normal heat-treatment temperatures, their presence acts as a barrier to grain growth. If the steel is over-heated so that the particles of carbide and nitride go into solution, then grain growth immediately takes place rapidly.

13.63   Vanadium has a strong carbide-stabilising tendency and forms the carbide VC. It also stabilises martensite and bainite on heat-treatment and increases the depth of hardening. Since it combines readily with oxygen and nitrogen it is often used as a 'scavenger' or 'cleanser' in the final stage of deoxidation to produce a gas-free ingot. One of the most important effects of vanadium is that it induces resistance to softening at high temperatures provided that the steel is first heat-treated to absorb some of the vanadium carbide into solid solution. Consequently vanadium steels are used for hot-forging dies, extrusion dies, die-casting dies and other tools operating at elevated temperatures. Most high-speed steels (14.10) contain vanadium, sometimes in amounts up to 5.0%. Some steels containing vanadium are shown in Table 13.5.

Heat-resisting Steels

13.70   Steels required for service at high temperatures are used for such components as exhaust valves for aero-engines; racks for enamelling stoves; conveyor chains; furnace arch and floor plates; recuperator tubes; rotors for gas turbines; annealing boxes; pyrometer sheaths; retorts; super-heater element supports; burner nozzles; etc. The main requirements of such steels are:

   (i) resistance to oxidation by gases in the working atmosphere;

   (ii) a sufficiently high strength (creep resistance) at the working temperature

   (iii) freedom from structural changes at the working temperature such as the formation of grain-boundary precipitates which would cause embrittlement.

13.71   The surface of iron scales readily at temperatures above 700°C and the layer of oxide produced adheres only loosely to the surface. Moreover the mobility of metallic and non-metallic ions through the oxide film is high (21.21) and increases with temperature, resulting in continual destructive oxidation. In a number of oxides, notably those of beryllium,

Table 13.5 *Steels Containing Vanadium*

Type of steel	Composition (%)						Condition	Hardness (VPN)	Heat-treatment	Uses
	C	Si	Mn	Cr	Mo	V				
¼% V steel	0.7 – 1.0	0.25	0.25	—	—	0.2	—	—	Water-quenched from 850°C. Temper as required.	Cold-drawing dies, etc.
							—	—		Coining and embossing dies (12.70—Part 2)
¾% Cr–V steel	0.57	0.25	0.8	0.75	—	0.2	Hardened and tempered	575	Pre-heat to 700°C and then oil-quench from 850°C. Temper as required.	Die-casting dies for zinc-base alloys.
5% Cr–Mo–V steel	0.4	1.0	0.3	5.0	1.35	0.45	Hardened and—(a) Tempered at 550°C	600	Pre-heat to 850°C and then to 1000°C; soak for 10–30 minutes and air-cool; temper at 550–650°C for 2 hours.	Hot-forging dies for steel and copper alloys where excessive temperatures are not reached. Extrusion dies and mandrels for aluminium alloys. Pressure and gravity dies for aluminium die-casting.
	0.35	1.0	0.3	5.0	1.5	1.0	(b) Tempered at 650°C	375		
High C–high Cr steel	1.6	0.2	0.3	13.0	1.0	0.5	Hardened and—(a) Tempered at 200°C	800	Pre-heat to 850°C and then to 1000°C; soak for 14–45 minutes and quench in oil or air; temper at 200–400°C for 30–60 minutes.	Fine press tools. Deep-drawing dies and forming dies for sheet metal. Hobbing dies. Thread-rolling dies. Wire-drawing dies and wortle plates. Blanking dies, punches and shear blades for hard metals.
	1.9	0.3	0.2	12.5	0.8	0.25	(b) Tempered at 400°C	700	Pre-heat to 800°C and then 1030°C air-cool; temper at 180–450°C; for 2–3 hours.	
Vanadium tool steel	1.4	—	0.4	0.4	0.4	3.6	—	—	Water-quench from 770°C. Temper at 150–300°C.	Cold-heading dies.
5% Cr–5% V steel	2.3	—	—	5.25	1.1	4.5	Hardened and tempered at 150°C	825	Oil- or air-quench from 940°C. Temper at 150–220°C.	Refractory moulds and other uses where resistance to abrasion is necessary.

Table 13.6 *Heat-resisting Steels*

Relevant specifications	Composition (%)						Condition	Typical mechanical properties at various temperatures				Maximum working temperature (°C)	Typical uses
	C	Si	Mn	Cr	Ni	Other elements		Testing temperature (°C)	Yield stress (N/mm²)	Tensile strength (N/mm²)	Elongation (%)		
BS 3100: 1398 Grade A	0.2	0.3	0.8	—	—	Mo 0.5	As cast	20	278	463	18	450	Steam-turbine and other castings.
BS 3100: 1398 Grade D	0.1	0.2	0.6	0.4	—	Mo 0.5 V 0.25	As cast	20	309	510	17	510	A creep-resisting steel for turbine castings working at high pressures.
AISI Series 311	0.15	—	—	20.0	25.0	—	Forged or rolled	—	—	—	—	820	Conveyor chairs and skids, heat-treatment boxes, recuperator tubes, exhaust valves and other furnace parts.
									0.2% Proof stress				
BS 970 (Pt. 4): 401S45	0.45	3.5	0.5	8.0	—	—	Oil-quenched 1050°C; tempered 800°C.	20 200 400 600	801 708 585 185	1001 909 816 339	23 20 20 45	850	A low-cost, heat-resisting steel used mainly for valves.
BS 970 (Pt. 4): 352S52	0.5	0.2	9.0	21.0	4.0	Nb 2.5 N 0.45	Water-quenched 1200°C, then air-cooled from 800°C. (Precipitation treatment)	20 200 400 600 800	585 447 385 323 216	1001 847 755 616 385	16 20 23 24 28	850	A valve steel which can be precipitation-hardened.

Material	C	Si	Mn	Cr	Ni	—	Condition	Temp °C	0.1% Proof stress		Elong. %	Max. service temp °C	Remarks
AISI Series 302B	0.1	1.5	1.0	19.0	11.0	—	Air-cooled from 1100°C	20 200 400 600 800 900	347 262 248 217 139 85	698 543 527 465 248 155	55 43 40 37 45 50	1000 (air) 950 (flue gases)	A fairly cheap grade of heat-resisting steel with a good combination of properties.
AISI Series 309	0.3	1.2	0.8	23.0	12.5	—	As cast	800 1000 1050	— — —	48.3 15.4 11.0	— — —	1050	Oil- and gas-fired installations —heat-treatment furnaces, nozzle burners, salt pots, super-heater supports. Used in the cast, forged or rolled form.
—	0.45	1.2	0.8	15.0	35.0	—	As cast	800 1000 1050	— — —	69.0 20.7 15.9	— — —	1050	A casting alloy. Not suitable for use when highly sulphurous gases are present. Useful for conditions involving cyclic thermal shock.
—	0.35	2.0	0.8	13.0	60.0	—	As cast	800 1000 1050	— — —	45.5 15.4 11.7	— — —	1050 1100 (air)	A casting alloy suitable for carburising and quenching equipment.

Material	C	Si	Mn	Cr	Ni	—	Condition	Temp °C	1.0% Proof stress		Elong. %	Max. service temp °C	Remarks
BS 1501 (Pt. 3): 310S24; AISI Series 310	0.07	0.8	1.0	25.0	20.0	—	Quenched from 1100°C to retain carbides in solution	20	270	543	40	1100	A wrought steel. Hearth plates, trays and muffles.
—	0.05	0.8	0.8	28.0	—	Co 50.0	As cast	—	—	—	—	1200	Quenching baskets, etc. Good wear-resistance at high temperatures, hence used in cement-production equipment.

aluminium, silicon and chromium, ionic mobility is low and since these oxides tend also to be dense they offer good protection against further oxidation. The first three of these elements tend to have deleterious effects on the mechanical properties of steels in the *quantities necessary to impart corrosion resistance* so that chromium—sometimes assisted by small amounts of silicon—is used. These elements form barrier films of oxide which adheres tenaciously to the surface.

13.72 The considerable amount of chromium required to effect oxidation resistance unfortunately also favours rapid grain growth. Hence nickel is often added to counteract this as in many other steels already dealt with here. However, nickel must be absent from those steels which are in contact with atmospheres containing sulphur gases, since brittle grain boundary films of nickel sulphide would be formed (21.20):

$$Ni \rightarrow \boxed{Ni^{++}} + 2e^-$$
$$S + 2e^- \rightarrow \boxed{S^{--}}$$

nickel sulphide

13.73 The chromium present contributes to some extent to the high-temperature strength by ordinary substitutional solid solution. However, creep strength is generally increased by 'stiffening' the alloy with small amounts of tungsten, molybdenum, vanadium, titanium, niobium, aluminium or nitrogen. Reduction in creep at high temperatures is then achieved by dispersion hardening provided by the small particles (8.62) of the carbides of niobium, tungsten, molybdenum, vanadium or titanium; the nitrides of aluminium, titanium or niobium; and also $NiAl_3$.

Tungsten, molybdenum and cobalt also render steels resistant to tempering effects at high temperatures as they do in high-speed steel (14.10), and thus raise the limiting creep stress.

In some cases precipitation hardening can be used to strengthen the alloy provided that at the working temperature, non-coherent precipitation (9.92) does not occur since this would result in loss of strength and an increase in brittleness. Generally, steels working at high temperatures are either austenitic or austenitic/ferritic in structure so that brittle martensite will not form each time the component cools, otherwise cracking in the martensitic areas would be likely due to internal stress arising from non-uniform cooling.

13.74 Short-time creep tests (5.72) are often employed in evaluating heat-resistant alloys. Design is frequently based on the stress which will produce a creep rate of 0.0001% per hour (1% extension in the gauge length in 10 000 hours) at the maximum operating temperature. Safe working stresses are often based on 50% of this stress. However, such a value can only be applied under conditions of direct axial static loading and relatively uniform temperature. When impact loading or rapid temperature changes are involved higher safety factors must be used, ie 33⅓% or even 25% of the creep stress as derived above. The compositions and properties of some typical heat-resisting steels are given in Table 13.6.

Manganese Steels

13.80 Although the CIS is the largest producer of manganese, South Africa also mines considerable quantities of the metal, followed by Brazil, Gabon, Australia, India, Mexico, China and Ghana. The bulk of Britain's imports are from South Africa, Brazil and Ghana. Interest is currently focused on the large deposits of ferromanganese nodules which occur in the main basins of the Pacific, Atlantic and Indian Oceans at depths between 4 and 5 km. The origin of these nodules appears to be biological as well as chemical.

13.81 Manganese is very widely used in steel production both for deoxidation and desulphurisation of the molten steel. It is generally added as ferromanganese (80Mn; 6C; bal.Fe) or as spiegeleisen (5–20Mn; 5C; bal.Fe) and eliminates most of the oxygen and sulphur:

$$FeO + Mn \rightarrow Fe + MnO \text{ (volatilises or joins the slag)}$$
$$FeS + Mn \rightarrow Fe + MnS \text{ (insoluble in iron—so joins slag)}$$

In order that these reactions proceed as shown an excess of manganese must be added and the excess remains in the steel. Such amounts are usually less than 1.0%, and it is only when the manganese content exceeds this amount that it is regarded as a deliberate alloying element. Of recent years, however, there has been a tendency to increase the manganese content of ordinary carbon steels at the expense of a little carbon, thus giving improvements in respect of ductility and Izod values, in both the normalised and heat-treated conditions. The susceptibility of low-carbon steels to brittle fracture is reduced by raising the manganese carbon ratio, using up to 1.3% manganese (5.53). Similarly free-cutting steels contain up to 1.7% manganese so that sulphur will be present as manganese sulphide (MnS) globules which aid machining (6.63).

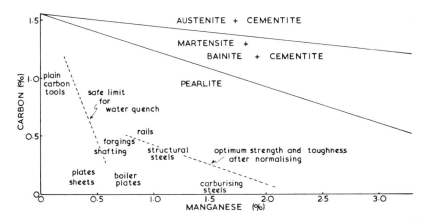

Fig. 13.12 A Guillet-type diagram illustrating the relationship between manganese and the structure and uses of low-manganese steels.

13.82   Like nickel, manganese stabilises austenite by raising the A_4 temperature and depressing the A_3 temperature; but, unlike nickel, manganese has a very powerful stabilising effect on carbides without itself being a very strong carbide former. It does however form stable Mn_3C. Manganese also has a considerable strengthening effect on the ferrite and also increases the depth of hardening to a useful degree. Despite these advantageous properties, the low-carbon, low-manganese steels are not widely used, though manganese is used to some extent to replace nickel in low-alloy steels.

13.83   As recorded at the beginning of this chapter, the development of high-manganese steel in 1882 by Sir Robert Hadfield was one of the landmarks in the history of alloy steels. Hadfield's steel contained approximately 12.5% manganese and 1.2% carbon, and, both as far as composition and heat-treatment are concerned, this is basically the same steel as that used to-day. It is austenitic in type, though if the steel is allowed to cool slowly to room temperature some precipitation of carbides will take place. The steel is therefore water-quenched from 1050°C in order to keep the carbon in solution.

Being austenitic, high-manganese steel is extremely tough and shock-resistant, and although relatively soft, it nevertheless wears extremely well. The reason for this apparently paradoxical situation is the mechanical disturbance causes the surface layers to become extremely hard so that the resultant component has a soft, though shock-resistant core and a very hard, wear-resistant case. Abrasion of the surface will cause the Brinell figure to rise from 200 to 550. The reason for this phenomenon is still uncertain. Some claim that mechanical disturbance of the austenite causes martensite to form at the points of high stress concentration, whilst others have suggested that the cause is simply work-hardening of the austenite. There is little positive evidence of martensite formation and it seems most probable that strain-hardening occurs as a result of mechanical disturbance.

High-manganese steel is available as castings, forgings or hot-rolled sections, but is very difficult to machine because of its tendency to harden as

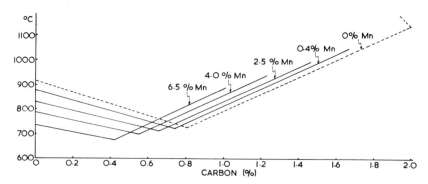

Fig. 13.13   The effect of manganese in displacing the eutectoid point in carbon steel. Like nickel, manganese depresses both A_1 and A_3 temperatures.

Table **13.7** Manganese Steels

B.S. 970 designation	Composition (%) C	Mn	Condition	Typical mechanical properties Yield point (N/mm²)	Tensile strength (N/mm²)	Elongation (%)	Reduction area (%)	Izod (J)	Brinell	Heat-treatment	Uses
150M28	0.25	1.5	Normalised	355	587	20	—	—	—	Oil-quench from 860°C (water-quench for sections over 38 mm dia.). Temper as required.	Automobile axles, crankshafts, connecting rods, etc., where a cheaper steel is required.
150M36	0.35	1.5	28.5 mm bar, oil-quenched and tempered at 600°C	510	711	27	64	68	—	Oil-quench from 840–860°C, temper at 550–650°C and cool in oil or air.	Automobile and general engineering where a carbon steel is not completely satisfactory, but where the expense of a nickel–chromium steel is not justified.
—	1.0	2.0	Hardened and tempered	—	—	—	—	—	620	Oil-quench from 770°C; temper at 200–300°C.	Plug gauges; thread and precision gauges; blanking dies; press tools; reamers; taps; plastics moulds; cams; general purpose low-cost, oil-hardening steel.
—	1.2	12.5	Water-quenched from 1050°C	—	849	40	—	—	Case-550 Core-200	Finish by quenching from 1050°C to keep the carbides in solution.	*Jaw-crushing machinery*: liner plates for ball- and rod-mills, parts for gyratory crushers. *Dredging equipment*; tumblers, buckets, bucket lips, heel plates. *Track work*; points and crossings. Digger teeth, crane track wheels, elevator chains, axle-bar liners, bottom plates for magnets and where non-magnetic properties are required.
—	0.6	14.0 Cr-2.0 Ni-2.0	Water-quenched	345	835	69	55	145	190	Heat at 700°C for 2 h and then water-quench	Japanese Nonmagne 30—an austenitic steel of low magnetic permeability. Used in development of nuclear fusion reactors & magnetic levitation railways.

soon as this is attempted. Despite this drawback, no really suitable substitute has been found for Hadfield steel, for applications where toughness combined with resistance to conditions of extreme abrasion are required, eg rock-crushing machinery, dredging equipment and track-work points and crossings. The austenitic core can be further toughened by the addition of carbide-forming elements, such as chromium and vanadium.

Some manganese steels are shown in Table 13.7.

Steels Containing Tungsten

13.90   Up to the middle of the eighteenth century the name 'tungsten' (or 'heavy stone') implied the mineral scheelite, but in 1781 the renowned chemist Scheele showed that scheelite contained a peculiar acid which he called tungstic acid. Two years later metallic tungsten was isolated by J. J. y Don Fausto d'Elhuyar. To-day tungsten is mined in many countries, the leading producers being China, the CIS, Bolivia, South Korea, the USA, North Korea, Thailand, Australia, Portugal and Canada. The bulk of British imports are from Portugal. The principal ore of tungsten has always been known as 'wolfram' but IUPAC (International Union of Pure and Applied Chemistry) suggests that in future this name should be used to describe the metal itself. However, in this edition we have decided to retain the original name of 'tungsten'.

13.91   Tungsten dissolves in both γ- and α-iron but, having a BCC structure, tends to stabilise α (ferrite). It therefore raises the A_3 temperature and, like chromium, forms a closed γ-loop in the phase diagram. Unlike chromium, however, it inhibits grain growth and therefore has a grain-refining effect. It also reduces decarburisation during working and heat-treatment. Both of these features are useful since high temperatures are involved during the heat-treatment of all tungsten steels.

Tungsten has a high affinity for carbon forming the extremely hard and very stable carbides W_2C and WC and, in steel, a double carbide Fe_4W_2 C. These carbides dissolve very slowly in steel when the latter is heated, and then only at very high temperatures. Once in solution the dissolved tungsten renders transformations extremely sluggish. Therefore, when successfully hardened, steels containing tungsten have a great resistance to tempering conditions and can be heated in the range 600–700°C before carbides begin to precipitate and softening sets in. For this reason tungsten is an important constituent of most high-speed steels (14.10) and hot-working die steels in which it develops high-temperature ('red') hardness following suitable heat-treatment.

13.92   Since the carbides of tungsten are so hard, tungsten is commonly used in small amounts in other tool steels (Table 13.8). It is also used in heat-resisting steels and alloys in which it assists in raising the creep strength at high temperatures.

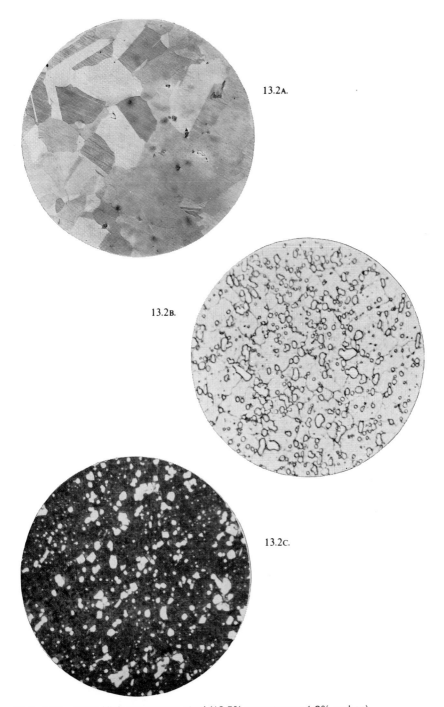

13.2A.

13.2B.

13.2C.

Plate 13.2 13.2A High-manganese steel (12.5% manganese; 1.2% carbon).
Water quenched from 1050°C. Austenite. × 100. Etched in 2% nital. *(Courtesy of Messrs Hadfields Ltd, Sheffield)*
13.2B High-speed steel (18% tungsten; 4% chromium; 1% vanadium).
Air quenched from 1250°C. Carbide globules in an austenite/martensite matrix, × 500. *(Courtesy of Messrs Edgar Allen & Co Ltd, Sheffield)*
13.2C High-speed steel (18% tungsten; 4% chromium; 1% vanadium).
Air quenched from 1250°C and then 'secondary hardened' at about 600°C. Austenite is transformed and the matrix etches darker. × 500. *(Courtesy of Messrs Edgar Allen & Co Ltd, Sheffield)*.

Table 13.8 Steels Containing Tungsten (Other than High-speed Steels)

Type of steel	BS 4659 designation	Composition (%)							Heat treatment	Uses
		C	Si	Cr	Mo	V	W	Other elements		
Tool steel	—	1.0	—	0.75	—	—	0.4	Mn 0.85	Oil-quench from 790°C and temper 200–250°C	Dies; stay-taps; delicate broaches; milling cutters; plugs; gauges; circular cutters; fine press tools and master tools
	BO1	0.9	0.2	0.5	—	—	0.5	Mn 1.20		
Hot-working die steel	BH12	0.35	1.0	5.0	1.5	0.4	1.35	—	Pre-heat to 800°C; soak and then heat quickly to 1020°C; then air-cool. Temper for 1½ hours at 540–620°C.	Extrusion dies, mandrels and noses for aluminium and copper alloys. Hot-forming, piercing, gripping and heading tools. Brass-forging and hot-pressing dies. Other hot-working operations.
Shock-resisting hot- and cold-working tool steel	—	0.5	0.6	1.1	—	—	2.25	—	Oil-quench from 880–920°C; temper at 200–300°C for cold-working tools; temper at 400–600°C for hot-working tools.	Hand and pneumatic chisels, caulking tools and general boilermaker's tools; punches; dies; blanking tools and shear blades. Coal-cutter picks. Cold-heading and nail-making dies. Engraving punches; hot-blanking tools. Hot-forging dies.
	—	0.53	—	1.8	—	0.2	1.9	—		

Type	Designation								Heat treatment	Applications
Hot-working tool steel	—	0.28	0.3	0.85	0.5	0.3	2.25	Ni 2.25	Pre-heat to 750°C and then to 1000°C. Quench in oil. Temper at 650°C for 30 minutes.	Hot-extrusion mandrels for steels and non-ferrous metals. Hot-forging tools. Die holders. Zinc die-casting dies.
Hot-working die steel	BH19	0.4	—	4.25	0.4	2.25	4.25	Co 4.25	Pre-heat to 850°C and then to 1150°C. Quench in oil. Temper at 550–680°C.	Extrusion dies. Hot-piercing punches. Brass-stamping dies. Die-casting dies.
Hot-working die steel	BH21	0.3	0.2	2.85	—	0.35	10.0	—	Pre-heat to 850°C and then heat rapidly to 1150–1200°C; oil-quench (or air-cool small sections); temper for 2–3 hours at 600–700°C	Hot-forging dies and punches for making bolts, nuts, rivets and similar small components where tools reach high temperatures. Hot-forging dies for copper alloys. Extrusion dies, mandrels, pads and liners for the extrusion of copper alloys. Die-casting dies for copper alloys and for pressure die-casting of aluminium alloys.
—		0.3	0.3	3.75	0.5	0.4	10.0	Ni 0.4		

Steels Containing Cobalt

13.100   Much of the World's supply of cobalt is mined in Zaire whilst significant amounts are produced in Morocco, Zambia and Canada. Britain's main sources of supply are Zaire and Zambia.

Cobalt is chemically very similar to nickel and the ores of these metals are generally associated with each other. Until recently cobalt was used principally in permanent-magnet alloys (14.36) and in 'super' high-speed steels (14.11). In the latter it gives a useful increase in red-hardness by promoting extreme sluggishness in transformation.

13.101   Cobalt is an essential constituent of most 'maraging' steels developed since the late nineteen-fifties. These are high-strength steels an important group of which contains 18Ni; 8–12Co; 3–5Mo and small amounts of titanium and aluminium. Carbon is present in very small amounts.

If such a steel is solution treated at 820°C to absorb precipitated compounds, uniform austenite is produced. On cooling in air, an iron–nickel martensite is formed, but unlike ordinary tetragonal martensite (12.22) this is of a lath-like BCC form and is much softer and tougher than ordinary martensite. If the structure is now 'age-hardened' at 480°C for three or more hours coherent precipitates of intermetallic compounds such as (Ti, Al or Mo)Ni_3 are formed and a high tensile strength up to 2400N/mm^2 is developed. Although the crystal structure of $TiNi_3$ is basically hexagonal there is a high degree of 'matching' between its (0001) planes and the (110) crystallographic planes of the martensite matrix (Fig. 13.14). Consequently

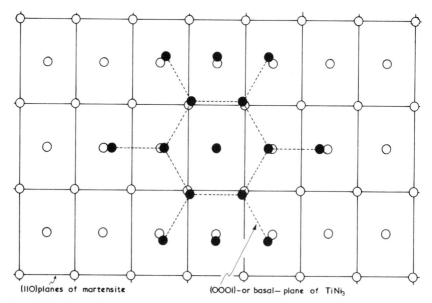

(110)planes of martensite   (0001)– or basal– plane of $TiNi_3$

Fig. 13.14   Illustrating the 'near-matching', and hence coherency, of the (0001) planes of $TiNi_3$ with the (110) planes of iron–nickel martensite.

Table 13.9 *Maraging Steels Containing Cobalt*

Type	Composition (%)						Heat treatment	Typical Mechanical Properties			
	Ni	Co	Mo	Ti	Al	C		0.2% Proof Stress (N/mm²)	Tensile strength (N/mm²)	Elongation (%)	Charpy Impact (J)
18Ni1400	18	8.5	3	0.2	0.1	0.01	Solution treated at 820°C for one hour, air cooled and then aged at 480°C for three hours.	1430	1465	9	52
18Ni1700	18	8	5	0.4	0.1	0.01		1740	1775	8	25
18Ni1900	18	9	5	0.6	0.1	0.01		1930	1965	7.5	21
18Ni2400	17.5	12.5	3.75	1.8	0.15	0.01	Solution treated at 820°C for one hour, air-cooled and aged at 480°C for twelve hours.	2390	2460	8	11
17Ni1600 (cast)	17	10	4.6	0.3	0.05	0.01	Solution treated at 1150°C for four hours, air-cooled and aged at 480°C for three hours.	1650	1730	7	18

a coherent precipitate is formed and the slight distortion produced in the martensite lattice impedes the movement of dislocations. The term 'maraging' is derived from this 'age-hardening' or 'ageing' (17.50) of martensite.

The main function of cobalt seems to be in producing more sites for the nucleation of (Ti, Al, Mo)Ni_3 precipitates. In these steels about half of the strength is due to the original iron–nickel martensite and the other half to the subsequent precipitation hardening. If the precipitation-treatment temperature exceeds 500°C, 'over-ageing' results and the strength begins to fall due to reversion of the structure to austenite.

13.102   These steels combine considerable toughness with high strength in the heat-treated condition and in this respect are much superior to conventional high-strength, low-alloy steels. This high strength is maintained up to 350°C. The wrought grades are very suitable for hot- or cold-working by most processes, whilst weldability is excellent presenting few problems. Heat-treatment involves few difficulties since there can be no decarburisation and no water quenches are required. An added advantage is that these steels, by virtue of composition, are very suitable for nitriding (19.40). They are, of course not stainless steels and under atmospheric conditions will rust at about half the rate of ordinary low-alloy steels. Hence some surface protection is required. Nevertheless stainless maraging steels are available and are based on chromium along with nickel, cobalt and molybdenum.

Among the more exotic uses of these very strong—but expensive—steels were aerospace components such as rocket-motor cases and parts for the suspension system in the Lunar Rover Vehicle. In general engineering, flexible drive shafts for helicopters, pressure vessels, barrels for rapid-firing guns, extrusion rams, cold-forming dies and die-casting dies are all examples which indicate the wide range of uses possible. Some high-strength maraging steels are shown in Table 13.9.

Steels Containing Boron

13.110   Most readers will be familiar with the non-metal boron in the form of the salt, borax. Pure boron, however, is a very hard grey solid with a high melting point (2300°C) and is added to some steels as ferro-boron. Though limited use is made of boron as an alloying element in Britain, boron steels are far more popular in the USA and Sweden.

13.111   Very small amounts—of the order of 0.0005 to 0.005%—are effective, when added to fully-deoxidised steels, in increasing their hardenability by displacing the TTT curves to the right, that is, by reducing the austenite → ferrite + pearlite transformation rates (Fig. 13.15). The bainitic transformation rate may also be reduced. The reduction in transformation rates is most evident in low-carbon steels (0.2–0.4C) but with steels in the region of the eutectoid composition the effect is negligible. A very important feature is that, in respect of hardenability, the amounts of other alloying elements in low-alloy steels can be reduced by as much as half, when small quantities of boron are added, without appreciable increase in

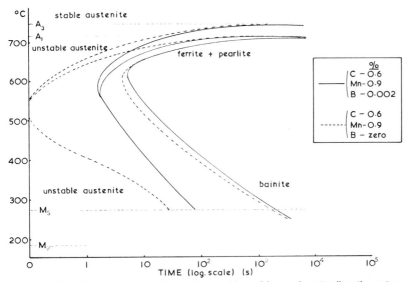

Fig. 13.15 The effects of extremely small quantities of boron in retarding the rates of transformation of austenite.

Table 13.10 *Steels Containing Boron*

Typical Composition (%)						Typical Mechanical Properties			
C	Mn	Ni	Cr	Mo	B	Yield point (N/mm^2)	Tensile strength (N/mm^2)	Elong. (%)	Uses
0.2	1.5	—	—	—	0.0005 min.	690	790	18	High-strength structural steel
0.17	0.6	—	—	0.6	0.003	690	860	18	High-tensile bainitic constructional steels.
0.15	1.3	—	—	0.25	0.003	690	830	18	
0.18	0.85	—	1.0	0.2	0.003	690	900	15	Extra high-strength constructional steels.
0.18	0.85	0.55	1.0	0.25	0.003	690	900	15	
0.45	0.85	0.3	0.45	0.12	0.002	700	1000	15	Automobile and general engineering.
0.16	1.4	—	0.6	0.25	0.002	630	930	14	Welded structures— bridges, submarines, off-shore structures, cranes, penstocks, lorries, trucks, etc.
0.25	1.2	—	0.5	0.20	0.002	1300	1550	8	An abrasion resistant steel— conveyors, gears, dumptrucks, bulldozers, excavators, presses, etc.

transformation rates. In addition to reduction in cost this generally means that cold formability and machinability are improved.

Low-carbon manganese steels containing boron are used in Britain for the manufacture of high-tensile bolts, self-tapping screws and thread-rolled wood screws. Small additions of nickel, chromium and molybdenum are also made to some of these boron steels (Table 13.10).

Steels Containing Silicon

13.120   Silicon is commonly present in amounts up to 0.35% in steels for sand castings, since it improves casting fluidity. However, in most wrought medium-carbon and high-carbon steels silicon is limited to a maximum of 0.25% because of its graphitising effect on cementite. It has a high affinity for oxygen and is used as a deoxidant in many steels.

The increase in high-temperature corrosion-resistance imparted by silicon has been mentioned (13.71) and silicon is a constituent of some heat-resistant valve steels (Table 13.6). Silicon–iron alloys have a high magnetic permeability but a very low hysteresis (14.38) making them particularly suitable for transformer cores and the like.

13.121   Silicon dissolves in ferrite thus increasing its strength and hardness. Since it stabilises ferrite (Fig. 13.16(i)) it raises both A_1 and A_3 temperatures. Low-alloy steels containing silicon as the principal addition are relatively inexpensive, but because silicon has a graphitising effect these steels also contain up to 1% manganese as a carbide stabiliser. Both elements combine in strengthening the ferrite and in increasing hardenability, so that silicon-manganese steels respond to oil-quenching. Subsequent tempering provides a good combination of strength and impact toughness. These steels (Table 13.11) have been widely used for coil and leaf-type springs, as well as for a variety of tools such as punches and chisels where shock-resistance is necessary.

Table 13.11 *Steels Containing Silicon*

BS 970 Part. 5.:	Typical Composition (%)					Type of steel
	C	Si	Mn	Cr	Mo	
250A53	0.53	1.9	0.85	—	—	
250A61	0.61	1.9	0.85	—	—	Spring steels.
925A60	0.60	1.9	0.85	0.3	0.25	
—	0.07	4.3	0.08	—	—	Low-hysteresis iron for transformer cores.

Steels Containing Copper

13.130   Although steels containing copper are little used in Britain they are fairly popular in the USA. Since copper has a FCC structure it tends to stabilise austenite when added to steel. It is sparingly soluble in ferrite

(Fig. 13.16(ii)); this solubility decreasing from 1.8% at 835°C to about 0.3% at 0°C. Copper does not form carbides and in fact has a graphitising effect. Consequently copper is added to low-carbon steels only and then in amounts of no more than 1.5%.

In steel copper has the following effects:
  (i) it improves corrosion resistance;
  (ii) it produces an alloy which can be precipitation hardened enabling an increase of some 175 N/mm² in tensile strength to be obtained.

The steep slope of the solvus line *SV* (Fig. 13.16(ii) indicates that precipitation hardening will be possible for a steel containing, say, 1% copper if it is solution treated at some temperature above X and then quenched. If the carbon content is very low martensite is not produced even if the steel is quenched from the γ-phase region. Some precipitation hardening steels available in the USA are shown in Table 13.12.

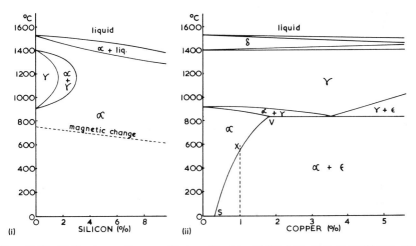

Fig. 13.16 (i) The iron–silicon equilibrium diagram; (ii) the iron–copper equilibrium diagram. The phase ε is solid solution which is, however, almost pure copper.

Table 13.12 *Steels Containing Copper*

Typical Composition (%)						Typical Mechanical Properties		
C	Mn	Cu	Ni	Nb	Others	Yield point (N/mm²)	Tensile strength (N/mm²)	Elongation (%)
0.25	1.3	0.4	—	—	—	340	490	20
0.20	1.0	1.0	1.9	—	—	450	620	20
0.06	0.5	1.2	1.3	0.02	—	550	620	18 Precipitation hardening steels.
0.20	0.7	0.4	0.7	—	Cr 0.3	350	500	22
0.20	1.0	1.25	1.5	0.01	Mo 0.25	550	700	18

In those steels which are not precipitation hardened a small increase in yield strength is obtained over those steels containing no copper, though little improvement in tensile strength is noticed.

HSLA and Other 'Micro-Alloyed' Steels

13.140 Recent decades have seen a continuous improvement in the chemical purity of dead-mild steels. This in turn has meant that 'micro' additions of suitable alloying elements can be effective in producing considerable and *reproducible* improvements in mechanical properties. Previously, varying amounts of impurities would have masked and made less predictable such improvements in properties. Increase in strength can be achieved whilst retaining inherent ductility and toughness.

Very small amounts of vanadium, niobium, titanium and aluminium are now used in high-strength low-alloy (HSLA) steels employed widely for bodywork in the automotive industries as well as in the construction of buildings, bridges and pipelines. The function of these alloying elements is to combine with small amounts of carbon and nitrogen present to form minute carbo-nitride precipitates, typically TiC, TiN and NbCN. These precipitates severely restrict crystal growth which would otherwise take place during necessary hot-working and annealing processes. The strength of HSLA steels tends to increase as grain-size decreases, presumably due to dispersion hardening (8.62) associated with these precipitates. Niobium additions of 0.06–0.1% will develop yield strengths of 350 N/mm^2 in cold-worked annealed products whilst hot-rolled HSLA steels with yield strengths of 550 N/mm^2 are produced. A typical off-shore pipeline composition might be C 0.09; Si 0.3; Al 0.03; Mn 1.35; Nb 0.035; and V 0.07. Such a 'steel' can be more correctly described as a 'dispersion strengthened ferrite' alloy and is very different from an ordinary ferrite/pearlite mild steel.

Within the automotive industry developments in cold-reduced annealed steels are not confined to high-strength products. For example autobody pressings of difficult shapes require steel with very low strength and high ductility not obtainable in normal mild steel. Vacuum degassed steels containing very small amounts of titanium and/or niobium are known as 'interstitial free' steels, since these additions combine with what would otherwise remain as interstitially-dissolved atoms of carbon and nitrogen and instead form *separate precipitates* of TiC, TiN and NbCN. Since no carbon or nitrogen now remain in *solid solution* in the ferrite the product has a very low yield stress combined with a high ductility and hence excellent formability. These 'steels' are useful for severe pressings and are commonly supplied as hot-dipped metallic-coated sheet (21.82). Consumer pressure and (in the USA) legislation now demand that motor car manufacturers give extended anticorrosion warranties of at least five or seven years.

Exercises

1. Estimate the eutectoid composition of a steel containing 5% nickel. What would be the lower-critical temperature of such a steel? Say why such a steel is not manufactured. (Fig. 13.3; Fig. 13.4 and 13.12)
2. Estimate the eutectoid composition and temperature of a steel containing 10% tungsten. What properties would such a steel possess? (Fig. 13.3; Fig. 13.4 and 13.90)
3. Estimate the tensile strength of a steel containing 0.35% carbon; 0.2% silicon; 0.8% manganese and 1.0% nickel in the normalised condition. (13.17)
4. A piece of steel sheet containing 0.05% carbon; 8.0% nickel; 23.0% chromium and 1.0% titanium was heated to 1100°C and cooled in still air. What phases are likely to be present in the structure? (13.18)
5. Discuss the statement 'Carbon is the most important alloying element in steel'.
6. What metallurgical reasoning must be applied when alloy steels are being specified for machine parts?
7. (i) What effect does a substitutional element have upon the iron–carbon thermal equilibrium diagram if (a) it is a γ-stabiliser (b) it is an α-stabiliser? Relate these ideas to the ability of alloy steels to be heat-treated.
   (ii) What effects do these substitutional elements have upon the other properties of steels? (13.11 to 13.18)
8. Discuss the use of nickel and chromium on the heat-treatment properties and microstructures of hardenable steels. (13.40)
9. What is meant by the term 'temper brittleness'?
   Show how the mechanical properties of the steel are affected and indicate a way in which the effect may be minimised. (13.50)
10. Name the main groups of stainless steel, indicating those which are capable of being hardened by heat-treatment.
    Discuss difficulties which may arise with welded stainless steels. (13.35; 13.44 and 13.46)
11. Explain, with the aid of constitutional diagrams (where necessary), the effects of the following alloying elements when added singly to a plain carbon steel: (i) nickel; (ii) chromium; (iii) manganese.
    Why are nickel and chromium often added in conjunction to many low-alloy steels? (13.21 to 13.24; 13.31 to 13.33; 13.81 to 13.83; 13.40)
12. A tool is intended for forging hot steel (forging temperature 1250°C).
    What conditions will the tool have to withstand? Suggest a material suitable for this application and justify your choice. (13.91)
13. Suggest a typical application for each of the steels shown below and explain the role of the additions. Where appropriate, describe the heat-treatments required and the microstructures obtained.
    (a) 1.2% carbon.
    (b) 0.35% carbon; 1.5% manganese; 0.45% molybdenum.
    (c) 0.3% carbon; 12% chromium.
    (d) 0.03% carbon; 18% chromium; 8% nickel. (Table 7.1; Fig. 11.10; 13.50 and Table 13.4; 13.36; 13.44)
14. Give the approximate compositions of alloy steels which would be suitable for the following:
    (i) a rustless kitchen sink unit;
    (ii) an aero-engine connecting rod;
    (iii) the lip of a dredger bucket;
    (iv) a rustless fruit-knife blade;

(v) an extrusion die for copper alloys;
(vi) a high-tensile steel bolt.
Give reasons in each case for the suitability of the steel and outline the heat-treatment necessary for each steel you specify. (13.35; Table 13.4; 13.83; 13.36; Table 13.8; 13.111)

15. What is the derivation of the term 'maraging' as applied to steels?
Outline the principles involved in the heat-treatment of these steels.
Under what circumstances would these expensive materials be used? (13.101)

Bibliography

Bain, E. C. and Paxton, H. W., *Alloying Elements in Steels*, American Society for Metals, 1961.

Cobalt Monograph Series, *Cobalt-containing High-strength Steels*, Centre d'Information du Cobalt, Brussels, 1974.

Hume-Rothery, W., *The Structures of Alloys of Iron*, Pergamon, 1966.

Lula, R. A., *Stainless Steel*, American Society for Metals, 1986.

Peckner, D., and Bernstein, I. M., *Handbook of Stainless Steels*, McGraw-Hill, 1978.

Pickering, F. B., *Physical Metallurgy and Design of Steels*, Applied Science, 1978.

Roberts, G. A. and Carey, R. A., *Tool Steels*, American Society for Metals, 1980.

Thelning, K-E., *Steel and its Heat-treatment: Bofors Handbook*, Butterworths, 1984.

United States Steel, *Isothermal Transformation Diagrams*, 1963.

Wilson, R., *Metallurgy and Heat-treatment of Tool Steels*, McGraw-Hill, 1975

BS 970:1987 and 1988 *Wrought Steels in the Form of Blooms, Billets, Bars and Forgings (Pt. 2: Direct-hardening Alloy Steels; Pt. 4: Stainless, Heat-resisting and Valve Steels; Pt. 5: Carbon and Alloy Spring Steels).*

BS 4490:1969 *Methods for the Determination of the Austenitic Grain Size of Steel.*

14

Complex Ferrous Alloys

High-speed Steels

14.10 The development of high-speed steel had its origins in 1861 when Robert Mushet of Sheffield was investigating the value of manganese in the production of Bessemer steel. He made Henry Bessemer's process viable by deoxidising the 'wild' steel with manganese, and discovered that one composition containing rather a large amount of manganese hardened when it was cooled in air. Analysis showed approximately 6% tungsten to be present in addition, and this alloy subsequently found its way on to the market under the name of air-hardening steel. It was discovered soon afterwards that cooling from 1100°C in an air blast gave better results for tools. Some twenty years later Messrs. Maunsel White and Fredrick Taylor, both of the Bethlehem Steel Company, replaced the manganese in Mushet's steel by chromium, and also increased the tungsten content. These changes in composition enabled the steel to be forged readily and produced a superior tool steel. Production engineers will already be familiar with the name of Taylor, who devoted a great deal of attention to time-study and machine-shop methods.

Modern high-speed steel was first introduced to the public at the Paris Exposition in 1900. The tools were exhibited cutting at a speed of about 0.3 m/s with their tips heated to redness. It was found later that maximum efficiency was obtained with a chromium content of 4% and about 18% tungsten, the carbon content having been reduced from 2.0%, in the original Mushet steel, to between 0.6 and 0.8%. In 1906 vanadium was first added to high-speed steel. Molybdenum was investigated as a substitute for some of the tungsten, but increases in the price of molybdenum in the 1920s suspended this development, particularly since molybdenum steels were regarded as being inferior in properties. Only in more recent years has molybdenum replaced much of the tungsten in many grades of high-speed steel. This is particularly true in the USA, the World's leading producer of molybdenum, where the greatest annual tonnage of high-speed steel is of the molybdenum type.

14.11 The main features of a high-speed steel are its great hardness in the heat-treated condition, and its ability to resist softening at relatively high working temperatures. Thus, high-speed steel tools can be used at cutting speeds far in excess of those possible with ordinary steel tools, since high-speed steel resists the tempering effect of the heat generated.

In high-speed steels ordinary cementite, Fe_3C, is replaced by single or double carbides of three different groups based on the general formulae M_6C; $M_{23}C_6$ and MC, where 'M' represents the total metallic atoms. Thus M_6C is represented by Fe_4W_2C and Fe_4Mo_2C, whilst $M_{23}C_6$ is present as $Cr_{23}C_6$ and MC as WC and VC. Vanadium Carbide is very abrasion-resistant and so improves cutting efficiency with abrasive materials.

Since vanadium is also an important grain-refiner, this effect is very useful in high-speed steels because of the very high heat-treatment temperatures involved. It also increases the tendency towards air hardening by retarding transformation rates. Vanadium tends to stabilise the δ-ferrite phase (13.11) at high temperatures and this leads to carbide precipitation so that high-vanadium steels may be somewhat brittle. Nevertheless vanadium is now added in amounts up to 5.0% to modern high-speed steels.

Up to 12.0% cobalt is also added to 'super' high-speed steels. Not being a carbide former it goes into solid solution and raises the solidus temperature, thus allowing higher heat-treatment temperatures to be used with a consequent increase in the solution of carbides and, hence, hardenability and wear-resistance. Since cobalt promotes excellent red-hardness these super-high-speed steels are useful for very heavy work at high speeds.

As mentioned above, high-speed steels containing greater proportions of molybdenum are now widely used. Generally these alloys require greater care during heat-treatment, being more susceptible to decarburisation than the tungsten varieties. With modern heat-treatment plant, however, this is not an unsurmountable difficulty. Molybdenum high-speed steels are considerably tougher than the corresponding tungsten types and are widely used for drills, taps and reamers.

All of the metallic carbides mentioned above are harder than ordinary cementite, Fe_3C, but the most important feature of high-speed steel is its ability to resist softening influences once it has been successfully hardened. This 'red' hardness, or resistance to softening at temperatures approaching red-heat, is due to tungsten, molybdenum, cobalt, etc being taken into solid solution in the austenite along with carbon before the steel is quenched. Further transformations in the resultant martensite are very sluggish as a result of the large quantities of alloying elements in solid solution, so that the steel can be raised to quite high temperatures before softening sets in due to carbide precipitation.

14.12 In the cast condition the structure of a high-speed steel resembles that of a cast iron. This cast structure is broken up by forging at temperatures between 900 and 1150°C, followed by reheating to about 750°C to relieve working stresses. If required in the soft condition high-speed steel is usually annealed by soaking for about four hours at 850–900°C, followed by a very slow rate of cooling (10 to 20°C per hour) down to 600°C, when it can be cooled to room temperature in still air. At this stage the ferrite present contains relatively little dissolved alloying elements. These are present as massive carbide particles which have precipitated during annealing.

14.13 Obviously, when we are dealing with a complex alloy containing up to six different elements, we cannot represent the system by a simple binary diagram (9.100). However, we can construct a two-dimensional diagram showing the relationship between microstructure, temperature and carbon content for a steel containing a *fixed* quantity of other elements (in this case 18% tungsten, 4% chromium and 1% vanadium). Such a diagram (Fig. 14.1) is generally called a pseudo-binary diagram and, whilst not being a true equilibrium diagram, serves as a useful constitutional chart.

As is shown in the pseudo-binary diagram, austenite forms when the steel is heated to 840°C. Since the eutectoid point has been displaced so far to the left by the influence of tungsten and other added elements this austenite contains little more than 0.2% dissolved carbon. If the steel were quenched from 840°C, the martensite ultimately produced would temper easily and show little advantage over that in ordinary types of tool steel.

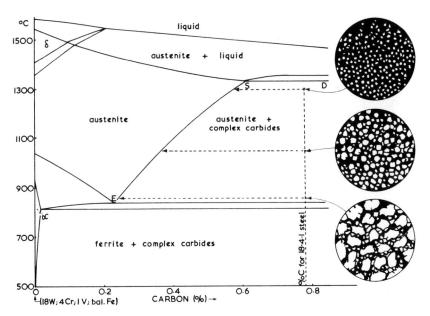

Fig. 14.1   A 'pseudo-binary' diagram representing the structure of 18-4-1 high-speed steel. Here the tungsten–chromium–vanadium content is kept constant but carbon remains variable. The sketches of microstructures show the effects of solution temperature upon the extent of solubility of the complex carbides (light) in the matrix (dark) following quenching. The higher the quenching temperature the greater the solubility of carbides (indicated by the slope of *ES*).

High-speed steel is therefore heated further, so that more carbide is absorbed by the austenite in accordance with the increase in solubility of the carbide, as indicated by the sloping phase boundary ES in Fig. 14.1. Quenching from temperatures above 840°C will produce increasing red-hardness with rise in temperature, due to more carbon and *associated alloying elements* being taken into solution and retained in solution, thus rendering the resultant martensite more sluggish to tempering influences. Even when the solidus temperature is reached at S (approximately 1330°C), no more than 0.6% carbon has been absorbed and the remainder exists as isolated globules of complex carbides (Pl. 13.2). During the heating process most of the chromium carbide has gone into solution by the time 1100°C has been reached and at the quenching temperature of, say, 1250°C, carbides amounting to 10% of the total remain undissolved. These are principally the double carbides Fe_4W_2C and Fe_4Mo_2C and, because vanadium has a very strong carbide-forming tendency, practically all of the VC.

14.14   Therefore, in order to obtain the highest cutting efficiency in a high-speed steel, it must be hardened from a temperature little short of the solidus SD, at which fusion commences. This temperature will vary with steels of different composition but is usually between 1170° and 1320°C. Heat-treatment at such temperatures has its attendant difficulties,

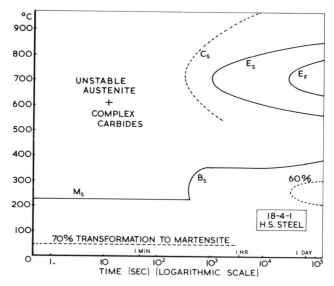

Fig. 14.2 TTT curves for an 18-4-1 high-speed steel.
The initial structure consists of saturated austenite together with some undissolved carbides. The line C_s indicates the commencement of precipitation of primary carbides (since the steel is not of eutectoid composition). E_s and E_F indicate the start and finish of eutectoid ('pearlite') formation. B_s shows the start of bainite formation whilst the associated broken line indicates the time required for 60% transformation to bainite. Note that at room temperature the transformation of austenite to martensite is incomplete (70% at 50°C so that air- or oil quenching will produce a mixture of martensite and retained austenite.

and requires some skill and experience in management, as well as adequate plant. Grain growth and oxidation can be excessive at high temperatures unless the conditions under which the steel is heated are suitably controlled.

Oxidation is effectively prevented by heating the tool in a slightly carburising atmosphere; whilst grain growth is minimised by using a two-chamber furnace. The tool is first pre-heated in the lower-temperature chamber to about 850°C, and then transferred to the higher-temperature chamber (or to a high-temperature salt bath) in which it is heated rapidly to the hardening temperature. The time for which the tool is maintained at the higher temperature must be strictly controlled if grain-growth and decarburisation are to be avoided. Sections below 10 mm in thickness should not be in the high-temperature compartment for more than four minutes whilst sections of 50 mm should be heated for no more than ten minutes. As soon as it has been in the high-temperature chamber for the allotted time it is oil-quenched or cooled in an air blast according to the composition of the steel (Table 14.1). Pre-heating reduces the time during which the surface of the tool would otherwise be in contact with the high-temperature conditions.

Modern trends are towards the use of vacuum furnaces for the large-scale heat-treatment of some high-speed steels, particularly 'molybdenum 562'. The tools are heated to 1250°C in a chamber maintained at high vacuum, the resistance-heating elements being of graphite which will not oxidise under such conditions. When the quenching temperature is reached the vacuum chamber is purged with high-velocity *inert* gas so that hardening of the charge is achieved. Such a process has obvious advantages over traditional methods of heat-treatment:
1) a bright, oxide-free finish is retained;
2) the tool surface suffers neither loss nor gain of carbon;
3) repeatability and uniformity of control are possible;
4) hazardous fumes and toxic wastes are absent.

14.15   Due to the sluggishness produced in the austenite → martensite transformation by the presence of alloying elements, the quenched high-speed steel will contain considerable amounts of retained austenite (Fig. 14.2). It is necessary to transform this to martensite, and this can be accomplished by sub-zero treatment or, more generally, by tempering. Low-temperature tempering between 300 and 400°C reduces the hardness slightly but increases the toughness. On tempering between 400 and 600°C the hardness increases again, often to a figure higher than the original (Fig. 14.3), though this will be accompanied by some reduction in toughness. This phenomenon, known as 'secondary hardening', is associated with the presence of vanadium. Secondary hardening is usually carried out by heating the tool for one-half to three hours at 550–600°C according to the composition of the steel. Some austenite may still be present following tempering, and for this and other reasons double tempering is now generally used. The retained austenite is 'conditioned' during the first tempering process and some of it changes to martensite on cooling. During the second tempering process the transformation goes still further towards completion

Table 14.1 *High-speed Steels*

Type of steel	BS4659 designation	Composition (%)						Heat-treatment		Hardness (VPN)	Characteristics and uses
		C	Cr	W	V	Mo	Co	Quench in oil or air blast from:	Secondary-hardening treatment		
14% Tungsten	—	0.65	4.0	14.0	0.25	—	—	1250–1300°C.	Double temper at 565°C for 1 hour	800–860	General shop practice—all-round work, moderate duties. Also for some cold-working tools (blanking tools and shear blades).
18% Tungsten (18–4–1)	BT1	0.75	4.2	18.2	1.2	—	—	1290–1310°C.	" "	800–890	Lathe, planer and shaping tools; millers and gear cutters. Router bits; hobs; reamers; broaches; taps; dies; drills; slitting discs. Hacksaws; bandsaws; etc. Roller-bearings at high temperatures, e.g. gas turbines.
22% Tungsten	BT20	0.78	4.5	21.5	1.4	0.5	—	1280–1320°C.	" "	830–950	Large hobs and milling cutters, automatic lathe tools, accurate form-tools, large broaches, etc. Holds its edge longer than the 18% tungsten type of steel.
6% Cobalt	BT4	0.8	4.5	18.5	1.2	0.5	5.5	1290–1320°C.	" "	800–900	Lathe, planer and shaper tools on very hard materials and for extra heavy work, e.g. high-tensile steels, hard-cast irons, new and brake-hardened railway tyres, etc. Austenitic steels which

		C	Cr	W	V	Mo	Co	Hardening temp.	Tempering	Hardness	Uses
12% Cobalt	BT6	0.8	4.75	21.5	1.5	0.5	12.0	1300–1320°C.	"	900–950	Lathe, planer and shaper tools, milling cutters, hobs, twist drills, etc., for exceptionally hard materials. Has maximum red-hardness and toughness. Suitable for severest machining duties—manganese steel, steels of over 1200 N/mm². T.S., close-grained cast irons.
		0.78	4.5	18.8	2.0	0.7	10.0		"		
12% Tungsten–5% vanadium	BT15	1.5	4.5	12.5	5.0	—	5.0	1230–1300°C.	Double temper at 560°C for 1 hour	900–1000	Extremely hard due to high vanadium %, but also tough. Excellent red-hardness and edge retention. High accuracy and finish—form tools, turning, milling, shaping. Also cutting gas-turbine alloys.
18% Tungsten–5% vanadium	—	1.5	5.0	17.5	5.0	0.5	8.75	1250°C.	Double temper at 560°C for 2 hours	900–990	For very hard materials—gas-turbine alloys, etc. Tool bits, form tools, milling cutters, etc.
Molybdenum "562"	BM2	0.83	4.25	6.5	1.9	5.0	—	1250°C.	Double temper at 565°C for 1 hour	850–900	Roughly equivalent to standard 18–4–1 high-speed steel, but tougher. Used for drills, reamers, taps, milling cutters, punches, threading dies, cold-forging dies.
9% molybdenum	BM1	0.8	3.75	1.6	1.25	9.0	—	1200–1230°C.	"	830–900	A general-purpose Molybdenum-type high-speed steel—drills, taps, reamers, cutters. More susceptible to decarburisation during heat-treatment than ordinary 18–4–1.
									"		
9% molybdenum–8% cobalt	BM42	1.0	3.75	1.65	1.1	9.5	8.25	1180–1210°C.	Triple temper at 530°C for 1 hour	830–935	Similar uses to 12% Cobalt–21% Tungsten high-speed steel.

as more of the retained austenite changes to martensite. This second tempering process also relieves internal stresses which have been set up as austenite transformed to martensite during the first tempering process. In the case of high-speed steel the initial quench from a high temperature tends not to produce internal stresses, since an adequate cushion of soft austenite is retained. The first tempering process causes much of the retained austenite to transform to hard martensite and this may lead to the introduction of internal micro-stresses.

The final structure of a correctly heat-treated high-speed steel will consist of isolated globules of complex carbide, in a matrix of martensite which contains sufficient dissolved alloying elements to render it sluggish to tempering influences. The traditional type of 18-4-1 high-speed steel, when correctly hardened, will retain a high hardness to about 550°C whilst the more expensive cobalt steels will retain a similar hardness for a further 100°C (Fig. 14.3). Although high-speed steels are generally noted for their use in machining operations under severe conditions, their high strength at elevated temperatures makes them useful for hot-working tools of other types, eg hot shears and punches for the hot-forming of nuts. The wear of such steels can be further improved by nitriding (19.40) after normal heat-treatment has been completed.

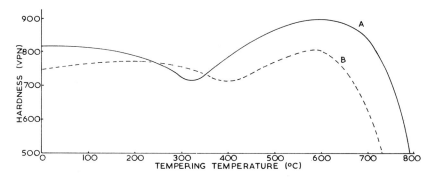

Fig. 14.3 The relationship between hardness and tempering temperature for high-speed steels.
(A) Represents the 12% cobalt type, and (B) the ordinary 18% tungsten type.

Cemented Carbide and Other 'Cermet' Cutting Materials

14.20 Mention has already been made (6.50) of the production of cutting tools by powder metallurgy, and though, strictly speaking, not ferrous alloys, these materials will be discussed here because of their close association with high-speed steels.

Cemented carbide tools are prepared by sintering at high temperature and pressure, in an atmosphere of hydrogen, a mixture of tungsten carbide powder with powdered cobalt. Sometimes tungsten carbide is replaced by powdered carbides of other refractory metals, tantalum, molybdenum, titanium and also by silicon carbide. In each case cobalt (or some other tough metal) forms a ductile bonding matrix between the carbide particles.

The resultant structure resembles that of a high-speed steel, in that spherical particles of carbide are held in a tough matrix, but whereas the properties of a high-speed steel are developed by heat-treament, the structures of cemented carbide materials are produced by mechanical methods.

14.21   In other 'cermets', which are sintered products similar in principle to the cemented carbides mentioned above, the object is to combine the high-temperature strength and hardness of suitable metallic oxides, nitrides or borides with sufficient ductility provided by a metallic bonding element, thus giving a material with considerable resistance to thermal and mechanical shock and suitable for jet-engines or other high-temperature applications as well as for cutting tools. Typical hard-phase/ductile-bond cermets are shown in Table 14.2.

Table 14.2 *Some Typical Cermets*

Group	Hard Phase	Bonding Metal or Alloy	Uses
	Tungsten carbide, WC	Cobalt	
Carbides	Titanium carbide, TiC	Molybdenum, cobalt or nickel/ chromium alloy	Cutting tool bits.
	Molybdenum carbide, Mo_2C	Cobalt or nickel	
	Silicon carbide, SiC	Cobalt or chromium	
	Chromium carbide, Cr_3C_2	Nickel	Gauge blocks, slip gauges, wire-drawing dies, jet-engine parts
	Aluminium oxide, Al_2O_3	Cobalt, iron or chromium	'Throw-away' tool bits, sparking-plug bodies, rocket motor and jet-engine parts.
Oxides	Magnesium oxide, MgO	Magnesium, aluminium, iron, cobalt or nickel	
	Chromium oxide, Cr_2O_3	Chromium	
	Titanium boride, TiB_2	Cobalt or nickel	
Borides	Chromium boride, Cr_3B_2	Nickel	Cutting-tool tips.
	Molybdenum boride, Mo_2B	Nickel or nickel/chromium	

Magnetic Properties and Materials

14.30   In the vicinity of either a magnetised material or an electrical conductor carrying a current, a compass needle is acted upon by a couple which tends to make it rotate. This couple is attributed to the presence of a magnetic field of force surrounding the magnetised material (or the current-carrying conductor). The *strength of the magnetic field* at any point is denoted by H (Fig. 14.4). This magnetic field will give rise to *magnetic induction* in any material upon which it acts and may be conveniently produced either by a 'permanent magnet' or by a current-carrying solenoid which, as Ampère illustrated, is equivalent to a series of magnets. The quantity, magnetic induction, is denoted by B and is related to the applied field strength by the equation:

$$B = \mu H$$

where μ is the *magnetic permeability* and is generally not constant for a material. If the value of μ is high then a relatively smaller applied field, H, is required to induce a given field, B, within the material.

Whilst iron is the metal generally associated with magnetism three other transition metals—cobalt, nickel and the 'rare earth' gadolinium—are also very strongly magnetic, or *ferromagnetic* as it is termed. Nevertheless other elements possess weak magnetic properties. Many metals are *paramagnetic*, that is they are very weakly attracted by a strong magnetic field. Consequently a paramagnetic metal like chromium when placed in a magnetising field carries a slightly greater magnetic flux than would a corresponding vacuum. Some metals, and all non-metals, are *diamagnetic*, that is they are weakly repelled by a magnetic field. Thus diamagnetic copper would carry a slightly smaller magnetic flux than the corresponding vacuum

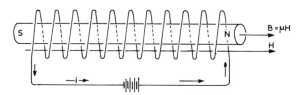

Fig. 14.4   The magnetic flux in a material is $B = \mu H$, where H is the magnetic field in the solenoid and μ is the permeability of the material.

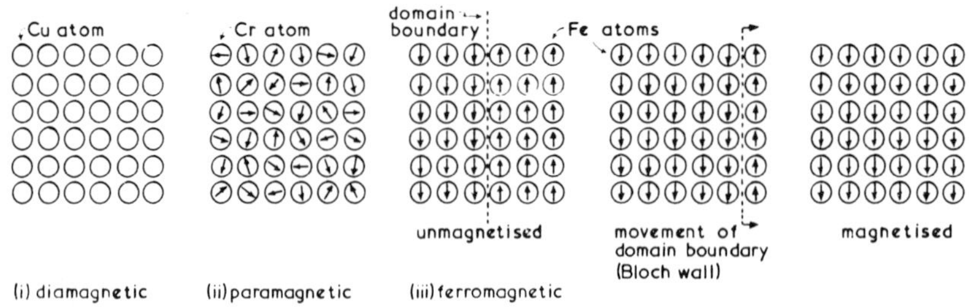

,Cu atom	, Cr atom	domain boundary	,-Fe atoms	
(i) diamagnetic	(ii) paramagnetic	(iii) ferromagnetic unmagnetised	movement of domain boundary (Bloch wall)	magnetised

Fig. 14.5   The three types of magnetism.
In diamagnetic substances like copper the *atoms* have no resultant moment whilst in paramagnetic substances (like chromium) there is no resultant field although the individual atoms have magnetic moments. In ferromagnetic materials domains are re-aligned during magnetisation.

under the influence of a magnetising field. Only those metals, alloys and mixtures which exhibit ferromagnetism are useful as magnetic materials.

14.31 Ferromagnetism   The relationship between an electric current and a magnetic field was demonstrated by Ampère and it is now considered that this magnetic field originates from the 'spin' of electrons. In a stable atom or ion the electrons can be thought of as particles moving in orbitals around the nucleus. At the same time each electron behaves as though it were spinning on its own axis and it is this spin which generates the magnetic field.

Pauli's Exclusion Principle indicates that in a stable atom not more than two electrons are able to occupy the same energy level. These two electrons will possess opposite 'spins' so that in effect their fields cancel out (Fig. 14.6(i)). In a large number of a elements electrons are 'paired' in their orbitals in this way so that such elements will be diamagnetic. Atoms of some of the transition elements, however, have only partly-filled *d* sub-shells (Table 1.4) so that unpaired electrons will be present. Consequently in an atom more electrons will spin in one direction than in the other (Fig.14.6(ii)) and such an atom will have a magnetic moment. For this reason the ferromagnetic elements and the strongly paramagnetic elements are found in the transition groups in the Periodic Table. In the elements iron, cobalt, nickel and gadolinium the magnetic moments are large enough and the ions suitably spaced in the crystal structure so that when they become magnetically aligned (during 'magnetisation') a powerful magnetic field is produced.

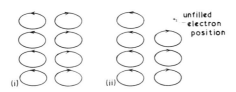

Fig. 14.6   Unfilled 'd' sub-shells as the basis for ferromagnetism.

14.32   Magnetic properties are very dependent upon atomic spacing. If the atoms are too far apart the forces between them are too small to resist the effects of thermal agitation in throwing spins out of alignment so that there is no resultant field. If the atoms are very close together then inter-atomic forces are too great to permit alignment of spins. Favourable conditions appear to be obtained when the atomic radius is between about 1.5 and 2.0 times the radius of the sub-shell containing the un-paired electrons. Thus iron, cobalt, nickel and gadolinium fulfil these requirements whilst other transition metals such as titanium, chromium and manganese are only just outside the range. In fact when manganese contains sufficient interstitially-dissolved nitrogen to increase its atomic spacing by an adequate amount it becomes ferromagnetic. Similarly the interatomic spacings in some ceramic materials like $BaO.6Fe_2O_3$ make them suitable as ferromagnetic materials. Modern magnetic alloys contain varying proportions of iron, cobalt, aluminium, nickel, copper, niobium, chromium and titanium; whilst sintered ceramics contain boron, barium, samarium and other 'rare earths' such as neodymium. Currently magnets based on neodymium are replacing those containing samarium because of the greater abundance of the former and its consequent lower cost.

14.33   Within the crystal of a ferromagnetic material are small regions known as *magnetic domains*. These are about 0.01 mm in width and microscopic examination using a colloidal suspension of magnetite (which behaves like minute iron filings) shows them to be of a shape as indicated in Fig. 14.7. Inside each domain the electron spins are aligned, but since the alignment varies from one domain to the next the resultant fields cancel and the material appears un-magnetised. If an external magnetic field is applied then the energy of these domains which are aligned *against* the applied field is raised so that they begin to reverse direction. In this way the favourably oriented domains grow at the expense of the others by a movement of boundaries which separate them. The boundary zones (called 'Bloch walls') move so that domains become similarly oriented and a resultant field is set up (Fig. 14.8)

If this alignment is retained when the magnetising field is removed the material is said to be *magnetically hard* but in many cases the magnetism is not retained and the material is said to be *magnetically soft*. Whether or not a material is magnetically hard or soft depends to a large extent on its physical condition. Magnetic hardness generally coincides with physical hardness, possibly because residual strains oppose randomisation of domains once the material has been magnetised. The presence of inclusions or holes gives rise to regions of low energy and these tend to anchor Bloch walls. Since they are difficult to move the material is more difficult to magnetise but, once magnetised, the magnetism is retained.

High magnetic 'coercivity' can also be promoted by using very small needles which have a thickness less than the natural domain wall thickness of the material. These needles are spontaneously magnetised and the direction of magnetisation cannot be changed because domain walls are absent. Some of the modern permanent-magnet materials consist of suitably aligned very small needles of iron which are then permanently embedded in a plastics matrix.

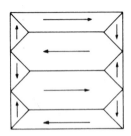

Fig. 14.7   Magnetic domains in an 'unmagnetised' ferromagnetic material.

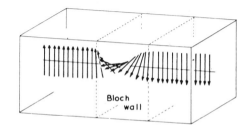

Fig. 14.8   Re-alignment within a Bloch wall.

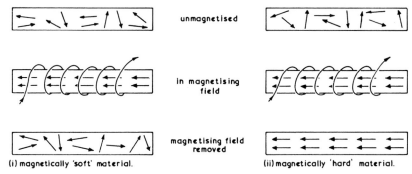

Fig. 14.9 The effects of a magnetic field on magnetically 'soft' and 'hard' materials. The arrows represent domains.

When a ferromagnetic material has been magnetised to saturation increase in temperature causes a corresponding increase in randomisation so that the magnetism progressively decreases reaching zero at the Curie point (769°C, 1120°C and 358°C for iron, cobalt and nickel respectively) when the material ceases to be ferromagnetic and becomes paramagnetic. This change is reversible on cooling, ie

$$\alpha\text{-iron} \overset{769\,°C}{\rightleftharpoons} \beta\text{-iron}$$

14.34 A permanent magnet is designed to retain a high intensity of magnetisation for an indefinite period provided it is not subjected to unusually high demagnetising influences. The actual magnetic strength, or flux, in a permanent magnet is dependent upon its design and the steel or other alloy from which it is made. The essential characteristics of permanent-magnet materials are:

(a) coercive force;
(b) residual magnetism or remanence;
(c) energy product value.

Coercive force refers to the ability of the magnet to withstand demagnetising influences. Remanence is the amount of magnetism remaining in the magnet after the magnetising field has been switched off. The energy product value is the quantity of magnetic energy possessed by the magnet, and is calculated from the demagnetising curve *RU* (Fig. 14.10). The magnetising force is expressed in 'ampere-turns per metre', whilst the SI unit of induced flux density is the tesla (T), symbolised by *H* and *B* respectively.

14.35 We will now consider what is known as the hysteresis loop. Fig. 14.10 shows a typical hysteresis loop (*PRUQST*) and magnetisation curve (*OP*) for a permanent magnet. When a piece of steel is placed in a solenoid with a current passing through it, magnetic flux is induced in the steel. The amount of this flux can be measured in webers (Wb) and the flux density in teslas (T), where Wb = Vs and T = Wb/m². In Fig. 14.10 magnetic flux

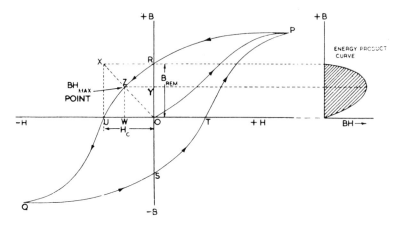

Fig. 14.10 Magnetic hysteresis loop.

values (B) are plotted vertically, whilst the magnetising force values (H), calculated from the current passing through the solenoid and the number of turns of wire on it, are plotted horizontally.

Beginning with a small value of H and gradually increasing it, corresponding values of B and H are plotted, and the curve OP produced. This is the magnetisation curve, which is continuous until it runs almost parallel to the horizontal axis at P. This state represents complete magnetic saturation of the specimen.

If we now gradually reduce the magnetising field to zero, the corresponding fall in induced magnetic flux is represented by the line PR. The intercept OR represents the value of the magnetic flux remaining in the steel when all external magnetising influence is removed. This is termed the remanent magnetism or remanence B_{rem}.

To demagnetise the steel, and at the same time to measure the amount of force required to achieve this, a current is once more passed through the solenoid but in the reverse direction, and gradually increased. This produces the curve RU, which is called the demagnetisation curve. The intercept OU corresponds to the amount of force required to completely demagnetise the steel, and is called, as already stated, the coercive force. To complete the loop the operations outlined are carried on to negative saturation and then repeated in the reverse sense.

The hysteresis loop, therefore, represents the lagging of magnetic flux produced behind the magnetising force producing it. (The word 'hysteresis' was in fact derived by Ewing from a Greek word meaning 'coming after'.) The area of the loop is proportional to the amount of energy lost when the specimen it represents is magnetised or demagnetised, and for permanent magnets the area of the loop must be as large as possible.

The ultimate criterion of a permanent magnet is the maximum product of B and H obtained from a point on the demagnetisation curve. In Fig. 14.10 it is shown that by completing the rectangle $RXUO$ and drawing the diagonal OX, the point Z at which the diagonal cuts the curve RU gives an approximation to the point at which the product of YZ and WZ is a maximum for the curve. This is called BH_{max}, and corresponds to the maximum energy the magnet can provide in a circuit external to itself.

14.36 Fully hardened carbon steels containing about 1.0 % carbon were originally used for permanent magnets. These were later replaced by alloy steels containing chromium and tungsten, and later cobalt. The cobalt steels are the only ones of any practical value today and these have been superseded by alloys of the Alni-Alnico-Alcomax series. These were developed from magnetic aluminium–nickel alloys discovered by Japanese workers in 1930.

The Alni-Alnico series have a lower remanence than cobalt steels but a much higher coercive force. Moreover, they retain magnetism much better under influences of shock and of heat (Fig. 14.11). The most modern alloys of the Alcomax-Hycomax series are superior also in remanence. They are *anisotropic* materials, that is they have better magnetic properties along a preferred axis. This effect is obtained by heating the alloy to a high temperature and allowing it to cool in a magnetic field. Groups of atoms become orientated as the magnet cools and this orientation is frozen in.

It is estimated that the current usage of permanent magnets in domestic applications averages fifty per household in the Western World. A large number of these are permanent magnets of the rare-earth type developed in recent years. Most of them are produced by powder-metallurgy methods. Amongst the earliest were magnets of the samarium–cobalt ($SmCo_5$) type followed by alloys containing copper as well as the rare earths and cobalt. This eventually led to the development of high-energy magnets of the type $Sm(Co–Cu–Fe.x)$ where x is Zr, Ti or Hf.

In more recent years commercial production of permanent magnets based on the intermetallic compound $Nd_2Fe_{14}B$ has begun. Such new materials combine very high polarisation coercivity with a very high value of BH_{max}, these values being the most important indications of permanent magnet performance.

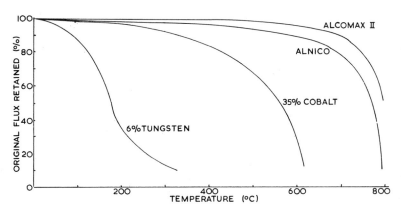

Fig. 14.11 The retention of magnetic flux by representative magnet alloys at high temperatures.

Table 14.3 Permanent Magnet Materials

Type of Material	Composition (%)							Special Details	Magnetic Properties			Characteristics and uses
	Fe	Co	Al	Ni	Cu	Ti	Other elements		B_{rem} (mT)	H_c (kA/m)	BH_{max} (kJ/m³)	
Carbon steel	99	—	—	—	—	—	C 1	—	900	4.4	1.5	Obsolete
35% Co steel	53	35	—	—	—	—	C 0.9; Cr 6; W 5.	—	900	20	7.5	Largely obsolete—a few 'semi-hard' applications eg hysteresis motors
'Alnico'	57	12	9	16	5	1	—	Cast or sintered	725	45	13.5	General purposes
'Alcomax 3 SC'**	50	24	8	14	3	—	Nb 0.6; Si 0.3	Chill cast; part grain orientation	1320	56	49	Widely used in electrical, communications and engineering industries. All are anisotropic alloys which can only be produced with a linear axis of magnetisation.
'Columax'**	50	24	8	14	3	—	Nb 0.6; Si 0.3	Full grain orientation	1350	60	60	
'Hycomax 3'**	35	34	7	15	4	5	—	Cast or sintered	900	128	43	
'Hycol 3'**	35	34	7	15	4	5	—	Full grain orientation	1040	130	60	
Rare-earth/cobalt*	—	66	—	—	—	—	Sm 34 (as the compound SmCo5)	Sintered in argon	800	600	127	High performance/low inertia motors for NC machine tools; high performance and/or very small stepper motors for watches, camera apertures, rocket guidance systems, etc.
'Neodure'*	76	—	—	—	—	—	Nd 16; B 8 (as the compound Nd2Fe14B)	Sintered ceramic	1100	800	215	Magnetic bearings, switches, servomotors, actuators, audio-visual and similar consumer products, disc file systems, etc.

Material				Composition	Condition				Applications
'Feroba 1'	—	—	—	BaO.(Fe₂O₃)₆	Sintered ceramic	220	135	8	Cheap general purpose magnets. Cycle dynamos. Magnetic repulsion bearings on Kwh meters. Small magnetic glandless couplings for water and gas meters.
'Feroba 2'*	—	—	—	BaO.(Fe₂O₃)₆	Sintered ceramic	380	160	29	Loudspeakers, door catches, computer track shifters on disc store devices.
'Feroba 3'*	—	—	—	SrO.(Fe₂O₃)₆	Sintered ceramic	370	240	26	Magnetic separators; DC motors for car heaters and wipers. Larger motors for washing machines, lawn mowers, outboard motors, electrical cars and industrial drives.
'Crofeco'*	Bal.	10 to 20	—	Cr 20 to 30	Cold-worked	1200	50	40	Mechanical properties similar to steel—tough and strong. Used for rotors, speedometers, general purpose round bar magnets, etc.
'Lodex'*	20	11	—	Pb 69	Fe-Co alloy particles in a lead matrix.	650	70	25	An 'elongated single domain' material—iron/cobalt alloy needles in a lead matrix.
Magnetic rubber I	—	—	—	BaO.(Fe₂O₃)₆	Ceramic in a rubber matrix	160	96	4	Magnetic display boards, eg BBC's former Weather Forecast Chart; refrigerator door seals and field magnets for cheap low performance motors (toys and windscreen washers).
A*						230	160	10	

*Anisotropic alloys and materials, properties measured along the preferred axis.

14.37 Special mention must be made of those magnetic materials which give a 'square' hysteresis loop of the type shown in Fig.14.12(ii). Such materials are used in computer memory units. Magnetic saturation to a moderate value of B is reached quickly and a large proportion of the flux is retained when the magnetising field is removed; thus it can be used as a 'memory' to activate systems when called upon to do so. Even with a small reverse field the strong flux is still available for 'read-out' and when it is desired to 'scrub' the information this can be done cleanly, leaving the device ready for remagnetisation. Such a material has a high remanence, B_{rem}, and a small but definite coercive force, H_c. Ceramics based on Mn–Mg–Fe oxides and Co–Fe oxides are used.

14.38 Components like transformer cores and dynamo pole pieces need to be made from a material which is magnetically 'soft' but of high permeability. Permeability (μ) changes with the magnetic field, rising to a maximum and then falling again. Since a transformer core works on alternating current, the core is first magnetised, then demagnetised and then remagnetised with reverse polarity. This type of cycle produces the closed curve *PRUQST* (Fig. 14.10) and the area which it encloses is a measure of the amount of energy wasted in overcoming remanent magnetism each time the current is reversed (which occurs many times per second). This energy is converted into heat, and it is therefore necessary to use materials, for transformer cores, which give very narrow hysteresis loops, and hence, a small 'hysteresis loss' in order to reduce energy loss and avoid overheating of the core. The differences in properties between a material suitable for a permanent magnet and one suitable for a transformer core is indicated by the different shapes of the hysteresis loops as shown in diagrams (i) and (iii) in Fig. 14.12.

Soft iron and iron–silicon alloys are most useful in respect of low hysteresis. The iron–silicon alloys contain up to 4.5 % silicon and practically no carbon, and are usually supplied in the form of dead-soft sheet from which transformer-core laminations can be stamped. The use of laminations

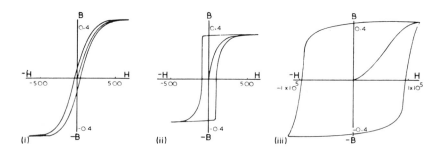

Fig. 14.12 Hysteresis loops for representative magnetic materials.
(i) A magnetically 'soft' material (transformer cores). (ii) A 'square loop' material (computer memory units). (iii) A modern magnetically 'hard' material (permanent magnets). (Note the compressed *H*-scale for the latter).

reduces losses from eddy currents, which would be more prevalent in a solid core.

Nickel–iron alloys are also of importance as low-hysteresis, high-permeability alloys. 'Permalloy' contains Ni78 and Fe22 with carbon and silicon kept as low as possible. 'Mumetal' contains Ni74; Cu5; Mn up to 1; Fe–bal. Both alloys are used in communications engineering, where the high permeability will utilise to the full the small currents involved. In addition to being used for transformer cores, these alloys are used as shields for submarine cables. Like the iron–silicon alloys, they are used in the dead-soft condition, the best treatment consisting of heating to 900°C, followed by slow cooling, and then heating to 600°C, followed by cooling in air. This treatment avoids the ordered structure which would otherwise form in these alloys. Since the magnetic properties are very sensitive to work-hardening these alloys must always be used in the annealed state.

In all magnetically soft materials properties depend upon the ease of domain movement and so work-hardening and the presence of impurities must be avoided. In iron–silicon alloys the crystal orientation of the sheet is also controlled. Since magnetisation is easiest in the <100> directions, the sheet is prepared with the (100) planes parallel to the plane of the sheet. This leads to a lower power loss in transformer cores.

A number of ceramic-type magnetically-soft materials are now used in electronics and radio-communications equipment. Most of these materials are based on the formula Fe_3O_4 ('magnetite' which also occurs naturally as a magnetic mineral—the 'lodestone' of the ancient mariners). In this formula the Fe^{++} ions are replaced by various mixtures of Mn^{++}, Zn^{++}, Ni^{++} and Cu^{++} ions and are generally known as 'ferrites'. This term has no direct connection with 'ferrite' which signifies α-iron but is a reference to the general formula $M.Fe_2O_4$. They are powder metallurgy products.

Exercises

1. Discuss reasons for and against the use of molybdenum as a replacement for tungsten in high-speed steels. (14.10; Table 14.1)
2. By reference to Fig. 14.1 estimate the % carbon in solution in austenite at (i) 900°C; (ii) 1300°C, for a typical high-speed steel.
   Show how this affects the quenching temperature for such a steel. (14.13; 14.14)
3. Discuss the significance of Fig. 14.2 on the heat-treatment programme of a high-speed steel. (14.15)
4. (a) What do you understand by the term 'tool steel'?
   (b) Give typical approximate chemical compositions for tool steels suitable for thread rolls or other components that need great resistance to wear and abrasion.
   (c) Describe how you would heat-treat such a tool, mentioning the precautions that would be necessary to produce a satisfactory article. (Table 14.1; 14.14; 14.15)
5. What are the basic requirements of a tool steel? Describe the heat-treatment of a high-speed tool steel, giving reasons for each step. Explain how the

microstructure is related to the service requirements of the tool. (14.11; 14.14; 14.15)

6. Distinguish between *ferromagnetism*, *paramagnetism* and *diamagnetism*. Outline the basic theory which seeks to explain ferromagnetic properties. (14.30 to 14.33)

7. Explain why the value BH_{max} is important in assessing the suitability of a permanent magnet material for service.
   Outline developments which have taken place in the formulation of permanent magnet materials in recent years. (14.35 to 14.37)

8. Define *magnetic permeability* (μ) and sketch representative hysteresis curves for both magnetically 'hard' and 'soft' materials.
   For what purposes are magnetically 'soft' materials used? (14.38)

9. Some magnetic materials are said to give a 'square' hysteresis loop.
   Explain what this means and say for what purposes such materials are used. (14.39)

Bibliography

Berkowski, A. E. and Kellner, E., *Magnetism and Metallurgy*, Academic Press, 1970.

Hoyle, G., *High Speed Steels*, Butterworths, 1988.

Rosenberg, H. M., *The Solid State*, Oxford University Press, 1978.

Thelning, K-E., *Steel and its Heat-treatment: Bofors Handbook*, Butterworths, 1984.

Wilson, R., *Metallurgy and Heat-treatment of Tool Steels*, McGraw-Hill, 1975.

15

Cast Irons and Alloy Cast Irons

15.10 In Victorian times almost anything was likely to be made of cast iron—street lamps, domestic fireplaces, railings, mock-gothic church window frames and the ornamental water fountain in the local park. Sadly many of these relics of the nineteenth century foundryman's art are no more since most fell victim to the urgent need for steel during the Second World War. Even as late as the author's childhood toy 'six-shooters' were of white cast iron, whereas to-day's children play with 'space-age' guns made of plastics materials. One important reason for the widespread use of cast iron in Victorian times was the fact that in those days not all pig iron produced was suitable for conversion into merchantable steel. In particular those pig irons high in phosphorus and sulphur were suitable only for ornamental castings which required little strength. Today cast iron is used exclusively for engineering purposes and its technology, like that of other alloys, continues to be developed at highly sophisticated levels. 'Spheroidal graphite' cast iron and, more recently, 'compacted graphite' irons are of this type.

15.11 Ordinary cast iron is very similar in composition and structure to the crude pig iron from the blast furnace. In most foundries the 'pigs' are melted, often with the addition of low-grade steel scrap, in a cupola, any necessary adjustments in composition being made during this melting process. The relative scarcity and consequent high cost of metallurgical coke has led some foundries to adopt electric melting, usually by high-frequency induction furnaces. An obvious advantage of electric melting is its chemical cleanliness and lack of sulphur contamination as compared with cupola melting where the iron is in intimate contact with the burning coke. Consequently many of the high quality grey cast irons, as well as spheroidal-graphite irons and alloy cast irons are now melted electrically.

Because production costs of pig iron are relatively low as compared with other alloys, and since no expensive refining process is necessary, cast iron is a cheap metallurgical material which is particularly useful where a casting

requiring rigidity, resistance to wear or *high compressive strength* is necessary. Other useful properties of cast iron include:
  (i)   good machinability when a suitable composition is selected;
  (ii) high fluidity and the ability to make good casting impressions;
  (iii) fairly low melting range (1130–1250°C) as compared with steel;
  (iv) the availability of high strengths when additional treatment is given
        to suitable irons, eg spheroidal-graphite iron, compacted-graphite
        irons or pearlitic malleable irons.

The structure and physical properties of a cast iron depend upon both chemical composition and the rate at which it solidifies following casting.

The Effects of Composition on the Structure of Cast Iron

15.20   Ordinary cast iron is a complex alloy containing a total of up to 10% of the elements carbon, silicon, manganese, sulphur and phosphorus; the balance being iron. Alloy cast irons, which will be dealt with later in this chapter, contain also varying amounts of nickel, chromium, molybdenum, vanadium and copper.

15.21 Carbon   can exist in two forms in cast iron, namely as free graphite or combined with some of the iron to form iron carbide (cementite). These two varieties are usually referred to as graphic carbon and combined carbon respectively, and the total amount of both types in the specimen of iron as total carbon. In ordinary engineering cast iron the form in which carbon is present depends largely upon the cooling rate during solidification and upon the quantity of silicon present. In alloy cast irons those elements mentioned above will also affect the resultant structure. In fact the effects of various elements on the microstructure of cast iron is generally the same as in steels (13.12).

Cementite is a hard, white, brittle compound, so that irons which contain much of it will present a white fracture when broken, and will have a low resistance to shock. At the same time they will possess a high resistance to wear. Such irons are called white irons. The fractured surface of a cast iron containing graphite, however, will appear grey, and the iron will be termed a grey iron.

Although the form of the iron–carbon diagram shown in Fig. 11.1 is of great value to the practical metallurgist it is not a true *equilibrium* diagram because in fact cementite is not an 'equilibrium phase'. Cementite is said to be a *metastable* phase, that is, it has a natural tendency to decompose forming a mixture of iron (ferrite) and carbon (graphite). In ordinary steels this decomposition almost never occurs because in iron supersaturated with carbon the nucleation of cementite takes place much more readily than the nucleation of graphite. Once cementite has formed it is quite stable and for practical purposes can be regarded as an equilibrium phase.

In a simple iron–carbon alloy containing, say, 3% carbon, solidification would begin with the formation of primary austenite and at 1131°C a

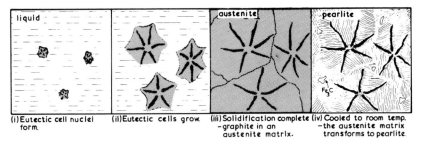

liquid		austenite	pearlite
(i) Eutectic cell nuclei form.	(ii) Eutectic cells grow.	(iii) Solidification complete – graphite in an austenite matrix.	(iv) Cooled to room temp. – the austenite matrix transforms to pearlite.

Fig. 15.1 The formation of 'flake' graphite resulting from the growth of eutectic cells. In (iv) some free cementite will also be present as the solubility of carbon in austenite decreases from 2.0% at 1131°C to 0.8% at 723°C.

eutectic consisting of austenite and cementite would form. However, engineering cast irons contain sufficient silicon to increase the instability of cementite to the point where graphite is precipitated from solution instead. Therefore primary austenite will separate out first until the eutectic temperature (1131°C) is reached, but at that temperature the eutectic which forms consists of austenite and *graphite*. This eutectic develops from nuclei and is in the form of approximately spherical particles known as *eutectic cells*. The layers of austenite and graphite develop in roughly radial form, both being deposited directly from the liquid phase (Fig. 15.1). In Fig. 15.1 and Plate 15.1A graphite appears to be in the form of separate *flakes*, but in fact the eutectic cells are three-dimensional and roughly spherical in shape so that the graphite layers can be regarded as something like those shown in Fig. 15.2. That is, rather like a cluster of potato crisps growing out from a central nucleus.

As in the case of the eutectoid point in steels (13.14) the eutectic point in cast irons may be displaced to the left by the presence of other elements, notably silicon. Thus, a cast iron containing less than 4.3% carbon may be

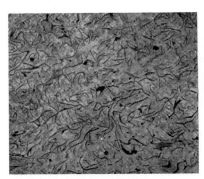
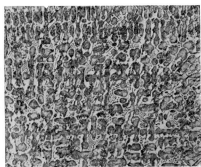

Plate 15.1   15.1A Grey cast iron.
Graphite flakes in a matrix of pearlite. × 60. Etched in 4% picral.
15.1B White cast iron.
Primary cementite (white) and pearlite. × 100. Etched in 4% picral. *(Courtesy of BCIRA)*.

Fig. 15.2   A three-dimensional impression of graphite flakes in a eutectic cell.

hyper-eutectic in composition and so deposit primary graphite (instead of cementite) before the eutectic cells start to form. This primary graphite is generally coarse and was known as 'kish'.

Rapid cooling, which produces a greater degree of undercooling, initiates the formation of a greater number of eutectic cells and also more frequent branching in the eutectic graphite 'leaves', giving much finer graphite flakes. The smaller the eutectic cells, the finer the graphite flakes and hence the better the mechanical properties.

The presence of graphite gives a softer iron which machines well because of the effect of the graphite flakes in forming chip cracks in advance of the edge of the cutting tool (6.61). Moreover, since graphite occupies a greater volume in the solid structure than does carbon in solution in the liquid iron, it tends to counteract the effects of shrinkage during solidification.

15.22 Silicon   dissolves in the ferrite of a cast iron and is the element which has the predominant effect on the relative amounts of graphite and cementite which are present. Silicon tends to increase the instability of cementite (13.12) so that it decomposes, producing graphite and hence, a grey iron. The higher the silicon content, the greater the degree of decomposition of the cementite, and the coarser the flakes of graphite produced.

Thus, whilst silicon actually strengthens the ferrite by dissolving in it, at the same time it produces softness by causing the cementite to break down to graphite. When, however, silicon is present in amounts in excess of that necessary to complete the decomposition of all the cementite, it will again cause hardness and brittleness to increase. Both the direct and indirect effects of silicon must therefore be considered.

The presence of silicon in a cast iron is beneficial in so far as it increases the fluidity of the molten iron, and so improves its casting properties.

15.23 Sulphur   The possible effect of the silicon present in an iron cannot be completely estimated without some reference to the sulphur content, since sulphur has the opposite effect, in that it tends to stabilise cementite. Sulphur, then, inhibits graphitisation and so helps to produce a hard, brittle, white iron. Moreover, its presence as the sulphide FeS in cast iron will also increase the tendency to brittleness.

During the melting of cast iron in the cupola there is a tendency for some silicon to be oxidised and lost in the slag. At the same time some sulphur will always be absorbed from the coke in the cupola. Both of these changes in composition tend to make the iron more 'white' as they are opposed to the formation of graphite. The foundryman therefore makes allowances for these changes during melting in the cupola.

15.24 Manganese   The effect of sulphur is governed, in turn, by the amount of manganese present. Manganese combines with sulphur to form manganese sulphide, MnS, which, unlike iron(II) sulphide, is insoluble in the molten iron and floats to the top to join the slag. The indirect effect of manganese, therefore, is to promote graphitisation because of the reduction of the sulphur content which it causes. Manganese has a stabilising effect on carbides, however, so that this offsets the effect of sulphur reduction in promoting graphitisation. The more direct effects of manganese include the hardening of the iron, the refinement of grain and an increase in strength.

15.25 Phosphorus   is present in cast iron as the phosphide Fe_3P, which forms a eutectic with the ferrite in grey irons, and with ferrite and cementite in white irons. These eutectics melt at about 950°C, so that high-phosphorus irons have great fluidity. Cast irons containing 1% phosphorus are, therefore, very suitable for the production of castings of thin section. High phosphorus contents should be avoided in heavy sections however, since Fe_3P is brittle and lowers strength considerably. Its presence tends to promote increased shrinkage.

Phosphorus has a negligible effect on the stability of cementite, but its direct effect is to promote hardness and brittleness due to the large volume of phosphide eutectic which a comparatively small amount of phosphorus will produce. Phosphorus must therefore be kept low in castings where shock-resistance is important.

The Effect of Rate of Cooling on the Structure of Cast Iron

15.30   A high rate of cooling during solidification tends to favour the formation of cementite rather than graphite. That is, the higher the rate of cooling for any given cast-iron composition the 'whiter' and more brittle the casting is likely to be. This effect is important in connection with the choice of a suitable iron for the production of castings of thin section. Supposing an iron which, when cooled slowly, had a fine grey structure containing small eutectic cells were chosen for such a purpose. In thin sections it would cool so rapidly that cementite would form in preference to graphite and a thin section of completely white iron would result. Such a section would be brittle and useless.

15.31   This effect is illustrated by casting a 'stepped bar' of iron of a suitable composition (Fig. 15.3 and Pl. 15.2). Here, the thin sections have

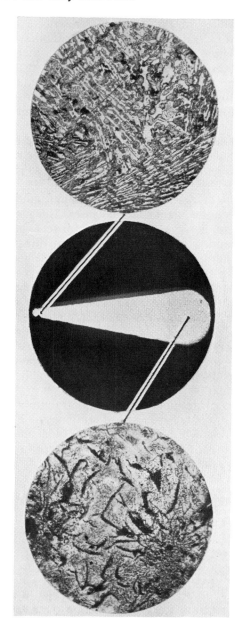

Plate 15.2   Illustrating the effects of thickness of section, and hence rate of cooling on the structure of a grey iron. The thinnest part of the section has cooled quickly enough to produce a white iron structure, whilst the core of the thickest part has a grey iron structure. The relationships between sectional thickness and microstructure are similar to those indicated in Fig. 15.3 on the opposite page. Both micrographs × 300 and etched in 2% nital. Macrosection × 3.

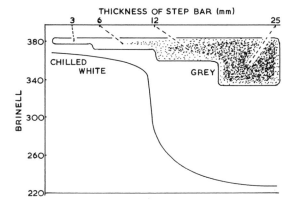

Fig. 15.3   The effect of thickness of cross-section on the rate of cooling, and hence upon the microstructure of a grey cast iron.

cooled so quickly that solidification of cementite has occurred, as indicated by the white fracture and high Brinell values. The thicker sections, having cooled more slowly, are graphitic and consequently softer. Due to the chilling effect exerted by the mould, most castings have a hard white skin on the surface. This is often noticeable when taking the first cut in a machining operation.

In casting thin sections, then, it is necessary to choose an iron of rather coarser grey fracture than is required in the finished casting. That is, the iron must have a higher silicon content than that used for the production of castings of heavy section.

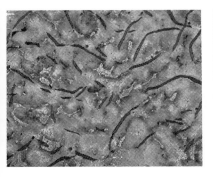
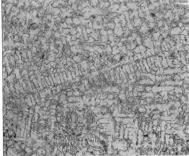

Plate 15.3   15.3A A phosphoric grey cast iron.
Flakes of graphite (dark) and patches of phosphide eutectic (light) in a matrix of pearlite. × 100. Etched in 4% picral. *(Courtesy of BCIRA)*
15.3B An inoculated grey cast iron.
Note the refinement produced in the structure. × 60. Etched in 4% picral. *(Courtesy of BCIRA)*.

The Microstructure of Cast Iron

15.40   Neglecting any patches of phosphide eutectic already mentioned, the following structures are possible for cast irons of different compositions and treatments:

(a) *Primary Cementite and Pearlite*. This structure is typical of the hard, white, low-silicon (and possibly high-sulphur) irons and is found also in other types of iron which have been chilled. A cementite-austenite eutectic forms at 1131°C and as the alloy cools to 723°C the austenite transforms to pearlite.

(b) *Primary Cementite, Graphite and Pearlite*. These are 'mottled' irons of such compositions that localised small changes in either composition or cooling rate will favour the formation of either cementite *or* graphite. The remaining austenite present at 723°C transforms to pearlite.

(c) *Graphite and Pearlite*. This structure is typical of a high-duty grey iron which has solidified as a graphite-austenite eutectic and in which the austenite has then transformed to pearlite at 723°C.

(d) *Graphite, Pearlite and Ferrite*. This will generally be a coarser grey iron which will be weaker and softer. The silicon content will be high. Here the graphite-austenite eutectic has cooled slowly enough through the eutectoid temperature (723°C) for some of the carbon which was dissolved in the austenite to deposit on to the existing graphite flakes leaving a matrix of ferrite. Some of the austenite has also transformed to pearlite.

(e) *Graphite and Ferrite*. This variety of iron usually has a very high silicon content. The graphite-austenite eutectic has cooled slowly enough for the *whole* of the carbon dissolved in the austenite to diffuse on to existing graphite flakes leaving a matrix entirely of ferrite. Such a cast iron is very soft and easily machined. The ferrite present will contain dissolved silicon and manganese.

In addition to the phases enumerated above, iron phosphide, Fe_3P, as previously mentioned, may be present in the microstructure. In white or mottled iron it will be present as a ternary eutectic with cementite and ferrite, whilst in coarse grey irons of the type (e) above, a binary eutectic with ferrite will be formed. This binary eutectic contains only 10% phosphorus, so that an overall 1% phosphorus in the iron may produce sufficient phosphide eutectic to account for 10% by volume of the resulting structure. The embrittling effect of the phosphide eutectic network will then be evident.

'Growth' in Cast Irons

15.50   If an engineering cast iron is heated for prolonged periods above 700°C the pearlitic cementite tends to break down to give a mixture of ferrite and graphite. This leads to an increase in volume of the iron. Inevitably some warping will follow and surface cracks will form. As hot gases penetrate these cracks internal oxidation will occur and lead to the progressive deterioration of the casting. Cast iron moulds used for the casting of non-ferrous ingots develop surface cracks in this way, giving a 'crazy paving' like surface. Trouble will then arise due to dressing oil collecting in these cracks. Fire bars in ordinary domestic grates often break up due to the combined action of 'growth' and internal oxidation.

Certain alloy cast irons have been developed to resist 'growth' at high temperatures. Silal is relatively cheap and contains about 5% silicon with low carbon. This very high silicon content will favour the formation of graphite rather than cementite during solidification so that its structure consists entirely of ferrite and graphite, and no cementite is present which can cause growth. Unfortunately Silal is rather brittle, so that, where the higher cost is justified, the alloy Nicrosilal can be used with advantage. This is an austenitic nickel–chromium cast iron (see Table 15.5).

Varieties and Uses of Ordinary Cast Iron

15.60 Foundry irons can be classified in the following groups according to their properties and uses:

15.61 Fluid Irons which can be used where mechanical strength is of secondary importance, and in which high fluidity is obtained by means of high silicon (2.5–3.5%) and high phosphorus (up to 1.5%) contents. Both silicon and phosphorus improve fluidity, and a high silicon content will further ensure that thin sections will have a reasonably tough grey structure, since silicon prevents deep chilling. Such irons were used extensively at one time for the manufacture of lamp-posts, fireplaces and railings, but they have now been replaced by ceramic materials for uses such as these.

15.62 Engineering Irons must have mechanical strength and the best type of microstructure is one with a small eutectic cell size (small graphite flakes in a pearlite matrix). An iron of this type will possess the best all-round mechanical properties and also good machinability. The silicon content will depend on the thickness of section to be cast, but in general it will not exceed 2.5% for castings of thin section, and may be as low as 1.2% for castings of heavy section.

The phosphorus content must be kept low where a shock-resistant casting is necessary, though up to 0.8% phosphorus may be present in some cases in the interests of fluidity. Sulphur must also be kept low (below 0.1%), as it leads to segregation, hard spots and general brittleness.

Irons of this type can have hard, wear-resistant, chilled surfaces introduced at various parts of the casting if desired. This is done by inserting chilling fillets at appropriate points in the sand mould. These fillets cause rapid cooling, and hence a layer of white iron on the surface of the casting at these points. At the same time the core of the casting remains grey and tough.

Table 15.1 *Compositions and Uses of Some Typical Cast Irons*

| | Composition (%) | | | | |
C	Si	Mn	S	P	Uses
3.50	1.15	0.8	0.07	0.10	Ingot moulds and heat-resisting castings
3.30	1.90	0.65	0.08	0.15	Motor brake drums
3.25	2.25	0.65	0.10	0.15	Motor cylinders and pistons
3.25	2.25	0.50	0.10	0.35	Light machine castings
3.25	1.75	0.50	0.10	0.35	Medium machine castings
3.25	1.25	0.50	0.10	0.35	Heavy machine castings
3.60	1.75	0.50	0.10	0.80	Light and medium water pipes
3.40	1.40	0.50	0.10	0.80	Heavy water pipes
3.50	2.75	0.50	0.10	0.90	Ornamental castings requiring low strength—now obsolete.

15.63 Heavy Castings   do not require a high silicon content, as there is little danger of chilling. Usually, silicon is no higher than 1.5%, with up to 0.5% phosphorus for such castings as columns and large machine frames. In castings, such as ingot moulds, which are exposed to high temperatures a close-grained, non-phosphoric iron should be used.

High-strength Cast Irons

15.70   Over the years a number of modifications of grey cast iron were made available to give enhanced properties. Thus 'semi steel' was made by melting a high proportion of steel scrap along with pig iron in the cupola, whilst Lanz Perlit iron was a fine-grained product made by casting a low-silicon iron into heated moulds in order to retard the cooling rate. Various 'trademarked' fine-grained irons were produced by 'inoculating' a suitable molten iron with materials such as calcium silicide. Modern

Table 15.2 *Typical Mechanical Properties of Engineering Grey Irons (to B.S. 1452)*.*

BS 1452: Grade	0.1% proof stress (N/mm^2)	Tensile strength (N/mm^2)	Total strain at failure (%)	Compressive strength (N/mm^2)	Fatigue limit (N/mm^2)
150	98	150	0.68	600	68
180	117	180	0.60	672	81
220	143	220	0.52	768	99
260	169	260	0.57	864	117
300	195	300	0.50	960	135
350	228	350	0.50	1080	149
400	260	400	0.50	1200	152

*The values are for guidance only—chemical compositions cannot be suggested since these will depend largely on sectional thickness of castings.

high-strength irons rely almost entirely on treatments which replace the flake-graphite of ordinary grey iron with graphite in the form of either spherical particles or 'compacted' flakes.

15.71 Spheroidal-graphite (SG) Cast Iron   (also known as **Nodular Iron** or **Ductile Iron**). The production of ordinary cast iron has fallen considerably in recent years due largely to such influences as the development of welding fabrication methods and the use of other materials like concrete and plastics. Nevertheless the introduction of SG iron in 1946 was without doubt the most important event during this century in the iron trade since in SG iron the engineer finds a worthy competitor to cast steel but often at much lower cost.

In an ordinary grey cast iron graphite is present as 'flakes' which tend to have sharp-edged rims. Since these flakes have negligible strength they act as wide-faced discontinuities in the structure whilst the sharp-edged rims introduce regions of stress-concentration. In SG cast iron the graphite flakes are replaced by spherical particles of graphite (Plate 15.4A), so that the metallic matrix is much less broken up, and the sharp stress raisers are eliminated.

The formation of this spheroidal graphite is effected by adding small amounts of cerium or magnesium to the molten iron just before casting. Since both of these elements have strong carbide-forming tendencies, the silicon content of the iron must be high enough (at least 2.5%) in order to prevent the formation, by chilling, of white iron in thin sections. Magnesium is the more widely used, and is usually added (as a nickel–magnesium alloy) in amounts sufficient to give a residual magnesium content of 0.1% in the iron. SG cast irons produced by the magnesium process have tensile strengths of up to 900 N/mm^2 or even higher in some heat-treated irons. The term 'SG iron' really describes a family of cast irons, which include some alloy irons, but in all cases treatment by inoculants is

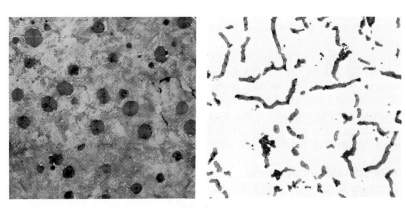

Plate 15.4   15.4A A spheroidal-graphite cast iron.
Here the graphite has been made to precipitate in nodular form by adding a nickel–magnesium alloy. × 100. Etched in 4% picral. *(Courtesy of BCIRA)*
15.4B A compacted graphite cast iron.
Unetched to show the rounded edges of the graphite flakes. × 100. *(Courtesy of BCIRA).*

Table 15.3   *Grades of Spheroidal-graphite Iron to BS 2789**

Grade	0.2% proof stress (N/mm²)	Tensile strength (N/mm²)	Elongation (%)	Typical hardness (H_b)	Structure—spheroidal graphite in a matrix of:
370/17	230	370	17	175	Ferrite
420/12	250	420	12	200	Ferrite
500/7	310	500	7	205	Ferrite-pearlite
600/3	350	600	3	230	Ferrite-pearlite
700/2	400	700	2	265	Pearlite
800/2	460	800	2	300	Pearlite, or a heat-treated structure.

*No 'typical compositions' are included since composition, related to the structure in the last column, will vary with sectional thickness. It will be noted that the 'Grade' designation indicates both tensile strength and % elongation.

employed to produce spheroidal-graphite particles. Some patents claim the production of SG iron using the following substances instead of cerium or magnesium: calcium, calcium carbide, calcium fluoride, lithium, strontium, barium and argon. In all cases high-quality raw material, as free as possible from carbide-stabilising elements is required. BS 2789 grades SG irons according to their mechanical properties and some of these are shown in Table 15.3.

Those irons consisting of graphite nodules in a ferrite matrix will have high ductility and toughness whilst those consisting of graphite nodules in a pearlite matrix will be characterised by high strength. Some of these irons are heat-treated to give even better mechanical properties. Thus, sections of the American motor industry harden some of their SG iron gears by the use of 'interrupted austempering'. This involves austenitising the gears at 900°C for 3.5 h in a nitrogen atmosphere followed by quenching to 235°C and holding at that temperature for 2 h. Since transformation from austenite occurs isothermally at 235°C there is little distortion in shape. It is claimed that SG iron hypoid ring and pinion gears are comparable with those of steel in terms of fatigue and also have a greater torsional strength.

Tensile strengths of the order of 1600 N/mm² (with an elongation of 1%) can be obtained by austempering SG iron at 250°C, following an initial austenitising at 900°C; whilst higher austempering temperatures up to 450°C will yield bainitic structures of lower strengths (900–1200 N/mm²) but elongations up to 14%. SG iron crankshafts cast to near final shape are less expensive and some 10% lighter than equivalent forged components. They are heat-treated in a similar way to the gears mentioned above.

15.72 Malleable Cast Irons   are irons which have been cast to shape in the ordinary way, but in which ductility and malleability have been increased from almost zero to a considerable amount by subsequent heat-treatment. The names of the two original malleabilising processes, the Blackheart and the Whiteheart, refer to the colour of a fractured section after heat-treatment has been completed. A more recent process for the manufacture of pearlitic malleable iron aims at the production of castings which, when suitably treated, will have tensile strengths of up to 775 N/mm². In all three processes the original casting is of white iron, which will be very brittle before heat-treatment. This white structure is achieved by keeping the silicon content low, usually not much more than 1.0%, whilst in the Whiteheart process some sulphur, too, is permissible, thus increasing still further the stability of the cementite.

15.72.1. Blackheart Malleable Iron castings are manufactured from white iron of which the following composition is typical:

Carbon	2.5%
Silicon	1.0%
Manganese	0.4%
Sulphur	0.08%
Phosphorus	0.1%

In the original process the castings were fettled and placed in white-iron containers along with some non-reactive material, such as gravel or cinders, to act as a mechanical support. The lids were luted on to exclude air and the containers loaded into an annealing furnace which was fired by coal, gas, fuel oil or pulverised fuel. Treatment temperatures varied between 850 and 950°C and were dependent upon the desired quality of product and the analysis of the original casting. The duration of heat-treatment was between 50 and 170 hours, again depending upon the type of casting and the analysis of the iron. The development of modern annealing furnaces in which castings need no longer be packed in an insulating material has resulted in a reduction in heating and cooling-off times, so that malleabilising can now be effected in 48 hours or less. Such furnaces are of the continuous type in which a moving hearth carries the castings slowly through a long furnace of small cross-section and in which a controlled non-oxidising atmosphere is circulated.

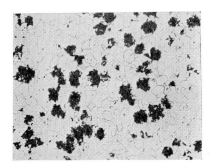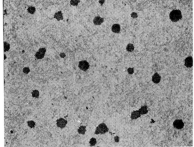

Plate 15.5   15.5A Blackheart malleable cast iron.
'Rosettes' of temper carbon in a matrix of ferrite. × 60. Etched in 4% picral. *(Courtesy of BCIRA)*
15.5B Whiteheart malleable cast iron.
Micrograph taken from the *core* of the material and showing rosettes of temper carbon in a matrix of pearlite. × 60. Etched in 4% picral. *(Courtesy of BCIRA).*

The effect of this prolonged annealing is to cause the cementite to break down, but instead of forming coarse graphite flakes, the carbon is precipitated in the form of small 'rosettes' of 'temper carbon'. A fractured section will thus be black, hence the term 'blackheart'. The final structure, which consists entirely of ferrite and finely divided temper carbon, is soft, readily machinable and almost as ductile as cast steel. Blackheart malleable castings find particular application in the automobile industries because of the combination of castability, shock-resistance and ease of machining they afford. Typical uses include rear-axle housings, wheel hubs, brake shoes, pedals, levers and door hinges.

15.72.2. Whiteheart Malleable Iron castings are manufactured from white iron of which the following composition is typical:

Carbon	3.3%
Silicon	0.6%
Manganese	0.5%
Sulphur	0.25%
Phosphorus	0.1%

In this process the castings are heated in contact with some oxidising material, such as haematite ore, for between 70 and 100 hours at a temperature of about 1000°C.

During annealing, the carbon at the surface of the casting is oxidised by contact with the hematite ore, and lost as carbon dioxide. This causes more carbon to diffuse outwards from the core, and, in turn, this is lost by oxidation. Thus, after treatment a thin section may be completely ferritic, and on fracture present a steely white appearance; hence the name 'whiteheart'. Heavier sections will not be completely decarburised, and so, whilst the outer layer is ferritic, this will merge into an inner zone containing some pearlite, and at the extreme core some nodules of temper carbon. The surface layer may exhibit some oxide penetration.

Thin-sectioned components requiring high ductility are often made in the form of whiteheart malleable castings. Examples include fittings for gas, water, air and steam pipes; bicycle- and motorcycle-frame fittings; parts for agricultural machinery; switchgear equipment; and parts for textile machinery.

15.72.3. Pearlitic Malleable Iron is produced from raw material similar in composition to that of blackheart malleable iron, though in some cases alloying elements may be used to stabilise the required pearlitic matrix in the final structure. When an unalloyed iron is used it is first malleabilised either fully or partially at about 950°C to cause adequate breakdown of the primary cementite. It is then reheated to 900°C so that carbon will dissolve in the austenite present at that temperature. Subsequent treatment consists either of air-cooling (to produce a pearlitic matrix) or some other form of heat treatment designed to give a bainitic or tempered-martensite type of matrix.

If an alloyed iron is used it is often given a normal malleabilising treat-

Table 15.4 *Mechanical Properties (Minimum Values) of Malleable Cast Irons.*

Type	Grade	Diam. of test piece (mm)*	0.5% proof stress (N/mm²)	Tensile Strength (N/mm²)	Elongation (%)	Hardness (H_B) (not mandatory)
White-heart malleable BS 309	W-410/4	9	190	350	10	229 max.
		12	230	390	6	
		15	250	410	4	
	W-340/3	9	—	270	7	229 max.
		12	—	310	4	
		15	—	340	3	
Black-heart malleable BS 310	B-340/12	15	200	340	12	149 max.
	B-310/10	15	190	310	10	
	B-290/6	15	170	290	6	
Pearlitic malleable BS 3333	P-690/2	15	540	690	2	241–285
	P-570/3	15	420	570	3	197–241
	P-540/5	15	340	540	5	179–229
	P-510/4	15	310	510	4	170–229
	P-440/7	15	270	440	7	149–197

*Whiteheart malleable castings of greater cross-section are likely to be stronger since the core will contain more pearlite.

ment, the presence of carbide-stabilising elements causing retention of the pearlitic matrix during this process.

The final structure will consist of rosettes of temper carbon in a matrix of pearlite, bainite or tempered martensite according to the final heat-treatment given. Tensile strengths in the region of 775 N/mm² are possible and in many respects the product can compete with cast steel, despite the extra cost of the heat-treatment processes.

15.73 Compacted-graphite (CG) Irons   These materials have their origins in work carried out by Morrogh at BCIRA* many years ago and are characterised by graphite structures (Pl. 15.4B)—and consequently also physical properties—which are intermediate between those of ordinary grey flake-graphite irons and those of SG iron. With appropriate chemical treatment the graphite flakes produced are short and stubby and have *rounded* edges. This has led to the term 'vermicular iron' being used in the USA, though the flakes are not quite 'wormlike' as the title implies.

CG iron is produced when molten iron of near-eutectic composition is first de-sulphurised in the ladle and then treated at 1400°C with a single alloy containing appropriate amounts of magnesium, titanium and cerium so that the resultant iron contains Mg (0.015–0.03%), Ti (0.06–0.13%) and Ce (a trace). The presence of magnesium tends to produce spheroidal graphite (as it does in SG iron) and this is controlled by the restraining tendency of titanium which makes the amount of the magnesium less critical.

The mechanical properties are roughly intermediate between those of grey iron and those of SG iron, whilst the resistance to scaling and 'growth' at high temperatures are very good. Since in cast irons generally graphite

* British Cast Iron Research Association, Alvechurch, Birmingham.

Table 15.5 *Compositions, Properties and Uses of Some Typical Alloy Cast Irons*

Type of iron	Composition (%)								Mechanical properties		Properties and uses
	C	Si	Mn	S	P	Ni	Cr	Other elements	Tensile strength N/mm^2	Brinell	
Chromidium	3.2	2.1	0.8	0.05	0.17	—	0.32	—	278	230	Cylinder blocks, brake drums, clutch casings, etc.
Ni-Tensyl	2.8	1.4	—	—	—	1.5	—	—	355	230	An 'inoculated' cast iron.
Wear-resistant	3.6	2.8	0.6	0.05	0.50	—	—	Vanadium 0.17	—	—	Piston rings for aero, automobile and diesel engines. Has wear-resistance and long life.
Ni–Cr–Mo iron	3.1	2.1	0.8	0.08	0.09	0.5	0.9	Molybdenum 0.9	371	300	Automobile crankshafts. Hard, strong and tough.
Heat-resistant	3.4	2.0	0.6	0.05	0.10	0.35	0.65	Copper 1.25	278	220	Good resistance to wear and heat-cracks. Used for brake drums and clutch plates.
Wear- and shock-resistant	2.9	2.1	0.7	0.05	0.10	1.75	0.10	Molybdenum 0.8 Copper 0.15	448	300	Crankshafts for diesel and petrol engines. Good strength, shock-resistance and vibration-damping capacity.

Material											Description
Ni-hard	3.3	1.1	0.5	—	—	4.5	1.5	—	—	600	A martensitic iron used to resist severe abrasion, e.g. chute plates in coke plant
Ni-resist	2.8	1.3	—	—	—	21.0	2.0	—	170	140	Chemical plant handling sulphur or chloride solutions. Austenitic, non-magnetic and corrosion-resistant.
Ni-resist	2.9	2.1	1.0	0.05	0.10	15.0	2.0	Copper 6.0	216	130	Pump castings, handling concentrated brine; an austenitic, corrosion-resistant alloy.
Nicrosilal	2.0	5.0	1.0	—	—	18.0	5.0	—	263	330	An austenitic, corrosion-resistant alloy, pump components, flue gas dampers.
No-mag	2.8	1.5	6.0	—	—	11.0	—	—	232	210	An austenitic, corrosion-resistant alloy.
Silal	2.5	5.0	—	—	—	—	—	—	170	—	Resistant to high temperatures.
High Chromium BS 4844: Grade 3A	2.7	0.5	1.0	—	—	15.0	—	Molybdenum 1.3	—	450	
Grade 3C	2.6	0.5	1.0	—	—	20.0	—	Molybdenum 1.5	—	450	Abrasion-resistant
Grade 3E	3.0	0.5	1.0	—	—	25.0	—	Molybdenum 0.75	—	500	White irons

Table 15.6 Some Flake and SG Austenitic Alloy Cast Irons covered by BS3648

Grade* Flake (L) / S.G. (S)	Typical Composition (%)					Typical mechanical properties					Typical uses
	C	Si	Mn	Ni	Cr	0.2% proof strength (N/mm²)	Tensile strength (N/mm²)	Elong. (%)	Compressive strength (N/mm²)	Hardness (HB)	
L–NiMn 13–7	3.0	3.25	7.0	13.0	—	—	180	—	735	135	Non-magnetic pressure covers for turbine generator sets; housings for switch gear; insulator flanges, terminals and ducts.
S–NiMn 13–7						235	425	20	—	150	
L–NiCr 20–2	3.0	1.9	1.0	20.0	2.0	—	190	2.5	770	170	Pumps handling alkalis; vessels for alkalis—in soap, food and plastics manufacture
S–NiCr 20–2						230	420	14	—	170	Pumps; valves; compressors; turbo supercharger housings; gas exhaust manifolds.
L–NiSiCr 30–5–5	2.5	5.0	1.0	30.0	5.0	—	205	—	560	180	Good corrosion resis. even to dilute sulphuric acid—pump parts; flue dampers
S–NiSiCr 30–5–5						275	440	2.5	—	210	Pump parts and flue gas dampers subjected to high mechanical stress.
L–Ni 35	2.4	1.5	1.0	35.0	—	—	150	2.0	630	130	Low thermal expansion—parts requiring high dimensional accuracy. Machine tools, scientific instruments, glass moulds.
S–Ni 35						225	390	30.0	—	155	

*These particular compositions, selected from B.S.3648, can be used as 'flake-graphite' irons or can be treated to give spheroidal-graphite structures. The considerable improvement in mechanical properties obtained by the use of S.G. iron is apparent both in terms of strength and ductility.

is a better conductor of heat than is the iron matrix, it follows that CG irons have a thermal conductivity only slightly less than grey iron but better than nodular graphite irons. All of these factors make CG iron attractive as a heat-resisting material and it was developed originally for the manufacture of ingot moulds and vehicle brake components. In modern engineering there are many other applications, eg gear pumps, eccentric gears, fluid and air cylinders, discharge manifolds, etc.

15.74   The choice between ordinary grey iron, SG iron and malleable iron for a specific application is, as with most materials problems, dictated by both economics and technical requirements. Grey iron is, of course, lowest in cost and also the easiest in which to produce sound castings. With regard to the relative merits of the two high-strength competitors, SG iron and malleable iron, the choice may be less easy. Malleable iron is the more difficult to cast since at this stage it will have a white structure. This also involves limitations as to cross-section as compared with SG iron. Moreover, the cost of the malleabilising process makes the product a little more expensive than SG iron. However, it is sometimes necessary to anneal SG iron in order to improve its machinability, and in this case malleable iron may prove more attractive cost-wise.

Alloy Cast Irons

15.80   The microstructural effects which alloying elements have on a cast iron are, in most cases, similar to the effects these elements have on the structure of a steel. Alloying elements can therefore be used to improve mechanical properties, to refine grain, to increase the hardness by stabilising cementite and forming other harder carbides; and to stabilise, when desirable, martensitic and austenitic structures at ambient temperatures. Improvements in resistance to atmospheric corrosion and also deterioration at high temperatures are also attainable.

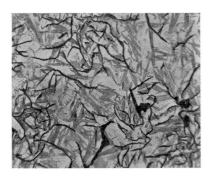 

Plate 15.6   15.6A A martensitic alloy cast iron.
Graphite in a matrix of martensite. × 600. Etched in 4% picral. *(Courtesy of BCIRA)*
15.6B An austenite alloy cast iron (Ni-resist).
Graphite and carbides in an austenite matrix. × 100. Etched in 4% picral. *(Courtesy of BCIRA)*.

15.81 Nickel is a common alloying element and, as in steel, it tends to promote graphitisation. At the same time it has a grain-refining effect, so that whilst nickel will help to prevent chilling in thin sections, it will also prevent coarse grain in thick sections. Nickel also reduces the tendency of thin sections to crack.

15.82 Chromium is a strong carbide stabiliser, and so inhibits the formation of graphite. Moreover, the carbides formed by chromium are more stable and less likely to graphitise under the application of heat than is iron carbide. Irons containing chromium are therefore less susceptible to growth.

As in the low-alloy steels, the disadvantages attendant on the use of either nickel or chromium separately are overcome by using them in conjunction in the ratio of two to three parts of nickel to one part of chromium.

15.83 Molybdenum, when added in small amounts, dissolves in the ferrite, but in larger amounts forms double carbides. It increases the hardness of thick sections and promotes uniformity of microstructure. Impact values are also improved by amounts in the region of 0.5% molybdenum.

15.84 Vanadium promotes heat-resistance in cast irons, in so far as the stable carbides which it forms do not break down on heating. Strength and hardness are increased, particularly when vanadium is used in conjunction with other alloying elements.

15.85 Copper is only sparingly soluble in iron, and has a very slight graphitising effect. It has little influence on the mechanical properties, and its main value is in improving the resistance to atmospheric corrosion.

15.86 Martensitic irons, which are very useful for resisting abrasion, usually contain 4–6% nickel and about 1% chromium. Such an alloy is Nihard, included in Table 15.5. It is martensitic in the cast state, whereas alloys containing rather less nickel and chromium would need to be oil-quenched in order to obtain a martensitic structure. Austenitic irons usually contain between 10 and 30% nickel and up to 5% chromium. These are corrosion-resistant, heat-resistant, non-magnetic alloys. Some of them are treated to produce structures containing spheroidal instead of flake graphite. Table 15.6 indicates the degree of improvement in both strength and ductility which arise from this additional treatment.

Exercises

1. What factors control the structure of cast iron? Explain how these principles are used in the foundry to control the strength of grey-iron castings. (15.20 and 15.30)
2. Discuss the influence of the following elements on the structure and properties of cast iron: (i) silicon; (ii) manganese; (iii) sulphur; (iv) phosphorus. (15.22 to 15.25)
3. Describe briefly the methods available to control the size and shape of graphite in cast iron, and explain how the mode of occurrence of the graphite affects the properties. (15.22; 15.30; 15.71; 15.73)
4. Explain fully what is meant by the term *growth* as applied to cast irons. Discuss methods which are used to overcome this phenomenon. (15.50)

5. Cast iron is a cheap, easily produced engineering material, but ordinary grey cast iron has relatively poor mechanical properties. Discuss, in general terms, the methods employed to overcome these limitations. (15.71; 15.72; 15.73)
6. Discuss the effect of chemical composition and cooling rate on the structure and properties of cast irons. Briefly describe one method for producing (a) malleable iron and (b) nodular iron. (15.20; 15.30; 15.72; 15.71)
7. Most cast irons consist of a dispersion of graphite in a steel-like matrix. Illustrate this statement by sketches of microstructures and discuss the variations in properties that can arise, by reference to: (a) pearlitic grey cast iron; (b) blackheart malleable iron; (c) SG iron; (d) compacted-graphite iron. Indicate briefly the treatment necessary to produce blackheart malleable iron, SG cast iron and compacted-graphite cast iron. (15.40; 15.72; 15.71; 15.73)
8. How are malleable castings produced? What are the most important engineering properties of these materials? (15.72)
9. Discuss the role of microstructure in determining the general properties of cast iron. (15.20; 15.30; 15.70)
10. Discuss the uses of nickel, chromium and molybdenum in cast irons. (15.50; 15.81 to 15.83)

Bibliography

Angus, H. T., *Cast Iron: Physical and Engineering Properties*, Butterworths, 1960.
Copper Development Association, *Copper in Cast Iron*, 1964.
Hume-Rothery, W., *The Structures of Alloys of Iron*, Pergamon, 1966.
BS 1452: 1977 *Specification for Grey Iron Castings*.
BS 6681: 1986 *Specifications for Malleable Cast Iron*.
BS 2789: 1985 *Iron Castings with Spheroidal or Nodular Graphite*.
BS 1591: 1983 *Corrosion-resistant High Silicon Iron Castings*.
BS 3468: 1986 *Austenitic Cast Iron*.

16

Copper and the Copper-Base Alloys

16.10   Copper was undoubtedly the first useful metal to be employed by Man. In many countries it is found in small quantities in the metallic state and, being soft, it was readily shaped into ornaments and utensils. Moreover, many of the ores of copper can easily be reduced to the metal, and since these ores often contain other minerals, it is very probable that copper alloys were produced as the direct result of smelting. It is thought that bronze was produced in Cornwall by accidentally smelting ores containing both tin- and copper-bearing minerals in the camp fires of ancient Britons.

As a schoolboy the author was taught that the very protracted period of Man's history when progress was extremely slow, namely the 'Stone Age', was followed by a 'Bronze Age' when primitive technology developed relatively quickly. It is now fairly certain that the Bronze Age was preceded by a comparatively short Copper Age but, because copper corrodes more quickly than bronze, few artefacts of that period have been recovered by archaeologists. It is fairly certain that the Egyptians used copper compounds for colouring glazes some 15 000 years ago and they may have been the first to extract the metal. In Europe copper production seems to have begun in the Balkans some 6000 years ago, and very soon metal workers were hardening copper by the addition of small amounts of arsenic for the manufacture of knives and spear heads. This may well have been the first instance, in Europe at least, of the deliberate manufacture of an alloy.

16.11   At the end of the eighteenth century practically all the world's requirement of copper was smelted in Swansea, the bulk of the ore coming from Cornwall, Wales or Spain. Deposits of copper ore were then discovered in the Americas and Australia, and subsequently imported for smelting in Swansea. Later it was found more economical to set up smelting plants near to the mines, and Britain ceased to be the centre of the copper industry.

Climbers and ramblers who ascend Snowdon by the popular Pyg track will pass the ruins of an old copper-mine, perched on the mountain side below the Crib Goch ridge. This small mine was one of many located in the mountainous areas of Britain, but all have long since ceased production. Their output was insignificant compared with that of a modern mine like that at Chuquicamata, in Chile. Nevertheless supplies of the higher grades of copper ore are diminishing and the demand for yet higher outputs has meant that low-grade, less economical sources have to be worked. Until a couple of decades ago the world tonnage of copper produced annually was second only to that of iron but now copper ranks third, having been overtaken by aluminium.

Although the USA is still the largest producer of copper, as of many other metals, her output is being closely approached by those of the CIS and Chile. Canada, Zambia, Zaire and Peru are also leading producers, whilst in Europe the outputs of Poland and Yugoslavia are significant. The bulk of Britain's supply of copper comes from Zambia and Canada.

Properties and Uses of Copper

16.20   A very large part of the world's production of metallic copper is used in the unalloyed form, mainly in the electrical industries. Copper has a very high specific conductivity, and is, in this respect, second only to silver, to which it is but little inferior. When relative costs are considered, copper is naturally the metal used for industrial purposes demanding high electrical conductivity.

16.21 The Electrical Conductivity of Copper   The International Annealed Copper Standard (IACS) was adopted by the International Electrochemical Commission as long ago as 1913, and specified that an annealed copper wire 1 m long and of cross-section 1 mm^2 should have a resistance of no more than 0.017241 ohms at 20°C. Such a wire would be said to have a conductivity of 100%. Since the standard was adopted in 1913 higher purity copper is now commonly produced and this explains the anomalous situation where electrical conductivities of up to 101.5% are frequently quoted.

As indicated in Fig. 16.1, the presence of impurities reduces electrical conductivity. To a lesser degree, cold-work has the same effect. Reduction in conductivity caused by the presence of some elements in small amounts is not great, so that up to 1% cadmium, for example, is added to telephone wires in order to strengthen them. Such an alloy, when hard-drawn, has a tensile strength of some 460 N/mm^2 as compared with 340 N/mm^2 for hard-drawn, pure copper, whilst the electrical conductivity is still over 90% of that for soft pure copper. Other elements have pronounced effects on conductivity; as little as 0.04% phosphorus will reduce the electrical conductivity to about 75% of that for pure copper.

16.22 The Commercial Grades of Copper   include both furnace-refined and electrolytically refined metal. High-conductivity copper (usually referred to as OFHC or oxygen-free high conductivity) is of the highest purity, and contains at least 99.9% copper. It is used where the highest electrical and thermal conductivities are required, and is copper which has been refined electrolytically.

Fire-refined grades of copper can be either tough pitch or deoxidised according to their subsequent application. The former contains small amounts of oxygen (present as copper(I) oxide, Cu_2O) absorbed during the manufacturing process. It is usually present in amounts of the order of 0.04–0.05% oxygen (equivalent to 0.45–0.55% copper(I)oxide). This copper(I) oxide is present as tiny sky-blue globules which were originally part of a Cu/Cu_2O eutectic. Hot-working of the copper ingots breaks down the Cu_2O layers of this eutectic into globules. These globules have a negligible effect as far as electrical conductivity, and most other properties, are concerned. The presence of copper(I) oxide is, in fact often advantageous, since harmful impurities, like bismuth, appear to collect as oxides associated with the copper(I) oxide globules, instead of occurring as brittle intercrystalline films, as they would otherwise do.

For processes such as welding and tube-making, however, the existence of these globules is extremely deleterious, since reducing atmospheres containing hydrogen cause gassing of the metal. Hydrogen is interstitially soluble in solid copper so that it comes into contact with subcutaneous globules of copper(I) oxide, reducing them thus:

$$CU_2O + H_2 \rightleftharpoons 2Cu + H_2O$$

Equilibrium proceeds to the right according to the Law of Mass Action because the concentration of dissolved hydrogen is high relative to that of copper(I) oxide. The water formed is present as steam at the temperature of the reaction. Since steam is virtually insoluble in solid copper it is precipitated at the crystal boundaries, thus, in effect, pushing the crystals apart and reducing the ductility by as much as 85% and the tensile strength by 30–40%. Under the microscope gassed tough-pitch copper is recognised by the thick grain boundaries, which are really minute fissures, and by the absence of copper(I) oxide globules.

For such purposes as welding, therefore, copper is deoxidised before being cast by the addition of phosphorus, which acts in the same way as the manganese used in deoxidising steels. A small excess of phosphorus, of the order of 0.04%, dissolves in the copper after deoxidation, and this small amount is sufficient to reduce the electrical conductivity by as much as 25%. So, whilst copper which is destined for welding or thermal treatment in hydrogen-rich atmospheres should be of this type, copper deoxidised by phosphorus would be unsuitable for electrical purposes, where either electrolytic copper or good-quality tough-pitch copper must be used.

16.23 Mention was made (16.10) of the use of arsenic in hardening copper by Balkans metal workers some 6000 years ago. In more recent times up to 0.5% arsenic was added to much of the copper used in the construction of locomotives. This addition considerably increased the strength at elevated temperatures by raising the softening temperature from about 190°C for the pure metal to 550°C for arsenical copper. This made arsenical copper useful in the manufacture of steam locomotive fire-boxes, boiler tubes and rivets, since the alloy, whilst being stronger at high temperatures still had a high thermal conductivity.

The addition of 0.5% lead or tellurium imparts free-cutting properties to copper (6.64) and provides a material which can be machined to close tolerances whilst still retaining 95% of the conductivity of pure copper.

16.24 Although copper has only a moderate tensile strength (16.21) it is a metal with very high malleability and ductility. It is very suitable for both hot- and cold-working by the main processes. Mention has been made of 'deformation twins' (4.18) but another type of twinning occurs when certain of the FCC metals—particularly copper and its alloys and austenitic steels—are annealed *following cold-work*. The presence of these annealing twins is indicated by what appear to be pairs of parallel straight lines crossing individual crystals (Plates 16.1c and 16.3c). Like deformation twins, these regions represent a part of the crystal in which a change in direction of the crystal lattice occurs and in this instance are formed during re-crystallisation. The growth of annealing twins is related to the amount of internal energy associated with the formation of dislocation faults within the crystals during previous cold-work. In aluminium and its alloys this form of internal energy is high so that new crystal boundaries tend to form whereas in copper alloys this energy is low and twinning occurs instead. Thus copper and its alloys show twinned crystals in the cold-worked/annealed state whilst aluminium alloys do not.

16.25 The Effects of Impurities   on the electrical properties of copper

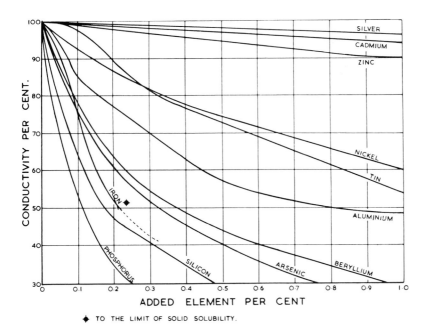

Fig. 16.1   The effect of impurities on the electrical conductivity of copper.

have already been mentioned. Quite small amounts of some impurities will also cause serious reductions in the mechanical properties.

Bismuth is possibly the worst offender, and even as little as 0.002% will sometimes cause trouble, since bismuth is insoluble in amounts in excess of this figure, and, like iron(II) sulphide in steel, collects as brittle films at the crystal boundaries. Antimony produces similar effects and, in particular, impairs the cold-working properties.

Selenium and tellurium make welding difficult, in addition to reducing the conductivity and cold-working properties; whilst lead causes hot-shortness, since it is insoluble in copper and is actually molten at the hot-working temperatures.

The Copper-base Alloys

The ever-present demand by the electrical industries for the World's diminishing resources of copper has led industry to look for cheaper materials to replace the now expensive copper alloys. Whilst the metallurgist has been perfecting a more ductile mild steel, the engineer has been developing more efficient methods of forming metals so that copper alloys are now only used where high electrical conductivity or suitable formability coupled with good corrosion resistance are required. The copper-base alloys include brasses and bronzes, the latter being copper-rich alloys containing either tin, aluminium, silicon or beryllium; though the tin bronzes are possibly the best known.

The Brasses

16.30 The brasses comprise the useful alloys of copper and zinc containing up to 45% zinc, and constitute one of the most important groups of non-ferrous engineering alloys.

As shown by the constitutional diagram (Fig. 16.2), copper will dissolve up to 32.5% zinc at the solidus temperature of 902°C, the proportion increasing to 39.0% at 454°C. With extremely slow rates of cooling, which allow the alloy to reach structural equilibrium, the solubility of zinc in copper will again decrease to 35.2% at 250°C. Diffusion is very sluggish, however, at temperatures below 450°C, and with ordinary industrial rates of cooling the amount of zinc which can remain in solid solution in copper at room temperature is about 39%. The solid solution so formed is represented by the symbol α. Since this solid solution is of the disordered type, it is prone to the phenomenon of coring, though this is not extensive, indicated by the narrow range between liquidus and solidus.

If the amount of zinc is increased beyond 39% an intermediate phase, β', equivalent to CuZn, will appear in the microstructure of the slowly-cooled brass. This phase is hard, but quite tough at room temperature and plastic when it changes to the modification β above 454°C. Unlike copper, which is FCC and zinc, which is CPH, β' has a structure which is often

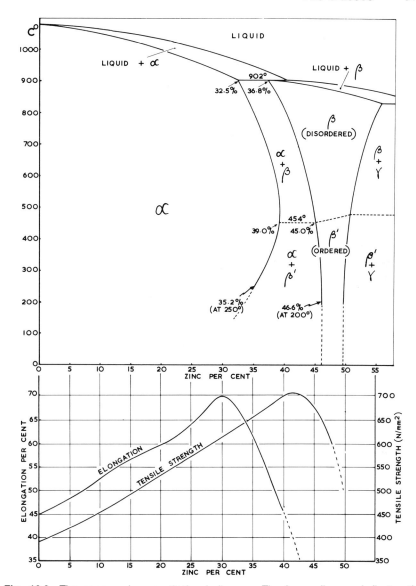

Fig. 16.2 The copper–zinc constitutional diagram. The lower diagram indicates the relationship between composition and mechanical properties.

loosely described as BCC. However this is not strictly correct since the term 'body-centred cubic' implies a structure in which all atoms are similar. The structure of β′ is in fact of the 'caesium chloride' type in which two interlacing simple cubic lattices are involved (Fig. 16.3). Each atom occupying a 'body-centred position' is surrounded by four atoms of the other metal. Since the copper and zinc atoms occupy fixed positions in the lattice,

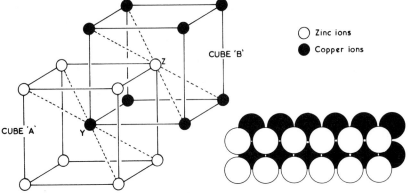

Fig. 16.3 The crystal structure of β'-brass.
Atom 'Y' occupies a body-centred position in cube 'A', whilst atom 'Z' occupies a body-centred position in cube 'B'. This is in fact the 'caesium chloride' type of structure and not really BCC as it first appears.

β' crystals are not cored. Recent research suggests that if the β' phase is allowed to cool *extremely* slowly it undergoes a eutectoid transformation at about 240°C to produce an α + γ structure. However when a brass containing β' is cooled at ordinary industrial rates this transformation never occurs and β', which can contain between 48 and 50% copper, persists at ambient temperatures. For this reason we have omitted the β' → α + γ transformation from the *constitutional* diagram (Fig. 16.2). Further increases in the zinc content beyond 50% cause the appearance of the phase γ in the structure. This is very brittle, rendering alloys which contain it unfit for engineering purposes.

Due to coring effects, an alloy which is nominally α-phase in structure may contain some β' particles at the boundaries of the cored α-crystals when in the as-cast condition. This coring will depend upon the rate of cooling and the nearness of the composition of the alloy to the α/α + β' phase boundary. Such β'-phase will usually be absorbed fairly quickly during hot working. A Widmanstätten structure (11.53) is formed in cast α + β' alloys as the temperature falls, due to the manner in which the needle-like crystals of α precipitate within the β crystals as the alloy cools from out of the β-phase area.

16.31 As mentioned above, the α-phase is quite soft and ductile at room temperatures, and for this reason the completely α-phase brasses are excellent cold-working alloys. The presence of the β'-phase, however, makes them rather hard and with a low capacity for cold-work; but since the β-phase is plastic at red heat, the α + β' brasses are best shaped by hot-working processes, such as forging or extrusion. The α-phase tends to be rather hot-short within the region of 30% zinc and between the temperatures of 300 and 750°C, and is therefore much less suitable as a hot-working alloy unless temperatures and working conditions are strictly controlled. The α-phase would also introduce difficulties during the extrusion of the α + β' alloys were it not for the fact that it is absorbed into the β-phase when the 60–40 composition (one of the most popular alloys of this group) is heated to a point above the α + β/β phase boundary in the region of 750°C, thus producing a uniform plastic structure of β-phase only. The α-phase is usually in process of being precipitated whilst hot-working is taking place, so that, instead of the Widmanstätten structure being formed again as the temperature falls, it is replaced by a refined granular α + β' structure which possesses superior mechanical properties to those of the directional Widmanstätten structure. The needle-shaped crystals of the α-phase are prevented from forming by the mechanical disturbances which accompany the working process.

Thus the brasses can conveniently be classified according to whether they are hot-working or cold-working alloys.

16.32 The Cold-working α-brasses   These are generally completely α-phase in structure, though a limited amount of cold-work may also be applied to those α + β' alloys which contain only small amounts of the β'-phase. The α + β' alloys proper are, however, shaped by hot-working processes in the initial stages, and such cold-work as is applied is merely for finishing to size or to produce the correct degree of work-hardening for subsequent use.

The α-brasses are useful mainly because of their high ductility, which reaches a maximum at 30% zinc, as shown in Fig. 16.2. Such alloys need to be of very high purity, since the inclusion of even small amounts of impurity will lead to a big loss in ductility. The need to use high-purity copper and zinc in the manufacture of 70–30 brass makes it a very expensive alloy. For this reason it has been replaced in many instances in engineering design by BOP mild steel which has a high ductility because of its low nitrogen content. The α-brasses are also rather sensitive to annealing temperatures and, since grain growth is rapid at elevated temperatures, it is easy to burn the alloy. (This trade term should not be confused with oxidation of the metal. It is widely used industrially to signify overheating.) α-Brasses should be annealed at about 600°C. If overheated to 750°C, grain growth is so rapid that on subsequent pressing an orange peel effect is apparent on the surface. This is due to coarse grain being large enough to be visible on the surface.

16.33   Cold-worked α-brasses are subject to 'season cracking'. Dislocations become piled-up at crystal boundaries as a result of cold-work, making these regions zones of high energy. Therefore, any corrosion which takes place tends to be intercrystalline since the high-energy zones are 'anodic' (21.40) to their surroundings. As a result of corrosion the grain boundaries become weakened and fracture occurs there because of the locked-up stresses which are present. The term 'season-cracking' was originally used to describe the spontaneous cracking found in stored cartridge cases in India during the monsoon season. It was particularly prevalent when the damp atmosphere contained ammonia emanating from nearby cavalry stables. Season cracking can be avoided by giving the components a low-temperature, stress relief anneal at about 250°C after fabrication.

16.34   Other elements may be added in small amounts to the α-brasses in order to improve either corrosion-resistance or mechanical properties.

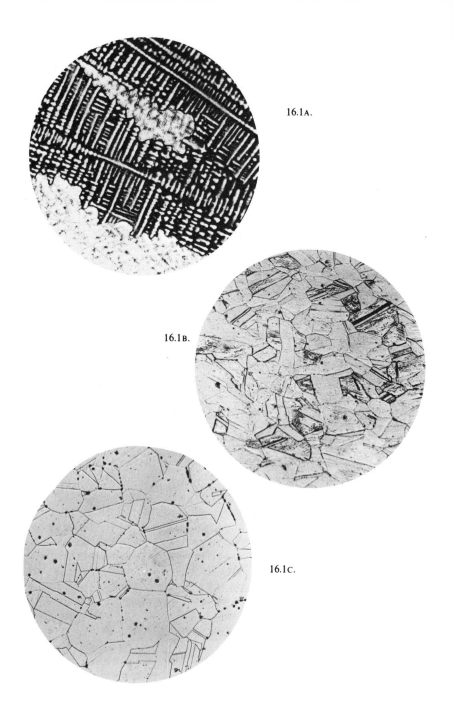

16.1A.

16.1B.

16.1C.

Plate 16.1   16.1A 70/30 Brass, as-cast.
Cored crystals of α solid solution throughout. These crystals reflect light at different intensities depending upon the angles their crystallographic planes make with the surface. × 60. Etched in ammonia/hydrogen peroxide.
16.1B 70/30 Brass, extruded and cold-worked.
Distorted α crystals showing strain bands. Coring has been removed by the hot-working process. × 60. Etched in ammonia/hydrogen peroxide.
16.1C 70/30 Brass, extruded, cold-drawn and then annealed.
Twinning α crystals. × 60. Etched in ammonia/hydrogen peroxide.

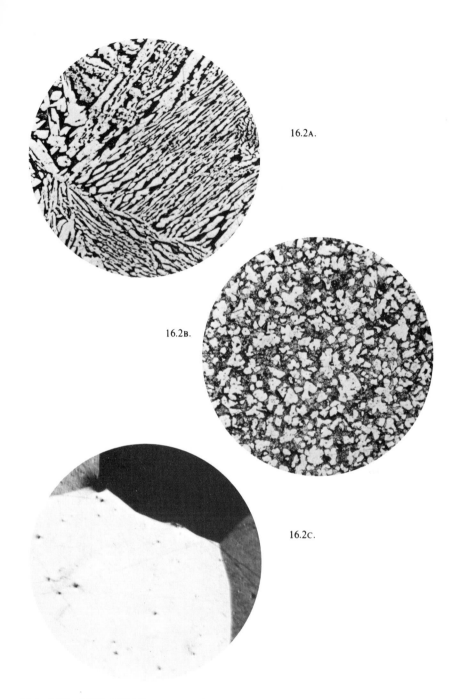

16.2A.

16.2B.

16.2C.

Plate 16.2   16.2A 60/40 Brass, as-cast.
Widmanstätten structure of α (light) and β′ (dark). × 150. Etched in acid iron(III) chloride.
16.2B 60/40 Brass, extruded.
Extrusion has stimulated recrystallisation of the coarse as-cast structure and produced a
fine granular structure of α (light) in a matrix of β′ (dark). × 150. Etched in acid iron(III)
chloride.
16.2C 50/50 Brass, as-cast.
Crystals of β′ only. These are not cored since β′ is an intermediate phase of fixed composition
(CuZn). The individual crystals reflect light at different intensities because their crystallographic planes cut the surface at different angles. × 15. Etched in acid iron(III) chloride.

Table 16.1 *Typical Commercial Brasses*

BS 2870/5 desig- nation	Composition (%)					Condition	Typical mechanical properties				Characteristics and uses
	Cu	Zn	Pb	Sn	Other elements		0.1% Proof stress N/mm²	Tensile strength N/mm²	Elong- ation (%)	Hard- ness (VPN)	
CZ101	90	10	—	—	—	Annealed Hard	77.2 463	278 510	55 4	60˚ 150	*Gilding Metal.* Used for architectural metal-work, imitation jewellery, etc., on account of its gold-like colour and its ability to be brazed and enamelled.
CZ106	70	30	—	—	—	Annealed Hard	77.2 510	324 695	70 5	65 185	*Cartridge Brass.* Deep-drawing brass having the maximum ductility of the copper–zinc alloys. Used particularly for the manufacture of cartridge and shell cases.
CZ107	65	35	—	—	—	Annealed Hard	92.7 510	324 694	65 4	65 185	*Standard Brass.* A good general-purpose cold-working alloy useful where the high ductility of the 70–30 quality is not necessary. Used for press-work and limited deep-drawing.
CZ108	63	37	—	—	—	Annealed Hard	92.7 541	340 726	55 4	65 185	*Common Brass.* A general-purpose alloy suitable for simple forming operations by cold-work
CZ123	60	40	—	—	—	Hot-rolled	108	371	40	75	*Yellow or Muntz Metal.* Hot-rolled plate used for tube plates of condensers and similar purposes. Can be cold-worked to a limited extent. Also as extruded rods and tubes.

Alloy	Cu	Zn			Other	Condition					Description
CZ120	59	39	2.0	—	—	Annealed Hard	92.7 463	371 618	45 5	75 190	*Clock Brass.* Used for the plates and wheels in clock and instrument manufacture. Lead imparts free-cutting properties.
CZ121	58	39	3.0	—	—	Extruded rod	139	448	30	100	*Free-cutting Brass.* Most suitable material for high-speed machining, but can be only slightly deformed by bending, etc. A 61% Cu alloy has greater impact strength.
CZ111	70	29	—	1.0	0.02–0.06 As	Annealed Hard	77.2 432	340 587	70 10	65 175	*Admiralty Brass.* A standard composition for condenser tubes. Tin gives improved corrosion resistance over plain 70–30 brass. Arsenic inhibits dezincification.
CZ112	62	37	—	1.0	—	Extruded	154	417	35	100	*Naval Brass.* For structural applications and forgings. Tin reduces corrosion, especially in sea-water.
CZ110	76	22	—	—	0.02–0.06 As 2.0 Al	Annealed Hard	108 463	371 618	70 8	65 175	*Aluminium Brass.* Possesses excellent corrosion-resistance, and is a popular alloy for condenser tubes.
CZ114	58	Rem.	1.0	0.75	1.5 Mn 1.0 Al 0.7 Fe	Extruded	280	500	15	150	*High-tensile Brass.* Stronger than plain brasses of similar copper content.
CZ132	61.5	Rem.	2.5	—	0.1 As	Hot stamping	210	380	25	120	*Dezincification-resistant Brass.* Water fittings for use where the water supply dezincifies plain $\alpha + \beta$ brasses. After hot stamping the alloy is annealed at 525°C and water quenched to achieve dezincification resistance.

Tin is added in amounts up to 1.0% in order to improve corrosion-resistance, particularly in naval brass and Admiralty brass for condenser tubes. Such small quantities of tin are retained in solid solution. Alternatively, small amounts of *arsenic* (0.01–0.05%) may be added to 70–30 brass destined for the manufacture of condenser tubes, as it is said to improve corrosion-resistance and inhibit dezincification. *Lead*, in amounts of the order of 2.0%, is added to improve machinability. It is insoluble in brass, and exists as small globules, which cause local fractures during machining (6.66). *Aluminium* is sometimes added, in amounts up to 2.0%, to brass for the manufacture of naval condenser tubes, since it imparts excellent corrosion-resistance, particularly to 'impingement' attack. *Nickel* is retained in solid solution, and small amounts may be added to brass to improve corrosion-resistance. The now obsolete twelve-sided threepenny pieces were made from a brass containing 20% zinc, 1% nickel and the balance copper.

16.35 The Hot-working α + β′ brasses Whilst the α-brasses are specifically cold-working alloys they are generally hot-worked in the 'breaking-down' stages; but α + β′ alloys, containing not more than 60% copper are shaped almost entirely by hot work.

The only important 'straight' brass in this group is 60–40 brass, formerly known as 'Muntz metal'. As already mentioned, the α-phase is entirely absorbed into the β-phase when the alloy is heated to about 750°C, so that the best hot-working temperature range is while cooling between 750 and 650°C, during which range the α-phase is being deposited. The mechanical-working process breaks down the α-phase into small particles as it is deposited, and prevents the reintroduction of the coarse Widmanstätten structure.

16.36 Free-cutting Brasses of the 60–40 type contain about 2.0% lead, whilst a similar alloy of higher purity is used extensively for hot-forging where a machining operation is to follow. Up to 0.15% *arsenic* may be added to those brasses destined for the manufacture of water fittings as it is known to improve corrosion-resistance and inhibit dezincification.

16.37 High-tensile Brasses are misleadingly called Manganese Bronze, possibly because the manganese they often contain produces an oxidised-bronze effect on the surface of extruded rod. These brasses contain 54–62% copper, up to 7.0% other elements and the balance zinc. In addition to being hot-working alloys, they are also used in the cast form for such applications as marine propellers, water-turbine runners, rudders, gun mountings and sights, and locomotive axle boxes. The wrought sections are used for pump rods, and for stampings and pressings for automobile fittings and switch gear. The tensile strength is increased, by the addition of these other elements, to as much as 700 N/mm^2 in the chill-cast or forged condition. The additions usually in amounts up to 2.0% each, include manganese, iron and aluminium; whilst up to 1.0% tin may also be included to improve corrosion-resistance.

Details of some of the more important wrought brasses are given in Table 16.1.

The Tin Bronzes

16.40 Bronzes containing approximately 10% tin were probably the first alloys to be used by Man. In Britain bronze articles almost four thousand years old have been found, and during the Roman occupation of Britain the copper-mines of Cumberland and Wales were in a state of rapidly increasing production. Centuries before the Roman invasion, however, Phoenician traders from Tyre and Sidon brought their ships to Cornwall in search of tin.

One of the significant factors in the early Roman conquests was undoubtedly the bronze sword, and it is thought that in even earlier times metal-workers realised that a high tin content, in the region of 10%, produced a hard bronze whilst less tin gave a softer alloy. The relatively high cost of copper—and particularly tin—coupled with competition from other new and improved materials has led to a decline in the wide use of bronzes in the modern world.

16.41   The relationship between the equilibrium diagram and the actual microstructure produced for a given alloy is rather more complex in the case of tin bronzes than with the brasses. The rate of diffusion of copper and tin into each other is much lower than it is with copper and zinc. This is indicated by the wide range of composition at any temperature, between the liquidus AL and the solidus AS (Fig. 16.4), and leading to a high

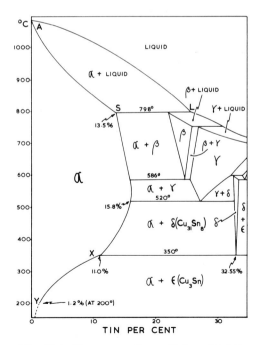

Fig. 16.4 The copper–tin equilibrium diagram.

degree of coring during the actual solidification process. Moreover, structural changes below approximately 400°C take place in copper–tin alloys with extreme sluggishness. Both of these factors mean that a cast bronze, cooled to ambient temperature under normal industrial conditions, will rarely exhibit the structure indicated by the equilibrium diagram.

In short, whilst the peritectic reaction ($\alpha \rightarrow \beta$) at 789°C and the eutectoid transformations ($\beta \rightarrow \alpha + \gamma$) at 586°C and ($\gamma \rightarrow \alpha + \delta$) at 520°C will take place as indicated by the diagram during ordinary rates of cooling, the eutectoid transformation ($\delta \rightarrow \alpha + \varepsilon$) at 350°C would occur only under conditions of extremely slow cooling, such as would never be encountered industrially. Hence the phase ε (Cu_3Sn) is never seen in the structure of a cast bronze containing more than 11.0% tin. Further, due to the slow rate of diffusion of copper and tin atoms below 350°C, the precipitation of ε from α in alloys containing less than 11.0% tin, in accordance with the phase boundary XY, will not occur. For practical purposes the reader can therefore ignore that part of the equilibrium diagram below 400°C and assume that whatever structure has been attained at 400°C will persist to room temperature under normal industrial rates of cooling.

16.42   As with brasses, the α-phase, being a solid solution, is tough and ductile, so that α-phase alloys can be cold-worked successfully. The δ-phase, however, is an intermetallic compound of composition equivalent to $Cu_{31}Sn_8$, and is a hard, brittle, blue substance, whose presence renders the $\alpha + \delta$ bronzes rather brittle. The δ-phase must, therefore, be absent from alloys destined for any degree of cold-work.

Due to heavy coring arising from slow rates of diffusion, cast alloys with as little as 6.0% tin will show particles of δ at the boundaries of the cored α crystals. The skeletons of the α-phase crystals will be much richer in copper than the nominal 94%, thus making the outer fringes correspondingly richer in tin to such an extent that the δ-phase is formed. In order to make such an alloy amenable for cold-work, the δ-phase can be absorbed by prolonged annealing (say six hours at 700°C), which will promote diffusion so that equilibrium is attained in accordance with the equilibrium diagram and a uniform α-phase structure produced. Subsequent air-cooling —or even furnace-cooling at the usual industrial rates—will be too rapid to permit precipitation of any of the ε-phase when the phase boundary XY is reached; and so the uniform α structure will be retained at room temperature. By using such initial heat-treatment to produce a uniform α structure it is possible to cold-work bronzes containing as much as 14% tin, though in general industrial practice only alloys with up to 7% tin are produced in wrought form.

The tin bronzes can be classified as follows:

16.43 Plain Tin Bronzes   These comprise both wrought and cast alloys, the former usually containing up to 7% tin and the latter as much as 18% tin. The wrought alloys contain none of the δ-phase, the absence of which makes them amenable to shaping by cold-working operations. These alloys are usually supplied as rolled sheet, drawn rod or drawn turbine blading.

The cast alloys are used mainly for bearings, since the structure fulfils the requirements for that duty, namely, hard particles of the δ-phase, which will resist wear, embedded in a matrix of the α-phase, which will resist shock.

16.44 Phosphor-bronzes   Most of the tin bronzes mentioned above contain small amounts of phosphorus (up to 0.05%) residual from deoxidation which is carried out before casting. They are often incorrectly called phosphor-bronzes. True phosphor-bronze contains phosphorus as a deliberate alloying addition, generally present in amounts between 0.1 and 1.0%.

Wrought phosphor-bronzes contain up to 8.0% tin and up to 0.3% phosphorus, and, like the plain tin bronzes, are supplied in the form of rod, wire and turbine blading. The phosphorus not only increases the tensile strength but also, it is claimed, improves the corrosion-resistance.

Cast phosphor-bronzes contain up to 13.0% tin and up to 1.0% phosphorus, and are used mainly for bearings and other components where a low friction coefficient is desirable, coupled with high strength and toughness. The phosphorus is usually present in these alloys as copper phosphide, Cu_3P, a hard compound which forms a ternary eutectoid with the α- and δ-phases.

16.45 Bronzes Containing Zinc   These, again, comprise both wrought and cast alloys. The wrought alloys are used chiefly for the manufacture of coinage and contain up to 3.0% tin and up to 2.5% zinc. In recent years the tin content of British 'copper' coinage has been reduced from the pre-war figure of 3.0% to as low as 0.5%. This began as a war-time measure when the Japanese occupation of Malaya cut off our main supplies of tin; and has continued since because of the high prices which have prevailed for the metal. The replacement of tin by some zinc cheapens the alloy, zinc being only about a tenth the price of tin. A subsidiary function of zinc is that, like phosphorus, it acts as a deoxidiser and forms zinc oxide, ZnO, which floats to the top of the melt. The wrought bronzes containing zinc are all α-phase alloys with similar structures to those of straight tin bronzes of like compositions.

The best known of the cast alloys is 'Admiralty gunmetal', or 88–10–2 gunmetal, containing 10% tin and 2% zinc. It is no longer used in naval ordnance, but is still widely employed where a strong, corrosion-resistant casting is required. Its structure is similar to that of a straight tin bronze containing 11% tin, so that, due to considerable coring, a lot of $\alpha + \delta$ eutectoid will be present. The zinc not only cheapens the alloy and acts as a deoxidiser but also improves casting fluidity.

16.46 Bronzes Containing Lead   Up to 2.0% lead is sometimes added to bronzes as well as to brasses in order to improve machinability. Larger amounts are added for some special bearings; and such bronzes permit 20% higher loading than do lead-base or tin-base white metals. The thermal conductivity of these bronzes is also higher, so that they can be used at higher speeds, since heat is dissipated more quickly. With normal lubrication they have excellent wear resistance but, equally important, the seizure resistance is high since lead will function temporarily as a lubricant should normal lubrication fail. Leaded bearing bronzes contain up to 20% lead

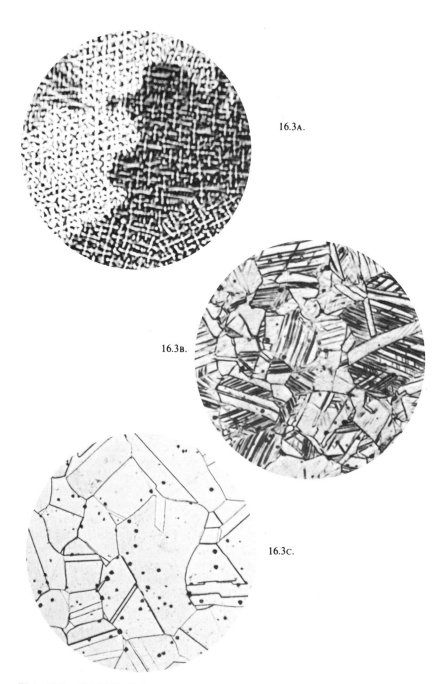

16.3A.

16.3B.

16.3C.

Plate 16.3   16.3A 5% Tin bronze, as-cast.
Large crystals of cored α solid solution. × 100. Etched in ammonia/hydrogen peroxide.
16.3B 5% Tin bronze, cold-worked, annealed and cold-worked again.
Uniform α crystals showing large numbers of strain bands. × 250. Etched in ammonia/
hydrogen peroxide.
16.3C 5% Tin bronze, cold-worked and annealed.
Twinned α crystals with the effects of strain removed by annealing. × 250. Etched in
ammonia/hydrogen peroxide.

Table 16.2 *Tin Bronzes, Phosphor-bronzes and Gunmetal*

Relevant specifications	Composition (%)					Condition	Typical mechanical properties				Characteristics and uses
	Cu	Sn	P	Zn	Other elements		0.2% Proof stress N/mm²	Tensile strength N/mm²	Elongation (%)	Hardness (VPN)	
—	95.5	3	—	1.5	—	Annealed / Hard	120 / 600	324 / 726	65 / 5	60 / 200	*Coinage Bronze.* Used for British 'copper' coinage.
BS 2870: PB101	96	3.75	0.1	—	—	Annealed / Hard	150 / 620	340 / 741	65 / 5	60 / 210	*Low-tin Bronze.* Used where good elastic properties combined with resistance to corrosion and corrosion fatigue are necessary. Widely used as springs and instrument parts.
BS 2870/4: PB102	94	5.5	0.1	—	—	Annealed / Hard	150 / 620	355 / 695	65 / 15	65 / 180	*Drawn Phosphor-bronze.* Generally used in the work-hardened condition. Useful for engineering components subjected to friction. Also for steam-turbine blading and other corrosion-resisting applications.
BS 1400: CT1/B	Rem.	10	0.1	—	—	Sand-cast	140	278	15	90	*Cast Phosphor-bronze.* A suitable alloy for general sand-castings. With phosphorus raised to 0.5% (BSS 1400/PB1/C) the alloy is a standard phosphor-bronze for bearings. It is often supplied as cast sticks for turning small bearings, etc.
—	Rem.	18	Up to 1.0	—	—	Sand-cast	154	170	2	—	*High-tin Bronze.* Used for bearings subjected to heavy compression loads—bridge and turntable bearings, etc.
BS 1400: G1/C	88	10	—	2	—	Sand-cast	140	293	16	85	*Admiralty Gunmetal.* Widely used for pumps, valves and miscellaneous castings, particularly for marine purposes because of its corrosion-resistance. Also used for statuary because of good casting properties.
BS 1400: LG2/A	85	5	—	5	Pb 5	Sand-cast	110	216	13	65	*85–5–5–5 Leaded Gunmetal or Red Brass.* Often used as a substitute for Admiralty gunmetal and also where pressure tightness is required.
BS 1400: LB5/B	Rem.	5	—	0.5	Pb 20	Sand-cast / Chill-cast	80 / 95	175 / 200	8 / 9	65 / 70	Useful where lubrication is likely to fail. Also in aero and automobile construction. For carrying heavy loads a steel backing is necessary.

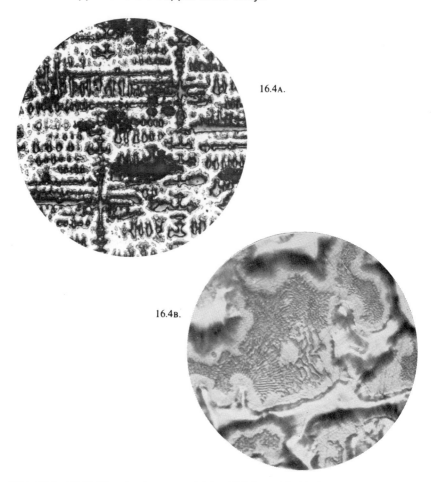

Plate 16.4   16.4A Phosphor bronze containing 10% tin and 0.5% phosphorus, as-cast. Cored α (dark) with an infilling of α, δ and Cu_3P eutectoid. The irregular dark regions are shrinkage cavities. × 150. Etched in ammonia/hydrogen peroxide.
16.4B The same bronze at higher magnification showing the nature of the eutectoid. At this higher magnification the primary α appears light. × 1100. Etched in ammonia/hydrogen peroxide.

(LB5) and are used for aero and automobile crankshaft bearings. 'Red brass' (LG2) is also used occasionally as a bearing metal but more often for pressure-tight castings (Table 16.2.)

Aluminium Bronze

16.50   Part of the copper–aluminium constitutional diagram is shown in Fig. 16.5. As in the case of the copper–zinc alloys, further structural changes occur in some of these alloys if they are cooled *very slowly indeed*

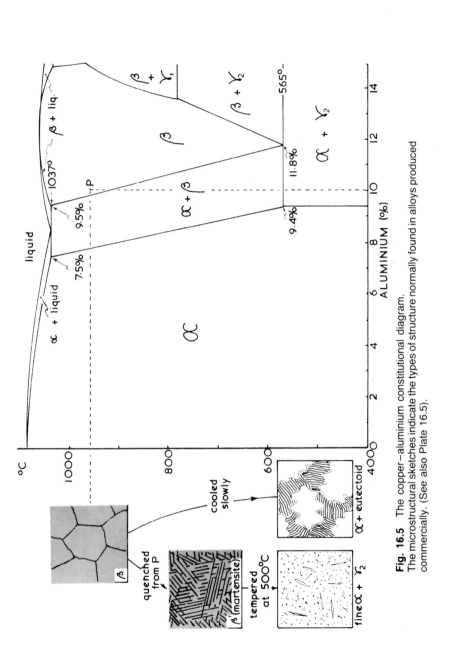

Fig. 16.5 The copper–aluminium constitutional diagram. The microstructural sketches indicate the types of structure normally found in alloys produced commercially. (See also Plate 16.5).

under laboratory conditions below 400°C. However, such very slow cooling rates are never encountered in industrial production methods and we can assume that the $\alpha + \gamma_2$ structure persists in alloys containing between 9.4 and 16.2% aluminium which are cooled reasonably slowly ('furnace cooled') to ambient temperatures. Consequently that part of the constitutional diagram below 500°C can be ignored for our purposes and is not shown here.

Like the brasses the aluminium bronzes can be divided into two main groups; the cold-working and the hot-working or casting alloys respectively. The constitutional diagram indicates that a solid solution (α) containing up to 9.4% aluminium at room temperature, is formed. Like the other α solid solutions based on copper, it is quite ductile. With more than 9.4% aluminium the phase γ_2 is formed. This is an intermetallic compound of the formula Cu_9Al_4 and, in common with compounds of this type, is very hard and brittle, resulting in an overall brittleness of alloys containing the γ_2-phase.

16.51 Further inspection of the constitutional diagram reveals similarities between it and the iron–carbon diagram. The two α-phases are analogous; the β-phase solid solution of the copper–aluminium diagram corresponds to the γ (austenite) phase of the iron–carbon diagram; and the $\alpha + \gamma_2$ eutectoid is similar to the ferrite + cementite eutectoid (pearlite) of the steels. As a result of these similarities in the positions of the phase fields in the respective diagrams, a 10% aluminium bronze can be heat treated in a manner parallel to that of steel so that a martensite-type transformation occurs. Nevertheless it should be realised that the crystallography of the aluminium bronzes is different from that of the corresponding steels.

Consider a 10% aluminium bronze; this will consist entirely of the phases α and γ_2 if it is allowed to cool slowly in a furnace to ambient temperature. If it is reheated the $\alpha + \gamma_2$ eutectoid is transformed to the solid solution β when the eutectoid temperature (565°C) is reached, and as the temperature rises further, the α-phase is absorbed until at about 900°C the structure consists entirely of the solid solution β. Water-quenching from this temperature produces a structure consisting of the phase β'. This is not shown in the equilibrium diagram, since, like martensite in steels, it is not an equilibrium phase. The β'-phase is hard and brittle like martensite, and is in fact very similar in microstructural appearance. Tempering this β'-phase at 500°C causes the precipitation of a fine agglomerate of the phases α and γ_2, closely resembling the tempered martensite of a steel treated in a parallel manner.

In fact the thermal equilibrium characteristics of 10% aluminium bronze resemble more closely those of an air-hardening alloy steel than those of a plain carbon steel. For example, if the 10% aluminium bronze is air cooled from the β-phase field ('normalised') then the resultant structure is likely to be either β' ('martensitic') or a bainitic type of structure containing finely precipitated γ_2. The $\alpha + \gamma_2$ ('pearlitic') structure will only be obtained by annealing, followed by furnace cooling to ambient temperature. Extremely slow cooling under controlled laboratory conditions may cause a

Table 16.3 *Aluminium Bronzes*

Relevant specifications	Composition (%)				Condition	Typical mechanical properties				Characteristics and uses
	Cu	Al	Fe	Other elements		0.2% Proof stress N/mm²	Tensile strength N/mm²	Elonga-tion (%)	Hard-ness (VPN)	
BS 2870/5: CA101	Rem.	5	—	Mn and/or Ni up to 4.0	Annealed Hard	123 587	386 772	70 4	80 220	Cold-worked for decorative purposes, imitation jewellery, etc. Also useful in various engineering applications, especially in tube form. Excellent resistance to corrosion and to oxidation on heating. Tubes for heat exchrs.
BS 2870/5: CA102	Rem.	7	—	Fe, Mn and Ni up to 2.0	Hot-worked	154	432	45	100	Suitable for chemical-engineering applications, especially at moderately elevated temperatures.
BS 2872: CA103	Rem.	9.5	Fe + Ni = 4.0		Forged	247	556	30	180	These alloys are cold-worked to a limited extent only and are shaped by hot-working processes, including forging. The properties quoted are for the hot-worked condition. The materials can be heat-treated by quenching and tempering (16.51).
BS 2872: CA104	Rem.	10	5	Ni 5	Forged	463	726	20	215	
BS 2872: CA107	Rem.	6.2	0.6	Mn up to 0.5 Si-2.2	Forged	400	600	30	200	Low magnetic permeability imparted by Si—developed for components for Navy mine counter-measure vessels. Also for commercial uses.
BS 1400: AB1	Rem.	9.5	2.5	Ni & Mn up to 1.0 each.	Cast	200	575	20	115	Popular die-casting alloy.
BS 1400: AB2	Rem.	9.5	4.5	Ni-5.5 Mn up to 1.5	Cast	300	690	16	160	Most common alloy for aluminium-bronze castings.
BS 1400: AB3	Rem.	6.2	0.6	Si-2.2	Cast	185	480	25	—	The casting version of CA107.

further structural change to take place at 385°C with the formation of another intermediate phase (χ) but as mentioned earlier that need not concern us here.

16.52.   In spite of these heat-treatment phenomena, the main industrial uses of aluminium bronzes depend upon other features such as:

(a) the ability to retain strength at elevated temperatures, particularly when certain other elements are present;
(b) the high resistance to oxidation at elevated temperatures;
(c) good corrosion-resistance at ordinary temperatures;
(d) good wearing properties;
(e) the pleasing colour which makes some of these alloys useful for decorative purposes, particularly as a substitute for gold in imitation jewellery.

A serious drawback to the wider use of aluminium bronze has always been present in the difficulties arising during casting. Since aluminium oxidises so readily at the high casting temperature necessary (above 1100°C), oxide skin and dross is persistently formed and will be drawn into the mould during casting unless special non-turbulent methods of casting are employed (2.33–Pt2). Such skilled casting techniques inevitably cost more money.

16.53 The α-phase Cold-working Alloys   contain between 4.0 and 7.0% aluminium and occasionally up to 4.0% nickel, which improves corrosion-resistance still further. This latter type of alloy is particularly useful in the manufacture of condenser tubes and heat-exchangers, where high strength and corrosion resistance are required up to temperatures of about 300°C. Since the composition of these α-phase alloys can be adjusted to give a colour similar to that of 18-carat gold, rolled sheet is used in the manufacture of imitation-gold cigarette-cases and for other decorative articles of this type. In less permissive times large quantities of the alloy were used for cheap 'wedding rings'. Many products originally made from aluminium bronze are now produced in anodised aluminium the surface of which is further treated to give a gold-like finish at lower cost.

16.54 The $\alpha + \gamma_2$ Hot-working and Casting Alloys   with between 7.0 and 12.0% aluminium, may also contain other elements, such as iron, nickel or manganese. The hot-working alloys contain from 7.0 to 10.0% aluminium and are shaped by forging or by hot-rolling according to their subsequent application. These alloys may also contain up to 5.0% each of iron and nickel, the iron acting as a grain-refiner. Such alloys are used for chemical-engineering purposes (especially for components exposed to high temperatures); also for a variety of purposes where a corrosion-resistant forging is required.

The alloys used for sand- and die-casting contain between 9.5 and 12.0% aluminium with varying amounts of iron and nickel up to 5.0% each, and manganese up to 1.5%. These alloys are used widely in marine engineering, eg for pump rods, valve fittings, propellers, propeller shafts and bolts. They are also used for valve seats and sparking-plug bodies in internal-combustion engines and for brush-holders in generators; for bearings to work at heavy duty; for gear wheels; for pinions and worm wheels; and in the manufacture of non-sparking tools such as spanners, wrenches, shovels and hammers in the potentially 'dangerous' gas, paint, oil and explosives industries, though they are softer and hence inferior to beryllium bronze in this direction.

Copper–Nickel Alloys

16.60   As the thermal-equilibrium diagram (Fig. 9.9) shows, the metals copper and nickel are soluble in each other in all proportions, forming a continuous series of solid solutions. In the cast condition cored crystals will be present, but, though coring may be extensive, it can never lead to the precipitation of a brittle secondary phase, as may happen with the other copper alloys dealt with. On annealing, all copper–nickel alloys consist of uniform solid solutions.

16.61   Being completely solid solution in structure, the copper–nickel alloys are all ductile. They are also very accommodating with regard to the methods of manufacture, and can be hot or cold-worked by rolling, forging, stamping, pressing, drawing and spinning. The absence of any second phase in the microstructure eliminates the possibility of electrolytic corrosion (21.70) in the presence of an electrolyte, and this contributes to the high corrosion-resistance of copper–nickel alloys. This corrosion-resistance reaches a maximum with Monel Metal. Cladding of ships' hulls and parts in off-shore structures with 90–10 cupro-nickel is now widely used as a protection against marine corrosion. Details of the more important copper–nickel alloys are included in Table 16.4.

16.62 Nickel Silvers   are alloys containing from 10 to 30% nickel, 55 to 63% copper and the balance zinc. They are all completely solid solution in structure, and comparable with similar brasses as far as mechanical properties are concerned. They are, however, white in colour, which makes them admirably suitable for the manufacture of spoons, forks and other table-ware. Being ductile, the alloys can be cold-formed for these applications and are usually silver-plated (the stamp EPNS means 'electroplated nickel silver'). The current fashion for stainless steel table-ware has meant a decline in the popularity of the more expensive EPNS. A lot of brassware carrying a very thin 'flashing' of silver also now masquerades as EPNS. The addition of 2.0% lead makes nickel silvers easy to engrave and such alloys are used in the manufacture of Yale-type keys where the presence of lead increases the ease with which the blank can be cut.

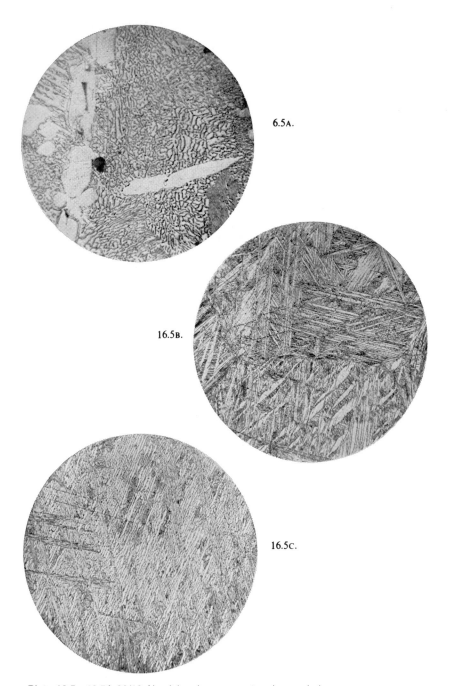

6.5A.

16.5B.

16.5C.

Plate 16.5 16.5A 90/10 Aluminium bronze, cast and annealed.
Crystals of primary α (light) in a matrix of $\alpha + \gamma_2$ eutectoid. $\times$ 1100. Etched in acid iron(III) chloride.
16.5B 90/10 Aluminium bronze, water quenched from 900°C.
Martensitic type of structure (β'). $\times$ 100. Etched in acid iron(III) chloride.
16.5C 90/10 Aluminium bronze, water quenched from 900°C and tempered at 500°C.
Martensitic β' transforming to a fine $\alpha + \gamma_2$ structure. $\times$ 100. Etched in acid iron(III) chloride.

Copper Alloys which can be Precipitation Hardened

16.70   It is often assumed that 'age hardening' or, more properly, 'precipitation hardening' (9.92), is a phenomenon confined to the light alloys. Some of the copper-base alloys can nevertheless be so treated, though these alloys have on the whole rather specialised uses.

16.71 Beryllium Bronze   is one of the most outstanding of these alloys. It contains 1.75–2.5% beryllium and up to 0.5% cobalt, the balance being copper. It is solution-treated at 800°C (Fig. 16.6(i)), quenched, cold-worked (if necessary) and then precipitation-hardened at 275°C for one hour. After such treatment the alloy attains the remarkable tensile strength of 1300–1400 N/mm^2 due to the formation of the coherent precipitate β'. It is particularly suitable for the manufacture of tools where non-sparking properties are required (hammers, chisels and hacksaw blades in gas, oil, paint and explosives industries and in 'dangerous' mines), since it is both hard and tough. It also has a high fatigue strength and is therefore used for the manufacture of springs and for diaphragms and tubes of pressure-recording instruments.

16.72 Chromium Copper   containing 0.4–0.5% chromium (Fig. 16.6(ii) is another heat-treatable alloy which finds use in electrical switchgear and as pressure (spot and seam) welding electrodes (20.61), since it combines a reasonable strength of 540 N/mm^2 with an electrical conductivity of 80% in the hardened state. It is quenched from 1020°C (*P* in Fig. 16.6(ii)) and precipitation hardened at 450°C.

16.73 Zirconium Copper   The solid solubility of zirconium in copper increases from about 0.01% at 0°C to 0.15% at the eutectic temperature of 965°C. Consequently copper-base alloys containing approximately

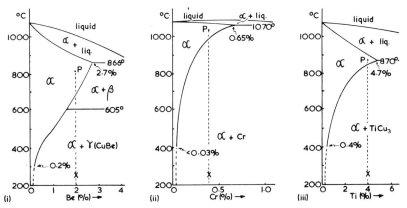

Fig. 16.6 Equilibrium diagrams for some copper-base alloys which can be precipitation hardened. In each case the commercial alloy is of approximate composition 'X' and is solution treated at 'P' at which temperature the microstructure is completely of unsaturated solid solution $\propto$.

Table 16.4 Copper–Nickel Alloys

Relevant specifications	Composition (%)			Typical mechanical properties					Characteristics and uses
	Cu	Ni	Other elements	Condition	0.1% Proof stress N/mm²	Tensile strength N/mm²	Elonga- tion (%)	Hardness (VPN)	
BS 2870/5: CN101	95	5	Fe 1.2 Mn 0.5	Annealed Hard	— —	263 463	50 5	65 130	Has slightly better mechanical properties and corrosion-resistance than pure copper. Can be regarded as toughened copper.
BS 2870/5: CN103	85	15	Mn 0.25	Annealed Hard	— —	324 494	45 5	70 145	The alloy of lowest nickel content which has a more or less white colour
BS 2870/5: CN104	80	20	Mn 0.25	Annealed Hard	108 463	340 541	45 5	75 165	Used for manufacture of bullet envelopes because of its high ductility and corrosion-resistance. Will withstand severe cold-working.
BS 2870/5: CN105	75	25	Mn 0.25	Annealed Hard	— —	355 602	45 5	80 170	Used mainly for coinage, eg the present British silver coinage.
BS 2870/5: CN106	70	30	Mn 0.4	Annealed Hard	108 541	355 649	45 5	80 175	Used for condenser and cooler tubes where good resistance to corrosion is required.

					0.2% Proof stress				
—	60	40	—	Annealed Hard	— —	386 649	45 5	90 190	*Constantan.* Largely used as a resistance material and for thermocouples. Has a high specific resistance and a very low temperature coefficient. Eureka is a similar alloy.
BS 3072/6: NA13	29	68	Fe 1.25 Mn 1.25	Annealed Hard	216 571	541 726	45 20	120 220	*Monel Metal.* Combines good mechanical properties with excellent corrosion–resistance. Used in chemical–engineering plant, etc.
BS 3072/6: NA18	29	66	Al 2.75 Fe 0.9 Mn 0.4 Ti 0.6	Annealed Hard Heat-treated	340 571 788	680 757 1070	40 25 22	175 225 310	*K Monel.* A heat-treatable modification of monel metal. Used for motor-boat propeller shafts.
NES* 824	Rem.	30	Cr–1.8 Fe–0.75 Mn–0.75 Si–0.3 Zr–0.1 Ti–0.15	Sand cast	310	510	22	190	Submarine sea-water systems—valves, pump casings, pipe fittings.

* Naval Engineering Standard

0.12% Zr can be solution treated at 950°C followed by quenching to give a super-saturated solid solution which is then precipitation treated at 450°C to form a coherent precipitate (probably Cu_3Zr). Mechanical properties are similar to those of chromium copper.

Copper containing small amounts of both zirconium and chromium (0.7 Cr; 0.1 Zr) have slightly better strength after heat-treatment. These materials are used in commutators because of their good thermal and electrical conductivities; rocket nozzles and resistance-welding electrodes.

16.74 Titanium Copper   alloys containing 2.5 and 4.0% titanium are in production. The 4% alloy is solution treated at 850–900°C (Fig. 16.6(iii)) for one hour, quenched and then cold-worked up to 40% reduction. It is then precipitation hardened in the range 400–525°C to give a tensile strength of 1076 N/mm^2. It has similar applications to beryllium bronze, ie bellows and diaphragms; springs; electrical contacts and non-sparking tools.

16.75 K Monel   The mechanical properties of ordinary monel metal (16.61) can be further improved by the addition of 2.0–4.0% aluminium to give an alloy designated K Monel. This alloy can be solution treated, cold-worked and then precipitation-hardened to give a tensile strength as high as 1160 N/mm^2.

Some brasses and bronzes which contain small quantities of nickel and aluminium can be precipitation hardened. Thus an alloy containing 20 Zn; 6 Ni; 1.5 Al; bal.Cu is quenched from 850°C and then precipitation hardened at 500°C. These alloys are used for machine parts such as gear wheels; instrument pinions and pressings where strength, hardness and corrosion resistance are important.

16.76 'Narcoloy'   an aluminium bronze, contains 7 Al; 0.5 Co and 0.25 Sn. It is an α-type alloy and its heat-treatment does not depend upon the formation of martensite as with the α + $γ_2$ aluminium bronzes (16.50), but upon precipitation hardening. It is solution treated at 800°C, cold-worked up to 25% and then precipitation hardened at 450°C to give a tensile strength of 580 N/mm^2 and a hardness of 250 H_v. Its high impact value coupled with excellent resistance to stress corrosion make it particularly useful in marine atmospheres for cold-forged bolts and thread-rolled studding.

Since it has an attractive gold-like appearance, takes a high polish and resists corrosion by all household detergents and vinegar, it is suggested that it could be used for tableware such as forks and spoons.

'Shape Memory' Alloys

16.80   'Shape memory' is a phenomenon associated with a limited number of alloy systems and was first investigated in 1951 in some gold–cadmium alloys. The fundamental characteristics of these alloys is the ability to exist in two distinct shapes or crystal structures above and below a critical transformation temperature. Below the critical temperature a martensitic type of structure forms and develops as the temperature falls further. This martensitic change is unlike that observed in steels in that it is reversible, and as the temperature is raised again the martensite is absorbed and ultimately disappears. This reversible change in structure is linked to a change in dimensions and the alloy exhibits a 'memory' of the high and low temperature shapes.

16.81 The best know commercial 'memory metals' are in the copper–zinc–aluminium system—the SME ('shape-memory-effect') brasses—the compositions of which vary in the range 55–80% copper with 2–8% aluminium, the balance being zinc. By choosing a suitable composition, transition temperatures between -70 and $+130°C$ can be achieved. The usable force associated with the change in shape can be employed to operate a range of temperature-sensitive devices; snap on/off positions often being obtained by the use of a compensating bias spring. Following progress with the SME brasses other alloys are being developed in this field, eg a Cu–Al–Ni alloy which operates in the region of 250°C. SM alloys also exhibit other interesting characteristics, including superelasticity where large recoverable strains can be introduced into the material. These strains disappear as the alloy is heated past its transition point. Among such alloys is a nickel–titanium series of which *Nitinol* (55Ni–45Ti) is possibly the best known. It has a transition temperature of 115°C.

16.82 The SM alloy element is generally in the form of a heavy-gauge coiled-spiral spring and such units are used in automatic greenhouse ventilators, thermostatic radiator valves, liquid-gas safety switches, de-icing switches on air-conditioning plant, valves in solar heating systems, electric kettle switches, boiler damper safety switches and warning devices to indicate the presence of ice on roads.

Many other more sophisticated uses are currently being developed for these alloys. Thus *Nitinol* is used in a number of aerospace projects, for example in the form of a coupling ring for joining stainless steel and titanium tubes. The ring contracts on reaching its transition temperature and four knife-like protrusions lock mechanically on to the tube surfaces to form a hermetic seal. Another SM nickel–titanium alloy is used for the manufacture of spectacle frames. The martensitic alloy used results in the frame being soft and comfortable to wear, and if it is accidentally bent the frame can be restored to its original shape by immersion in warm water. A more unexpected use is in the engineering of underwired 'bras'. The low 'yield' strength and superelastic straining at relatively constant stress apparently makes these garments very comfortable to wear and virtually impossible to bend out of shape. This contrasts with conventional underwires which are either comfortable or durable but not both.

Exercises

1. By reference to Fig. 16.2 estimate as accurately as you can, the temperature at which the Widmanstätten structure *begins* to develop as a cast 60–40 brass cools slowly.

   Assuming that the brass continues to cool slowly what will be the approximate composition of the phases present at 300°C and in what proportions will they exist?
2. By reference to Fig. 16.4 describe, step by step, the phase changes which will take place in a bronze containing 22% tin as it cools *extremely slowly* from 900°C to 350°C.

   In what proportions will the phases α and δ be present at 350°C?
3. Describe the microstructures which would be produced when an alloy containing 89.5 Cu–10.5 Al is water quenched from (i) 950°C; (ii) 700°C, after being slowly heated to the quenching temperature. Account for the structures you describe. (16.51)
4. By reference to Fig. 9.9 estimate as accurately as possible the freezing range of the British copper–nickel coinage alloy.
5. Describe the effects of alloying on the following properties of copper: (*a*) electrical conductivity; (*b*) machinability; (*c*) formability; (*d*) corrosion resistance; (*e*) mechanical strength. (16.21; 16.23; 16.31; 16.34; 16.37; 16.44; 16.46; 16.70)
6. Give an account of the effects of small amounts of oxygen and other elements on the properties and uses of copper. (16.20)
7. Pure copper is weak. How has it been found possible to increase the strength of the metal and still retain a relatively high conductivity? (16.21)
8. By reference to the appropriate thermal equilibrium diagram, show how the composition and structure of a commercial brass dictates the method by which it will be shaped and the subsequent uses to which it will be put. (16.30)
9. With reference to the equilibrium diagram for copper–tin alloys, give an account of the structure, properties and uses of tin bronzes. Discuss the reasons for adding: (*a*) phosphorus; (*b*) zinc; (*c*) lead to these alloys. (16.41; 16.42; 16.44; 16.45; 16.46)
10. Discuss the constitution, structure, properties, heat-treatment and uses of the industrially important aluminium bronzes. (16.50 to 16.54)
11. The basic requirements of a precipitation-hardening alloy system is that the solid solubility limit should decrease with decreasing temperature.

    Discuss this statement with reference to copper alloys containing beryllium, chromium and titanium in small quantities.

    How can further strengthening be obtained with Cu–Be alloys? At what stage should this further treatment be applied? (16.70)

Bibliography

American Society for Metals, *Source Book on Copper and Copper Alloys*, 1980
Higgins, R. A., *Engineering Metallurgy (Part II)*, Edward Arnold, 1986.
West, E. G., *Copper and its Alloys*, Ellis Horwood, 1982.

A large number of books and leaflets dealing with the properties and uses of copper and its alloys are available from Copper Development Association.
BS 1400: 1985 *Copper Alloy Ingots and Copper and Copper Alloy Castings*.

BS 2870: 1980 *Specifications for Rolled Copper and Copper Alloys: Sheet, Strip and Foil*.
BS 2871: 1971 and 1972 *Copper and Copper Alloys: Tubes*.
BS 2872: 1969 *Copper and Copper Alloys: Forging Stock and Forgings*.
BS 2873: 1969 *Copper and Copper Alloys: Wire*.
BS 2874: 1986 *Copper and Copper Alloys: Rods and Sections*.
BS 2875: 1969 *Copper and Copper Alloys: Plate*.

17

Aluminium and its Alloys

17.10 The properties of aluminium which chiefly dictate its use as an engineering material are its low relative density coupled with a reasonably high tensile strength when used in one of its alloyed forms. Since its relative density is only about one third that of steel its alloys are widely used in aero-, automobile and constructional engineering. A combination of alloying and heat-treatment can produce alloys which, weight-for-weight, are in the same class as many steels.

17.11 The high affinity of aluminium for oxygen is both disadvantageous and useful. It is disadvantageous in so far as it increases the cost of extraction of the metal by making necessary a relatively expensive electrolytic extraction process. Usually a metal is extracted by heating its oxide ore with a cheap reducing agent, such as carbon (in the form of coke), and the resulting crude metal is refined by allowing the bulk of the impurities present in it to be oxidised. This is the basis for the production of pig iron and its subsequent conversion to steel. The very high chemical affinity of aluminium for oxygen means that aluminium cannot be reduced by carbon by ordinary chemical means. Obviously any other reducing agents able to separate aluminium from oxygen will be thermodynamically far too expensive, so that aluminium can only be produced economically by electrolytic means.

It was using an expensive reducing agent in the form of metallic potassium which enabled the Danish physicist and chemist H. C. Oersted to produce the first samples of aluminium in 1825. Consequently aluminium was worth about £250 per kg in those days, and, even later, it is said that the more illustrious foreign guests to the Court of Napoleon III were privileged to use forks and spoons made from aluminium, whilst the French nobility had to be content with mere gold plate and silver cutlery. In the Gay Nineties aluminium was still regarded as being in the nature of a curious precious metal, though it was in 1886 that C. M. Hall, a twenty-two-year-old student, had discovered a relatively cheap method for producing aluminium by electrolysing a fused mixture of aluminium oxide and the mineral cryolite.

Aluminium cannot be purified by blowing air through it, as in the case of iron. This treatment would oxidise the aluminium and leave behind the impurities. Therefore the ore must be purified before being electrolysed, and this involves an expensive chemical process. The ore of aluminium is bauxite, which is named after Les Baux, in the south of France, where it was discovered in 1821. Although France is still an important producer of bauxite, Jamaica and the Republic of Guinea mine the largest quantities of the ore followed by the CIS, Surinam and Guyana. In Europe Hungary, Greece and Jugoslavia, in addition to France, all produce considerable quantities of the ore. Britain is wholly dependent upon imported ore to maintain a token small-scale production at the Kinlochleven and Lochaber works, both of which are sited with access to water transport and relatively cheap hydroelectric power in the Scottish Highlands; but most aluminium is imported as the metal.

17.12 Although aluminium has a great affinity for oxygen, its corrosion-resistance is relatively high. This is due to the dense impervious film of oxide which forms on the surface of the metal and protects it from further oxidation. The corrosion-resistance can be further improved by anodising (21.91), a treatment which artificially thickens the natural oxide film. Since aluminium oxide is extremely hard, wear-resistance is also increased by the oxide layer; and the slightly porous nature of the surface of the film allows it to be coloured with either organic or inorganic dyes. In this respect high oxygen-affinity is an asset.

The high affinity of aluminium for oxygen also makes it useful as a deoxidant in steels and also in the Thermit process of welding (20.52).

17.13 The fact that aluminium has over 50% of the specific conductivity of copper means that, weight for weight, it is a better conductor of electricity than is copper. Hence it is now widely used, generally twisted round a steel core for strength, as a current carrier in the electric 'grid' system. Large increases in the price of copper a few years ago led to the use of aluminium as a domestic current carrier, though this type of wire is often copper coated to increase the conductivity of the surface skin.

17.14 Unlike copper which is normally available as a high-purity metal because of the ease of its electrolytic refinement, commercial grades of 'pure' aluminium may contain no more than 99.0% of the metal. This is due to the difficulty in refining crude aluminium already mentioned. Although electrolytic refining methods are now practised producing aluminium of 99.99% purity this is obviously costly. Available commercial qualities covered by BS specifications contain 99.99%, 99.8%, 99.5%, 99.0% aluminium respectively. The lower commercial grades are used fairly widely for drawing, pressing and spinning large numbers of cooking and other kitchen utensils to which non-stick coatings of PTFE (polytetrafluoroethene) can be applied where necessary. Aluminium is suitable for both hot- and cold-working processes but unlike copper, another FCC metal, it does not exhibit annealing twins following recrystallisation.

Pure aluminium is relatively soft and weak with a tensile strength of no more than 90 N/mm^2 in the annealed condition, so that for most other engineering purposes it is used in the alloyed form.

Alloys of Aluminium

17.20   The addition of alloying elements is made principally to improve mechanical properties, such as tensile strength, hardness, rigidity and machinability, and sometimes to improve fluidity and other casting properties.

17.21   One of the chief defects to which aluminium alloys are prone is porosity due to gases dissolved during the melting process. Molten aluminium will dissolve considerable amounts of hydrogen if, for any reason, this is present in the furnace atmosphere. When the metal is cast and begins to solidify the solubility of hydrogen diminishes almost to zero, so that tiny bubbles of gas are formed in the partly solid metal. These cannot escape, and give rise to pinhole porosity. The defect is eliminated by treating the molten metal before casting with a suitable flux, or by bubbling nitrogen or chlorine through the melt. A more practical method is to use specially prepared tablets of hexachloroethane which decompose liberating small quantities of chlorine when the tablet is plunged beneath the surface of the melt.

17.22   Aluminium alloys are used in both the cast and wrought conditions. Whilst the mechanical properties of many of them, both cast and wrought, can be improved by precipitation hardening, a number are used without any such treatment being applied. Forty years ago there was a bewildering multitude of aluminium alloys in the manufacturers' lists. Fortunately obsolescence and rationalisation has reduced the number considerably, though some new compositions have been introduced as a result of expansion of the market for aluminium alloys outside the field of aerospace.

17.23   Most of the aluminium alloys available commercially are supplied according to British Standards Specifications and are covered by General Engineering Specifications BS 1470/1475 (wrought materials) and BS 1490 (cast alloys). Specialised applications are covered under a number of supplementary specifications, whilst a few manufacturers produce alloys not covered by a BS number. For wrought alloys the general specification number, ie BS 1470/1475, is followed by a four-digit designation identifying a specific alloy. The first of the four digits indicates the alloy group as follows:

Aluminium (99.0% min. and greater)		– 1xxx
Aluminium alloy groups (by major alloying elements):	Copper	– 2xxx
	Manganese	– 3xxx
	Silicon	– 4xxx
	Magnesium	– 5xxx
	Mg + Si	– 6xxx
	Zinc	– 7xxx
	Other elements	– 8xxx
	Unused series	– 9xxx

In the first group (aluminium containing more than 99.0% Al) the *last two* digits are the same as the two digits to the right of the decimal point in the 'minimum % aluminium' when expressed to the nearest 0.01%. The second digit in the designation indicates modifications in impurity limits or alloying elements. When the second digit is zero it indicates unalloyed aluminium having natural impurity limits; integers 1 to 9, assigned consecutively as required, indicate special control of one or more impurities or alloying elements.

In Groups 2xxx to 8xxx the last two digits in the designation have no special significance and are used only to identify different alloys in the group. The second digit in the designation indicates modifications. Thus, if the second digit is zero it indicates the original alloy; integers 1 to 9 assigned consecutively indicate modifications. National variations of wrought aluminium alloys registered by another country are identified by a serial letter *after* the four-digit designation.

The condition of the material may be denoted by suffix letters as follows:

M	–	material 'as manufactured'.
O	–	fully annealed to give lowest strength.
H1 to H8	–	work-hardened. Material cold-worked after annealing. Designations are in ascending order of strength and hardness.
TB	–	solution-treated and 'aged' naturally. (Material receives no cold-work after solution treatment.)
TE	–	cooled from an elevated-temperature shaping process and precipitation hardened.
TF	–	solution treated and precipitation hardened.
TH	–	solution treated, cold-worked and then precipitation hardened.
TD	–	solution treated, cold-worked and 'aged' naturally (tubes).

The cast aluminium alloys are covered by BS 1490 and here the suffix letters indicate the condition of the material as follows:

M	–	'as cast' with no further treatment
TS	–	castings 'stress relieved' only
TE	–	precipitation treatment after casting
TB	–	solution treated only
TB7	–	solution treated and 'stabilised'
TF	–	solution treated and precipitation hardened
TF7	–	solution treated, precipitation hardened and then 'stabilised'.

It is convenient to classify aluminium alloys into the following four main groups according to the condition in which they are employed.

Table 17.1 Wrought alloys which are not heat-treated

Relevant specifications (BS)	Composition (%)†						Condition	Typical mechanical properties			Typical uses
	Cu	Mg	Si	Fe	Mn	Other elements		0.2% Proof stress (N/mm^2)	Tensile strength (N/mm^2)	Elongation (%)	
1470:1080A 1475:1080A	0.03	0.02	0.15	0.15	0.02	Al-99.80	0 H4 H8	— — —	30 90 130	40 10 5	High-purity aluminium, very high ductility, corrosion resistance and electrical conductivity.
1470:1200 1471:1200 1474:1200	0.05	—	Si + Fe 1.0		0.05	Al-99.0	0 H4 H8	— —	85 125 155	35 5 4	Panelling and moulding, lightly stressed and decorative assemblies especially in architecture, holloware, electrical conductors, equipment for chemical, food and brewing industries, packaging, automobile radiator parts.
1475:4047A	0.3	0.1	11.0 – 13.0	0.6	0.15	—	M	—	—	—	Wire for welding (this alloy is also used in the sheet form for panelling and light marine construction.)
1475:4043A	0.3	0.2	4.5 – 6.0	0.6	0.15	—	M	—	—	—	Welding wire and rods.
1470:3103 1475:3103	0.1	0.3	0.5	0.7	0.9 – 1.5	Zr + Ti* 0.10	H4 H8	— — —	110 155 200	30 7 4	Metal boxes, bottle caps, domestic and commercial food containers and cooking utensils; roofing sheets; panelling of land-transport vehicles; wire; automobile radiators.

Specification		Mg				Others	Temper	0.2% Proof stress	Tensile strength	Elong. %	Typical uses
1470:5251 1471:5251 1472:5251 1474:5251 1475:5251	0.15	1.7 – 2.4	0.4	0.5	0.1 – 0.5	—	0 H3 H6	60 130 175	185 220 265	20 8 4	Marine superstructures, life boats, panelling exposed to marine atmospheres, fencing wire, chemical plant, panelling for road and rail vehicles, now used in automobile radiators.
1470:5154A to 1475:5154A	0.1	3.1 – 3.9	0.5	0.5	0.5	Ti – 0.2* Mn + Cr 0.1 – 0.5	0 H2 H4	85 165 225	240 270 300	18 7 4	Shipbuilding and other marine structures, pressure-vessels, storage tanks, welded structures, deep-pressings for racing-car bodies.
1473:5056A 1475:5056A	0.1	4.5 – 5.6	0.4	0.5	0.1 – 0.6	Mn + Cr 0.1 – 0.6	0 H2	— —	255 290	18 8	Mainly for corrosion-resistant rivets.
1475:5556A	0.1	5.0 – 5.5	0.25	0.4	0.6 – 1.0	Cr 0.05 – 0.2	M	—	—	—	Welding electrodes, rod and wire.
1470:5083 1471:5083 1472:5083 1474:5083	0.1	4.0 – 4.9	0.4	0.4	0.4 – 1.0	Cr 0.05 – 0.25	0 H2 H4	125 235 270	315 345 375	15 8 6	Unheated welded pressure vessels, marine, auto and aircraft parts, TV towers, drilling rigs, transportation equipment, missile components.
(Alcan:2004)	6.0	—	—	—	—	Zr – 0.4	M	120	200	7	A 'superplastic alloy'—many components in the aerospace industries.
———	—	—	—	2.0	—	—	—	—	—	—	In the rolled condition iron has a dispersion-hardening effect—electrical purposes and household foil.

†Single values are maxima unless otherwise stated.
*Optional

Wrought Alloys Which Are Not Heat-treated

17.30   The main requirements of alloys in this group are sufficient strength and rigidity in the work-hardened state, coupled with good corrosion-resistance. These properties are typical of the alloys shown in Table 17.1. As will be seen, these alloys are widely used in the manufacture of panels for land-transport vehicles. Here the high corrosion-resistance of the aluminium–magnesium alloys is utilised, those with a higher magnesium value having an excellent resistance to sea-water and marine atmospheres, so that they are used extensively for marine superstructures. The desired mechanical properties are produced by the degree of cold-work applied in the final cold-working operation, and these alloys are commonly supplied as soft (0), or having undergone varying degrees of work-hardening as denoted by the BS Specification suffixes H1, H2, H3, etc. up to H8 (full hard). The main disadvantage is that, once the material has been finished to size, no further variation can be made in mechanical properties (other than softening by annealing), whereas, with the precipitation-hardening alloys, the properties can be varied, within limits, by heat-treatment.

17.31   As would be expected, most of these alloys (the main exception being those of aluminium and silicon) have structures consisting entirely of solid solutions. This is a contributing factor to their high ductility and high corrosion-resistance. In the case of aluminium–magnesium alloys the solubility of magnesium falls sharply from 14.9% at 451°C to about 1.5% at 0°C (Fig. 17.1). Consequently care must be taken to cool these alloys rapidly in order to prevent the intercrystalline precipitation of the intermediate phase β which would lead to intercrystalline corrosion and embrittlement. Some of the alloys listed are occasionally used in the cast condition.

17.32   Recently a number of superplastic alloys have been developed for use by stretch forming techniques. Early development work was carried out on the eutectic alloy (Al–33Cu) (4.64) but currently the most widely used alloy is 'Alcan 2004' (Al–6Cu–0.4Zr). Some of these alloys can be further strengthened by precipitation treatment.

17.33   Aluminium–iron alloys containing up to 2% iron are now in

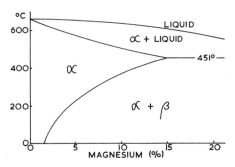

Fig. 17.1   The aluminium–magnesium thermal equilibrium diagram.

production. These alloys can be rolled to foil which has higher strength and good electrical conductivity, the rolling technique being such that nearly all of the iron remains out of solid solution. In this condition it has a considerable dispersion hardening effect whilst at the same time, since little iron is in solution, electrical conductivity remains high.

Cast Alloys which are not Heat-treated

17.40 Alloys in this group are widely used in the form of general-purpose sand castings and die castings. They are used where rigidity, good corrosion resistance and fluidity in casting are of greater importance than strength.

17.41 Undoubtedly the most widely used alloys in this class are those containing between 9.0 and 13.0% silicon, with occasionally small amounts of copper. These alloys are of approximately eutectic composition (Fig. 17.2), a fact which makes them eminently suitable as die-casting alloys, since their freezing range will be small. The rather coarse eutectic structure can be refined by a process known as modification. This consists of adding

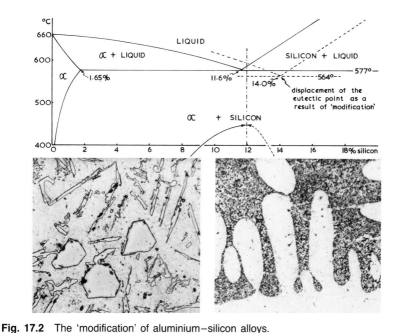

Fig. 17.2 The 'modification' of aluminium–silicon alloys.
In (i) the structure consists of primary silicon + eutectic because the composition is just *to the right* of the eutectic point (11.6% Si). Note that the eutectic is very coarse and consists of needles of silicon in a matrix of α (aluminium rich).
(ii) Here 'modification' has displaced the eutectic point to the right (14% Si) so that the composition of the alloy under consideration is now *to the left* of this point and the structure consists of primary α (light) + eutectic. The latter is now extremely fine-grained and gives superior mechanical properties.

Table 17.2 Cast Aluminium Alloys Which Are Not Heat-treated

BSS. 1490:	Composition (%)					Condition	Typical mechanical properties		Characteristics and uses
	Si	*Cu*	*Mg*	*Mn*			Tensile strength (N/mm^2)	Elongation (%)	
LM.2.	9.0–11.5	0.7–2.5	–	–	M	Sand cast Chill cast	– 150	– 2	Pressure die-castings. General purposes. Withstands moderate stresses. Good fluidity, hence useful for thin-walled castings.
LM.4.	4.0–6.0	2.0–4.0	–	0.2–0.6	M TF	Sand cast Chill cast Sand cast Chill cast	140 160 230 280	2 2 – –	Sand-castings; gravity and pressure die-castings. Good foundry characteristics and inexpensive. General-purpose alloy where mechanical properties are of secondary importance. Can be heat-treated. The most widely used and versatile casting alloy.
LM.5.	–	–	3.0–6.0	0.3–0.7	M	Sand cast Chill cast	140 170	3 5	Sand-castings and gravity die-castings. Suitable for moderately stressed parts. Good corrosion-resistance for marine work. Takes an excellent polish. Food processing.
LM.6.	10.0–13.0	–	–	–	M	Sand cast Chill cast	160 190	5 7	Sand-castings; gravity and pressure die-castings. Excellent foundry characteristics. Large castings for general and marine engineering. Radiators; sumps; gear-boxes; etc. One of the most widely used aluminium alloys.
LM.18	4.5–6.0	–	–	–	M	Sand cast Chill cast	120 140	3 4	Good fluidity but lower cost than LM.6. Good corrosion resistance—hence used in cooking and food utensils.
LM.24	7.5–9.5	3.0–4.0	–	–	M—Chill cast		180	1.5	A very popular alloy for pressure die-castings. Good strength and machinability.
LM.27	6.0–8.0	1.5–2.5	–	0.2–0.6	M	Sand cast Chill cast	140 160	1 2	Sand- and gravity-die castings, e.g. large cambox (sand) and small steering housing (gravity). Also good general purpose alloy.
LM.30	16.0–18.0	4.0–5.0	0.4–0.7	–	M—Chill cast TS—Chill cast		150 160	– –	Very low coefficient of thermal expansion. High hardness. Special applications such as cylinder blocks with unlined bores.

small amounts of sodium (about 0.01% by weight of the charge) to the melt just before casting. The effect is to delay the precipitation of silicon when the normal eutectic temperature is reached, and also cause a shift of the eutectic composition towards the right of the equilibrium diagram. Therefore, as much as 14% silicon may be present in a modified alloy without any *primary* silicon crystals forming in the structure (Fig. 17.2). It is thought that sodium collects in the liquid at its interface with the newly-formed silicon crystals, inhibiting and delaying their growth. Thus, undercooling occurs and new silicon nuclei are formed in large numbers resulting in a relatively fine-grained eutectic structure. Remelting tends to restore the original structure due to a loss of sodium by oxidation. The modification process raises the tensile strength from 120 to 200 N/mm^2 and the elongation from 5.0 to over 15.0%. The relatively high ductility of this cast eutectic alloy is due to the fact that the α-solid-solution phase in the eutectic constitutes nearly 90% of the total structure. As will be seen (Fig. 17.2) it is therefore continuous in the microstructure and acts as a cushion against much of the brittleness arising from the hard silicon phase.

These aluminium–silicon alloys can therefore be produced in the wrought form, though they are better known as materials for die-casting and other types of casting where intricate sections are required, since high fluidity and low shrinkage, as well as the narrow freezing range mentioned above, make them very suitable for such purposes. Their high corrosion-resistance makes these alloys useful for marine work, and the fact that they are somewhat lighter than the aluminium–copper alloys makes them suitable for aero and automobile construction.

17.42 Aluminium–copper alloys containing up to 10% copper were popular in the early days of the aluminium industry. These contained considerable amounts of the brittle compound CuAl$_2$ (see Fig. 17.3) and were useful when rigidity and good casting properties were required in components not subjected to shock. They also machined well but since their corrosion-resistance was poor they have become completely obsolete. However, small amounts of copper are added to some of the aluminium–silicon alloys in order to improve strength and machinability at the cost of some losses in castability and corrosion resistance. One of these alloys (LM4) though generally used in the 'as-cast' form is responsive to heat-treatment.

17.43 The main feature of the aluminium–magnesium–manganese alloys is good corrosion resistance which enables them to receive a high polish. Since many of these alloys are also rigid and shock-resistant they are very suitable for moderately-stressed parts in marine constructions.

Wrought Alloys which are Heat-treated

17.50 Without doubt the most metallurgically-important feature of the aluminium alloys is the ability of some of them to undergo a change in properties when suitably heat-treated. Although this phenomenon is common to many alloys in which a change in solubility of some constituent

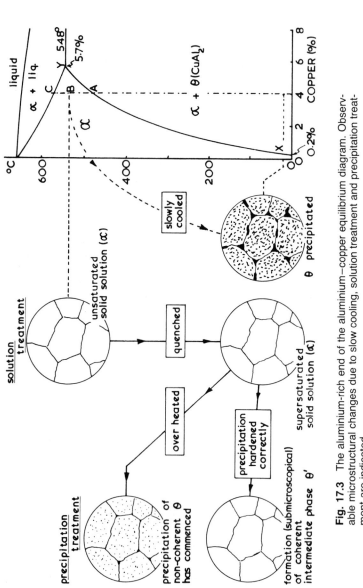

Fig. 17.3 The aluminium-rich end of the aluminium–copper equilibrium diagram. Observable microstructural changes due to slow cooling, solution treatment and precipitation treatment are indicated.

in the solvent metal takes place with variation in temperature (9.90) it is more widely used in suitable aluminium-base alloys than in any others.

The phenomenon was first observed by a German research metallurgist, Dr. Alfred Wilm who, in 1906, noticed that an aluminium–copper alloy which had been water-quenched from a temperature of about 500°C, subsequently hardened unassisted at ambient temperature over a period of several days. The strength increased in this way, reaching a maximum value in just under a week, and the effect was subsequently known as age-hardening. About four years later Wilm transferred the sole rights of his patent to the Dürener Metal Works in Germany, and the alloy produced was named Duralumin. Although this name is often used to describe any wrought aluminium alloy of this type, it is, strictly speaking the proprietary name given to a series of alloys produced by certain companies both here and abroad. The first significant use of duralumin was during the First World War, when it found application in the structural members of the airships bearing the name of Graf von Zeppelin.

17.51 The 'age-hardening' process can be accelerated and higher strengths obtained if the quenched alloy is heated at temperatures up to 180°C. Such treatment was originally known as artificial age-hardening but both of these descriptions have been replaced in metallurgical nomenclature by the general term precipitation hardening. Much speculation took place over the years as to the fundamental nature of this phenomenon but metallurgists now generally agree that the increases in strength and hardness produced are directly connected with the formation of coherent precipitates (9.92) within the lattice of the parent solid solution. We will consider the application of this principle to the precipitation-hardening of aluminium–copper alloys.

17.52 Let us assume that we have an aluminium–copper alloy containing 4.0% copper. At temperatures above 550°C this will consist entirely of α solid solution as indicated by the equilibrium diagram (Fig. 17.3). If we now allow the alloy to cool *very slowly* to room temperature, equilibrium will be reached at each stage and particles of the intermetallic compound $CuAl_2$ (θ) will form as a non-coherent precipitate. This precipitation will commence at A and continue until at room temperature only 0.2% (X) copper remains in solution in the aluminium. The resulting structure will lack strength because only 0.2% copper is left in solution, and it will be brittle because of the presence of coarse particles of $CuAl_2$.

If the alloy is now slowly reheated, the particles of $CuAl_2$ will be gradually absorbed, until at A we once more have a complete solid solution α (in industrial practice a slightly higher temperature, B, will be used to ensure complete solution of the $CuAl_2$). On quenching the alloy we retain the copper in solution, and, in fact, produce a supersaturated solution of copper in aluminium at room temperature. In this condition the alloy is somewhat stronger and harder because there is more copper actually in solid solution in the aluminium, and it is also much more ductile, because the brittle particles of $CuAl_2$ are now absent. So far these phenomena permit of a straightforward explanation, forthcoming from a simple study of the microstructure. What happens subsequently is of a sub-microscopical nature, that is, its observation is beyond the range of an ordinary optical microscope.

17.53   If the quenched alloy is allowed to remain at room temperature, it will be found that strength and hardness gradually increase (with a corresponding reduction in ductility) and reach a maximum in about six days. After this time has elapsed no further appreciable changes occur in the properties. The completely α-phase structure obtained by quenching is not the equilibrium structure at room temperature. It is in fact super-saturated with copper so that there is a strong urge for copper to be rejected from the solid solution α as particles of the non-coherent precipitate $CuAl_2$ (θ). This stage is never actually reached at room temperature because of the sluggishness of diffusion of the copper atoms within the aluminium lattice. However, some movement does occur and the copper atoms take up positions within the aluminium lattice so that nuclei of the intermediate phase (θ') are formed. These nuclei are present as a coherent precipitate *continuous* with the original α lattice, and in this form, cause distortion within the α lattice. This effectively hinders the movement of dislocations (9.92) and so the yield strength is increased. The sub-microscopical change within the structure can be represented so:

$$\text{Cu} \quad + \quad 2\,\text{Al} \quad \rightarrow \quad \theta'$$

An improvement in properties over those obtained by ordinary natural age-hardening can be attained by tempering the quenched alloy at temperatures up to nearly 200°C for short periods. This treatment—called precipitation treatment—increases the amount of the intermediate coherent precipitate θ' by accelerating the rate of diffusion, and so strength and hardness of the alloy rise still further. If the alloy is heated to a higher temperature, a stage is reached where the structure begins to revert rapidly to one of equilibrium and the coherent intermediate phase θ' precipitates fully as non-coherent particles of θ ($CuAl_2$):

$$\theta' \quad \rightarrow \quad \theta$$

[Intermediate
coherent
precipitate]

[Non-coherent
precipitate
$CuAl_2$]

When this occurs both strength and hardness begin to fall. Further increases in temperature will cause the θ particles to grow to a size making them easily visible with an ordinary optical microscope, and this will be accompanied by a progressive deterioration in mechanical properties. Fig. 17.4 illustrates the general effects of variations in time and temperature during post-quenching treatment for a typical alloy of the precipitation-hardening variety. At room temperature (20°C) the tensile strength increases slowly and reaches a maximum of about 390 N/mm² after approximately 100 hours. Precipitation-treatment at temperatures above 100°C

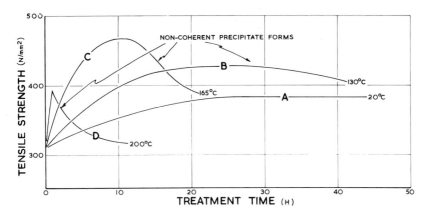

Fig. 17.4 The effects of time and temperature of precipitation-treatment on the structure and tensile strength of a suitable alloy.

will result in a much higher maximum tensile strength being reached. Optimum strength is obtained by treatment at 165°C for about ten hours, after which, if the treatment time is prolonged, rapid precipitation of non-coherent particles of θ (CuAl₂) will cause a deterioration in tensile strength and hardness as shown by curve *C*. Treatment at 200°C, as represented by curve *D*, will give poor results because the rejection from solution of non-coherent θ is very rapid such that precipitation will overtake any increase in tensile strength. This process of deterioration in structure and properties due to faulty heat-treatment is generally termed 'reversion'. Time and temperature of precipitation-treatment differ with the composition of the alloy, and must always be controlled accurately to give optimum results.

17.54 Precipitation-hardening is not confined to the aluminium-base alloys, but can be applied to alloys of suitable composition from many systems in which a sloping phase boundary such as *X A Y* (Fig. 17.3) exists. As far as aluminium alloys are concerned, those containing copper and those containing magnesium and silicon (which cause hardening due to the formation of Mg₂Si) are the most important. In addition, precipitation-hardening is utilised in a number of magnesium-base alloys (18.11), titanium alloys (18.62), some copper-base alloys (16.70) and certain steels.

17.55 A number of alloys are sold under the general trade name of Duralumin. Whilst some of them rely on the presence of approximately 4% copper to effect hardening, others contain magnesium and silicon so that hardening will be assisted by the coherent precipitation of Mg₂Si. The original Duralumin contained 4.0 Cu; 0.5 Mg; 0.5 Si; 0.7 Fe; and 0.7 Mn, the main function of the manganese being as a grain refiner.

The initial hot-working of duralumin, either by hot-rolling or extrusion, is done between 400 and 450°C. This treatment breaks up the coarse eutectic to some extent, and the alloy can then be cold-worked. It is subsequently solution-treated at 480–500°C for one to three hours in order to absorb the $CuAl_2$ and Mg_2Si and is then water-quenched. Accurate temperature control is essential, as the temperature of treatment must be maintained between A and C (Fig. 17.3). After quenching, cold-working operations can be carried out on the resulting solid-solution structure if desired. The alloy is subsequently hardened, either by allowing it to remain at room temperature for about five days or by heating it to some temperature between 100 and 160°C for up to ten hours, according to the properties required. Heating at above 160°C may cause visible particles of $CuAl_2$ to appear in the microstructure, with consequent losses in strength and hardness.

17.56   Just as the application of heat will accelerate precipitation-hardening so will refrigeration impede the process. This fact was utilised extensively in aircraft production during the Second World War. An alloy rivet must be of such a composition that it will 'age' at ambient temperature, since it is inappropriate to precipitation-harden a completed airframe structure by any sort of furnace treatment. Unfortunately once solution-treated such a rivet will soon begin to harden at ambient temperature and if this hardening had begun any attempt to drive the rivet in this hard condition would cause it to split. Hence, immediately after quenching from the solution-treatment temperature, rivets were transferred to a refrigerator at about -20°C. By this means age-hardening was considerably slowed down and rivets could be stored at sub-zero temperatures until required.

17.57   Although the addition of copper forms the basis of many of the precipitation-hardening aluminium alloys, copper is absent from a number of them which rely instead on the presence of magnesium and silicon. Such alloys have a high electrical conductivity approaching that of pure aluminium, so that they can be used for the manufacture of overhead conductors of electricity. Most commercial grades of aluminium contain iron as an impurity, but in some of these alloys it is utilised in greater amounts in order to increase strength by promoting the formation of $FeAl_3$, which assists in precipitation-hardening. Titanium finds application as a grain refiner, whilst other alloys containing zinc and chromium produce tensile strengths in excess of 620 N/mm^2 in the heat-treated condition.

17.58   In the 1920s aluminium–lithium alloys were investigated as being potentially useful in aircraft construction, since the low relative density of lithium (0.51) could mean reductions of 10% in the density of alloys containing the metal. Subsequently interest was revived in lithium as an alloying element and the US Navy's 'Vigilante' contains the alloy '2020', though here lithium was used to increase the temperature range over which existing aluminium–copper alloys could be used rather than to reduce weight. More recent research has been based on alloys containing 2.5% lithium along with small amounts of magnesium, copper and—as a grain refiner—zirconium. The development of RSP (9.110) has led to experiments with alloys containing 4% lithium since under these conditions more lithium is retained in solution.

Since lithium is one of the chemically reactive 'alkali metals' in Group I of the Periodic Classification (Fig. 1.2) there are difficulties encountered in melting and casting these alloys. Moreover, early aluminium–lithium

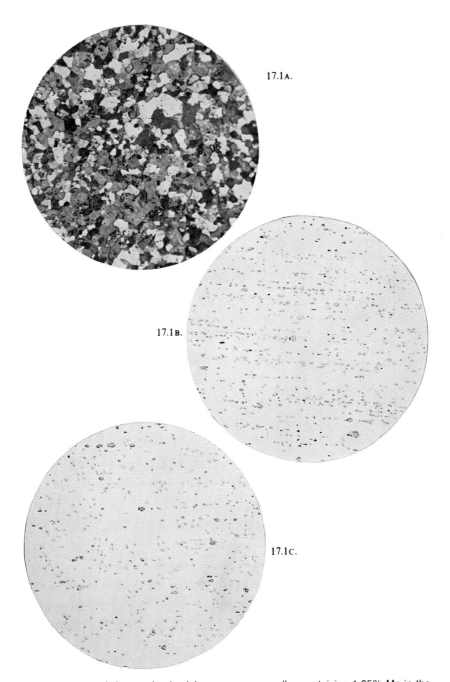

17.1A.

17.1B.

17.1C.

Plate 17.1   17.1A A wrought aluminium–manganese alloy containing 1.25% Mn in the annealed condition, showing uniform crystals of solid solution.
Etched electrolytically and photographed under polarised light which emphasises grain contrast. All of the crystals, however, are of the same uniform solid solution. × 100.
17.1B An alloy of the duralumin type, in the hot-rolled condition, showing particles of $CuAl_2$ in a matrix of solid solution. × 100. Unetched.
17.1C The same alloy in the heat-treated condition.
Much of the $CuAl_2$ is now in solution. × 100. Unetched.

Table 17.3 Wrought Aluminium Alloys Which Are Heat-treated.

Relevant Specification (BS)	Composition (%)‡						Heat-treatment	Condition	Typical Mechanical Properties			Characteristics and uses
	Cu	Mg	Si	Mn	Fe	Other elements			0.2% Proof Stress (N/mm²)	Tensile strength (N/mm²)	Elongation (%)	
1471:6063 1472:6063 1474:6063 1475:6063	0.1	0.45 – 0.9	0.2 – 0.6	0.1	0.35	Cr-0.1* Ti-0.1*	Solution-treated at 520°C, quenched and precipitation-hardened at 170°C for 10 hours.	O TB TF	– 100 180	150 180 250	– 15 17	Good corrosion-resistance. Glazing bars and window sections, windscreen and sliding roof sections for automobile industry. Good surface finish available with this alloy.
1472:2031 2L83†	1.0 – 2.8	0.6 – 1.2	0.5 – 1.3	0.5	0.6 – 1.2	Ni-0.6-1.4 Ti-0.2*	Solution-treated at 530°C, quenched and precipitation-hardened at 170°C for 10–20 hours.	TB TF	150 295	310 385	13 6	Air-screw forgings, sheets, tubes
1470:2014A to 1475:2014A 3L63; 2L77†	3.9 – 5.0	0.2 – 0.8	0.5 – 0.9	0.4 – 1.2	0.5	Zr + Ti-0.2*	Solution-treated at 510°C, quenched and precipitation-hardened at 170°C for 10 hours	TB TF	245 375	450 510	10 8	Highly-stressed components in aircraft, e.g. stressed-skin construction. Engine parts such as connecting rods. Also supplied in Alclad form, i.e. covered with a thin layer of pure aluminium which protects the alloy from corrosive media.
1471:6061 1473:6061 1474:6061 1475:6061	0.15 – 0.4	0.8 – 1.2	0.4 – 0.8	0.15	0.7	Cr-0.04-0.35 Ti-0.15*	Solution-treated at 520°C, quenched and precipitation-hardened at 170°C for 10 hours	H4 TB TF	160 115 235	200 215 300	5 13 8	Bars and sections for marine use; tubes; bolts and wire.
1470:6082 to 1474:6082	0.1	0.6 – 1.2	0.7 – 1.3	0.4 – 1.0	0.5	Cr-0.25* Ti-0.1*	Solution-treated at 510°C, quenched and precipitation-hardened at 175°C for 10 hours.	O TB TF	– 117 250	155 220 325	17 15 8	Structural members, for road, rail and sea transport vehicles, general architectural work, ladders and scaffold tubes. High electrical conductivity – used for overhead lines, etc.

Spec						Other	Condition	Temper				Applications
2L84†	1.0 – 2.0	0.5 – 1.2	0.8 – 1.3	1.0	0.7	Ti-0.3*	Solution-treated at 525°C, quenched and precipitation-hardened at 170°C for 10 hours.	TF	400	430	12	Structural members for aircraft and road vehicles, tubular furniture.
3L86†	1.5 – 3.0	0.2 – 0.5	0.7	0.5	0.7	Ti-0.3*	Solution-treated at 495°C, quenched and aged at room temperature for 5 days	TB	150	300	20	Rivets for built-up structures.
2L77†	3.5 – 4.7	0.4 – 1.2	0.2 – 0.7	0.4 – 1.0	0.7	Cr + Ti-0.3*	Solution-treated at 480°C, quenched and aged at room temperature for 4 days	TB	275	400	10	General purposes, stressed parts in aircraft and other structures. The original Duralumin.
2L88; 2L95	1.0 – 2.2	2.0 – 3.0	0.5	0.3	0.5	Zn-5.0–7.5 Cr-0.15 Ti-0.3*	Solution-treated at 465°C, quenched and precipitation-hardened at 120°C for 24 hours	TF	580	650	11	Highly-stressed aircraft components such as booms. Other military equipment requiring the highest strength/weight ratio. The strongest aluminium alloy produced commercially.
Alcoa (USA) '2020'	4.5	–	–	0.5	–	Li 1.1 Cd-0.2	–	–	–	–	–	Upper wing skins in aircraft—good creep properties up to 175°C
(RAE 8090)	1.3	0.8	–	–	–	Li 2.5 Zr-1.2	–	TF	485	570	6	Military aircraft construction. Low relative density (2.53). Can be shaped 'superplastically'.
(RAE 8091)	1.8	0.9	–	–	–	Li 2.6 Zr-0.12	–	TF	490	550	4	High specific modulus due to low relative density (2.54)

‡ Single values are maxima unless otherwise stated.
* Optional
† BS Aerospace Series, Section L (aluminium and light alloys).

alloys failed because of poor ductility and impact values. However, in recent British alloys, these values have been maintained at satisfactory levels along with a relative density of only 2.54, and hence an increase in *specific modulus* (2.36) of up to 25%.

Precipitation hardening is due mainly to the phase δ' (Al$_3$Li) a precipitate coherent with the matrix. Magnesium is added to reduce density further and also confer limited solid-solution hardening, but unfortunately this leads to the formation of detrimental grain-boundary precipitates of Al$_2$MgLi. However the addition of some copper causes the precipitation of δ' and the phase T$_1$ (Al$_2$CuLi) instead. The simultaneous precipitation of two phases gives rise to an appreciable improvement in impact toughness when precipitation of the two phases is suitably controlled.

A selection of typical wrought heat-treatable alloys of aluminium is given in Table 17.3.

Cast Alloys which are Heat-treated

17.60   German air-raids on England during the First World War were carried out by 'Zeppelins', the airframes of which were constructed from Duralumin. Britain's defence against these raids was provided by heavier-than-air flying machines, the airframes of which were largely of non-metallic construction. Nevertheless light-weight internal combustion engines were required to power these aircraft and the National Physical Laboratory carried out much of the development work on new aluminium alloys in those days.

One of the very successful alloys developed by NPL is still known by the series letter used to identify it during development—Y Alloy. Like Duralumin this is of the 4% copper type but contains also about 2% nickel and 1.5% magnesium. Precipitation-hardening is due to the combined effects of coherent precipitates based on both CuAl$_2$ and NiAl$_3$, and whilst fundamentally a casting alloy it can be used in the wrought condition. As a cast alloy it is useful where reasonable strength at high temperatures is required, as in high-duty pistons and cylinder heads for internal combustion engines. Its relatively low coefficient of expansion also makes it useful in this direction.

17.61   There are other casting alloys similar in composition to 'Y Alloy', but containing rather less copper. Some of these alloys (eg RR50) can be precipitation hardened by heating at 155–170°C without the usual preliminary solution-treatment at a high temperature. This is due to sluggishness in precipitation of phases like CuAl$_2$ and NiAl$_3$ as the casting cools, following the initial pouring process. The 5.0 and 13.0% types of aluminium–silicon alloy are brought into this group by the addition of small amounts of either copper and nickel or magnesium and manganese, which render the alloys amenable to precipitation-hardening.

17.62   The use of aluminium–silicon alloys for automobile cylinder blocks has obvious advantages in terms of weight saved as compared with the usual iron casting. Other advantages include higher thermal conductivity which assists heat transfer in the water cooling and lubrication systems; and enhanced corrosion resistance. Moreover, die-casting can be used in the production of aluminium alloys further reducing production costs when large numbers of castings are involved. Unfortunately the wear-resistance of normal 11–13% silicon alloys is inadequate and cast iron liners need to be fitted resulting in increased costs.

Recently a range of hyper-eutectic aluminium–silicon alloys was developed. These contain 18–25% silicon and because of the presence of large amounts of *primary* silicon the wear resistance of these alloys is

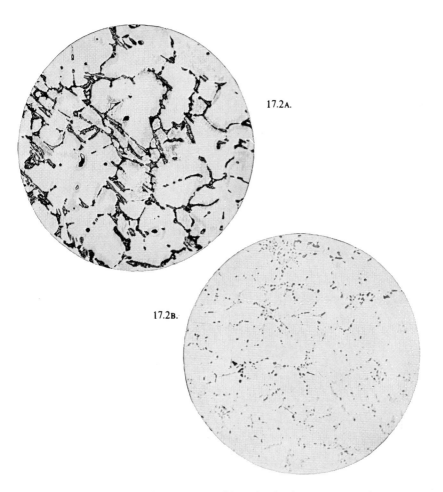

17.2A.

17.2B.

Plate 17.2   17.2A ∨ alloy in the as-cast condition, showing large amounts of intermetallic compounds in the structure. × 100. Etched in 0.5% hydrofluoric acid. *(Courtesy of Aluminium Laboratories Ltd)*
17.2B The same alloy in the solution-treated and precipitation-hardened condition. Much of the intermetallic compounds are now in solution. × 100. Etched in 0.5% hydrofluoric acid. *(Courtesy of Aluminum Laboratories Ltd).*

Table 17.4 Cast Aluminium Alloys Which Are Heat-treated

Relevant BS Specifications	Composition (%)					Heat-treatment	Typical mechanical properties				Characteristics and uses
	Si	Cu	Mg	Mn	Ni		Condition	Tensile strength (N/ mm^2)	Elongation (%)	Hardness (H$_b$)	
1490:LM 4.	4.0–6.0	2.0–4.0´	—	—	—	Solution-treated at 520°C for 6 hours; quenched in hot water or oil; precipitation-hardened at 170°C for 12 hours.	TF Sand cast Chill cast	230 280	—	—	General purposes (sand, gravity and pressure die-casting). Withstands moderate stresses, shock and hydraulic pressure. Also used in the non-heat-treated condition.
1490:LM9.	10.0–13.0	—	0.2–0.6	0.3–0.7	—	Solution-treated at 530°C for 2–4 hours; quenched in warm water; precipitation-hardened at 150–170°C for 16 hours.	M Sand cast Chill cast TE Sand cast Chill cast TF Sand cast Chill cast	— 190 170 230 240 295	— 3 1.5 2 — —	— — — — — —	Suitable for intricate castings, stressed but not subjected to shock. Good corrosion-resistance like the plain high-silicon alloy.
1490:LM 10.	—	—	9.5–10.5	—	—	Solution-treated at 525°C for 8 hours; cooled to 390°C and then quenched in oil at 160°C or boiling water.	TB Sand cast Chill cast	280 310	8 12	— —	High combination of mechanical properties and high resistance to corrosion. Initially tough but suffers progressive embrittlement and susceptibility to stress corrosion cracking. Not recommended for sustained tensile stresses (applied or residual).
1490:LM 13.	10.0–13.0	0.7–1.5	0.8–1.5	—	1.5 Opt	Solution-treated at 520°C for 4–12 hours; oil-quenched; precipitation-hardened at 165–185°C for 6–12 hours.	TE Sand cast Chill cast TF Sand cast Chill cast TF7 Sand cast Chill cast	— 210 170 280 140 200	— — — — — —	90–130 100–150 65–90	Notable for low thermal expansion. Widely used for automobile pistons.

Alloy						Condition / HB (Brinell)		HB	Uses
4L35 ('Y Alloy')	0.3	4.0	1.5	—	2.0	Solution-treated at 510°C; precipitation-hardened in boiling water for 2 hours or aged at room temperature for 5 days.	TF Chill cast 270	—	Pistons and cylinder-heads for liquid- and air-cooled engines. Heavy duty pistons for diesel engines. General purposes.
1490:LM 16; 3L78.	4.5–5.5	1.0–1.5	0.4–0.6	—	—	Solution-treated: 520° for 12 hours; water-quenched; precipitation-hardened at 150–160°C for 8–10 hours.	TB Sand cast 170 / Chill cast 230; TF Sand cast 230 / Chill cast 280	2, 3, —, —	Useful for intricate shapes and pressure-tightness. Cylinder-heads; valve bodies; water jackets.
3L51 (RR 50)	2.5	1.8	0.2	—	1.2	No solution-treatment required; precipitation-hardened at 155–170°C for 8–16 hours.	TE-Chill cast 200	3	A good general-purpose alloy for sand-casting and gravity die-casting. High rigidity and moderate shock-resistance.
1490:LM 25.	6.5–7.5	—	0.2–0.6	—	—	Solution treated at 540°C for 4–12 hours; quenched; precipitation-hardened at 165°C for 8–12 hours.	M Sand cast 150 / Chill cast 160; TE Sand cast 150 / Chill cast 190; TF Sand cast 250 / Chill cast 280	2, 3, 1, 2, —, 2	General purpose die-casting alloy. Used where good corrosion-resistance coupled with good mechanical properties are necessary. Also used for sand castings.
1490:LM 26	8.5–10.5	2.0–4.0	0.5–1.5	—	—	Precipitation-treated at 180–190°C for 8–12 hours.	TE-Chill cast 210	—	90–120 Gravity die cast pistons.
1490:LM 29.	22.0–25.0	0.8–1.3	0.8–1.3	—	0.8–1.3	Precipitation treatment and microscopic examination to ensure primary silicon is evenly distributed in the eutectic (size shall not exceed 40 μm average and 70 μm maximum.)	TE Sand cast 120 / Chill cast 190; TF Sand cast 120 / Chill cast 190	—	100–140 Very low coefficient of thermal expansion. High hardness. Special pistons for internal combustion engines.

adequate for their use in cylinder blocks without the inclusion of cast-iron liners. Both the wear-resistance and mechanical strength of these alloys can be improved considerably by controlling the size and distribution of the primary silicon. The finer and more evenly distributed the primary silicon the better the wear resistance. A number of methods have been proposed to refine this primary silicon and include the addition of small amounts of phosphorus, arsenic or sulphur to the melt just prior to casting. Since copper, magnesium and nickel are also present (LM29) in sufficient quantities these alloys can be further hardened and strengthened by pre-cipitation treatment. However, despite the apparent advantages of the high-silicon alloys production difficulties have presumably prevented them from becoming established even though they are adequately covered in BS 1490.

A selection of the alloys representative of this group is given in Table 17.4.

Exercises

1. By using Fig 17.2 estimate the proportions of primary α to eutectic in an unmodified sand cast alloy containing 6% silicon which is just below its solidification temperature.

2.

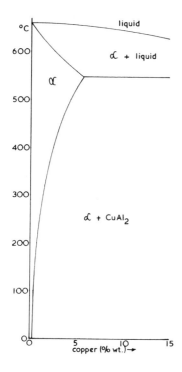

The figure alongside shows the aluminium-rich end of the aluminium–copper thermal equilibrium diagram.
(i) What are the % compositions of phases present in an alloy containing 12% copper which has cooled slowly to 0°C?
(ii) In what proportions by mass will the phases in (i) exist?
(iii) What will be the temperature range over which an alloy containing 4% copper could be solution treated?
(iv) What, therefore, would be an appropriate practical solution-treatment temperature for this alloy? Give reasons.
(v) Explain, step by step, what happens if this 4% Cu alloy is allowed to cool *slowly* from its solution-treatment temperature to ambient temperature.
(vi) What mechanical properties will characterise the resultant structure in (v)?
(Atomic masses: Cu–63.53; Al–26.98)

3. What characteristics of aluminium and its alloys account for its versatility in uses? Suggest suitable competitors for two possible applications.
4. What is understood by the terms *modification* and *age-hardening* as applied to the alloys of aluminium? What is the structural basis of each of these procedures and why are they carried out? (17.41; 17.50)
5. What is meant by the terms solution-treatment and precipitation-treatment? Illustrate your reference to a suitable aluminium-base alloy. (17.50)
6. What are the essentials of precipitation-hardening treatments? Give three distinct examples of alloys in which precipitation-hardening is found. (9.92; 13.101; 16.70; 17.50; 18.10)
7. Describe two methods of hardening aluminium alloys. Give the principles underlying each method. (17.30; 17.50)
8. Which types of alloy can be hardened *(a)* by cold-working; *(b)* by precipitation-hardening; *(c)* by a combination of *(a)* and *(b)*? Why is it necessary to exercise close control of heat-treatment variables in precipitation-hardening heat treatment? (17.30; 17.50; 17.53; 17.56)
9. Outline the basic theory which seeks to explain the precipitation hardening of an aluminium alloy containing 4% copper. (9.92; 17.52)
10. Suppose a batch of aluminium-alloy aircraft rivets had been on the shop-floor over the week-end. Would they be usable? If not could they be salvaged— and, if so, how? (17.56)

Bibliography

Higgins, R. A., *Engineering Metallurgy (Part II)*, Edward Arnold, 1986.

King, F., *Aluminium and its Alloys*, John Wiley, 1987.

Martin, J. W., *Precipitation Hardening*, Pergamon, 1968.

Martin, J. W., *Micromechanisms in Particle-hardening Alloys*, Cambridge University Press, 1979.

Mondolfo, L. E., *Aluminium Alloys, Structure and Properties*, Butterworth, 1976.

Polmear, I. J., *Light Alloys (Metallurgy of the Light Metals)*, Edward Arnold, 1989.

BS 1490: 1988 *Aluminium and Aluminium Alloy Ingots and Castings*.

BS 1470: 1987 *Wrought Aluminium and Aluminium Alloys for General Engineering Purposes: Plate, Sheet and Strip*.

BS 1471: 1972 *Wrought Aluminium and Aluminium Alloys for General Engineering Purposes: Drawn Tube*.

BS 1472: 1972 *Wrought Aluminium and Aluminium Alloys for General Engineering Purposes: Forging Stock and Forgings*.

BS 1473: 1972 *Wrought Aluminium and Aluminium Alloys for General Engineering Purposes: Rivet, Bolt and Screw Stock*.

BS 1474: 1987 *Wrought Aluminium and Aluminium Alloys for General Engineering Purposes: Bars, Extruded Round Tubes and Sections*.

BS 1475: 1972 *Wrought Aluminium and Aluminium Alloys for General Engineering Purposes: Wire*.

18

Other Non-ferrous Metals And Alloys

Magnesium-base Alloys

18.10   It is perhaps surprising that magnesium-base alloys should be suitable for engineering purposes, since in composition they are similar to the incendiary bombs used by the RAF and the Luftwaffe during the Second World War. Incendiary-bomb cases contained about 7% aluminium (which was added to introduce better casting properties) and small amounts of manganese; the balance being magnesium. Fortunately, combustion of such an alloy takes place only at a relatively high temperature, and this was generated by filling the bomb case with thermit mixture (20.52). Ordinarily, magnesium alloys can be melted and cast without mishap, provided that the molten alloy is not overheated.

18.11   Although pure magnesium is a relatively weak metal, its alloys containing suitable amounts of aluminium, zinc or thorium can be strengthened considerably by precipitation-hardening. The thermal-equilibrium diagrams (Fig. 18.1) indicate the changes in solid solubility of these three metals in the respective α-phases; changes which are necessary if precipitation-hardening is to be effected (9.92).

18.12   Small amounts of manganese, zirconium and the rare earth metals are also added to some magnesium alloys. Manganese improves corrosion-resistance, whilst zirconium is an effective grain refiner. Thorium and the rare-earth metals (in particular, cerium) give further increase in strength whilst allowing the alloy to retain a somewhat higher ductility, but their most important function is to give improved creep-resistance at high working temperatures.

The greater strength obtained by alloying, coupled with their low relative density of about 1.8, makes these alloys particularly useful where weight

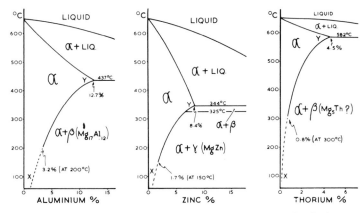

Fig. 18.1 Changes in solid solubility along *XY* make precipitation-hardening possible in each case.

is a limiting factor. Such uses include various castings and forgings used in the aircraft industry—landing-wheels, petrol tanks, oil tanks, crankcases and airscrews, besides many engine parts in both piston and jet engines. Those alloys which can be precipitation hardened are first solution treated at temperatures between 300°C and 565°C—410°C for those alloys containing aluminium; 320°C for those containing zinc; and 565°C for those containing thorium as indicated by the equilibrium diagrams (Fig. 18.1). Precipitation treatment is usually carried out at between 170° and 200°C for up to 12 hours. Examples of both cast and wrought alloys are given in Tables 18.1 and 18.2.

18.13   The melting and casting of these alloys present some difficulty in view of the ease with which molten magnesium takes fire, particularly when overheated. They are generally melted under a flux containing the fluorides of calcium and sodium and the chlorides of potassium and magnesium. During the casting operation flowers of sulphur is often dusted on to the stream of molten alloy. The sulphur takes fire in preference to the magnesium, and thus protects the latter with a surrounding atmosphere of sulphur dioxide gas. Nevertheless modern developments such as fluxless melting and mechanical stirring have improved commercial production in recent years.

Zinc-base Die-casting Alloys

18.20   Zinc-base alloys containing 4% aluminium and small amounts of copper and magnesium have been used for many years for die casting a wide range of components both in engineering industries and for domestic appliances, and were sold initially under the trade name of 'Mazak'. These alloys have excellent casting properties as well as reasonable strength and rigidity and are used for many parts in washing machines, radio sets, alarm

Table 18.1 Cast Magnesium-base Alloys

Relevant specifications to BS 2970	Composition (%)						Condition	Typical mechanical properties.					
								0.2% Proof stress (N/mm^2)		Tensile strength (N/mm^2)		Elongation (%)	
	Al	Zn	Mn	Zr	Th	Rare Earths		Sand	Chill	Sand	Chill	Sand	Chill
General purpose alloys													
MAG.1 (Mg–Al8–Zn–Mn)	7.5 – 9.0	0.3 – 1.0	0.15 – 0.4	—	—	—	M TB	85 80	85 80	140 200	185 230	2 6	4 10
MAG.3 (Mg–Al10–Zn–Mn)	9.0 – 10.5	0.3 – 1.0	0.15 – 0.4	—	—	—	M TB TF	95 85 130	100 85 130	125 200 200	170 275 215	— 4 —	2 5 2
MAG.7 (Mg–Al8.5–Zn1–Mn)	7.5 – 9.5	0.3 – 1.5	0.15 – 0.8	—	—	—	M TB TF	85 80 110	85 80 110	125 185 185	170 215 215	— 4 —	2 5 2

Special alloys

Alloy							Condition						
MAG.5 (Mg–Zn4–RE–Zr)	—	3.5 – 5.0	—	0.4 – 1.0	—	0.75 – 1.75	TE	135	135	200	215	3	4
MAG.6 (Mg–RE3–Zn–Zr)	—	0.8 – 3.0	—	0.4 – 1.0	—	2.5 – 4.0	TE	95	110	140	155	3	3
MAG.8 (Mg–Th3–Zn2–Zr)	—	1.7 – 2.5	—	0.4 – 1.0	2.5 – 4.0	—	TE	85	85	185	185	5	5
MAG.9 (Mg–Zn5.5–Th2–Zr)	—	5.0 – 6.0	—	0.4 – 1.0	1.5 – 2.3	—	TE	155	155	255	255	5	5
(WE 43)	—	—	—	0.5	—	Y–4.0 Nd and others–3.5	TF	185	185	265	265	7	7
DTD.5035A	Ag 2.5	—	—	0.6	—	Nd and others–2.5	TF	185	185	240	240	2	2

M—as cast
TE—precipitation-treated
TB—solution-treated
TF—solution-treated and precipitation-treated.

Table 18.2 *Wrought Magnesium-base Alloys*

Relevant Specification (BS)	Composition (%)				Condi-tion	Typical mechanical properties		
	Al	Zn	Mn	Zr		0.2 % Proof stress (N/mm^2)	Tensile strength (N/mm^2)	Elonga-tion (%)
3370:MAG.S.101 3372:MAG.F.101 3373:MAG.E.101	—	—	1.0–2.0	—	M	70	200	5
3370:MAG.S.111 3373:MAG.E.111	2.5–3.5	0.6–1.4	0.15–0.40	—	M O	100 120	250 220	8 12
3372:MAG.F.121 3373:MAG.E.121	5.5–6.5	0.5–1.5	0.15–0.40	—	M	180	270	8
3370:MAG.S.131 3372:MAG.F.131 3373:MAG.E.131	—	1.5–2.3	0.6–1.3	—	M O	160 120	250 220	8 12
3370:MAG.S.141 3372:MAG.F.141 3373:MAG.E.141	—	0.75–1.5	—	0.4–0.8	M	170	250	8
3370:MAG.S.151 3372:MAG.F.151 3373:MAG.E.151	—	2.5–4.0	—	0.4–0.8	M	180	265	8
3372:MAG.F.161 3373:MAG.E.161	—	4.8–6.2	—	0.45–0.8	TE	230	315	8

BS3370—Plate, sheet and strip (S)
BS3372—Forgings and cast forging stock (F)
BS3373—Extruded bars, sections and tubes (E)

clocks, electric fires and accurately-scaled 'matchbox' toys; whilst the average motor car contains a multitude of zinc-base die-castings from the door handles to the wiper-motor frames.

18.21 Difficulty was experienced in the development of these zinc-base alloys owing to swelling of the casting during use, coupled with intercrystalline embrittlement. These faults were found to be due to intercrystalline corrosion caused by small amounts of cadium, tin and lead. For this reason high-grade zinc of four nines quality (ie 99.99% pure) is used for the manufacture of these alloys, and the impurities thus limited to the small quantities indicated in Table 18.3.

18.22 Normally, after casting, these zinc-base alloys undergo a slight shrinkage which is complete in about five weeks. When close dimensional limits are necessary a stabilising anneal in dry air at 100±5°C for six hours should be given before machining, in order to accelerate any volume change that is likely to take place.

In order to test these alloys for growth and intercrystalline corrosion, the steam test is applied. This accelerates the formation of these defects, and consists in suspending the test piece above water at 95°C in a closed vessel for ten days. The test piece is then examined for expansion, and

Table 18.3 Zinc-base Casting Alloys

Designation	Composition (%)					Condition	Typical mechanical properties				Relative density
	Zn	Al	Cu	Mg	Impurities (max.)		Tensile strength (N/mm^2)	Elongation (%)	Impact toughness (J)	Hardness (H_b)	
BS1004 (alloy A)	Bal.	4.1	0.10 max.	0.05		die cast	286	15	57	83	6.7
						'stabilised'	273	17	61	69	
BS1004 (alloy B)	Bal.	4.1	1.0	0.05		die cast	335	9	58	92	6.7
						'stabilised'	312	10	60	83	
ZA8	Bal.	8.4	1.0	0.03	Fe-0.10 Pb-0.004 Cd-0.003 Sn-0.002	sand cast	260	2	—	85	6.3
						pressure die-cast	375	8	40	103	
ZA12	Bal.	11.0	0.8	0.03		sand cast	300	2	—	100	6.0
						gravity die-cast	327	2	28	90	
ZA27	Bal.	26.5	2.2	0.02		sand cast	420	4.5	—	115	
						sand cast (heat-treated)	317	10	—	95	5.0
						gravity die-cast	424	1	13	115	

mechanically tested to reveal any intercrystalline corrosion which may have occurred.

18.23 Of the two casting alloys generally available, alloy A is used for engineering purposes since it has greater dimensional stability and responds consistently to the stabilising anneal. It is also more ductile and retains its impact strength at high temperatures much better than alloy B. The corrosion resistance of alloy A is also superior. Alloy B, however, has greater strength and hardness as well as better castability in thin sections and is therefore used for such components as zip-fastner slides and small gear wheels.

18.24 In recent decades the relatively low-strength die-cast alloys have been replaced for many purposes by plastics mouldings. This, together with a fashion change away from chromium-plated finishes, led to a big decline in the use of the 4% aluminium (BS1004) alloys. However a new series of zinc-base alloys containing greater quantities of aluminium and copper has been developed to give increased strength and hardness, and it is expected that these will find increasing use in the automobile industries as engine mountings and other parts where strength and toughness combined with a saving in weight are involved.

Three new alloys, designated by the trade ZA8, ZA12 and ZA27 contain respectively 8, 11 and 27% aluminium along with small amounts of copper and magnesium (Table 18.3). As with the BS1004 alloys the impurities lead, cadmium, tin and iron need to be strictly controlled in order to preserve dimensional stability. Whilst ZA8, like the BS1004 alloys, can be pressure die cast both ZA12 and ZA27 are more suitable for gravity die casting.

18.25   These alloys, like the BS1004 alloys, undergo changes both in mechanical properties and dimensions as a result of an ageing process subsequent to being cast. As is indicated in the zinc–aluminium equilibrium diagram (Fig.18.2) a eutectoid change occurs in these alloys at 275° when the solid solution β' transforms to a pearlite-like eutectoid of α and β. However the rapid cooling which prevails during die casting suppresses the eutectoid change. This is further aggravated by the presence of copper which also retards the eutectoid transformation.

Subsequently ageing occurs as the β' phase—unstable at ambient temperatures—gradually reverts to the $\alpha + \beta$ eutectoid with passage of time. This leads to a general deterioration in mechanical properties and at the same time a change in volume. The ZA27 alloy is sometimes heat treated by cooling down slowly from above the eutectoid temperature in order to produce a fine lamellar eutectoid (cf. normalising in steel—11.60) and as a result much improved toughness and ductility. The ductile/brittle transition temperature is also depressed from 0°C to −20°C. In order to stabilise it against volume changes the ZA27 alloy can be aged for a short time just below the eutectoid temperature.

18.26   The corrosion resistance of zinc-base alloys is adequate for most atmospheric environments. However, the availability of zinc anodising (21.90) and other protective finishes make them useful in aggressive marine environments such as oil platforms in the North Sea. One such treatment, *chromate passivation*, involves immersion of the castings in a solution containing sulphuric acid and sodium dichromate.

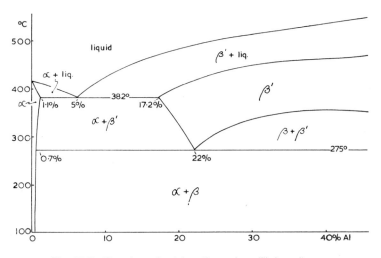

Fig. 18.2 The zinc–aluminium thermal equilibrium diagram.

Nickel–Chromium High-temperature Alloys

18.30   The main feature of the nickel–chromium alloys is their ability to resist oxidation at elevated temperatures. If further suitable alloy additions are made the strength is increased under conditions of stress at high temperatures. Alloys rich in nickel and chromium have a high specific resistance to electricity, which makes them admirable materials for the manufacture of resistance wires, and (because of their low rate of oxidation at high temperatures) heater elements of many kinds capable of working at temperatures up to bright red heat. Such alloys (the Brightray series) are included in Table 18.4.

18.31   A further group of high-temperature, nickel–chromium-base alloys are those containing iron and, sometimes, molybdenum and tungsten. Of these the Inconel series has been long established. Inconel 600 contains 76Ni; 16Cr; and 7Fe and is resistant to many acids and alkalis as well as oxidising atmospheres at high temperatures. Since it retains reasonable strength at high temperatures it is used for furnace equipment including retorts, muffles, heat-treatment trays, supports and nitriding

Table 18.4   *High-temperature Resistance Alloys*

Composition(%)				Tensile strength of annealed rod N/mm^2	Resistivity $(10^{-8}\,\Omega m)$	Maximum working temp. (°C)	Uses
Ni	Cr	Fe	Cu				
80	20	—	—	911	103	1150	Heaters for electric furnaces, cookers, kettles, immersion heaters, hair dryers, toasters, etc.
65	15	20	—	726	106	950	Similar uses to above, but for goods of lower quality—also for soldering-irons, tubular heaters, towel rails, laundry irons, etc.
34	4	62	—	664	91	700	Cheaper-quality heaters working at low temperatures, but mainly as a resistance wire for motor-starter resistances, etc.
45	—	—	55	402	48	300	Limited use for low-temperature heaters such as bed-warmers, etc. Mainly as a resistance wire for instrument shunts, field regulators and cinema arc resistances.

Table 18.5 Some High-temperature 'Superalloys'

Alloy	Relevant specifications (BS)	Composition (%) (Bal. -Ni)						Typical mechanical properties at 650°C			Uses in jet aircraft construction
		Cr	Co	Mo	Ti	Al	Other elements	0.2% Proof stress (N/mm²)	Tensile strength (N/mm²)	Elongation (%)	
Nimonic 75	HR5; HR203; HR403; 2HR504.	20	—	—	0.4	—	—	325	600	27	Casings and rings; sheet-metal fabrications (combustion chambers, hot-gas ducts, exhaust systems, silencers, thrust reversers, etc.)
Nimonic 80A	2HR1; 2HR201; 2HR401; 2HR601.	20	—	—	2.3	1.3	—	700	870	14	Discs; turbine blades and vanes; shafts; casings and rings.
Nimonic 90	2HR2; 2HR202; HR402; 2HR501; 2HR502; 2HR503.	20	17	—	2.5	1.4	—	740	900	15	Discs; blades and vanes for both compressors and turbines; shafts; casings and rings.
Nimonic 105	HR3	15	20	5	1.2	4.7	—	750	1100	27	Discs; turbine blades and vanes.
Nimonic 115	HR4	15	15	3.5	4	5	—	770	1110	22	Turbine blades and vanes.
Nimonic 263	HR10; HR206.	20	20	5.9	2.1	0.5	—	470	800	38	Casings and rings; sheet metal fabrications (combustion chambers, hot-gas ducts, exhaust systems, silencers, thrust reversers, etc.)

Alloy	Specification	Cr	Co	Mo	Ti	Al	Other				Applications
Nimonic 901	HR53; HR404.	12.5	1	5.7	2.9	0.3	Fe-35	810	1000	12	Compressor blades and vanes; discs; shafts; casings and rings.
Nimonic PE13	HR6; HR204.	22	1.5	9	—	—	Fe-18 W-0.6	310	620	45	Casings and rings; sheet-metal fabrications (combustion chambers, hot-gas ducting, exhaust systems, silencers, thrust reversers, etc.)
Nimonic PE16	HR55; HR207.	16	—	3.3	1.2	1.2	Fe-35	560	770	22	
Nimonic PK33	—	18	14	7	2.2	2.1	—	720	1060	23	Discs.
Nimonic AP.1	—	15	17	5	3.5	4	—	1000	1350	—	
Inconel 718	—	18.5	—	3.1	0.9	0.4	Fe18.5 Nb-5	950	1050	—	Compressor blades and vanes; discs; shafts; casings and rings.
Inconel MA754	—	20	—	—	0.5	0.3	Y_2O_3-0.6	—	—	—	Turbine blades and vanes.
Incoloy 903	—	—	15	—	1.4	0.7	Fe-42 Nb-3	—	—	—	Shafts; casings and rings.
Incoloy MA956	—	20	—	—	0.5	4.5	Y_2O_3-0.5	—	—	—	Sheet metal fabrications (combustion chambers; hot-gas ducting; exhaust systems; silencers; thrust reversers, etc.)

boxes, as well as for exhaust manifolds and electric heating elements for cookers. Hastelloy C (59Ni; 17Mo; 15Cr; 5Fe; and 4W) is resistant to corrosion by oxidising acids such as nitric and sulphuric, as well as to oxidising and reducing atmospheres up to 1000°C; whilst Hastelloy X (51Ni; 22Cr; 18Fe; 9Mo) is a high-temperature alloy maintaining good oxidation resistance and reasonable strength up to 1200°C making it useful in furnace equipment and combustion systems of jet aircraft.

18.32 By far the most spectacular of the nickel–chromium-base alloys, however, are those of the Nimonic series. Nimonic is a trade name coined by Messrs. Henry Wiggin, the manufacturers of these alloys, which have played a leading part in the construction of jet engines. They are basically nickel–chromium alloys which will resist oxidation at elevated temperatures, but which have been further stiffened for use at high temperatures by the addition of small amounts of titanium, aluminium, cobalt, molybdenum and carbon in suitable combinations. 'Nimonics' and 'super-alloys' generally must have a high creep resistance at high temperatures. Since creep proceeds by the thermally-activated movement of dislocations through the lattice, the basic requirements for high creep resistance are a high elastic modulus and low diffusion rates within the lattice at high temperatures. The presence of one or more of the elements mentioned above will provide solute atoms in which the atom size produces sufficient lattice distortion to increase the elastic modulus, and at the same time have low diffusion rates.

The most important strengthening mechanism depends upon the formation of *coherent* precipitates—generally designated γ'—within the lattice of the nickel-base matrix. These coherent precipitates are based on the intermetallic compounds $TiNi_3$, $AlNi_3$, or $NbNi_3$ which, like the nickel-base matrix have a FCC structure. However, there is a very small difference (less than 1%) in lattice dimensions between matrix and compound and since the mismatch is slight the compounds are able to form coherent precipitates within the nickel-base matrix which remain stable at high temperatures. Any carbides present form as finely-dispersed particles which strengthen the grain boundary regions and inhibit intergranular fracture.

Some of these alloys are dispersion strengthened (8.62) by the inclusion of particles of yttrium oxide Y_2O_3. Uniform distribution of such insoluble particles cannot be achieved by normal casting processes. Hence the powdered base alloy is ground with Y_2O_3 powder in a ball mill to promote uniform mixing and 'mechanical alloying' of the components. The blend is then hot extruded to consolidate the alloy which may be shaped by further hot-working processes.

The result is a series of alloys which will withstand oxidation and mechanical stresses at working temperatures in excess of 1000°C. In addition to their use in jet engines, they find application as thermocouple sheaths, tubular furnaces for the continuous annealing of wire, jigs for supporting work during brazing or vitreous enamelling, gas retorts, extrusion dies and moulds for precision glass manufacture. Some of the current Nimonic alloys together with other Wiggin 'super alloys' are shown in Table 18.5 whilst the relationship between strength and temperature for some of these alloys is shown in Fig. 18.3. It should be appreciated, however, that the true criteria to usefulness will be high creep strengths at elevated temperatures, available with these alloys.

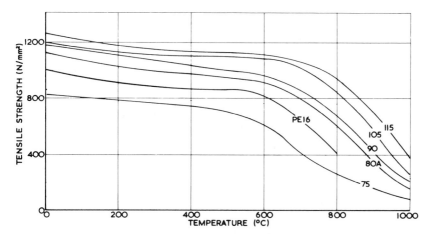

Fig. 18.3 The relationship between tensile strength and temperature for some important 'Nimonic' high-temperature alloys.

18.33 Nickel-base Corrosion-resistant Alloys containing molybdenum, iron, usually chromium and sometimes small amounts of tungsten and copper are used at ordinary temperatures in the chemical industries where they are in contact with concentrated acids and other extremely corrosive liquids. At ordinary temperatures up to 20°C molybdenum dissolves in nickel in the solid state and this increases both corrosion resistance and strength. The Hastelloy series of alloys is probably the most important in this group. Hastelloy B (65Ni; 28Mo; 5Fe) is used for transporting and storing hydrochloric acid, phosphoric acid and other non-oxidising acids; whilst Hastelloy C (59Ni; 17Mo; 15Cr; 5Fe; 4W) has a high resistance to a wide range of corrosive media including oxidising liquids and those containing chlorides. It is also reasonably resistant to concentrated sulphuric acid at ambient temperatures and to phosphoric acid both cold and boiling.

Bearing Metals

18.40 Bearings are devices used to transmit loads between relatively-moving surfaces. They may be either *plain* bearings involving sliding contact between surfaces or more *complex* bearings using costly ball- or roller-components. In this instance we shall consider alloys used for plain bearings.

If lubrication were always perfect and bearing surfaces accurately fitted so that 'running in' was unnecessary, then almost any metal would suffice as a bearing provided that it was strong enough to carry the load. In fact it is often difficult to align a long shaft accurately enough to run in a series of bearings. Consequently a relatively soft, malleable material is preferable so that it will align itself to the journal when under pressure. This prevents the build-up of stress at any 'high spots' on the surface so that the journal is less likely to be scored by grit in the lubricant. Nevertheless the bearing must be strong enough and hard enough, preferably with a low-friction, wear-resistant surface. This set of properties is generally not obtained in a single-phase alloy for, whilst solid solutions are tough, ductile and strong they are also rather soft; and whilst intermetallic compounds are hard, they are also brittle and weak. Hence, traditionally, bearing metals have been two-phase alloys, eg white metals and bronzes. In these materials harder, low-friction particles are held in a malleable, soft solid-solution (or eutectic) matrix. This matrix tends to wear, providing channels which assist the flow of lubricant.

Another desirable characteristic of a bearing metal is its ability to form 'soaps' with the lubricant, whilst low *overall* hardness will minimise damage to the shaft starting and stopping, when oil pressure may be low and the tendency towards pressure-welding of the journal to the bearing surface at its greatest.

Bearing materials are of three main types:

(i) Two-phase alloys of the traditional types, eg white metals and bronzes;

(ii) Two-phase alloys in which one phase can become a lubricant in extreme circumstances, eg leaded bronzes and aluminium–tin alloys;

(iii) Single-phase materials, eg nylon and PTFE (polytetrafluoro-ethene).

18.41 Copper-base Bearing Alloys   include the phosphor-bronzes (containing from 10 to 13% tin and 0.3 to 1.0% phosphorus) and the plain tin bronzes (containing from 10 to 15% tin). Both types of bronze satisfy the structural requirements of a bearing metal, since they contain particles of the hard intermetallic compound δ ($Cu_{31} Sn_8$) embedded in a tough matrix of the solid solution α. These alloys are very widely used for bearings of many types where the loading is heavy.

For small bearings and bushes of standardised sizes, sintered bronzes are often used. These are generally of the self-lubricating type, and are made by mixing copper powder and tin powder together, in the proportions of a 90–10 bronze, with the addition of graphite. The mixture is then compacted and sintered (6.53—Part I and 6.50—Part II) to produce a semi-porous, self-lubricating alloy suitable for low-duty bearings in sizes up to about 75 mm diameter shaft.

Leaded bronzes (Table 16.2) are used for main bearings in aero-engines, and for automobile and diesel crankshaft bearings. They have a high resistance to wear, and their good thermal conductivity enables them to keep cool when running. Should lubrication fail the lead is extruded under pressure, as overheating sets in, and forms a lubricating film which prevents seizure—a necessary feature with aircraft bearings. These alloys contain 5–30% lead and since lead is insoluble in molten bronze these bearings are often centrifugally cast to prevent excessive segregation of the lead. Alternatively, segregation is reduced by adding small amounts of nickel. Some of these bearings consist of sintered porous copper infiltrated with lead. Corrosion of the lead by sulphur compounds in the lubricant is prevented by overlaying the surface with a coating of lead–indium or tin–indium which is encouraged to diffuse into the bearing surface by low-temperature heat-treatment.

18.42 White Bearing Metals   may be either tin-base or lead-base. The former, which represent the better-quality white bearing alloys, are often called Babbitt metals, after Isaac Babbitt, their original patentee. All of these alloys contain between 3.5 and 15.0% antimony, much of which combines with some of the tin present to form an intermetallic compound SbSn. This compound forms cubic crystals, usually called cuboids (see Plate 18.1) which are hard and of low-friction properties.

Let us assume that we have a plain tin–antimony alloy containing 10% antimony, as being typical of a basic bearing-metal analysis. Fig. 18.4 shows the tin-rich end of the tin–antimony equilibrium diagram. Our 10% alloy will begin to solidify at X by depositing cuboids of the hard intermetallic compound SbSn. These SbSn crystals are of lower specific gravity than the liquid which remains, and they therefore tend to float to the surface of the melt. At 246°C a peritectic reaction takes place, and the solid solution α is formed as an in-filling between the cuboids. As the temperature falls, the solubility of antimony in tin decreases from about 10.3% at the peritectic temperature to 3.5% at room temperature, so that if cooling is slow enough, more cuboids of SbSn are precipitated uniformly within the solid solution α. Nevertheless, the segregation of the primary cuboids of SbSn at the surface of the cast bearing is a serious fault. It can, however, be minimised in one of the following ways:

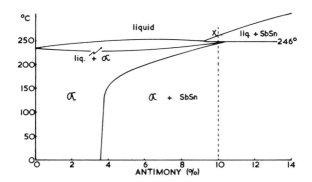

Fig. 18.4   Part of the tin–antimony thermal equilibrium diagram.

(*a*) By rapidly cooling the cast alloy when possible. The metal then solidifies completely before the cuboids have had an opportunity to rise to the surface.

(*b*) By adding up to 3.5% copper to the alloy. This combines with some of the tin, forming a network of needle-shaped crystals of Cu_6Sn_5 which precipitates from the melt *before* the cuboids of SbSn begin to form. This network traps individual cuboids as they develop and prevents their movement towards the surface, thus ensuring a uniform cast structure. Moreover, the compound Cu_6Sn_5, being hard, improves the general bearing properties.

Lead is added in the interests of cheapness, and this forms solid solutions of limited solubility with both tin and antimony. These solid solutions form a eutectic structure. In some of the cheap bearing metals, eg. Magnolia metal, tin is omitted completely. The cuboids then consist of almost pure antimony, embedded in a eutectic matrix of antimony-rich and lead-rich solid solutions. These lead-rich alloys are intended only for low pressures. The more important white bearing metals are indicated in Table 18.6.

18.43 Aluminium–tin Bearing Alloys   The more recent philosophy of using a duplex structure consisting of two interlocking networks—one strong and the other soft and capable of acting as a lubricant under occasional high contact pressures—led to the development of aluminium–tin alloys as bearings. These alloys contain approximately 80Al–20Sn. The two metals form a eutectic series (Fig. 18.5) in which the eutectic contains only 0.5% aluminium and solidifies at 228°C so that the alloy in question solidifies over a range of more than 400°C and segregation is a considerable danger. Nevertheless, by a combination of cold-rolling

Table 18.6   *White Bearing Metals*

BS 3332 designation	Composition (%)					Characteristics and uses
	Sn	Sb	Cu	Pb	Zn	
/1	90	7	3	–	–	High unit loads and high temperatures.
/2	87	9	4	–	–	Main bearings in automobile and aero engines.
/3	81	10	5	4	–	British Rail—moderately severe duties.
/4	75	12	3	10	–	
/5	75	7	3	15	–	Railway carriage work. Pumps, compressors and general machinery.
/6	60	10	3	27	–	
/7	12	13	1	74	–	Railway wagon bearings, mining machinery and electric motors.
/8	5	15	–	80	–	Magnolia Metal—Light loads and speeds at higher temperatures.
/9	68.5	–	1.5	–	30	Under-water bearings.

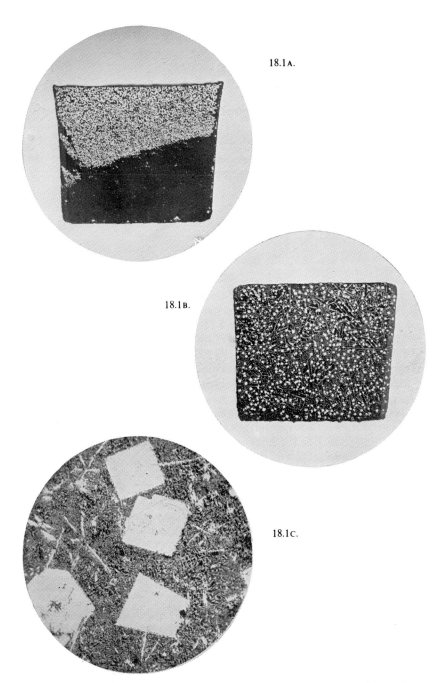

18.1A.

18.1B.

18.1C.

Plate 18.1 18.1A Vertical section through a small ingot of bearing metal (8% antimony; 32% lead; 60% tin).
A slow rate of cooling has enabled the cuboids of SbSn to float to the surface, where they have formed a layer. × 4. Etched in 2% nital.
18.1B Vertical section through a similar alloy (with the addition of 3% copper).
Although cooled at the same rate as the ingot in Plate 18.1A, segregation of the SbSn cuboids has been prevented by the needles of Cu_6Sn_5 which formed first. × 4. Etched in 2% nital.
18.1C Bearing metal containing 10% antimony; 60% tin; 27% lead; and 3% copper.
Cuboids of SbSn (light) and needles of Cu_6Sn_5 (light) in a matrix consisting of a eutectic of tin-rich and lead-rich solid solutions. × 100. Etched in 2% nital.

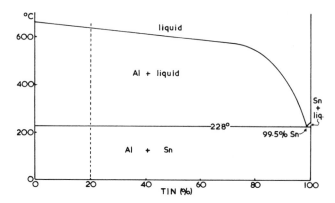

Fig. 18.5 The aluminium–tin equilibrium diagram.

and recrystallisation the tin network is broken up to some extent so that the aluminium forms a continuous strong matrix. Bearing shells of this type are generally carried on a steel backing strip and have been widely used in automobile design both for main and big-end bearings.

18.44 Polytetrafluoroethene (PTFE) and Nylon as non-metallic bearing materials are competing with metals in many fields. The coefficient of friction of PTFE ($\mu = 0.03$–0.1) is lower than for any other solid material. Unfortunately its mechanical properties are poor so that it has to be strengthened by the use of some 'filler' material. PTFE/bronze composites (Plate 18.2) are useful up to temperatures of 300°C—the upper working limit for PTFE. They are used in the automobile industry as bearings in windscreen wipers and steering system bushes. Nylon can be reinforced with bronze particles. An important feature of both materials is that they function well as 'dry' bearings. Hence they are useful in machinery for the preparation of food and textiles where contamination by a lubricant would be unacceptable. They are also useful in some cases where a bearing is so inaccessible as to make its lubrication difficult.

Plate 18.2 The microstructure of a PTFE filled sintered bronze bearing. *(Courtesy of The Glacier Metal Co Ltd, Northwood Hills, Middlesex)*.

Fusible Alloys

18.50   These contain varying amounts of lead, tin, bismuth and cadmium, and are generally ternary or quaternary alloys of approximately eutectic composition. A number of such alloys will melt below the boiling point of water, and in addition to such obvious uses as the manufacture of teaspoons for practical jokers, they are used for metal pattern work, for the fusible plugs in automatic fire-extinguishing sprinklers, for bath media in the low-temperature heat-treatment of steels and for producing a temporary filling which will prevent collapse of the walls during the bending of a pipe. Bismuth-rich alloys such as Cerromatrix expand slightly on cooling, and are useful for setting press-tool punches in their holders.

Table 18.7   *Fusible Alloys*

	Composition (%)				*Other elements*	*Melting point/range (°C)*
Type of Alloy	*Bi*	*Pb*	*Sn*	*Cd*		
Tinman's solder	–	38	62	–	–	183
Cerromatrix	48	28.5	14.5	–	Sb-9	102–225
Rose's Alloy	50	28	22	–	–	96
Wood's Alloy	50	24	14	12	–	71
Cerrolow 117	45	23	8	5	In-19	47
Ternary eutectic	–	–	12	–	Ga-82	17
					Zn-6	

Titanium and its Alloys

18.60   Mention of the use of small amounts of titanium in some steels and aluminium alloys has already been made. In recent years titanium has become increasingly important as an engineering material in its own right, both in the commercially pure and alloyed forms.

Titanium is not a newly discovered element (it was discovered in Cornwall in 1791 by W. Gregor, an English priest), and is only rare in the sense that it is a little-used metal, and one which is expensive to produce. In Nature it is very widely distributed, being the tenth element in order of abundance in the Earth's crust. Of the metallurgically useful metals only aluminium, iron, sodium and magnesium occur more abundantly (Table 1.2). Nevertheless it is unlikely that titanium will ever be an inexpensive metal because of its high chemical affinity for all non-metals except for the noble gases. For this reason it can only be produced by reducing its chloride, $TiCl_4$, with magnesium—obviously an expensive process—and the titanium so produced will react with all orthodox crucible materials. Hence it can only be remelted in vacuum by using an electric arc process in which the electrode consists of the titanium 'sponge' (produced by chloride reduction) and in which the crucible is virtually of titanium. This seemingly impossible feat is achieved by using a water-cooled copper crucible on to which a layer of titanium freezes—a thermodynamically inefficient but necessary situation.

Very strong titanium alloys can be produced and since the relative density of the metal is only 4.5 then these alloys have a very high specific strength (2.36). Its high melting point (1725°C) has contributed to its use in jet-engine construction. Moreover the creep properties and fatigue strengths of titanium alloys are very satisfactory so that they are finding increasing use in aerospace industries. As an example the four engines of Concorde contain some 16 tonnes of titanium alloys.

18.61   Although titanium is chemically very reactive it has an excellent corrosion resistance. Like aluminium it coats itself with a dense protective oxide skin. It is now very extensively used, generally in the 'commercially pure' form, in chemical plant where it will resist attack under extremely hostile conditions such as are provided by strong acids, chloride solutions and bromine. By adding 0.15% of the noble metal palladium the corrosion rate can be reduced by up to 1500 times because of 'anodic passivation'.

18.62   Titanium is a bright silvery metal and, when polished, resembles steel in appearance. The high-purity metal has a relatively low tensile strength (216 N/mm²) and a high ductility (50%), but commercial grades contain impurities which raise the tensile strength to as much as 700 N/mm² and reduce ductility to 20%. It is an allotropic element. The α-phase, which is hexagonal close-packed in structure, is stable up to 882.5°C, whilst above this temperature the body-centred cubic β-phase is stable. The polymorphic change point is affected by alloying in a similar manner to the A_3 point in iron (13.11). Thus, alloying elements which have a greater solubility in the α-phase than in the β-phase tend to stabilise α over a greater temperature range (Fig. 18.6(i)). Such elements include those which dissolve interstitially—oxygen, nitrogen and carbon—and also aluminium

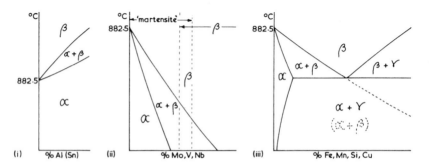

Fig. 18.6   The effects of alloying on the α → β transformation temperature in titanium.

which dissolves substitutionally. Elements which tend to dissolve in β, and consequently stabilise it, are usually those which, like β, are body-centred cubic in structure. These are mainly the transition elements iron, chromium, molybdenum, etc., and the resulting equilibrium diagrams are of the types (ii) or (iii) (Fig. 18.6). Those alloys represented by an equilibrium diagram of type (iii) could be expected to undergo a martensite-type transformation (12.22) if quenched from the β-range, and indeed they do. However, unlike the hard martensite produced by quenching steel the martensite formed here is relatively soft because the solute atoms are *substitutionally* dissolved and do not produce the same degree of lattice distortion as do the *interstitially* dissolved carbon atoms in steel. Due to considerable sluggishness the eutectoid transformation in type (iii) alloys is rarely complete and can be neglected so that most alloys can be regarded as belonging to group (i) or (ii) as far as subsequent heat-treatment is concerned.

18.63   The β-phase in alloys tends to be hard, strong and less ductile than the relatively soft α-phase. Nevertheless, the β-phase forges well so that most of the commercial alloys, which are of the α + β type, are hot-worked in the β range. Those elements like aluminium and tin which stabilise α tend to strengthen alloys by normal solid solution effects but the α + β alloys formed when molybdenum, vanadium, niobium, silicon and copper are added are strengthened by heat-treatment. Low alloy additions of these group (ii) and (iii) elements tend to give a martensitic structure when quenched from the β-range but with higher alloy additions a completely β structure is retained. Suitable ageing treatment produces a finely-dispersed precipitate of α within the β structure and the best all-round combination of mechanical properties. However, during the ageing treatment a very brittle intermediate phase ω is first precipitated but prolonged treatment causes this to disappear and be replaced by α. Consequently ageing treatment must be carried out in the region of 500°C for 24 hours to ensure the removal of all ω and its subsequent replacement by finely-divided α. Most of the heat-treatable titanium alloys are first solution treated in the range 850–1050°C according to composition, followed by ageing at 500°C for 24 hours.

18.64   Titanium alloys are now extensively used in both airframe and engine components of modern supersonic aircraft because of the attractive combination of specific strength and excellent corrosion resistance. Expensive precision cameras now contain titanium-alloy parts requiring low inertia such as shutter blades and blinds, whilst the high corrosion resistance of titanium has led to its use as surgical implants as well as in the chemical industries. Some alloys of titanium are given in Table 18.8.

Table 18.8 Some Titanium Alloys

Relevant specification (B.S.)	Composition (%)								Heat-treatment	Typical mechanical properties				Typical uses
	Al	V	Mo	Sn	Si	Zr	Other elements			0.2% Proof stress (N/mm^2)	Tensile strength (N/mm^2)	Elongation (%)	Stress to produce 0.1% strain in 100h (N/mm^2)	
2TA6	Commercially-pure titanium							Fe-0.2 max. H-0.0125 max.	Annealed at 650–750°C	460	650	15	—	In chemical plant for resistance to acids and chloride solutions.
2TA10 (Sheet and strip)	6	4	—	—	—	—	—		Annealed 700–900°C	900	1120	8	550 at 300°C	Air frame—gear box fittings, engine nascelles, fuselage skinning, wings, etc. Jet engines—blades and discs.
2TA28 (Forgings)	6	4	—	—	—	—	—		Heat to 870–950°C, quench and heat to 400–510°C for 2 hours.	970 min.	1200	14	—	Marine—heat exchangers, scrubbers, sonar, yacht fittings. Surgery—joint replacements, bone plates and screws.

							Heat treatment	0.2% proof stress	Tensile strength	Elongation %		Applications
TA48	4	—	4	2	0.5	—	Heat to 900°C and soak; cool in air. Heat at 500°C for 24 hours.	920	1125	9	465 at 400°C	Air frame—flaps and slat tracks, brackets in wings, engine pylons. Jet engines—discs and blades, fan casings, etc.
TA38	4	—	4	4	0.5	—	C-0.13	Heat to 900°C for 1 hour, cool in air. 'Age' at 500°C for 24 hr.	1095 min.	1350	14	—	Supplied as bars for machining components.
—	2.25	1	11	0.25	5	—	Heat to 1050°C for 30 minutes, quench in oil. Heat at 550°C for 24 hours.	970	1225	8	385 at 450°C	Jet engines—discs and blades.
(IMI 685)	6	—	0.5	0.25	5	—	Solution treated and precipitation-hardened.	At 520°C 850 / At 520°C 480	1020 / 700	6 / 9	300 at 520°C	Jet engines—rotary and static components, fan casings.
(IMI 834)	5.8	0.5	4	0.35	3.5	Nb-0.7 C-0.06	Solution-treated at 1025°C—air cooled or quenched according to section. Aged for 2 hours at 700°C.	960	1075	5	—	
TA52	—	—	—	—	—	Cu-2.5	Heated to 790–820°C, cooled in air.	550	810	10 min.	—	Sheet and strip.

Uranium

18.70   The mineral pitchblende was formerly thought to be an ore of tungsten, iron or zinc, until in 1789 M. H. Klaproth proved that it contained what he called 'a half-metallic substance' different from the three elements named. He named this new element Uranium in honour of Herschel's discovery of the planet Uranus a few years earlier. It was not until 1842, however, that E. M. Péligot, by isolating the metal itself, showed that Klaproth's 'element' was in fact the oxide.

18.71   In the solid form uranium is a lustrous, white metal with a bluish tint, and capable of taking a high polish. It is a very heavy metal, having a relative density of 18.7. Pure uranium is both malleable and ductile, and commercially it can be cast and then fabricated by rolling or extrusion, followed by drawing. It is a chemically reactive metal and oxidises at moderately high temperatures. Consequently, it must be protected during fabrication processes. For atomic energy purposes it is usually vacuum melted in high-frequency furnaces, and cast in 25 mm diameter rods.

Chemically, uranium is similar to tungsten and molybdenum. Like them, it forms stable carbides which led to the experimental use of uranium in high-speed steels. These steels have not survived, since they showed no advantage over the orthodox alloys.

18.72   Uranium is one of a number of elements which undergo natural radioactive disintegration. During such spontaneous disintegration three different types of emission proceed from the source—the so-called α, β and γ 'rays'. The phenomenon generally referred to as α-rays is in fact the effect produced by a stream of moving particles—α-particles. An α-particle is, in effect, of the same constitution as the nucleus of a helium atom. Similarly, β-rays consist of a stream of fast-moving β-particles (in actual fact these are electrons). Only γ-rays are in fact true electromagnetic radiation, and since this radiation is of short wavelength it is able to penetrate considerable thicknesses of metal.

All three forms of radiation are emitted at some stage in the natural radioactive disintegration of uranium, ^{238}U. Obviously the loss of an α-particle from the nucleus of a ^{238}U atom will cause a change in both its atomic mass number and atomic number and consequently in its properties. In this general way ^{238}U changes in a series of steps to radium, and finally to a stable, non-radioactive isotope of lead (^{206}Pb). Each stage of the radioactive decay is accompanied by the emission of one or more of the types of radiation described above. A somewhat simplified representation of this process is indicated in Fig. 18.7

18.73   Some stages in the process of radioactive disintegration take place more quickly than others, and each separate decay process is governed by the expression—

$$-\frac{dn}{dt} = \lambda n$$

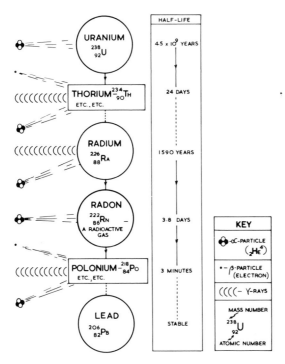

Fig. 18.7 Stages in the radioactive disintegration of uranium, $^{238}_{92}$U.

where n is the number of atoms of the species present and λ is a constant for that species. Integration of this expression and evaluation of the time necessary for half of the atoms present, initially, to decompose gives what is called the half-life period of the species.
   Thus—

$$t_{1/2} = \frac{0.6932}{\lambda}$$

Each radioactive species is characterised by its half-life period. Thus it takes 4.5×10^9 years for half of the ^{238}U atoms present in a mass of uranium to change to the next uranium isotope in the series, whilst it takes only 1590 years for half of the radium atoms present to change to the radioactive gas radon.
   18.74 Until its development as a source of fissionable material, uranium was little more than a chemical and metallurgical curiosity so that, before 1943, there was no commercial extraction process operating with the object of uranium as the sole product. Its industrial use, before the Second World War, was limited to the manufacture of electrodes for gas-

discharge tubes; whilst its compounds were used in the manufacture of incandescent gas mantles.

18.75   The reader will, no doubt, associate uranium with the production of nuclear power. This element has two principal isotopes (1.90), the atoms of which contain 92 protons and 92 electrons in each case. However, one isotope has 146 neutrons giving a total mass of 238, whilst the other has 143 neutrons giving a total mass of 235. These isotopes are generally represented by the symbols $^{238}_{92}U$ and $^{235}_{92}U$ respectively. Here the lower index represents the number of protons (92) in the nucleus (the *atomic number*, Z) and the upper index represents the total number of protons and neutrons (the *atomic mass*, A, of the isotope).

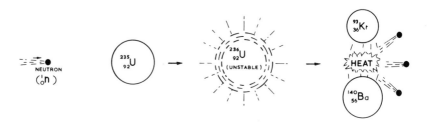

Fig. 18.8   The fission of an atom of uranium, $^{235}_{92}U$.
Here, isotopes of the metal barium, $^{140}_{56}Ba$, and the 'noble' gas krypton, $^{93}_{36}Kr$, have been produced together with three neutrons, $^{1}_{0}n$. About thirty similar reactions are known to take place during the fission of a $^{235}_{92}U$ atom, the two isotopes produced varying in mass number accordingly between 72 and 162. One such radioactive isotope is the notorious 'strontium ninety', ie $^{90}_{38}Sr$.

Since a neutron has no electrical charge, and hence suffers neither attraction nor repulsion by either the negatively charged electrons or positively charged protons, it can be fired into the nucleus of an atom. When a neutron enters the nucleus of an atom of $^{235}_{92}U$ the latter becomes unstable and splits into two approximately equal portions (Fig.18.8). At the same time a small reduction in the total mass occurs, for although the total number of particles (protons and neutrons) remains the same after fission, the sum of the masses of the resultant nuclei is slightly less. This is a measure of the *mass* equivalent of the nuclear binding *energy*. Einstein's Theory of Relativity states that mass and energy are not distinct entities but are really different manifestations of the same thing. The two are related to the velocity of light (c) by the expression:

$$E = mc^2$$

Since the velocity of light is 2.998×10^8 m/s it follows that even a small loss in mass can result in a great release of energy. As is often the case in engineering problems this energy is emitted in the form of heat. It can be calculated that the energy produced by the loss of one gramme of matter in this way is sufficient to heat about 200 tonnes of water from 0 to 100°C.

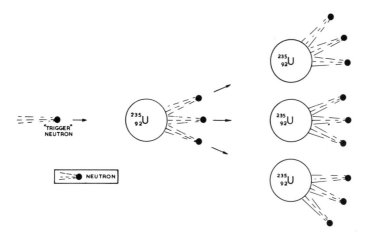

Fig. 18.9   Chain fission of $^{235}_{92}$U atoms.

Fission is also accompanied by the emission of neutrons and if the overall mass of $^{235}_{92}$U is great enough, other atoms will absorb some of these neutrons so that a chain reaction occurs leading to an 'atomic explosion'. There is a 'critical size' for a mass of $^{235}_{92}$U. Below this critical size neutrons will tend to escape rather than be absorbed by $^{235}_{92}$U nuclei, and a chain reaction will not, therefore, be promoted.

18.76   The rate of production of nuclear energy from this source can be controlled by introducing into an 'atomic pile' rods of some element which will absorb unwanted neutrons and thus control the rate of disintegration of other $^{235}_{92}$U atoms. In natural uranium total disintegration will not occur since only 0.7% $^{235}_{92}$U is present, mixed with the more common isotope $^{238}_{92}$U. Isotope $^{238}_{92}$ will however, absorb a neutron to give an atom of nuclear mass 239. This undergoes further change to produce plutonium, $^{239}_{94}$U, which can be used as a concentrated fuel in such 'fast reactors' as that at Dounreay.

Some Uncommon Metals

18.80   Of the sixty-nine metallic elements which occur naturally in the surface of our Earth only about half find any metallurgical applications. Some—like those of Group I—are far too chemically reactive, whilst others lack mechanical properties which interest the engineer. The technology of some 'new' metals has been developed because of properties which are of value in nuclear engineering. Others would be of greater interest to the aerospace industries were they not so expensive because of scarcity or high extraction costs. Several of these metals have very high melting points coupled with high corrosion resistance so that they and their alloys join the ranks of the expensive 'superalloys'.

18.81 Beryllium   was in fact discovered as long ago as 1797 by Vauquelin. The famous chemist Wöhler first produced the metal in quantity in 1828 but commercial development did not begin until 1916 in the USA. Beryllium bronze (16.71) first made its appearance commercially in the early 'thirties' but it is only in recent years that interest has been focused on the possibilities of using beryllium in the pure form or as an alloy base. For a 'light' metal beryllium has a very high melting point (1285°C) so that it can be used over a wide range of temperatures. It also has good strength and reasonable corrosion resistance. Since beryllium is less dense than aluminium but potentially stronger, the aerospace industries would be much more interested in the metal than they are, were it not for difficulties in forming but above all its extreme scarcity. The only significant source of beryllium is the mineral beryl, a semi-precious stone, mined in a number of countries but notably Brazil, Argentina and Central Africa. It is also present in the precious stone emerald.

In France beryllium is known as 'glucinium', a reference to the sweet taste of some of its compounds. The metal is, however, extremely toxic giving rise to a chronic form of poisoning called 'berylliosis'. This resembles septic pneumonia and constitutes a further hazard in fabricating the metal. Moreover, although *malleable* the metal lacks *ductility*, a fact which restricts the methods which can be used to shape it. Beryllium crystallises in the CPH form and metals with this type of structure are generally more brittle than those which crystallise in one of the cubic forms (4.12). Nevertheless beryllium is successfully extruded to produce bar, tube and various sections. Hot-working the metal in a mild steel sheath is a common method of shaping beryllium.

One property which interests the nuclear engineers is the low neutron absorption of beryllium. That is, it does not significantly impede the passage of neutrons. For this reason beryllium was initially popular as a canning material for nuclear fuel but partly because of its scarcity it has been largely replaced by zirconium alloys. Since it is transparent to X-rays beryllium is used for X-ray windows. Its low inertia has led to use in some expensive camera shutters whilst its high melting point has made it useful in neon-sign electrodes and targets for cyclotrons.

Beryllium has found limited use in aerospace structures, jet-engine turbine parts and also as a neutron source using α-particles from plutonium. Commercial quality beryllium (92Be; 4.5BeO; 0.2 max.Al; 0.5 max.C; 0.3 max.Fe) has a yield strength of approximately 560 N/mm^2 and a tensile strength of 890 N/mm^2.

18.82 Zirconium   In 1789 Klaproth identified a new metal in the mineral zircon, but it was Berzelius who, in 1824, succeeded in isolating zirconium. It was not until 1914 that the metal was produced in sufficient quantity to show that it was ductile. Then, during the First World War, rumours circulated that Germany was developing zirconium steels. Intensive research by the Allies, however, failed to produce an alloy of this type with useful properties. Eventually zirconium became known chiefly as an alloying addition to magnesium-base alloys (18.12), whilst its oxide zircona was developed as a high-temperature refractory. In more recent years it has been used in some of the titanium-base alloys (18.62).

In 1944 research was instituted with the object of producing high-purity zirconium and to-day it is probably the most useful nuclear-engineering material for use where low neutron absorption is necessary. In other respects too zirconium is an attractive proposition in nuclear engineering. As compared with beryllium it is relatively plentiful, and is also superior to beryllium in corrosion resistance. Moreover, zirconium and its alloys can be fabricated with comparative ease provided that the metal is not heated in contact with either oxygen, nitrogen or hydrogen. Each of these gases will form interstitial solid solutions in zirconium leading to its embrittlement.

Unalloyed zirconium tends to be mechanically weak at high temperatures. In addition, under such conditions it is rapidly corroded by water vapour and carbon dioxide. These difficulties have been largely overcome by alloying and most of the zirconium now used in nuclear engineering is supplied as alloys such as 'Zircaloy II' (1.5Sn; 0.12Fe; 0.05Ni; 0.1Cr; Bal.Zr) or 'Zircaloy IV' (1.5Sn; 0.2Fe; 0.1Cr). These alloys are used for fuel sheathing and structural components in pressurised-water reactors because of their combination of low neutron absorption, high strength and good corrosion resistance.

Since zirconium is chemically reactive in the powdered form it is compacted with lead to make lighter 'flints'. Before the days of 'electronic flash' it constituted the wire in many consumable photographic flash bulbs. In the form of ferro-zirconium or ferro-silico-zirconium it can be used as a 'scavenger' in steel making since it removes traces of both oxygen and sulphur. Any residue refines the grain of steel whilst the strength of some alloy steels is increased. Zirconium occurs chiefly as 'zircon sands' and is mined in Brazil, India, the USA and around the Australian coast. It is extracted from its minerals in a similar way to titanium.

18.83 Hafnium   One of the main difficulties in the use of zirconium as a nuclear canning material is the presence of small amounts of hafnium which materially increase neutron absorption. Unfortunately hafnium is always closely associated with zirconium in the mineral deposits of the latter, and the reduction of the hafnium content to the few parts per million which is all that can be tolerated in reactor-quality zirconium, is an expensive operation. Both zirconium and hafnium belong to the same 'transition group' in the Periodic Classification. Consequently their chemical properties are so alike that their separation is difficult and, hence, costly. In respect of neutron absorption, however, they are at opposite ends of the scale for, whilst zirconium has a very low neutron absorption, that of hafnium is extremely high.

For some time hafnium was used as control rods in nuclear reactors because of its high neutron absorption and because the 'daughter' atoms formed by such neutron bombardment were hafnium isotopes. This led to little dimensional change in the rods. Of recent years hafnium has been replaced by less expensive materials so that it now finds little industrial use.

The existence of hafnium was first suspected towards the end of the nineteenth century and it was discovered in rare-earth residues in 1911, though its properties were not verified until 1923. It appears to be named after Hafnia, the ancient name for Koben*havn* (Copenhagen).

18.84 Tantalum is important because it combines high ductility and exceptional corrosion resistance with a very high melting point. In the pure state tantalum is corrosion resistant to most acids and alkalis. Since, when pure, it is also very ductile, it is used in acid-proof equipment in chemical plant. Being impervious to acid attack, sections as thin as 0.3 mm can be used in heat-transfer equipment. It is used in the form of plugs to repair damage to vitreous lined steel tanks in distilleries and chemical plant.

Not only is pure tantalum unreactive towards animal fluids but it also provides a surface upon which tissue will grow. Consequently, it is used in bone surgery, plates to replace skull tissue and for other 'implants' in the human body. Woven tantalum gauze is used as a reinforcement to the abdominal wall in some hernia operations.

Tantalum can be anodically treated (21.91) and the film so formed is very stable and self-sealing. Moreover, this oxide film has excellent dielectric properties and, being impervious to electrolytes used, has made tantalum a very useful metal for the manufacture of very small sized electrolytic capacitors. Foil of the order of 0.012 mm thick can be produced resulting in electrolytic capacitors about one-tenth of the volume of ordinary aluminium-foil capacitors of equivalent capacitance. These small tantalum-foil capacitors are widely used in modern electronic circuitry.

In modern engineering tantalum alloys (Table 18.9) are used for steam-turbine blades, valves, nozzles, stills, agitators, containers and pipes in chemical industries; and for the tips of fountain-pen nibs. Its minerals are mined principally in Central and Southern Africa, Australia and Portugal. Though generally assumed to be a 'new' metal the presence of tantalum was first detected early in the nineteenth century. Because of the great difficulty in isolating it from its minerals the metal was named after Tantalus, son of Zeus, of Greek mythology. It was Tantalus who, for his alleged crimes, was stood in water up to his chin but as he attempted to drink the water always receded—hence to 'tantalize'. Although it was detected so early, a hundred years passed before the pure metal was isolated in 1905.

18.85 Niobium was first discovered near Connecticut in 1801 by a British chemist named Hatchett. In recognition of its source he called it 'columbium'. At about the same time Ekeburg in Sweden discovered a 'new' metal in close association with tantalum, and when later, in 1844, it was isolated from tantalum by Rose he appropriately named it 'niobium' after the sad Niobe, daughter of Tantalus. Soon afterwards 'columbium' and 'niobium' were identified as being the same metal, but the two names continued to be used on their respective sides of the Atlantic until some years ago, by international agreement, the name 'niobium' was officially adopted. Niobium was first produced in quantity in 1929 and has for a considerable time been used as a 1.0% addition to 18–8 stainless steels in which it induces resistance to 'weld decay' (20.93).

Like zirconium, niobium has a low neutron absorption. Moreover it has

Table 18.9 *Some Tantalum and Niobium High-temperature Alloys*

Proprietary names (Fansteel Metals, Chicago)	Composition (%)	Condition	Testing temperature (°C)	Tensile strength (N/mm^2)	10 hour rupture test (N/mm^2)
Tantalum (commercial)	—	Cold-worked	20	490	—
		Annealed	1100	105	16
'Fansteel 60'	10W; bal.-Ta	Annealed	1300	345	140
'Fansteel 63'	2.5W; 0.15Nb; bal.-Ta	Annealed	1100	130	—
Niobium (commercial)	—	Annealed	1100	70	30
SCb. 291	10Ta; 10W; bal.-Nb.	Annealed	1100	220	62
C. 103.	10Hf; 1Ti; 0.7Zr; bal.-Nb.	Annealed	1100	185	—
Cb. 132M.	20Ta; 15W; 5Mo; 1.5Zr, 0.1C; bal.-Nb	Heat-treated	1300	400	—
			1200	—	310

a very high melting point (2468°C) suggesting that it would be useful for nuclear-fuel sheathing where higher reactor temperatures are involved. Unfortunately niobium also reacts with the pile gases, as well as with other substances with which it is likely to come into contact in the reactor, at temperatures in excess of 500°C. However some of the alloys of niobium have very good resistance to oxidation and a number (Table 18.9) are used for high-temperature applications.

18.86 Cobalt   is second only to iron in terms of its strong ferromagnetic properties and finds use in permanent magnet alloys (14.30). It is also important as a constituent of super high-speed steels and as the binding matrix in some cemented carbide tool materials (14.20).

Although the corrosion-resistance of pure cobalt is poor that of some of its alloys is excellent, particularly at high temperatures. Most of these alloys are difficult to work and are therefore shaped by investment casting (5.40—Pt.2). They have good high-temperature strength. 'FSX.414' (52Co; 29.5Cr; 10.5Ni; 7W; 0.25C; 0.01B) is used for gas-turbine vanes; whilst 'MAR-M322' (60Co; 21.5Cr; 9W; 4.5Ta; 2.25Zr; 1C; 0.75Ti) is an investment casting alloy for jet-engine blades. 'NAS Co-W-Re' (67Co; 25W; 3Cr; 2Re; 1Zr; 1Ti; 0.4C) has a rupture strength of $45N/mm^2$ at 1300°C for 1000 hours and is used for high-temperature space applications.

Whilst the current tendency is towards the development of plastics materials and ceramics for surgical purposes because of their greater chemical inactivity, some cobalt-base alloys are used in orthopaedic implants. A popular cast alloy contains 65% cobalt, 28% chromium, 6% molybdenum and 1% nickel whilst a suitable wrought alloy contains 20% chromium, 10% nickel, 0.1% carbon and the balance cobalt. However even with alloys of the highest corrosion resistance some interaction may occur with the surrounding tissue, releasing potentially toxic substances.

18.90 As more becomes known of the rarer and more elusive of the metallic elements so are their properties exploited. Thus metals like yttrium (Table 18.5), samarium and strontium (Table 14.3) now find use in engineering alloys, whilst another 'rare earth' neodymium is used in a new generation of permanent magnets (14.30) as well as in the colouring of glass and the manufacture of some capacitors. Yet another 'rare earth', praseodymium, along with cerium, is used in the *de*colourising of some optical glasses, whilst lanthanum is used in high refractive index glasses. As long ago as the 1960s europium and yttrium compounds were developed for the coating of TV screens, whilst during the same period the nuclear energy programme was interested in europium, samarium, gadolinium and dysprosium as high neutron cross-section materials. The high thermal stability of cerium and yttrium hydrides led to their use as neutron moderators in atomic piles. In the field of medicine even the man-made element technetium (with a half-life of only a few hours) is used as a radioactive isotope.

Exercises

1. By reference to Fig. 18.1 estimate the maximum and minimum solution-treatment temperatures (prior to precipitation-hardening) for magnesium-base alloys which contain *singly*: (i) 10% aluminium; (ii) 5% zinc; (iii) 3% thorium.
2. By reference to Fig. 18.4 estimate the proportion of primary aluminium to eutectic in a slowly cooled bearing metal in the cast condition containing 80 Al–20 Sn.
3. Which of the following alloy compositions could be hardened by some form of heat-treatment: (i) 90Mg–10Al; (ii) 90Al-10Mg; (iii) 90Cu-10Zn; (iv) 90Cu-10Al; (v) 90Cu-10Ni? Give reasons for your answers.
4. What are the principal difficulties encountered during the manufacture of magnesium-base alloys?
   Where do most of these alloys find application? (18.10)
5. Write a short essay on light-weight non-ferrous alloys, their properties and uses (Ch. 17; 18.10)
6. Outline a process by which zinc-base alloys are generally shaped. What are (i) the attractive properties; (ii) possible difficulties, involved in the use of these materials? (6.20; 18.20)
7. Discuss the factors which influence the choice of nickel-base alloys for high-temperature engineering. Outline the mechanisms used to strengthen these materials. (18.30)
8. The development of materials to operate at elevated temperatures has been a most important aspect of metallurgical progress. Give a comprehensive account of the ferrous and non-ferrous alloys at present available for operation in the 600–1000°C range, discussing such factors as resistance to oxidation, resistance to corrosion and retention of strength and showing how these are obtained in particular alloys. (13.70; 18.30)
9. List, and briefly describe, the essential requirements of a bearing metal. Show,

by discussing the constitution and structure of materials, how the requirements are realised in: (i) certain copper-base alloys and (ii) white metals. Compare the applications of the two groups and give typical compositions.(18.40)

10. What particular properties of titanium and its alloys are important in considering them for aero-space design?
Outline the principles by which high strength is developed in these materials. (18.60)

11. Outline the roles played by (i) uranium; (ii) beryllium; (iii) zirconium; and (iv) hafnium, in 'nuclear engineering'. (18.70; 18.80)

Bibliography

Betteridge, W., *Cobalt and its Alloys*, Ellis Horwood, 1982

Betteridge, W., *Nickel and its Alloys*, Ellis Horwood, 1984

Betteridge, W. and Heslop, J., *The Nimonic Alloys and Other Nickel-base High-temperature Alloys*, Edward Arnold, 1974.

Hausner, H. H., *Beryllium, its Metallurgy and Properties*, University of California Press, 1966.

Kleefisch, E. W., *Industrial Applications of Titanium and Zirconium*, American Society for Testing and Materials, 1981.

Morgan, S. W. E., *Zinc and its Alloys*, Macdonald and Evans, 1977.

Polmear, I. J., *Light Alloys (Metallurgy of the Light Alloys)*, Edward Arnold, 1989.

Sims, C. T. and Hagel, W. C., *The Superalloys*, John Wiley, 1973.

BS 2970: 1972 *Magnesium Alloy Ingots and Castings*.

BS 3370: 1970 *Wrought Magnesium Alloys for General Engineering Purposes: Plate, Sheet and Strip*.

BS 3372: 1970 *Wrought Magnesium Alloys for General Engineering Purposes: Forgings and Cast Forging Stock*.

BS 3373: 1970 *Wrought Magnesium Alloys for General Engineering Purposes: Bars, Sections and Tubes, Including Extruded Forging Stock*.

BS 1004: 1985 *Zinc Alloys for Die Casting and Zinc Alloy Die Castings*.

BS 3072: 1983 *Specifications for Nickel and Nickel Alloys: Sheet and Plate*.

BS 3073: 1989 *Specifications for Nickel and Nickel Alloys: Strip*.

BS 3074: 1983 *Specifications for Nickel and Nickel Alloys: Seamless Tube*.

BS 3075: 1989 *Specifications for Nickel and Nickel Alloys: Wire*.

BS 3076: 1983 *Specifications for Nickel and Nickel Alloys: Bar*.

BS 115: 1987 *Metallic Resistance Materials for Electrical Purposes*.

BS TA 1 to TA 59: 1973/1986 *Specifications for Titanium and Titanium Alloys* (The TA series covers 'Aerospace' uses.)

BS 2TA 100: 1973 *Procedure for Inspection and Testing of Wrought Titanium and Titanium Alloys*.

19

The Surface Hardening of Steels

19.10 The service conditions of many steel components such as cams, gears and shafts make it necessary for them to possess both hard, wear-resistant surfaces and at the same time, tough, shock-resistant cores. In plain carbon steels these two different sets of properties exist only in alloys of different carbon content. A low-carbon steel, containing approximately 0.1% carbon, will be tough but soft, whilst a high-carbon steel of 0.8% or more carbon will be hard when suitably heat-treated but will also be relatively brittle.

19.11 The situation can best be met by employing a low-carbon steel with suitable core properties and then causing either carbon or nitrogen to penetrate to a regulated depth to produce a potentially hard surface skin; as in the principal surface-hardening processes of carburising and nitriding. Alternatively, a medium-carbon steel can be used, heat-treated first to produce desirable core properties and local hardness at the surface then introduced by one of the flame-hardening or induction-hardening processes. In the first case the hardenable material is localised, whilst in the second case it is the heat-treatment itself which is localised.

Rapid penetration of the surface of steel can only be effective if the solute element dissolves interstitially. This is the case with the elements used, viz. carbon, nitrogen and, to a small extent, boron. Once dissolved, the elements increase the hardness of the surface by forming interstitial compounds—carbides, nitrides or borides.

Case-hardening

19.20 The principles of case-hardening were used centuries ago in the conversion of wrought iron to steel by the 'cementation' process. Both this ancient process and case-hardening make use of the fact that carbon will diffuse into iron provided that the latter is in the FCC (γ) form which exists above 910°C. Thus carburising consists in surrounding the component with suitable carbonaceous material and heating it to above its upper critical temperature for long enough to produce a carbon-enriched layer of sufficient depth.

Solid, liquid and gaseous carburising media are used. The nature and scope of the work involved will govern which medium it is best to employ. The function of the carburising medium is to release atoms of carbon at the surface of the work piece so that, at the carburising temperature, they will be absorbed interstitially into the steel.

The rate at which the released carbon atoms diffuse beneath the surface of the work piece is governed by Fick's Law (8.25):

$$J = -S.D.\frac{\partial c}{\partial x}$$

where J = the amount of carbon which passes in unit time across area S
in a plane normal to the 'x' axis;

$\dfrac{\partial c}{\partial x}$ = the variation of carbon content with depth below the surface (the concentration gradient of the carbon);

D = the diffusion coefficient of carbon in γ-iron.

D is very dependent upon temperature but in this instance is of the order of 10^{-7} cm^2/s at the carburising temperature (900°C).

From the Diffusion Equation (8.25):

$$\frac{\partial c}{\partial t} = D\frac{\partial^2 c}{\partial x^2}$$

the *depth* of carbon penetration could be derived using one of the standard methods for solving differential equations. However, a simple formula (13.84—Pt.2), derived originally by Einstein from mathematical studies of diffusion, can be applied:

$$x = \sqrt{(2Dt)}$$

where D = the diffusion coefficient (cm^2/s);
and t = time of diffusion (s).
This can be written:

$$\text{Case depth} = k\sqrt{t}$$

where $k = \sqrt{2D}$ and gives a rough estimate of the time required to produce a case of given depth.

A very rough estimate of the depth of case may be made from a fractured section of a test piece carburised along with the work but a more accurate measurement may be made metallographically. A cross section of the test piece is polished and etched and its image projected on to the screen of a

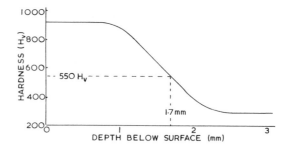

Fig. 19.1 The relationship between carbon penetration and hardness. Here the *useful* depth of case is 1.7mm, assuming a minimum hardness of 550 H_v.

camera microscope at known magnification. The depth of case can then be measured on the screen image using a rule. Some specifications quote a hardness index as indicating the limit of the useful case. Thus Fig. 19.1 represents a specification in which a hardness value of $550H_v$ is judged to coincide with the limit of the useful case—in this instance a case depth of 1.7 mm.

19.21 Carburising in Solid Media 'Pack-carburising' as it is usually called, involves packing the work into heat-resisting steel (25Cr; 20Ni) boxes along with the carburising material so that a space of approximately 50 mm exists between the components. Small-scale work may be carried out in mild-steel boxes but in either case the lids must fit tightly or be 'luted' on with fireclay. The boxes are then heated slowly to the carburising temperature (875–925°C) and maintained at this temperature for up to eight hours, according to the depth of case required. Fig. 19.2 indicates the relationship which exists between depth of case, carburising temperature and time of treatment.

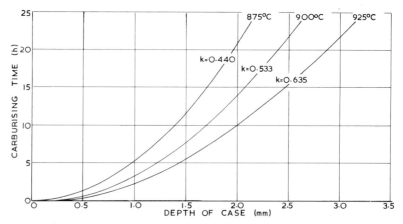

Fig. 19.2 The relationship between time and temperature of carburising and the depth of case produced.

Although a solid carbon-rich medium is packed around the work the actual carburising process depends upon the presence of the gas carbon monoxide, CO, to carry carbon to the surface of the work pieces. Oxygen in entrapped air forms this carbon monoxide:

$$2C + O_2 \rightarrow 2CO$$

At the surface of the work piece this carbon monoxide releases carbon atoms:

$$2CO \rightarrow CO_2 + C$$

The carbon atoms so released are dissolved interstitially at the surface of the steel.

As indicated above, charcoal alone could be used as the carburising medium but in practice the rate of carburisation is increased by adding an 'energiser', usually 10–15% barium carbonate. This dissociates at the carburising temperature:

$$BaCO_3 \rightarrow BaO + CO_2$$

and the carbon dioxide thus produced reacts with charcoal to form carbon monoxide:

$$CO_2 + C \rightleftharpoons 2CO$$

The above reaction is reversible and as the temperature is increased, pressure remaining constant, the proportion of carbon monoxide increases and so therefore does the rate of carburisation (Fig. 19.2). In commercially produced carburising media charcoal may be replaced by other carbon-rich substances such as petroleum coke.

If it is necessary to prevent any areas of the component from being carburised, this can be achieved by electro-plating these areas with copper to a thickness of 0.075–0.10 mm; carbon being insoluble in solid copper at the carburising temperature. An alternative method, which can be more conveniently applied in small-scale treatment, is to coat the area with a mixture of fireclay and ignited asbestos made into a paste with water. This is allowed to dry on the surface before the component is carburised.

When carburising is complete the components are quenched or cooled slowly in the box, according to the nature of the subsequent heat-treatment to be applied.

19.22 Carburising in a Liquid Bath   Liquid-carburising is carried out in fused mixtures of salts containing from 20 to 50% sodium cyanide, together with up to 40% sodium carbonate and varying amounts of sodium or barium chloride. This cyanide-rich mixture is melted in 'calorised' (21.84) pots to a temperature of 870–950°C and the work, which is contained in wire baskets, is immersed for periods varying from about five minutes up to one hour, depending upon the depth of case required. One of the main advantages of cyanide-hardening is that pyrometric control is so much more satisfactory with a liquid bath. Moreover, after treatment the basket of work can be quenched. This not only produces the necessary hardness but also gives a clean surface to the components. The process is particularly useful in obtaining shallow cases of 0.10–0.25 mm, though case depths up to 0.5 mm are used. Salt-bath carburising is used mainly for small parts requiring a shallow case depth. It is an economical process since the rate of heating is rapid due to the high heat-capacity of the liquid bath and the quick transfer of heat to the work. This reduces the total time of treatment.

It is believed that sodium cyanide, NaCN, is oxidised at the surface of the molten bath:

$$2NaCN + O_2 \rightarrow 2NaCNO$$

The sodium cyanate, NaCNO, thus formed diffuses into the bath and decomposes at the surface of the steel:

$$8\,NaCNO \rightarrow 4\,NaCN + 2\,Na_2CO_3 + 2\,CO + \boxed{4\,N}$$

$$\downarrow \qquad \text{dissolves in steel}$$

$$CO_2 + \boxed{C}$$

The carbon atoms dissolve interstitially in steel as they do in pack carburising, but in this process they are accompanied by nitrogen (which is released in the suitable atomic form). Nitrogen also increases the hardness of the case by forming nitrides as it does in the nitriding process (19.40).

Cyanides are, of course, extremely poisonous, and every precaution *must* be taken to avoid inhaling the fumes from a pot. Every pot should be fitted with an efficient fume-extracting hood. Likewise the salts should in no circumstances be allowed to come into contact with an open wound. Needless to say, the consumption of food by operators whilst working in the shop containing the cyanide pots should be *absolutely forbidden*. Disposal of cyanide wastes is also a problem. Effluents containing even small quantities of cyanides will destroy bacteria used in purifying sewage, consequently the disposal of these wastes must be rigorously controlled.

19.23 Gas Carburising is carried out in both batch-type and continuous furnaces (13.20—Pt. 2) and in recent years has become by far the most popular method for mass-production carburising, particularly when thin cases are required. Not only is it a cleaner process but the necessary plant is also more compact for a given output. Moreover the carbon content of the case can be controlled more accurately and easily than in either 'solid' or 'liquid' carburising.

The work is heated at about 900°C for three to four hours in an atmosphere containing gases which will deposit carbon atoms by decomposition at the work piece surfaces. This atmosphere is generally based on the hydrocarbons methane ('natural gas') CH_4; and propane, $C_3 H_8$, which are either partially burnt in the furnace or are diluted with a 'carrier' gas in

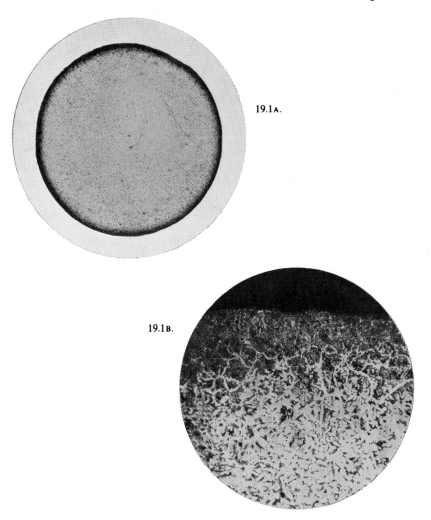

19.1A.

19.1B.

Plate 19.1   19.1A Cross-section through a carburised bar with a case depth of approx 1mm. × 3. Etched in 5% nital.
19.1B Structure in the region of the surface of a carburised bar, in the as-carburised condition. The original steel contains 0.15% carbon, but the carburised surface is now of eutectoid composition (0.8% carbon). × 35. Etched in 2% nital.

order to produce an atmosphere giving the required *carbon potential* at the work surface. This carbon potential is in practice the carbon content maintained in equilibrium in the surface film of the component whilst it is in contact with the gas atmosphere. A carbon potential of 0.8% is usually desirable (19.24). Carrier gases are generally of the 'endothermic' type made in a generator and consisting of a mixture of nitrogen, hydrogen and carbon monoxide. The active agent in the atmosphere is carbon monoxide as it is in so-called 'solid' carburising:

$$2CO \rightleftharpoons CO_2 + C \text{ (at the work surface)} \tag{1}$$

The atomic carbon thus released dissolves interstitially in the surface of the austenitic work pieces. Methane present in the atmosphere probably also releases carbon atoms:

$$CH_4 \rightleftharpoons 2H_2 + C \text{ (at work surface)} \tag{2}$$

The 'water gas reaction' may also be responsible for carburising:

$$CO + H_2 \rightleftharpoons H_2O + C \text{ (at work surface)} \tag{3}$$

Carbon dioxide formed in (1) probably reacts with methane:

$$CO_2 + CH_4 \rightleftharpoons 2H_2 + 2CO \text{ (away from the work surface)} \tag{4}$$

The concentration of carbon monoxide is thus maintained so that carburising continues.

The relative proportions of hydrocarbon and carrier are adjusted to give the desired carburising rate. Thus, the concentration gradient of carbon in the surface can be flattened by prolonged treatment in a less rich carburising atmosphere. Control of this type is possible only with gaseous media. There are a number of different methods by which the composition of the carburising atmosphere can be continually monitored, so that the carbon potential can be accurately controlled.

At present investigational work on 'plasma carburising' is taking place. This is similar in principle to 'ionitriding' or 'ion implantation' (19.43) but low-pressure nitrogen is replaced by low-pressure hydrocarbons. This process, if successfully developed, would lead to a big saving of natural gas. It is claimed that a case of one millimetre depth can be produced in thirty minutes.

19.24 Heat-treatment after Carburising If carburising has been correctly carried out, the core will still be of low carbon content (0.1–0.2% carbon), whilst the case should preferably have a carbon content of no more than 0.8% C (the eutectoid composition). If the carbon content of the case is higher than this then a network of primary cementite will coincide with the grain-boundary sites of the original austenite giving rise to intercrystalline brittleness and consequent exfoliation (or peeling) of the case during service. Even if this does not occur any cracks arising from the presence of primary cementite may initiate fatigue failure. Moreover a case containing 1.0% or more carbon may be soft at the surface after quenching due to retention of austenite.

After the component has been carburised heat-treatment will be necessary both to strengthen and toughen the core and to harden the case. At the same time prolonged heating in the austenitic range during carburising will have introduced coarse grain to the whole structure, so that a heat-treatment programme which will also refine the grain is desirable if the optimum properties are to be attained. For components which have been pack- or liquid-carburised to produce deep cases, a double heat-treatment is preferable to refine both core and case separately, as well as to harden the case and strengthen the core.

19.25 Refining the Core   The component is first heat-treated with the object of refining the grain of the core and consequently toughening it. This is effected by heating it to just above its upper critical temperature (about 880°C for the core) when the coarse ferrite/pearlite structure will be replaced by fine austenite crystals. The component is then water-quenched so that a fine ferrite/bainite/martensite structure is obtained in the core.

The core-refining temperature of 880°C is, however, still high above the upper critical temperature for the case, so that, at the quenching tempera-ture, the case may consist of large austenite grains. On quenching these will result in the formation of coarse brittle martensite. Further treatment of the case is therefore necessary.

19.26 Refining the Case   The component is now heated to about 760°C, so that the coarse martensite of the case changes to fine-grained austenite. Quenching then gives a fine-grained martensite in the case.

At the same time the martensite produced in the core by the initial quench will be tempered somewhat, and much will be reconverted into fine-grained austenite embedded in the ferrite matrix (point C in Fig. 19.3). The second quench will produce a structure in the core consisting of martensite particles embedded in a matrix of ferrite grains surrounded by bainite. The amount of martensite in the core is reduced if the component is heated quickly through the range 650–760°C and then quenched without soaking. This produces a core structure consisting largely of ferrite and bainite, and having increased toughness and shock-resistance.

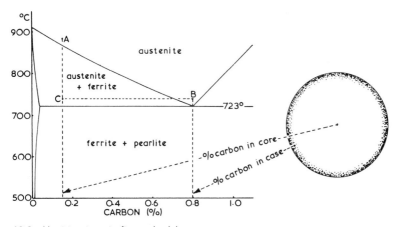

Fig. 19.3   Heat-treatment after carburising.
'A' indicates the temperature of treatment for the core and 'B' the temperature of treatment for the case.

Finally, the component is tempered at between 160°C and 220°C to relieve any quenching strains present in the case.

19.27 The above comprehensive heat-treatment may be regarded as the counsel of perfection and would be that applied to important components in which cases of considerable depth had been produced and in which the necessarily prolonged carburising cycle had given rise to coarse grain. The core-refining process would also cause some reduction in the high carbon content at the skin to a more acceptable value nearer 0.8%, thus reducing the chance of primary cementite networks at the surface. For thinner cases and lower quality work generally, modified heat-treatment is prevalent.

When thin cases (below 0.5 mm) are involved the component may be quenched direct from the carburising process with obvious advantages economically. The shorter carburising time will not have produced grain growth to the same extent as more prolonged treatment. With cases of intermediate thickness (0.5–1.25 mm) a component may be cooled slowly from the carburising temperature and given a single treatment by reheating to 820°C as a compromise between the upper and lower critical temperatures (for the core), water-quenched and finally tempered at between 160°C and 220°C. The case will then be hard though slightly coarse-grained whilst the core will consist of a ferrite/bainite/martensite structure. The ferrite will be rather coarse-grained, remaining from the original carburising and will not be completely redissolved by heating to 820°C. Hence the core will not attain maximum toughness.

Gas carburising gives greater flexibility in heat-treatment. Thus the actual carburising process may be carried out at 900–940°C to give a carbon content at the surface of 0.8%. When carburising is complete the furnace temperature can be lowered to 830°C and the work then quenched direct (Fig. 13.7—Pt. 2).

Case-hardening Steels

19.30 Plain Carbon Steels   used for case-hardening commonly contain up to 0.2% carbon when maximum toughness and ductility of the core are required. When higher core strength is necessary the carbon content may be raised to 0.3%, though it is generally more satisfactory to use an alloy steel in such circumstances.

Up to 1.4% manganese may be included since it aids carburisation by stabilising cementite. It also increases the depth of hardening but is liable to induce cracking during quenching. Silicon tends to graphitise cementite and, since it retards carburisation, is kept below 0.35%. Thus plain carbon steels used for case-hardening contain up to 0.3% carbon and up to 1.4% manganese.

19.31 Alloy Steels   for case-hardening contain, additionally, up to 4.5Ni; 1.5Cr; and 0.3Mo. Higher core-strength without serious loss of either toughness or ductility are obtained by alloying. Moreover, since the

Table 19.1 Case-hardening Steels

BS 970: Part 3:	Composition (%)						Typical mechanical properties of core			
	C	Mn	Ni	Cr	Mo	S	Tensile strength (N/mm²)	Elonga-tion (%)	Impact (J)	Characteristics and uses
045M10 ('10 carbon')	0.10	0.45	—	—	—	—	440	18	48	Good core toughness in thin sections—with economy.
080M15 ('15 carbon')	0.15	0.8	—	—	—	—	500	16	41	General purposes for medium-duty work—many types of gears and shafts.
130M15 ('15' carbon-manganese)	0.15	1.3	—	—	—	—	750	13	34	A carbon-manganese steel — High surface hardness where severe shock is unlikely.
214M15 ('15' carbon-manganese-free-cutting)	0.15	1.4	—	—	—	0.14	750	12	34	A free-cutting carbon-manganese steel
655M13 (3¼% nickel-chromium)	0.13	0.5	3.25	0.85	—	—	1005	9	41	High surface hardness combined with core toughness. High-duty gears, worm gears, crown wheels and clutch gears.
655M17 (1¾% nickel-molybdenum)	0.17	0.5	1.75	—	0.25	—	775	12	41	High hardness and severe shock—automobile gears, steering worm and quadrant, overhead valve mechanisms.
822M17 (2% nickel-chrome-molybdenum)	0.17	0.6	2.0	1.5	0.2	—	1315	8	27	Good combination of core strength, shock resistance and surface hardness along with economy—crown wheels, etc.
832M13 (3½% nickel-chrome-molybdenum)	0.13	0.5	3.5	0.8	0.2	—	1085	8	34	For severe shock and high stress—automobile gears, steering worm, overhead valve mechanisms.
835M15 (4% nickel-chrome-molybdenum)	0.15	0.4	4.0	1.2	0.2	—	1315	8	34	Best combination of surface hardness, core strength and shock resistance—crown wheels, bevel pins, aero-reduction gears. Intricate sections which require air hardening.

critical hardening rates are reduced oil quenching can be used to minimise the risk of distortion or cracking.

An important function of nickel is to retard grain growth during carburising and so give a comparatively fine-grained product. For this reason the core-refining stage in heat-treatment can be dispensed with in some applications of case-hardened nickel steels with obvious economic advantage, though this is generally restricted to those in which only a light case has been produced and in which the presence of a cementite network is unlikely.

Although nickel retards carburisation and tends to give a softer case than some plain carbon steels, the case is very wear resistant. Cracking and exfoliation are also less likely. The addition of chromium increases hardness and wear-resistance of the case and also stability of the cementite. It also improves strength of the core with little loss in ductility. Nevertheless chromium additions have to be limited because of the tendency of the metal to promote grain growth with its consequent loss of toughness.

Nitriding

19.40 Nitriding resembles carburising in so far as interstitial penetration of the solid surface of steel takes place during heating of the work in contact with the nitriding agent. Whilst in carburising the function of the carburising agent is to release atoms of carbon at the steel surface, in nitriding the nitriding agent releases single atoms of nitrogen at the steel surface so that, in this instance, hardness depends on the formation of hard nitrides instead of carbides. Whilst it is possible to nitride many types of steel, high surface hardness is only obtained when using special alloy steels containing aluminium, chromium, molybdenum or vanadium—elements which form hard and stable nitrides as soon as they come into contact with nitrogen atoms at the surface of the work piece.

Fig. 11.7 indicates that iron dissolves up to 0.1% nitrogen at 590°C and that above this amount it begins to form the hard nitride Fe_4N (γ'). This underlines a very important difference between carburising and nitriding. Whilst carburising must take place when mild steel is in its *austenitic* state at about 900°C, nitriding can be achieved with the work in its *ferritic* state at about 500°C.

19.41 Since it is conducted at a relatively low temperature, nitriding is made the final operation in the manufacture of the component, all machining and core-treatment processes having been carried out previously. The parts are maintained at 500°C for between 40 and 100 hours, according to the depth of case required, in a gas-tight chamber through which ammonia is allowed to circulate. Some of the ammonia dissociates according to the following equation:

$$NH_3 \rightleftharpoons 3H + N$$

Table 19.2. *Nitriding Steels*

BS 970 designation	Composition (%)						Typical mechanical properties					Heat-treatment (for core)	Uses
	C	Mn	Cr	Mo	V	Al	Yield point N/mm²	Tensile strength N/mm²	Elongation (%)	Izod (J)	V.P.N. (case)		
—	0.5	0.65	1.6	0.2	—	1.1	1000	1240	13	39	1050–1100	Oil-quench from 900°C and temper 550–700°C.	Where maximum surface hardness is essential coupled with very high core strength.
905M39	0.4	0.65	1.6	0.2	—	1.1	741	927	20	52	1050–1100	Oil quench from 900°C and temper 600–700°C.	For maximum surface hardness and high core strength.
905M31	0.3	0.65	1.6	0.2	—	1.1	556	741	24	65	1050–1100	Oil-quench from 900°C and temper 600–700°C.	For maximum surface hardness combined with reasonably high core strength.
—	0.2	0.65	1.6	0.2	—	1.1	463	618	30	72	1050–1100	Oil-quench from 900°C and temper 600–700°C.	For maximum surface hardness combined with ease of machining before hardening.
—	0.35	0.5	2.0	0.25	0.15	—	741	927	20	52	750–800	Oil-quench from 900°C and temper 600–700°C.	Moulds for plastics; other components requiring high hardness and good finish.
—	0.18	0.5	2.0	0.25	0.15	—	463	618	30	72	750–800	Oil-quench from 900°C and temper 600–700°C.	For ease of machinability and a high-class surface.
897M39	0.4	0.5	3.0	1.0	0.25	—	—	1390	14	39	850–900	Oil-quench from 900°C and temper 550–650°C.	Ball races, etc., where high core strength is necessary.
—	0.3	0.45	3.0	0.5	—	—	772	1000	20	52	800–850	Oil-quench from 900°C and temper 600–700°C.	Aero crankshafts, airscrew shafts, aero cylinders, crank-pins and journals.
722M24	0.2	0.45	3.0	0.5	—	—	618	772	24	65	800–850	Oil-quench from 900°C and temper 600–700°C.	Aero-engine cylinders.

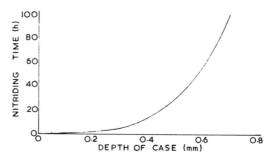

Fig. 19.4 The depth of nitriding for a 1% Al steel (905M39) in relation to time of nitriding at 500°C.

Part of the nitrogen, which is released as *single* atoms, is absorbed by the surface of the steel forming nitrides both with iron and with the elements mentioned above.

Nitriding of plain carbon steels would produce a case of only moderate hardness (about 400 H_v). This is largely because nitrogen diffuses fairly quickly beneath the surface forming Fe_4N (and so netimes Fe_2N) dispersed at greater depths so that surface hardness is reduced. Since aluminium— and to a lesser extent chromium, vanadium and molybdenum—has a higher affinity for nitrogen it prevents diffusion of the latter to a greater depth but instead forms very hard stable aluminium nitride near the surface, giving an extremely hard but shallow case up to 1.0 mm in depth. Chromium also helps to increase case hardness by the formation of chromium nitride. Since it forms at a greater depth than does aluminium nitride it prevents a sudden change in composition from hardened skin to soft core which would be likely to lead to spalling, or flaking, of the case. In addition to hardening the case, molybdenum also improves core toughness.

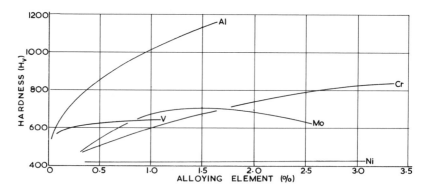

Fig. 19.5 The influence of alloying elements on the hardness of 'Nitralloy' type steels after nitriding.

In each case the steel contains 0.3% C and 0.65% Mn. The effect of nickel is slight because it does not form nitrides and such increase in hardness as is obtained is due only to solid solution effects.

When core strength is important aluminium is often omitted from the steel and a somewhat lower case hardness obtained, dependent upon the formation of chromium and vanadium nitrides (Fig. 19.5).

19.42   Before being nitrided the work is heat-treated to produce the required core properties. The normal sequence of operations is:

(a) oil-quenching from 850–930°C, followed by tempering at 550–700°C depending upon the composition of the steel and the core properties required;

(b) rough-machining, followed by a stabilising anneal at 550°C for five hours to remove stresses initiated by the cold-work;

(c) finish-machining, followed by nitriding at 490–520°C.

Any parts of the component which are required soft are protected by coating with a mixture of whiting and sodium silicate.

19.43 Ionitriding   The concept and principles of ionitriding (ion-nitriding)—also known as 'plasma nitriding' and 'ion implantation'—were first established as long ago as the early nineteen-thirties but only in recent years has the process become commercially established. The principles of ionitriding are similar to those of ion-plating (21.86) in that ions of the coating substance are attracted to the surface being coated. The work is made the cathode in a chamber containing nitrogen under near-vacuum conditions (1–10 m bar). Under a potential difference of 500–1000 volts (dc) the low-pressure nitrogen becomes ionised and the N^{+++} ions so produced are accelerated towards the negatively-charged work load where, on impact, they penetrate the surface of the steel. The kinetic energy of the ions is converted to heat so that the surface is quickly raised to the nitriding temperature (400–600°C). The work load is completely surrounded by a glow of ionised nitrogen so that uniform treatment is assured. Depending upon the type of steel and the depth of case required treatment times range from ten minutes to thirty hours. After treatment the load can be cooled under low pressure to prevent oxidation and also avoid the risk of distortion.

The degree of control is superior to that in other nitriding processes and the properties of the case can be effectively adjusted by alterations in the conditions of working. Maximum hardness in 'nitralloy-type' steels is achieved at a treatment temperature of 450°C. The higher the alloy content the thinner and harder the case since nitrides are formed near to the surface when aluminium is present.

Ionitriding is already widely used abroad and the throughput of an installation is about three times that of a comparable orthodox gas-nitriding plant. The process is currently being used to nitride components as large as 14 tonnes in mass down to the tiny balls of ball-point pens. Truck engine crankshafts and other automobile parts, as well as hot- and cold-working tools and dies are being ionitrided.

19.44 The Advantages of Nitriding   are, briefly, as follows:

(i) Since no quenching is required after nitriding, cracking or distortion are unlikely and components can be machine-finished before treatment.

(ii) A very high surface hardness of 1150 H_v is obtained with 'Nitralloy' steels containing aluminium.

(iii) Resistance to corrosion is good, particularly if the nitrided surface is left unpolished.

(iv) Resistance to fatigue failure is good.

(v) Hardness is retained to 500°C whereas in a carburised component hardness begins to fall at about 200°C.

(vi) The process is economical when large numbers of components are to be treated.

(vii) It is a very 'clean' process compared with cyanide-bath treatment where work must be rinsed to remove toxic salts. Such rinsing water, which must be disposed of, constitutes a threat to the environment.

19.45 The Disadvantages of Nitriding as compared with case-hardening are:

(*a*) The initial outlay for plant is higher than with case-hardening, so that the process is economical only when large numbers of components are to be treated.

(*b*) If the nitrided component is accidentally overheated the surface hardness will be lost completely, and the component must be nitrided again. A case-hardened component would only need to be heat-treated again (unless heating had been so excessive as to cause decarburisation).

19.46 Carbonitriding is a surface-hardening process in which both carbon and nitrogen are absorbed into the surface of the steel, nitrogen further increasing the hardness of the carburised layer. Although salt-bath carburising (19.22) also achieves this result, the term 'carbonitriding' is used to describe a process in which *gaseous* media are used. Treatment takes place at 800–875°C in a carbon monoxide/hydrocarbon atmosphere to which has been added 3% to 8% ammonia. The relative proportions of carbon and nitrogen dissolved may be controlled by varying both the concentration of ammonia and the temperature.

Since the steel must be in its austenitic condition to permit rapid solution of carbon this is in turn relatively unfavourable to the solution of nitrogen which dissolves fifty times more quickly in ferrite than it does in austenite. Nevertheless, solution of nitrogen in austenite is still considerable provided that the treatment temperature is kept below 900°C since the solubility of nitrogen in austenite falls as the temperature rises for a given concentration of ammonia.

The presence of more than 0.2% nitrogen, along with 0.8% carbon in the case, slows down transformation rates so that oil-quenching may sometimes be used to harden the case. However, the M_s temperature is also considerably reduced by the presence of more than 0.4% nitrogen so that large quantities of retained austenite may be present in the case after quenching. Nevertheless carbonitriding imparts increased hardenability and wear resistance compared with ordinary case-hardening.

Surface Hardening by Localised Heat-treatment

19.50   In these processes both core and case are of the *same* composition and it is the heat-treatment which is localised. Since the case must contain sufficient carbon to render it hardenable these treatments are applied to medium carbon steels containing 0.35–0.5% carbon. Low-alloy steels containing up to 1.0% chromium with, sometimes, 0.25% molybdenum and 0.5% nickel are also used and the processes are particularly suited to the treatment of components such as gear wheels, splined shafts and spindles where only limited areas need be hardened. The compressive stresses induced in the hardened surface layers improve the fatigue strength of the component.

The component as a whole is first heat-treated by quenching and tempering (or sometimes simply normalising) in order to achieve the necessary core properties. Its surface is then austenitised by being heated locally and immediately quenched to produce a hard martensitic structure. Thus the carbon content of the component is constant at about 0.4% throughout, but whilst the core structure will be tempered martensite (or ferrite/pearlite) that of the case will be martensite. Core and case will generally be separated by a 'cushion layer' of bainite which considerably reduces the risk of spalling.

Two methods of surface heating are available:

19.51 Flame Hardening   In this process the surface is heated by a gas flame derived from ethine (acetylene), propane or natural gas. Manually-operated torches are useful in treating small areas or localised surfaces such as the cutting edge of a blanking tool or the tip of a screw. As soon as the area has become austenitic the whole component is water-quenched. For the progressive hardening of larger areas a gas torch with built-in water jets can be used. The torch passes over the surface slowly enough to ensure its austenitisation and the water jets, following closely behind the flame, effect quenching.

Symmetrical components such as gears and shafts can be spun between centres within a ring burner. The work piece rotates quite slowly at about one revolution per second and as soon as the surface has reached its austenitic state it is quenched either by complete immersion in a bath or by water jets built into the burner ring. For the progressive surface hardening of long shafts a gas ring burner combined with a water-spraying system can be used.

19.52 Induction Hardening   is similar in principle to flame hardening in that only the surface is austenitised prior to being quenched. Heating of the surface is achieved by surrounding the component with an inductor coil carrying a high-frequency alternating current (in the range 2–500 kHz). The smaller the work pieces the higher the frequency used since this results in a shallower depth of penetration of the heated zone.

When an electric current passes through a coil a magnetic field surrounds the coil (Fig. 19.6 (i)) and a steel bar introduced into the field will carry a magnetic flux. Since the magnetic flux in this case is created by a high-frequency alternating current, 'eddy currents' are produced in the surface layers of the steel bar which consequently become heated (Fig. 19.6 (ii)). Additional heating is produced in the case of steel by hysteresis losses and the surface usually reaches its upper critical temperature in a few seconds. Since the copper coil tends to become heated by radiation from the work piece it is usually in the form of a tube which can be cooled by internal water flow.

As soon as the surface has reached the required temperature the component is quenched either by dropping it into a quenching bath or by lowering it automatically into a water spray. Long shafts, axles and similar components are hardened progressively by passing them through an inductor block which carries its own quenching spray attached to the block (Fig. 19.6 (iii)). The component is usually rotated to ensure even heating of the surface. Selective localised surface hardening is achieved by using an inductor 'probe' (Fig. 19.6 (iv)). In such cases hardening is usually attained by air cooling.

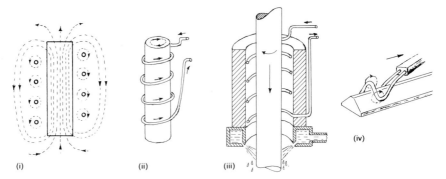

Fig. 19.6   The principles of induction hardening.

Friction Surfacing

19.60   This surface hardening process, known as *Fricsurfacing*, is a derivative of friction welding (20.67). The principles of the process are indicated in Fig.19.7.
The coating material in rod form is rotated under pressure against the surface to be coated so that a hot plastic layer is generated at the interface of the rod and the substrate. By moving the work piece across the face of the rotating rod a layer some 1–2 mm thick is produced. The surface of the coating is characterised by ripples but only 0.1 mm needs to be machined off after coating. During the coating cycle the applied layer reaches a temperature about 40°C below its melting point and severe plastic deformation ensures a uniform fine-grained microstructure as recrystallisation ensues.

19.61   Many of the hard wear-resistant alloys currently available can be deposited in this way. These include tool steels, high-speed steels and

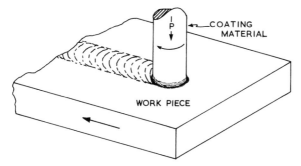

Fig. 19.7 The principles of friction surfacing.

cobalt-based *Stellites*. The method of deposition ensures a structure of finely divided carbides in a matrix of martensite which is formed as a result of the fairly rapid cooling which prevails in the substrate as heat is extracted by the cold work piece. Consequently a tempering process may be necessary or, in the case of high-speed steel, a secondary hardening treatment.

Exercises

1. Calculate the depth of case produced when a low-carbon steel is carburised for 20 hours at 875°C ($k = 0.440$). (Check your answer using Fig. 19.2)
2. Select a mass-produced steel component which is subject to heavy surface wear but requires only moderate strength to withstand operating conditions. Suggest the type of steel and describe the treatment necessary to produce such a component, giving reasons for your choice of material and method.
   (19.23 or 19.40)
3. By reference to the iron–carbon equilibrium diagram explain why the following sequence of treatments was specified for a case-hardened steel shaft 40 mm in diameter, made from plain-carbon steel:
   (i) heat to 950°C; oil-quench.
   (ii) reheat to 780°C; water-quench.
   (iii) heat to 180°C; air-cool.   (19.25;19.26)
4. Why has gas-carburising largely replaced other methods of carburising in mass-production operations ? (19.23)
5. Describe the processes of: (a) nitriding; and (b) gas-carburising and discuss the factors which influence the case and core properties, mentioning the types of steel used. What are the advantages and disadvantages of each process?
   (19.23;19.40)
6. Outline the principles underlying the surface-hardening of steel by methods which involve:
   (i) changes in both the composition and structure of the surface relative to the main body of the component;
   (ii) changes in structure only of the surface layer, the composition remaining constant throughout. (19.20; 19.40; 19.50)
7. Explain how you would produce a hard surface in the following:
   (i) a shaft made from 1%Al–1.5%Cr steel which has to operate at a temperature between 350 and 400°C;

(ii) a gear wheel made from a 0.5%C steel which requires a hard surface to improve its wear-resistant properties.

Give reasons for the choice you make and a brief description of the process chosen. (19.40; 19.50)

8. Outline the methods available for the surface-hardening of gears. Explain the principles underlying these processes. (19.20; 19.50)

9. Outline the process known as 'ionitriding' and show how it differs from the orthodox nitriding process. (19.41; 19.43)

Bibliography

Child, H. C., *Surface Hardening of Steel (Engineering Design Guide)*, Oxford University Press, 1980.

Thelning, K-E., *Steel and its Heat Treatment: Bofors Handbook*, Butterworths, 1984.

BS 970: 1988 *Wrought Steels in the Form of Blooms, Billets, Bars and Forgings. (Part 2 includes steels capable of surface hardening by nitriding.)*

20

Metallurgical Principles of the Joining of Metals

20.10 Apart from purely mechanical methods such as riveting the chief methods available for joining metals are soldering, brazing, welding or the use of resin-based adhesives. Industrial applications of these processes are many and varied and range from the soldering of sardine cans to the fabrication by welding of mass-produced ships by Henry J. Kaiser during the Second World War. Indeed the enormous progress in welding techniques since then has led to the replacement of riveted joints in steel structures by welded ones for almost every application. Modern welding produces a joint which saves up to 15% of the mass of the structure as compared with riveting. Moreover the joint is free from gaps and crevices and is easier to maintain by surface coating. The development of welding has made possible the replacement of many of the larger iron castings by weldments, resulting in tougher, lighter and sounder structures. At the same time it must be admitted that the relatively high labour costs, associated with welding and its allied processes, have led in recent years to the development of a host of metal fastening devices such as self-piercing rivets, self-clinching captive fasteners, thread-forming screws, toggle latches and pop-rivets which in many cases will provide an inexpensive alternative to a continuous metallic joint.

In soldering, brazing or welding complete or incipient fusion takes place at the surfaces of the two pieces of metal being joined so that a more or less continuous crystal structure exists as we pass across the region of the joint. Soldering and brazing are fundamentally similar processes in that the joining material always melts at a temperature which is lower than that of the work pieces, but the distinction between the two processes is imprecise. However, it is generally agreed that soldering—or *soft* soldering as it is often called—can be described as a process in which temperatures below 450°C are involved, whereas brazing temperatures are generally between 600°C and 900°C. Included in the brazing processes are *hard* soldering or *silver* soldering, terms tending to proliferate confusion.

20.11 In the most ancient welding operation—that used by the blacksmith—the two pieces of metal are hammered together whilst at a high temperature, so that crystal growth occurs across the surfaces in contact, thus knitting the halves firmly together. A welding process of this type was probably used by Tubal Cain. By the time the Parthenon was being constructed in Ancient Greece, on the initiative of Pericles, between 447 and 438 BC, welding played an important part in civil engineering. In the Parthenon the marble blocks used were not joined by mortar but by a system of iron dowels and double T-shaped clamps. Metallographic examination of surviving specimens has shown that the latter were made by welding the feet of two T-shaped pieces of iron together.

Soldering

20.20 A solder must be capable of 'wetting', that is, alloying with, the metals to be joined and at the same time have a freezing range appreciably lower so that the work itself is in no danger of fusion or deterioration of mechanical properties. The mechanical strength of the solder must also be adequate though it is often necessary to overlap workpieces considerably so that a joint of sufficient strength is obtained.

20.21 Alloys based on tin or lead fulfil most of these requirements for a wide range of metallurgical materials which need to be joined, since tin will alloy readily with iron and steel, copper and its alloys, nickel and its alloys and lead. At the same time tin–lead alloys possess mechanical toughness, and melt at temperatures between 183°C and 250°C, which is comfortably below the point at which deterioration in properties of the metals to be joined will take place. Those metals which coat themselves with a tenacious oxide skin are very difficult to solder. This is true of aluminium though some success is possible if a suitable flux and a tin–zinc solder are used. Zinc-base die-casting alloys containing aluminium are difficult to solder for the same reason, whilst the extremely dense oxide films on titanium and tantalum make them impossible to solder.

Solders are of two main types depending upon the freezing range required. Thus in the electronics industries and for the manufacture of tin cans, motor-car radiators and the like, a 'tinman's solder' is required which will solidify quickly over a narrow temperature range. Therefore an alloy is used which is as near to the eutectic mixture (62Sn–38Pb) as economic circumstances will allow, bearing in mind that the cost of tin is many times that of lead. 'Best' tinman's solder solidifies over the range 183–185°C whilst a cheaper compromise, 'coarse' tinman's solder (50Sn–50Pb) solidifies over the range 183–220°C. An increased awareness of the poisonous nature of lead has resulted in its prohibition for plumbing purposes. Tin–lead solders too are outlawed for domestic plumbing, even for the joining of copper pipes and fittings. Tin–silver solders or tin–copper alloys (Table 20.1) are now employed instead. Although these alloys are used by 'plumbers'—perhaps they should now be called 'coppersmiths'—they are not required to have the long freezing range of the original plumber's

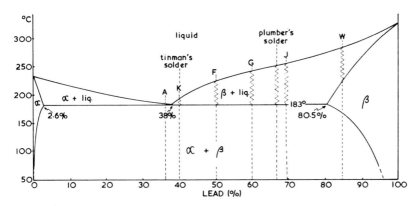

Fig. 20.1 The tin–lead thermal equilibrium diagram. The letters refer to corresponding designations of solders within BS 219 (Table 20.1).

solder. On the contrary they should freeze over as short a range, and hence as quickly, as possible in the manner of tinman's solder. For this reason these lead-free solders are of approximately eutectic composition (Fig. 20.2). Nevertheless freezing ranges involved are some 40°C higher than for equivalent tin–lead solders. They are also more expensive.

Although lead plumbing has been replaced for domestic use by copper or polythene, it is still often necessary to join lead pipes or cable-sheathing and for this purpose 'plumber's' solder is used. This is a mixture containing about 67% lead, and will consequently contain a mixture of liquid and solid phases between 183°C and about 250°C. This extended range over which the alloy will be in a pasty state is of advantage to the plumber, since it enables him to wipe joints in lead piping, a feat which would be almost impossible to accomplish with an alloy melting or freezing over a small range of temperatures.

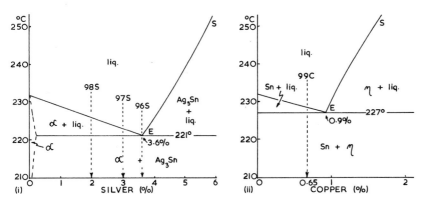

Fig. 20.2 Sections of the thermal equilibrium diagrams for (i) tin–silver; (ii) tin–copper, showing the freezing ranges for suitable lead-free soft solders. In either case an excess of silver or copper above the eutectic composition would give an unacceptable extension of the freezing range due to the steep slope of the liquidus line, *ES*.

Table 20.1 *Soft Solders*

| BS 219: | Composition(%) | | | Melting range (°C) | Uses |
	Sn	Pb	Other elements		
A	64	36	—	183–185	Mass soldering of printed circuits.
K	60	40	—	183–188	General soldering—sheet metal (steel, copper and tinplate); plumbing (capillary joints); printed circuits; electronics; food cans; electrical (hand soldering).
F	50	50	—	183–220	'Coarse tinman's solder'—general sheet-metal work (brass and galvanised sheet). Lower quality work in electrical trades.
G	40	60	—	183–234	Coating and pre-tinning; automobile radiators; refrigerators.
J	30	70	—	183–255	Electrical cable conductors; heat exchangers; automobile radiators; refrigerators.
W	15	85	—	227–288	Electric lamp bases.
C	40	57.6	Sb 2.4	185–227	Heat-exchangers; automobile radiators; refrigerators.
L	32	66.1	Sb 1.9	188–243	'Wiped joints' in lead-cable sheaths and lead pipes.
N	18	80.9	Sb 1.1	185–275	Dip-soldering—non-electrical.
96S	96.5	—	Ag 3.5	221	Eutectic alloy—high temperature service.
97S	96.75	—	Ag 3.25	221–223	For making capillary joints in all copper plumbing installations, particularly where the lead content is restricted in domestic and commercial installations.
98S	98	—	Ag 2	221–230	
99C	99.35	—	Cu 0.65	227–228	
—	—	97.5	Ag 2.5	304	Eutectic alloy—high temperature service; also soldering copper and its alloys.
—	92	—	Zn 8	200	Eutectic alloy—flux-cored wire for soldering aluminium.
62S	62	36	Ag 2	178–189	Soldering silver-coated surfaces.
—	48	—	In 52	117	Tin/indium eutectic —soldering glazed surfaces.
T	50	32	Cd 18	145	Making joints adjacent to other soldered joints which have used higher melting-point solders.

20.22 Solders are sometimes strengthened by the addition of small amounts of antimony which, within the limits in which it is added, remains in solid solution. When copper alloys are soldered it is always possible that at some point in the joint the concentrations of copper and tin will be such that one of the hard, brittle copper–tin intermetallic compounds, such as $Cu_{31}Sn_8$ or Cu_5Sn_6 will be formed in sufficient quantity to cause brittleness of the joint. This will be accentuated if the joint is subjected during service to high temperatures, which, though below the melting point of the solder, promote diffusion of copper and tin within the joint leading to an increase in the amount of compound formed. Such a joint will tear through at the interface containing these compounds revealing their characteristic bluish colour. This difficulty can be overcome by soldering copper with an alloy consisting of 97.5Pb–2.5Ag, for, whilst lead and copper are insoluble in each other, silver alloys with each, and thus acts as a metallic bond between the two without forming any brittle intermetallic compounds. A list of representative solders is given in Table 20.1.

20.23   In order that the solder shall 'wet', or alloy with, the surfaces of the work pieces, the latter must be clean and free from oxide film. Initial cleaning may involve pickling or some form of mechanical abrasion of the surface, but the thin layer of oxide which immediately forms on the cleaned surface of most metals must be dissolved by a suitable flux during the actual soldering process. The traditional soldering flux for iron and steel was 'killed spirits of salts'—hydrochloric acid to which had been added excess granulated zinc. Modern fluxes of this type contain zinc, ammonium, sodium or tin chlorides either separately or together. Orthophosphoric acid is a component of fluxes used for soldering stainless steel.

Whilst the above fluxes are very effective in action they leave behind a corrosive deposit. It is often inconvenient or impossible to wash this off, particularly in the case of electrical assemblies and some organic type of flux is then preferable. These are generally organic acids such as lactic, oleic or glutamic acids or their halogen compounds. Many of these volatilise during soldering or are easily water-soluble. They can be used for soldering steel, copper, brass and many electro-plated surfaces. For electrical work—including modern electronics—rosin-based fluxes are used. The least corrosive soldering flux is a solution of pure white rosin dissolved in propanol-2.

20.24   The reader will be familiar with simple hand-soldering using a 'soldering iron' and flux-cored solder. Whilst this method is still useful, mass production demands more sophisticated automatic methods. Thus, the solder may be 'pre-placed' in the form of washers, rings, discs, pellets or powder (generally a suspension in paste or liquid flux). The assembly, along with the pre-placed solder and flux is then heated on a hot-plate, or in some form of oven, or by high-frequency induction. Dip- or mass-soldering is also used where the work pieces are brought into contact with molten solder which also acts as the heat source. Printed-circuit soldering generally involves a modified form of dip-soldering, known as 'wave-soldering' in which the circuit board is passed over a wave-peak generated in the molten solder bath (Fig. 20.3).

20.25 Diffusion Soldering   The growth of the electronics industry has generated a need for soldering processes in which thermal damage to the workpiece is minimised. Such a process is diffusion soldering the elements of which are shown in Fig. 20.4. Here the workpieces to be joined are

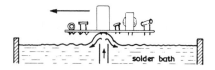

Fig. 20.3 The wave-soldering of printed circuit boards.

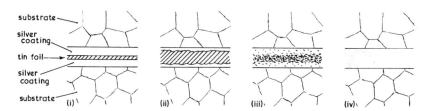

Fig. 20.4 Diffusion soldering, (i) a thin tin foil is sandwiched between the silver-metallised substrates; (ii) the tin foil melts and reacts with the silver film, forming Ag_3Sn; (iii) Ag_3Sn (m. pt. 830°C) solidifies forming a heterogeneous bond; (iv) further heating homogenises the bond forming a Ag/Sn solid solution (m. pt. about 900°C).

metallised* with silver and then separated by a thin layer of tin foil. The assembly is then heated to 250°C so that the tin melts and begins to dissolve the solid silver with which it reacts to form the intermetallic compound Ag_3Sn. Since Ag_3Sn has a melting point of 480°C it begins to solidify, and because the tin foil is thin compared with the silver layers it is soon used up so that the joint solidifies completely. Continued heating at 250°C causes the solid Ag_3Sn and the remaining solid silver to diffuse into each other more completely, so that Ag_3Sn transforms to Ag_5Sn (m.pt. 724°C) which in turn transforms to a silver–tin solid solution with a melting range in the region of 900°C.

Indium foil may be used instead of tin in which case a lower treatment temperature of 175°C can be used (m.pt. of indium 156.6°C). Here the intermetallic compound $AgIn_2$ is first formed, transforming to Ag_2In and ultimately to a solid solution with a freezing range in the region of 900°C. Using diffusion soldering a joint can be made using lower temperatures than those to which the device will be subjected in normal operation. Thus expansion mismatch stresses can be kept very low, a useful feature for example when joining a heat sink to a device.

Brazing

20.30 Metallurgically this process (17.80—Pt. 2) is similar to soldering in that the filler metal melts and alloys with the solid work pieces. Brazing is generally used where a tougher, stronger joint is required, provided that

---

* 'Metallising' involves coating some surface with metal by metal spraying, chemical or vacuum deposition.

the work will be neither melted nor otherwise damaged by the higher temperatures involved in melting the brazing solder. Most ferrous and non-ferrous alloys of sufficiently high melting point can be brazed. A traditional borax-type flux is used for solders which melt above 750°C, but below that temperature borax does not fuse readily and a mixture of borax and metallic fluorides is employed. Fluoride-base fluxes are also useful when brazing refractory materials, whilst alkali-halide fluxes are necessary for brazing aluminium.

Probably fifty-per-cent of industrial brazing is still carried out using a hand-held gas torch but mass production employs a wide range of heating methods in conjunction with pre-placed brazing alloy and flux. Stationary gas torches and batch or conveyor-hearth furnaces have been long established. Electric methods involve high-frequency induction and interface heating (Fig. 20.5), whilst superalloys are generally brazed in a near vacuum.

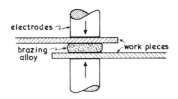

Fig. 20.5   The principles of resistance brazing by interface heating.

Ordinary brazing solder contains about 60Cu–40Zn. This would normally give a coarse Widmanstätten $\alpha + \beta'$ structure, but during brazing some of the zinc will volatilise and some may be absorbed by the work pieces. The resultant joint will therefore be of a tough ductile α structure. Higher-grade brazing compounds, or silver solders, contain up to 85% silver and melt over lower temperature ranges. Special brazing alloys are available for aluminium. Table 20.2 lists a few of the many brazing alloys dealt with in BS 1845.

20.31   The brazing of silicon units in modern electronics can be carried out using a brazing solder containing Al–12Si. This is of eutectic composition (Fig.17.2) and will also 'wet' silicon. If a lower brazing temperature is required an aluminium–germanium brazing solder containing 53% germanium is used and melts at 424°C. It alloys with silicon but has rather poor fluidity which is improved by the addition of silver.

Table 20.2 *Filler Alloys for Brazing*

BS 1845: 1977:	Composition (%)				Freezing Range (°C)	Suitable for brazing:
	Cu	*Zn*	*Ag*	*Other elements*		
AL1	4	—	—	Al—86; Si—10	535–595	Aluminium and some aluminium alloys.
AG1	15	16	50	Cd—19	620–640	Copper and copper-base alloys; mild-steel; carbon steels and alloy steels.
AG7	28	—	72	—	780 (eutectic alloy)	
AG17	41	34	25	—	700–800	
CP2	91.5	—	2	P—6.5	645–740	Copper and copper-base alloys.
CZ3	60	40	—	—	885–890	Copper; all types of steel; malleable and wrought iron; nickel alloys; cobalt alloys.
CZ7A	60	Rem.	—	Sn—0.3; Si—0.3; Mn—0.15	870–900	
CZ8	48	Rem.	—	Ni—9.5; Si—0.3	920–980	
NK5	—	—	—	Ni—rem.; Si—3.5; B—1.8	980–1070	Mild-, carbon- and alloy steels; stainless steels. Nickel and nickel alloys.
NK13	—	—	—	Co—rem; Cr—19; Ni—17; Si—8; W—4; B—0.8; C—0.4	1120–1150	

Welding

20.40   Some of the welding processes resemble both soldering and brazing in so far as molten metal is applied to produce a joint between the two pieces. However, in welding, the added metal is, more often than not, of similar composition to the metals being joined, and a more positive state of fusion exists at the metal surfaces. Thus, differences between the weld metal and the pieces being joined are structural rather than compositional.

The method of production of a weld calls for rather different technique to that employed in brazing and soldering. Since work pieces and filler metal are of similar compositions the speed of working is particularly important if fusion of the metal adjacent to the joint is to be avoided and this calls for a high 'temperature gradient' in and around the weld. However, some welding processes rely on pressure to effect joining of the two halves. In such cases no metal is added to form a joint, the weld metal being provided by the two halves being joined. Thus we have both *fusion* and *solid-phase* welding processes as indicated in Fig. 20.6.

Of recent years, welding has become one of the principal methods of fabricating and repairing metal products. It is an economical means of joining metals in almost all assembly processes and, assuming good welding technique, produces a very dependable result. There are now over thirty different welding methods available for specific applications so that it will be possible to mention only the more important processes here. A fuller description is given in Part 2.

Fusion Welding Processes

20.50 Either thermo-chemical sources, the electric arc or some form of radiant energy can be used to melt the weld metal. In each case the equipment must be capable of 'focusing' the high-energy heat source on to the weld area.

20.51 Gas Methods In these processes the surfaces to be joined are melted by the flame from a gas torch, the gases most commonly used being suitable mixtures of oxygen and ethine (acetylene), which produce temperatures up to 3000°C. Such a high temperature will quickly melt all the ordinary metals and alloys, and is necessary in order to overcome the tendency of sheet metals to conduct heat away from the joint so quickly that fusion cannot occur. Moreover, rapid melting will reduce distortion, overheating and oxidation of the surrounding metal.

In gas- and other fusion-welding processes a welding rod, or filler rod, is used to supply the necessary metal for the weld. The rod is held close to the work and melted by the flame, so that molten metal flows into the prepared joint between the pieces being welded. Since the edges of the pieces also fuse, a strong, metallurgically continuous joint is formed. Welding rods usually have a composition similar to that of the metals being welded, although in some cases they are richer in those elements which tend to volatilise during welding. Gas welding is now used mainly for jobbing and repair work in mild steel and carbon steels. The term 'bronze welding' refers to the joining of metals with high melting points, such as steel, cast iron, nickel and copper, by the use of a copper-alloy filler rod.

In gas welding the flame may be made oxidising, reducing or neutral but with those alloys which oxidise easily, such as those containing aluminium, some form of flux will be necessary. This is normally supplied as a coating to the filler rod. The flux should melt at a lower temperature than the metal being welded and thus dissolves oxide films which form before fusion of the metal begins. As the molten flux coats the surface it also protects the work from excessive oxidation.

20.52 Thermit Welding now finds only limited application, chiefly in the repair of large iron and steel castings, though it was the traditional method for joining rails on site. A mould is constructed around the parts to be joined, and above this is placed sufficient thermit powder. The parts to be welded are preheated and the powder then fired.

Thermit powder consists of an intimate mixture of powdered aluminium and iron(III) oxide in molecular proportions. The iron(III) oxide is reduced by aluminium to metallic iron:

$$Fe_2O_3 + 2Al \rightarrow Al_2O_3 + 2Fe$$

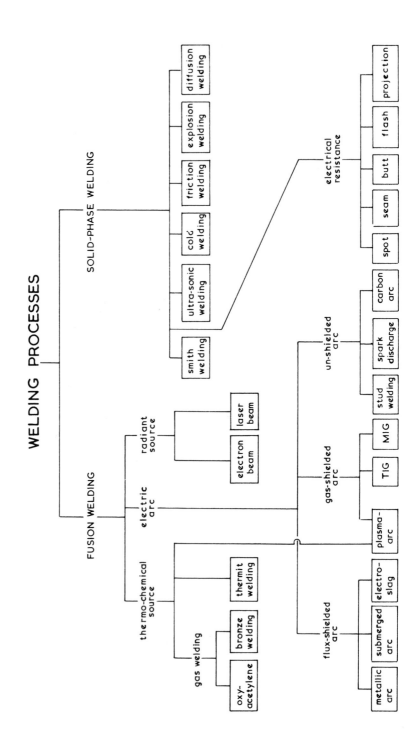

Fig. 20.6 A classification of the more important welding processes.

The heat of the reaction is so intense that the iron formed melts and runs down into the prepared mould, combining with the preheated surfaces and producing a weld. Variations of the process can be used to join non-ferrous cables on site.

20.53 Electric-arc Methods   In the earliest of these processes an electric arc was struck between two carbon electrodes or between a carbon electrode and the work itself. Filler metal was supplied to the arc separately. In modern processes the carbon electrode has been dispensed with and the filler rod itself becomes the electrode.

20.53.1. The metallic arc process is by far the most popular method (Fig.20.8(i)) of fusion welding. Basically it is similar to the carbon arc, but the electrode is a metal rod of suitable composition which serves also as the filler rod. The end of this rod melts and deposits on to the joint, whilst, at the same time, the heat generated melts the edges of the work, producing a continuous weld. The filler rod is usually coated with a suitable flux which not only supplies the weld but also assists in stabilising the arc. Although used chiefly for steel, the metallic-arc method also finds application in welding many of the non-ferrous alloys.

20.53.2. The submerged-arc process is essentially an automatic form of metallic-arc welding, which can be used in the straight-line joining of metals. A bare consumable electrode is used and powdered flux is fed into the prepared joint so that, on melting, it envelops the melting end of the electrode and so covers the arc. The weld area is thus coated by protective slag, allowing a smooth, clean weld surface to be produced. Very high welding currents can be used making very deep welds possible and the process is used in shipbuilding, structural engineering and for pressure vessels as well as in general engineering for mild and alloy steels.

20.53.3. Electro-slag welding is used to join plates generally more than 50 mm thick by a vertical 'run' (Fig. 20.7). A considerable pool of molten weld metal is maintained between the two work pieces by two copper 'dams' which travel upwards as the weld metal solidifies. One or more bare steel electrodes are fed into the slag bath which floats on top of the weld pool. The slag is sufficiently conductive to give rise to resistance heating and the electrodes melt providing the weld metal.

20.53.4. Gas-shielded arc processes are widely used for all materials

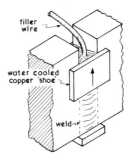

Fig. 20.7   The principles of electro-slag welding.

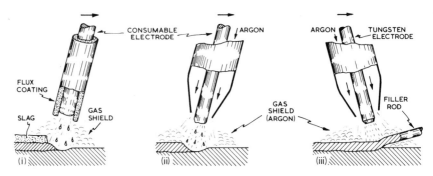

Fig. 20.8 Electric-arc welding processes.
(i) The metallic arc—here molten flux affords protection of the weld region, assisted to a small extent by a gas shield provided by burning carbonaceous matter in the flux coating.
(ii) The metallic inert gas-arc (MIG) in which a very effective gas blanket is obtained by a flow of argon or helium around the weld region. (iii) The tungsten inert-gas arc (TIG)—here a separate filler rod is used, the tungsten electrode being non-consumable.

particularly those in which oxidation is troublesome. In these processes a bare electrode protrudes from the end of a tube from which also issues an inert gas (argon, helium or carbon dioxide). The weld region is thus blanketed by a non-oxidising atmosphere making the use of flux unnecessary (Fig. 20.8(ii)). Carbon dioxide can act as an oxidising atmosphere when used as a gas shield at high temperatures. Nevertheless this can be overcome by using filler rods containing deoxidants such as manganese and silicon. It is a much cheaper gas than argon and during recent years the CO_2 process has become the most popular semi-automatic method for welding steel.

In the *tungsten inert gas*—or TIG—process (Fig. 20.8(iii)) the arc is struck using a non-consumable tungsten electrode and an atmosphere of argon or helium. Since filler metal is supplied from a separate rod this makes it a two-handed process requiring high skill when manual welding is employed. High quality welds are possible with most metals and the process is widely used in the aircraft and chemical industries, being particularly suitable for heavy-gauge stainless steel.

20.53.5. Plasma-arc welding is related to the TIG process in that argon passes through an electric arc formed within a very narrow anular orifice between the tungsten electrode and the water-cooled outer tube. The argon becomes ionised—or forms 'plasma'—and as these fast-moving ions strike the weld area to recombine with electrons, considerable heat is generated. Very deep penetration can be obtained.

Metallic inert gas—or MIG—welding employs a consumable electrode which is fed continuously from a coil within the hand-held 'gun'. Argon or helium are used as shielding gases for non-ferrous metals, but for iron and steel a mixture of argon and CO_2 or pure CO_2 are used. In the latter case the process is termed *CO_2 welding*, and the electrode wire will contain deoxidisers such as silicon, manganese, aluminium or titanium to offset the oxidising nature of CO_2 at high temperatures.

Gas-shielded welding using argon or helium is the only satisfactory method for welding aluminium alloys. An important feature, common to all gas-shielded arc methods, is that, since atmospheric nitrogen is kept away from the weld during fusion, nitrogen embrittlement is avoided. Gas-shielded arc welding will no doubt displace much of the gas- and metallic-arc welding now used, due to the superiority of the weld and the increased speed of manipulation which is possible.

20.53.6. Un-shielded arc processes are used for small scale work. Energy is supplied by *spark discharge* from a capacitor bank. Thermocouple wires are commonly joined in this way but the process is used for the mass production fixing of connecting wires to small components in the electronics industries. In *stud welding* by capacitor discharge a small protrusion is formed on the stud. When this is brought into contact with the work piece and the circuit thus closed the capacitor bank discharges and resistance heating melts the protrusion so that a weld is obtained.

20.54 Electron-beam Welding   In this process, fusion is achieved by focusing a beam of high-velocity electrons on the weld area. On striking the metal the kinetic energy of the electrons is converted into heat energy so that melting of the metal occurs. Welding must be carried out in a vacuum chamber, since the presence of oxygen and nitrogen molecules would lead to collisions with the electrons and their consequent scatter, as well as a loss of kinetic energy. Moreover, since the tungsten filament which emits the electrons is working at 2000°C, it would oxidise rapidly if exposed to the atmosphere.

The process is particularly useful for joining refractory metals, such as tungsten, molybdenum, niobium and tantalum, and also metals which oxidise easily, such as titanium, beryllium and zirconium.

20.55 Laser Welding   has been developed for the production mainly of spot welds in materials up to 3 mm thick. Unlike electron-beam welding a vacuum chamber is not required. Inert-gas shielding is provided by focusing the laser beam through the aperture of a nozzle carrying argon (Fig. 20.9). Very high temperatures and hence rapid welding can be achieved. Since the beam can be focused accurately to a point the process is adaptable to small-scale work for joining wires, foil and components in the electronics industries.

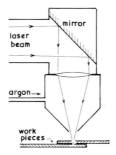

Fig. 20.9   The principles of laser welding.

Solid-phase Welding

These processes include the oldest of all welding operations, that practised by the blacksmith at his anvil. The smith heats the two parts to be welded to a temperature short of actual fusion, and then applies pressure in the form of hammer blows to cause the surfaces to unite. The sand, which he often uses as a flux, combines with the iron oxide produced by heating to a high welding temperature, and forms a fusible slag which is scattered by the hammer blows.

Electrical-resistance welding is similar in principle, and is used in the following modifications:

20.60 Spot Welding, in which the parts to be joined overlap and are firmly gripped between heavy metal electrodes (Fig. 20.10(i)). When the current is passed, local heating of the sheets to a plastic condition occurs, resulting in welding at a spot. This method is used mainly for the joining of sheets and plates and frequently in the production of temporary joints. It is a very widely used process, particularly when robot-controlled, in the motor industry.

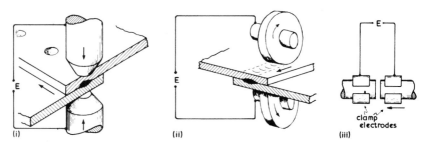

Fig. 20.10 The principles of resistance-pressure welding processes: (i) spot; (ii) seam; (iii) butt. In each case the electrical supply source (E) is the secondary of a 'step-down' transformer.

20.61 Seam Welding is similar in principle to spot welding but the electrodes are in the form of wheels which drive the work 'sandwich' forward (Fig. 20.10(ii)). The current is 'pulsed' to form a series of spot welds which either overlap or are regularly spaced according to the timing of motion and current pulsing. The pressure applied to the wheels is sufficient to produce a weld at the heated interface.

20.62 Butt Welding is used to join lengths of rod and wire. The ends are pressed together (Fig. 20.10(iii)) and an electric current passed through the work so that the ends are heated to a plastic state due to the higher electrical resistance existing at the point of contact. The pressure is sufficient to form a weld.

20.63 Flash Welding is similar to butt welding in that ends of sheets and tubes can be welded together. The process differs, however, in that the ends are first heated by striking an arc between them. This not only produces the necessary high temperature but also melts away any irregularities at the ends, which are then brought into sudden contact under pressure so that a sound weld is produced.

20.64 Projection Welding   is similar in some respects to stud welding except that here electrical resistance heating and pressure are supplied by the electrodes. A projection is punched into one of the work pieces so that point contact occurs when the work pieces are clamped between the electrodes. Localised heating occurs so that, under pressure, the projection collapses and a weld is formed.

20.65 Ultrasonic Welding   is a cold-joining process in which a bond is produced between the work pieces by ultrasonic vibratory energy. The work is gripped between an anvil and a 'sonotrode' tip which vibrates laterally at an ultrasonic frequency. Surface films of oxide are ruptured and welding then occurs between the slipping surfaces under the gripping pressure. This method is used for joining wires, foils and in printed circuits in the electronics industry.

20.66 Cold Welding   is used for lap joints in sheet and butt joints in wire and rod. Sheets are placed together and then punched so that slip occurs between their surfaces. This breaks up surface films of oxide so that intimate contact is achieved between the metal surfaces, leading to welding. Only malleable metals like copper and aluminium can be welded in this way.

20.67 Friction Welding   is used to butt-weld rods and bars. One bar is held in a rotating chuck whilst the other is held in a stationary but pressure-loaded chuck (Fig. 20.11). As the ends of the work pieces are brought together considerable friction between the interfaces causes softening and consequent welding. An advantage of this process is that dissimilar metals can be joined. Thus carbon steel can be welded successfully to high-speed steel, titanium, aluminium and zirconium. It has been used successfully in the motor car industry for joining transmission members, back axles and steering gear.

The underwater repair of cracked butt welds in off-shore structures by *friction stitch welding* is now employed. Essentially this involves rotating a tapered 'filler plug' in a prepared tapered cavity until welding occurs. Overlapping welds are made by 'stitching' along the length of the defect.

20.68 Explosive Welding   is useful in welding together sheets of dissimilar metals, eg cladding of steel. The sheets are placed at a slight angle to each other and a high-explosive charge detonated above the upper sheet. The surfaces virtually 'jet' together and welding occurs.

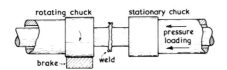

Fig. 20.11   The principles of friction welding.

20.69 Diffusion Bonding is also employed to join sheet material. The sheets are held face-to-face under light pressure in a vacuum chamber. The temperature is then raised sufficiently for diffusion to occur across the interface so that the surfaces become joined by a region of solid solution. To ensure that bonding occurs in less than two hours temperatures are usually in the range 0.5–0.8 T_m (where T_m is the melting point of the metal on the absolute scale). Applied pressures are sufficiently low so that deformation is less than 5%—above this the joining process is deemed to be pressure welding. Steel can be clad with brass in this way whilst titanium alloys such as Ti–6Al–4V can be diffusion bonded at $925°C$.

The Microstructure of Welds

20.70 Since high temperatures are necessary during welding, a weld will exhibit a coarse crystal structure in contrast to the parent metal around it which is generally in the wrought condition. Some elements such as chromium (in steel), accelerate grain growth, so that in these cases an enlarged grain may also be expected in that metal near to the weld.

A weld produced by one of the fusion methods will show an as-cast type of structure. Not only will the crystals be large but other as-cast features, such as coring, segregation and the like, may also be present, giving rise to intercrystalline weakness. The overall effect on grain size which will prevail in most metals and alloys is indicated in Fig. 20.12(i). Here the heat-affected zone will encompass those parts of the work pieces which have been heated in excess of the recrystallisation temperature long enough for recrystallisation to take place. In the arc welding of steel this heat-affected zone extends for only a few millimetres, but in oxyacetylene and electro-slag welding it may be considerably wider. When possible it is of advantage to work a weld mechanically by hammering it while hot. This produces a fine grain as recrystallisation takes place; it also minimises the effects of coring and segregation.

20.71 Changes other than simple recrystallisation may occur in the

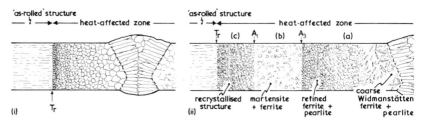

Fig. 20.12 (i) Indicates the relationship between the recrystallisation temperature (T_r) and the extent of the heat-affected zone for metals and alloys in which no phrase change occurs; (ii) shows the possible influences of the upper-critical temperature (A_3) and the lower-critical temperature (A_1) as well as the recrystallisation temperature (T_r) on the heat-affected zone in medium-carbon steel.

heat-affected zone near to a weld made in carbon steels containing more than 0.35% carbon (Fig. 20.12(ii)) and three separate sections may be present in the heat-affected zone:

(a) That part of the zone in which the temperature exceeds the upper-critical (A_3) will undergo complete transformation to austenite and the temperature gradient within the zone will result in variation in this austenite grain size. On cooling the austenite grain size will be reflected in that of the ferrite/pearlite structure produced so that very near to the weld a Widmanstätten structure may be present.

(b) The region in which the temperature reached was between the upper critical (A_3) and the lower critical (A_1 or 723°C) will transform only partially. Pearlite will transform to austenite containing 0.8% carbon and then begin to absorb primary ferrite as the temperature exceeds 723°C. However, subsequent cooling may be rapid enough for this austenite (being of *near-eutectoid* composition) to form martensite so that a brittle region is produced in that part of the heat-affected zone.

(c) In that section of the heat-affected zone where the temperature exceeds that of recrystallisation (about 500°C) the original cold-worked structure of the plate will tend to recrystallise without undergoing any phase-change.

The brittle martensite region may be more extensive in alloy steels of the oil- or air-hardening types since their lower transformation rates will be more likely to favour martensite formation.

The Inspection and Testing of Welds

20.80   Each fusion weld is produced individually and its quality will reflect the degree of skill of the welder. Even where some automation is possible in the welding process a number of variables will control the quality of the weld in so far as internal soundness, adequate fusion and extent of oxidation are concerned. Consequently for high-grade work inspection, coupled with non-destructive testing are necessary.

20.81 Visual Examination   The type of defect encountered may vary widely with the material being welded and the type of process used, but intelligent and careful examination of the resulting weld will often indicate whether it is likely to be satisfactory. Rough, burnt and blistered surfaces suggest overheating and gassing during welding, whilst spattering in arc welding indicates incorrect arc conditions. Insufficient fusion and fluxing are also defects which are relatively easy to detect.

In fusion welding the weld should be somewhat built-up above the parts being joined, to allow for subsequent trimming by grinding. Defects such as undercutting due to insufficient fusion, insufficient penetration and insufficient weld metal are shown in Fig. 20.13.

20.82 Non-destructive Tests   Cracks in or near welds can be detected by means of a *magnetic crack detector* (2.82), assuming of course that both weld metal and work pieces are of a ferromagnetic material (carbon steels

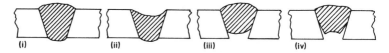

Fig. 20.13 Some welding defects revealed by simple visual inspection: (i) satisfactory weld; (ii) insufficient weld metal; (iii) incomplete penetration; (iv) undercutting and concavity of weld.

and ferritic alloy steels). An advantage of this method is that it will detect fissures just below the surface. Surface cracks in both ferrous and non-ferrous materials can be detected by *dye penetrant* (2.81) or *fluorescent dye penetrant* methods.

Internal defects such as slag inclusions, gas porosity or other cavities can be detected by using a *radiation method* (X-rays or γ-rays) or by an *ultrasonic* method. The penetrating power of radiation increases with its energy so that whilst normal X-ray machines (2.91) can be used for steel up to about 80 mm thick, greater thicknesses require the use of either special high-voltage X-ray tubes of the order of 1 MV or, alternatively, γ-rays derived from a suitable radioactive isotope (2.93). A particular advantage of the latter method is its portability as in site fabrication or where access is difficult.

Increasing use is made of ultrasonic testing (2.96) for detecting faults in both butt and fillet welds (radiation methods being inadequate for the latter). Ultrasonic methods can be adapted to the high speed mechanised testing of long straight welds as in seam-welded tubes. Here the height of the echo trace (which is a measure of the extent of the defect) is monitored on the cathode ray tube. When the height of the trace exceeds a predetermined maximum the monitor causes the defective region of the weld to be marked with paint.

20.83 Testing Specimen Welds Welded joints can be produced in the form of test pieces and tested to destruction. The result will provide useful information regarding the general quality of the process and indicate whether similar results may be expected during subsequent production.

Bend tests are often applied, whilst tensile tests and impact tests also provide useful information. Failure may occur in the weld, in the metal adjoining the weld or between the two. In the latter case, particularly when failure occurs at low stresses, inadequate fusion may be suspected and will often be clearly indicated by the appearance of the fracture, which, instead of being crystalline, is that of the smooth prepared surface. When failure occurs in the weld, visual examination will sometimes reveal such defects as porosity, slag globules and fissures.

Macro-examination of a suitably etched specimen will provide information on the soundness of a weld and also reveal the boundary between weld metal and base metal. A specimen should be cut from the welded joint so that it displays a complete transverse section of the weld. It can then be prepared as indicated in 10.40 and etched in one of the reagents mentioned in Table 10.6. The hydrochloric acid etch will reveal unsoundness by enlarging gas pockets and also by dissolving slag inclusions.

Microscopical examination of polished and etched sections will, in addition to revealing such defects as porosity, slag, cracks and oxide inclusions, also indicate such purely metallurgical defects as overheating, decarburisation and other microstructural defects associated with composition, temperature effects and rates of cooling. Special effects, such as carbide precipitation in stainless steels, can only be detected in this way.

The Weldability of Metals and Alloys

20.90   The ease with which an alloy can be welded depends upon a number of factors such as the tendency towards oxidation and volatilisation of one or more of its constituents, its ability to dissolve or react with other gases present and the coefficient of expansion which will influence crack formation in or near the weld during solidification. High thermal conductivity of the work pieces will demand a greater energy input during welding.

Fusion welding differs from brazing and soldering in that the filler metal is of a similar composition or at least contains a high proportion of the same elements as the work pieces themselves. Hence the problem of the filler metal being able to alloy with the work pieces does not arise. By developing suitable techniques most metals and alloys can be welded successfully by one or other of the standard processes whilst for the more refractory metals specialised methods such as electron or laser beam can be used.

20.91 Mild steel and structural steels   containing up to 0.3% carbon have been by far the most widely used materials in welded fabrications in all industries. Mild steel welds with ease using gas, arc or resistance methods. Electrodes are usually coated with a flux containing lime and silica along with ferromanganese or ferrosilicon to act as a deoxidant in the weld pool. Calcium fluoride may be included to increase flux fluidity.

As indicated in 20.71 some brittleness may occur in the heat-affected zone when steels containing 0.3% carbon are welded. Fortunately the improvement in quality of low-carbon structural steels in recent years has enabled carbon contents to be reduced to 0.1% or less, the required mechanical properties being attained by 'micro-alloying' (13.140). Thus weldability has been improved considerably as required by changes in construction methods where welding has replaced riveting as the main joining method over the last half century. 'Micro-alloy' steels suitable for welding contain 0.3% silicon and in excess of 1.0% manganese, in addition to 0.1% (or less) carbon and small amounts of niobium, vanadium or titanium; silicon to improve fluidity and manganese to assist in deoxidation in the weld pool.

20.92 High-carbon steels   are more difficult to weld because of reactions between carbon and any oxygen that enters the weld pool:

$$FeO + C \rightleftharpoons Fe + CO$$

Porosity in the weld is likely due to the gas, carbon monoxide, CO, evolved. This effect is minimised by using deoxidants such as ferromanganese or ferrosilicon in the flux coating when metallic-arc welding is employed.

In the CO_2 process the gas atmosphere can have a mildly oxidising effect as some dissociation occurs at the arc temperature:

$$2CO_2 \rightleftharpoons 2CO + O_2$$

Hence the bare electrodes used must contain sufficient manganese or silicon to act as deoxidants in the weld pool.

Brittleness arising from the formation of martensite during cooling (20.71) may be prevented by pre-heating (or post-heating) the work in order to induce a slower cooling rate. Carbon pick-up by the weld metal from the parent metal also causes brittleness, particularly since a coarse as-cast Widmanstätten structure can be expected to form, and this may lead to the formation of cracks in or near the weld during cooling. This is often overcome by a technique known as buttering. A layer of weld metal is deposited on each half of the work piece before they are joined. Some carbon is absorbed by this layer but, since the parts are not joined, cracking will be unlikely to occur. The two buttered sections can then be welded with safety since little carbon is likely to reach the actual weld.

Where applicable, the hammering of welds whilst still hot brings about an improvement in properties due to the refinement in crystal structure which it produces. Normalising will also help to reduce the size of the as-cast type of crystals present in ferrous welds made by fusion processes.

20.93 Alloy Steels   can generally be successfully welded, provided the correct type of filler rod is used. Low-alloy steels are often welded with plain steel welding rods containing 0.2% carbon, 1.4% manganese and 0.4% silicon, since there is generally sufficient pick-up of other alloying elements from the metals being welded.

Since low-alloy steels are basically of the oil- or air-hardening types it follows that martensite is very likely to form in the heat-affected zone during subsequent cooling giving rise to 'cold-cracking'. Where possible the whole weldment is preheated to between the M_s and M_f temperatures and held there for up to thirty minutes after welding is complete so that isothermal transformation of austenite to bainite will occur and frustrate the formation of martensite which would otherwise take place during air-cooling.

The filler rods used to weld stainless steels are usually of composition similar to the stainless steel being welded, thus limiting the possibility of electrolytic action at the weld during service. The problems associated with welding stainless steels depend upon whether the steels are of the ferritic or austenitic types (13.44). Ferritic high-chromium steels will suffer from grain growth in the heat-affected zone (Fig. 20.14 (i)) and also the formation of martensite if the steel contains sufficient carbon and is permitted to cool at a rate which will give rise to air-hardening. Alternatively, with slow cooling rates, the carbide Cr_7C_3 may be precipitated. Schaeffler dia-

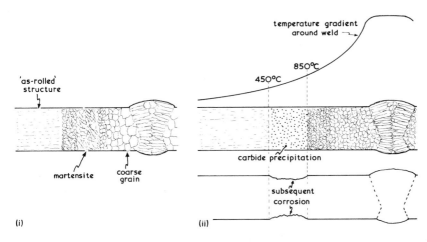

Fig. 20.14 Faults likely to occur during welding in (i) ferritic high-chromium stainless steel; (ii) austenitic 18/8 stainless steel which is not 'weld-decay proofed'.

grams (13.18) were originally devised with the object of predicting the structural changes likely to occur in stainless steels during welding.

Austenitic 18/8 stainless steels are likely to suffer from a defect commonly termed 'weld decay' if the carbon content exceeds 0.03%. Such a steel is normally cooled quickly from 1050°C after hot rolling is finished, in order to keep carbon in solution in the austenitic structure. If the steel is subsequently heated in the temperature range 450–850°C particles of the carbide $Cr_{23} C_6$ will precipitate at the grain boundaries of the austenite (Pl. 13.1c) and this will cause the region adjacent to the boundaries to be severely depleted in chromium. This region then becomes anodic (21.30) to the remainder of the crystal and so goes into solution in the presence of a strong electrolyte. During a fusion welding process some parts of the work pieces near the weld will inevitably be at temperatures between 450°C and 850°C for long enough to permit precipitation of carbides of chromium to take place and it is here that corrosion will occur (Fig. 20.14(ii)) particularly if the weld is in a stainless steel tank carrying corrosive liquids. The carbides of chromium particles can be re-dissolved by heating the complete weldment to 1050°C and quenching but this is not often possible for obvious reasons.

The most effective method of increasing the resistance of the steel to weld decay is to include about 1% of titanium or niobium. These metals have very high affinities for carbon and so effectively 'tie up' most of the carbon as titanium or niobium carbides. These do not dissolve, except in very small quantities, during the normal hot-working or annealing processes, so that the carbon content of the austenite is extremely low and carbides of chromium cannot therefore precipitate in significant amounts. Consequently most of the chromium remains distributed evenly within the austenite. Nevertheless small quantities of carbon can still precipitate as carbides of chromium and exposure to strong mineral acids or chloride solutions would lead to some corrosion. In such cases the carbon content of the steel is best kept below 0.03% in addition to the inclusion of 1% titanium or niobium.

20.94 Cast Iron   either of the grey or SG iron varieties are the ones most commonly welded. Each can be welded successfully when a suitable filler rod is used. For repair work by gas welding a cast-iron filler rod may be used but for arc welding a filler metal containing 55Ni–45Fe is necessary. Brittleness is likely to occur in the heat-affected zone. As the work pieces become austenitic on heating graphite flakes (or spheroids in S.G. iron) are dissolved and as the zone cools relatively rapidly this carbon is precipitated as an intercrystalline network of primary cementite instead of graphite giving a white-iron type of structure. At lower temperatures still, the remaining austenite, which contains 0.8% carbon, is likely to form some martensite thus adding to the brittleness introduced by the cementite network. This brittleness in the heat-affected zone may be reduced by preheating the work to 300°C before welding or by annealing the weldment at 650°C after welding. Alternatively, bronze welding may be used, though this introduces the danger of electrolytic action between the weld metal and the surrounding cast iron and may lead to the accelerated corrosion of the latter which will be anodic to the bronze weld (21.36).

Malleable cast iron of the whiteheart variety can be welded easily in the manner of mild steel. The temper carbon present in blackheart malleable castings, however, presents a problem similar to that arising from graphite in ordinary cast iron. This temper carbon dissolves in the austenite on heating and, on cooling, precipitates as cementite networks whilst some martensite may also form as in the case of grey iron. The best results are obtained by low-temperature bronze welding.

20.95 Copper and its Alloys   The high thermal conductivity of pure copper makes resistance welding of it by either spot or seam processes impracticable but, since most copper alloys have lower thermal conductivities, they can be joined successfully by these methods. The inadvisability of welding tough-pitch copper by gas methods has been mentioned (16.22) and, instead, phosphorus-deoxidised copper along with filler rods rich in deoxidants such as silicon and manganese or aluminium and titanium are used since phosphorus is inadequate as a deoxidant for the weld pool under these conditions. Tough-pitch copper may be welded by one of the inert gas shielded arc methods using similar filler rods to those just mentioned.

Those copper alloys which contain elements having a high affinity for oxygen, viz. aluminium, silicon, tin and zinc, are not subject to gas porosity and need no deoxidant additions to the filler metal. Brasses are difficult to weld because of volatilisation of zinc at the welding temperature. This may cause some porosity of the weld. For arc welding low-grade brasses and phosphor bronzes a filler rod of silicon bronze (96Cu; 4Si) is often used since silicon not only acts as a deoxidant but also increases fluidity of the weld metal.

Filler rods used for welding brasses sometimes contain small amounts of aluminium as a deoxidant. Welds in brass are improved by mechanical work which refines the grain. In the $\alpha + \beta$ phase brasses this work must be carried out whilst the weld is still hot, but in the ductile α-phase brasses it is best carried out in the cold and followed by a stress-relief annealing process which will lead to the formation of fine grain.

In the flame-welding of copper–nickel alloys care must be taken that sulphur is absent from the atmosphere, as this will cause a very serious deterioration in the properties of the metal around the weld. Nickel readily forms sulphides, for which reasons a sulphurous atmosphere always attacks alloys containing nickel.

20.96 Aluminium and its Alloys   have a high affinity for oxygen and the tenacious oxide skin formed on the surface of molten aluminium acts as a barrier in fusion welding. In oxyacetylene processes therefore a flux containing the fluorides and chlorides of lithium, sodium and potassium is used whilst for metallic arc welding potassium sulphate and cryolite are added to this mixture. These fluxes effectively dissolve aluminium oxide but, since they are potentially corrosive, must be washed off the completed weld. Consequently the bulk of fusion welding of aluminium alloys is now carried out by MIG or TIG processes.

Obviously attempts to fusion weld either work-hardened or precipitation-hardened aluminium alloys would lead to softening and loss of strength in the heat-affected zone. For this reason such alloys, which are widely used in aircraft construction, are generally joined by riveting. Some resistance welding of such alloys is carried out but this entails special surface cleaning and also the use of high welding currents because of the high thermal conductivity of aluminium.

The most common and troublesome defect in fusion welds in aluminium is porosity due to the presence of hydrogen. The gas is highly soluble in molten aluminium but only slightly soluble in solid aluminium—hence any dissolved during the welding process is rejected as minute bubbles during solidification of the weld. The main source of hydrogen is moisture, either in the electrode coating or in surface oxide. Thus porosity is at a minimum when welding clean work pieces by the TIG process.

20.97 Magnesium Alloys   like those of aluminium, have a high affinity for oxygen and the refractory oxide formed is only slightly less troublesome. In oxy-gas welding, fluxes containing the chlorides of sodium, potassium and calcium, along with sodium fluoride are used. These fluxes are very corrosive to the work pieces so that the TIG process is used for nearly all magnesium-alloy welding.

20.98 Nickel Alloys   may be welded by the majority of the common processes and the most frequent metallurgical problem is high-temperature embrittlement by sulphur, for which nickel has a high affinity forming weak intercrystalline films of nickel sulphide. Sulphur may be derived from drawing or machining oils, grease, paint or temperature-indicating crayons contaminating the surface of the work. Cleaning with a solvent such as trichloroethene is therefore advisable.

Porosity may arise from oxygen or nitrogen dissolved in the weld pool, so filler rods containing aluminium, titanium or niobium are used since both gases will combine with these metals.

20.99 Refractory and Reactive Metals   Beryllium, titanium, niobium, zirconium, tantalum and tungsten have specialised uses in nuclear and space technology. Since many of them are very reactive at the welding temperature special methods are necessary in welding them. The electron-beam method is particularly suitable since it operates in a vacuum and will also provide the necessary high temperature quickly. Spot, seam, pressure, ultrasonic and explosive welding are also employed but the TIG process, operated without a filler rod, is probably the most widely used method. Most of these metals are welded *autogenously*, that is, no filler metal is supplied, the weld metal being derived from the edges of the work pieces.

Exercises

1. What are the basic differences between *soldering* and *fusion welding*? Give the compositions of tin–lead solders used for: (i) soldering printed circuits; (ii) general sheet metal work, and say why these compositions are used for these purposes. (20.21)
2. Discuss the functions of fluxes used in both soldering and brazing. Why is it difficult to solder aluminium? (20.23; 20.30)
3. Describe, with the aid of a sketch, an electric-arc process in which the work is protected by means of a gas shield. For what metals and alloys is this process particularly suitable? (20.53.4)
4. Write a short account of fusion-welding processes. (20.50)
5. Why is spot welding such a widely used process? Name three common applications of spot welding. (20.60)
6. What is meant by 'solid-phase welding'? Describe two such processes. (20.60)
7. Two stainless-steel plates of the following composition: 0.25%C; 0.5%Mn; 12.0%Cr, are joined by a fusion-welding process. By reference to the Schaeffler diagram (Fig. 13.7) comment on the structure which is likely to be formed around the weld, assuming that the work receives no thermal treatment. How will this structure affect the mechanical properties? (20.93)
8. What difficulties might be experienced in the gas welding of: (i) tough-pitch copper; (ii) high-carbon steel; (iii) spheroidal-graphite cast iron; (iv) 13%Cr; 0.35%C steel; (v) aluminium alloys?
   What precautions would be taken to minimise those difficulties and what other methods of fusion welding would be more suitable for the materials mentioned? (16.22; 20.92; 20.93; 20.96)
9. What particular difficulties might be experienced in fusion welding: (i) a simple 18–8 stainless steel; (ii) ferritic stainless steel? Explain how these difficulties are overcome in practice. (20.93)
10. Sketch microstructures for the solders lettered A, G and W in Table 20.1, assuming them to be in the equilibrium state at room temperature. Calculate the proportions of α and β present in each case at 50°C.

Bibliography

Andrews, D. R. (Ed.)., *Soldering, Brazing, Welding and Adhesives*, Inst. Production Engineers, 1978.

Brooker, H. R. and Beatson, E. V., *Industrial Brazing*, Newnes-Butterworths, 1975.

Castro, R. J. and de Cadenet, J. J., *Welding Metallurgy of Stainless and Heat-resisting Steels*, Cambridge University Press, 1975.

Copper Development Association, *The Welding, Brazing and Soldering of Copper and its Alloys*.

Davies, A. C., *The Science and Practice of Welding*, Cambridge University Press, 1989.

Dawson, R. J. C., *Fusion Welding and Brazing of Copper and Copper Alloys*, Newnes-Butterworths, 1973.

Easterling, K., *Introduction to the Physical Metallurgy of Welding*, Butterworths, 1983.

Gourd, L. M., *Principles of Welding Technology*, Edward Arnold, 1986.

Gooch, T. G. and Willingham, D. C., *Weld Decay in Austenitic Stainless Steels*, The Welding Institute, 1975.

Higgins, R. A., *Engineering Metallurgy (Part II)*, Edward Arnold, 1986.

Houldcroft, P. T., *Welding Process Technology*, Cambridge University Press, 1977.

Lancaster, J. F., *Metallurgy of Welding*, Allen and Unwin, 1987.

Roberts, P. M., *Brazing (Engineering Design Guide)*, Oxford University Press, 1975.

Thwaites, C. J., *Soft-soldering Handbook*, International Tin Research Institute.

BS 499: 1980/1983 *Welding Terms and Symbols*.

BS 219: 1977/(1988) *Specifications for Soft Solders*.

BS 5245: 1985 *Phosphoric Acid Based Flux for Soft-soldering Joints in Stainless Steel*.

BS 1723: 1986/1988 *Brazing*.

BS 1845: 1984 *Specifications for Filler Metals for Brazing*.

BS 1724: 1959 *Bronze Welding by Gas*.

BS 4570: 1985 *Fusion Welding of Steel Castings*.

BS 1453: 1987 *Filler Materials for Gas Welding*.

BS 3019: 1960/1984 *TIG Welding*.

BS 2901: 1983 *Filler Rods and Wires for Gas-shielded Arc Welding*.

BS 5135: 1984 *Metallic-arc Welding of Carbon and Carbon-manganese Steels*.

BS 2926: 1984 *Chromium-nickel Austenitic and Chromium Steel Electrodes for Manual Metal-arc Welding*.

The above are only a selection of the more relevant British Standards Specifications covering metal joining processes, electrodes and tests.

21

Metallic Corrosion and its Prevention

21.10 Of all metallurgical problems which confront the engineer, few can be economically more important than the prevention of metallic corrosion. In Great Britain alone it is estimated that the cost of metallic corrosion is of the order of 3% of the annual gross national product which is equivalent to £10 billion or £150 per capita. This is perhaps less surprising when one considers, as an example, the team of painters permanently employed in protecting the steel of the Forth Railway Bridge from the ravages of air, rain and sea water. Nevertheless even more significant is the amount of steel which is allowed to rust away for lack of adequate protection. This amounts to some *1000 tonnes every single day*—surely an expensive way of 'saving jobs'. But enough of this cynicism.

Other metals, in addition to iron and steel, corrode when exposed to the atmosphere. The green corrosion-product which covers a copper roof, or the white, powdery film formed on some unprotected aluminium alloys is clear evidence of this.

Metals do not corrode where there is no atmosphere. The expensive Hasselblad camera left behind on the Moon by the American astronauts will remain in perfect condition as far as its metal parts are concerned, though radiation may damage some of its non-metallic components. Moreover little corrosion of most engineering alloys, including carbon steels, occurs in a *dry* atmosphere at ambient temperatures; but when the atmosphere is also moisture laden it is a very different matter and 'rusting' of steel is rapid. Fossil fuels, and in particular coal, contain varying quantities of sulphur and nitrogen compounds. On combustion these compounds yield sulphur dioxide and some oxides of nitrogen which, if not extracted from the flue gases, will escape to the atmosphere where they will combine with moisture to form the corrosive agents sulphurous, nitrous and nitric acids. Over much of Europe, both West and East, this accelerates the corrosion of metals, destroys ancient and modern buildings and kills large tracts of our forests. To extract the acidic oxides from flue gases is an expensive process which, in turn, will create some of its own environmental problems but it *must* be done. The only alternative seems to be the use of nuclear-derived power in our current state of technology.

The Mechanism of Corrosion

21.20 Oxidation or 'Dry' Corrosion   The metals of Group I and II of the Periodic Classification react readily with oxygen so that, apart from beryllium and magnesium which are useful because of their very low specific gravities combined with good strength, they are of little use as materials in constructional engineering. Most of the engineering metals are to be found in the 'transition' groups where affinity for atmospheric oxygen is rather less. Oxidation of many of these metals is extremely slow at ambient temperatures but occurs much more rapidly as the temperature rises as is demonstrated by the scaling of steel at red heat. When iron is heated in an atmosphere containing available oxygen it becomes coated with a layer of black oxide scale, FeO:

$$2Fe + O_2 \rightarrow 2FeO$$

The above equation is only a simplified way of expressing the reaction. In fact atoms of iron have been oxidised whilst atoms of oxygen have been reduced. These processes are related to a transfer of electrons from atoms of iron to atoms of oxygen:

$$Fe \rightarrow Fe^{++} + 2 \text{ electrons } (e^-) \quad \text{(Oxidation)}$$
$$O + 2 \text{ electrons } (e^-) \rightarrow O^{--} \quad \text{(Reduction)}$$

The terms 'oxidation' and 'reduction' have a wider meaning in chemistry and involve reactions in which oxygen takes no part. Thus, sulphur gases will attack nickel alloys at high temperatures giving rise to intercrystalline corrosion by the formation of nickel sulphide films. In this instance nickel has been 'oxidised' even though the element oxygen is not involved:

$$Ni \rightarrow Ni^{++} + 2e^- \quad \text{(Oxidation)}$$
$$S + 2e^- \rightarrow S^{--} \quad \text{(Reduction)}$$

Consequently, in its wider sense, oxidation can be described as a process where the atom involved *loses* electrons; whilst reduction can be described as a process where the atom *gains* electrons.

   21.21   Affinity for oxygen is not the sole criterion affecting the rate at which a metal oxidises. Although aluminium has a very high affinity for oxygen it is nevertheless corrosion resistant. This is due largely to the dense and impervious nature of the oxide film which forms on the surface and so protects it from further attack. The extent to which an oxide film will protect the metal beneath depends upon two factors:

(i) the continuity of the film and how effectively it bonds to the metallic surface. Some films are porous and offer poor protection whilst others crack or peel away with repeated heating and cooling of the metal. Such metals and alloys oxidise progressively.

(ii) the mobility of metal and non-metal ions within the oxide film. Iron at a high temperature will continue to oxidise even though coated with an oxide skin. Iron atoms at the metal interface ionise (Fig. 21.1), releasing electrons which travel quickly to the surface of the scale. There the electrons react with oxygen from the air forming ions, O^{--}. These oxide ions are then attracted towards the oppositely-charged Fe^{++} ions. Although the *equilibrium* composition of the oxide skin is $Fe^{++}O^{--}$ that near the metal surface will contain an excess of Fe^{++} ions and that adjacent to the atmosphere will contain more O^{--} ions. The rate at which new ions form depends upon their mobilities within the oxide film. That is, new Fe^{++} ions will form as existing ones diffuse away from the metal surface towards O^{--} ions. The more rapidly Fe^{++} ions move away, the more rapidly will iron atoms ionise to produce new Fe^{++} ions. Prolonged heating of large steel ingots for hot rolling produces a very thick layer of scale. The concentration gradients of Fe^{++} and O^{--} ions within this layer are often so great that three separate crystalline zones are found in the layer. The thin zone adjacent to the metal surface is of basic composition, FeO, whilst that in contact with the atmosphere is of composition Fe_2O_3. Between the two is a zone containing the intermediate oxide Fe_3O_4—or $FeO.Fe_2O_3$.

21.22   When two steel surfaces are in contact under fairly high pressure and at the same time subjected to alternating or vibrational stresses, *fretting corrosion* may occur. Differences in elastic properties between the surfaces may lead to localised welding of contacting high spots. The welds subsequently rupture and local high temperatures, set up as a result of friction, cause oxidation of the surface. Fretting is indicated by the presence of a reddish-brown powdery deposit of oxide, Fe_2O_3, sometimes referred to as cocoa powder. This is highly abrasive and so accelerates the attack. Fretting is prevalent in wire ropes and occurs in splines and other press-fitted components which are subjected to alternating stresses. In addition to general wear, such corrosion can lead to the initiation of fatigue cracks

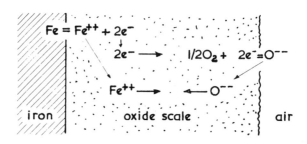

Fig. 21.1

particularly since the affected components are subjected to vibrational stresses. Though difficult to eliminate completely it can be minimised by excluding air by the use of a high-pressure grease or by the use of a solid high-pressure lubricant such as molybdenum disulphide.

21.23   As indicated above oxidation in industrial atmospheres is rarely a simple process involving oxygen only. Contaminants such as carbon dioxide, carbon monoxide, sulphur dioxide, oxides of nitrogen and water vapour are frequently responsible for the very rapid deterioration of metals at high temperatures. Under varying conditions carbon dioxide can promote both oxidation and carburisation in alloys. This phenomenon leads to *green rot* in Nimonic Alloys—precipitation of $Cr_{23}C_6$ leaves the matrix so depleted in protective chromium that extensive oxidation of the nickel to green NiO occurs rapidly.

Nimonic and other nickel-base superalloys are rapidly attacked in the presence of sulphur-rich combustion gases. This leads to the formation of black NiS so that the effect is aptly described as *black plague*. Solid ash from burning fuel, carried in the combustion gas stream, can also react chemically with surface oxidation products. As a result a fluxing action may take place resulting in the formation of fluid products at temperatures above 650°C. Fluid glassy slags so formed dissolve oxides rapidly so that a metallic surface is exposed to further oxidation. The very rapid corrosion of Fe–Ni–Cr alloys which occurs under such conditions is known as *catastrophic oxidation*.

21.30 Electrolytic Action or Wet Corrosion Involving Two Dissimilar Elements   Electrolytic action in one form or another is responsible for the bulk of corrosion which occurs in metals at ambient temperatures. In this particular instance it will occur when two dissimilar metals of different 'electrode potential' are in electrical contact with each other and with an 'electrolyte'. The term 'electrolyte' describes some substance which contains both positively- and negatively-charged ions, able to move about freely within it.

Much of this chemical action is similar to that which occurs in a simple Galvanic cell (Fig. 21.2), consisting of a copper plate and a zinc plate, immersed in dilute sulphuric acid (the electrolyte). When the external circuit is closed a current begins to flow through the ammeter. This current is composed of electrons which are released in the zinc plate and, as their concentration builds up there, are forced to flow to the copper plate. As a result of the loss of electrons zinc atoms become zinc *ions* (Zn^{++}) and pass into solution in the electrolyte:

$$Zn \rightarrow Zn^{++} + 2e^-$$

As the electrons, which have been forced round the external circuit by pressure of numbers, collect on the copper plate it becomes negatively charged so that hydrogen ions (H^+), present in the electrolyte from ionised sulphuric acid, are attracted to the copper plate where they combine with the available electrons to form ordinary atoms and, hence, molecules of hydrogen so that bubbles of the gas form on the copper plate:

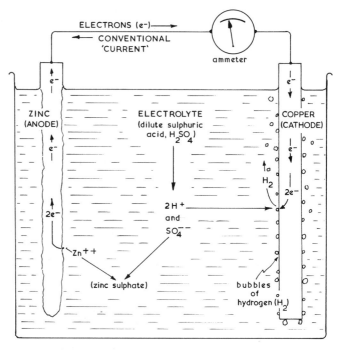

Fig. 21.2 The chemical reactions in the simple cell.
The conventional direction ascribed to the current in the early days of electrical technology is opposite to that in which electrons in fact flow.

$$2H^+ + 2e^- \rightarrow H_2 \text{ (bubbles on copper plate)}$$

ZN^{++} ions 'pair up' with SO_4^{--} ions in the electrolyte so that as the concentration of sulphuric acid falls that of zinc sulphate rises. Thus we obtain 'electrical energy' at the expense of a loss of chemical 'potential energy' by the zinc.

This electrolytic action occurs as the result of a difference in 'electrode potential' between copper and zinc (Table 21.1). The electrode potential of a metal is related to the amount of energy required to remove the valence electrons from its atoms. Thus zinc loses its valence electrons more readily than does copper and zinc is said to be *anodic* towards copper. The electrode which supplies electrons to the external circuit is called the *anode* whilst the electrode which receives electrons via the external circuit is called the *cathode*.

21.31 The reader may have been puzzled by the fact that Fig. 21.2 shows electrons as flowing in the opposite direction to that indicated for current in most elementary text books of electrical technology. The Greek derivation of the term anode suggests something which is being *built up* and in fact the anode was so named in 1800 because it was thought at that time that *positively*-charged 'particles of electricity' were passing through

Table 21.1   *The Electrochemical Series (Electrode Potentials for Some Metals)*

	Metal (ion)	Electrode Potential (volts) E_h
(Noble)	Gold (Au^{+++})	+1.50 (Cathodic)
	Platinum (Pt^{++++})	+0.86
	Silver (Ag^+)	+0.80
	Copper (Cu^+)	+0.52
	(Cu^{++})	+0.34
	Hydrogen (H^+)	0.00 (Reference)
	Iron (Fe^{+++})	−0.05
	Lead (Pb^{++})	−0.13
	Tin (Sn^{++})	−0.14
	Nickel (Ni^{++})	−0.25
	Cadmium (Cd^{++})	−0.40
	Iron (Fe^{++})	−0.44
	Chromium (Cr^{+++})	−0.74
	Zinc (Zn^{++})	−0.76
	Aluminium (Al^{+++})	−1.66
	Magnesium (Mg^{++})	−2.37
(Base)	Lithium (Li^+)	−3.04 (Anodic)

the external circuit from the cathode to the anode. We now know that the electric current in the external circuit is in fact a flow of *negatively*-charged particles (electrons) in the opposite direction. The original convention of current direction is, however, still retained.

21.32   The bulk of metallic corrosion is due to electrolytic action of this type. Electrolytic action is possible when two different metals or alloys are in electrical contact with each other and are also in common contact with an electrolyte. Action such as this occurs when a damaged 'tin can' is left out in the rain. If some of the tin coating has been scratched away so that the mild steel beneath is exposed, electrolytic action takes place between the tin and the mild steel when the surface becomes wet with rain water or condensation (Fig. 21.3A). Water containing oxygen or other dissolved gases is ionised to the extent that it will act as an electrolyte, and electrons flow from the iron (mild steel) to the tin, leading to the release of Fe^{++} ions into solution. Thus, once the coating is broken, the presence of tin accelerates the rusting of iron it was meant to protect, and the electrolytic action follows the same general pattern as that which prevails in the simple cell. The iron of the can corresponds to the zinc plate, the tin coating to the copper plate and the solution of oxygen in water to the dilute sulphuric acid used as the electrolyte in the simple cell. Industrial atmospheres accelerate corrosion because of the sulphur dioxide they contain. This gas is present due to the combustion of sulphur in coal and other fuels. Sulphur dioxide dissolves in atmospheric moisture forming sulphurous acid which has a much stronger electrolytic action than has dissolved oxygen. However, in the pages which follow electrolytic corrosion related only to the anion, $-OH^-$, is discussed; but it should be appreciated that in industrial atmospheres containing the anions $-SO_3^{---}$, $-NO_2^-$ and $-NO_3^-$, electrolytic corrosion will be accelerated by the presence of these strongly electronegative ions.

21.33   Metallic coatings are frequently used to protect steel from corrosion, but it is necessary to consider in each case how corrosion will be

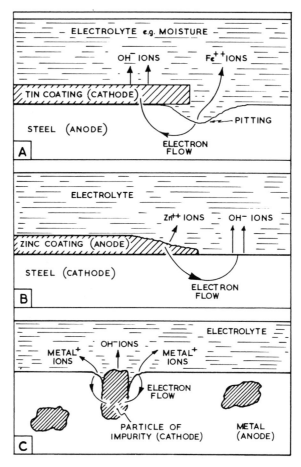

Fig. 21.3 The mechanisms of electrolytic corrosion.

affected if the coating becomes scratched or broken. Thus metals used to protect steel can be divided into two groups:

(*a*) Metals which are *cathodic* towards iron, such as copper, nickel or tin. These metals can be used only if a good-quality coating is assured, since protection offered is purely mechanical in that the coating isolates the surface of the steel from the corrosive medium. Suppose that at some point in a tin-coated iron sheet the tin film is broken (Fig 21.3A). Here the tin film acts as a cathode and the steel the anode. Fe^{++} ions go into solution and electrons, so released, travel to the tin cathode where they react with water and dissolved oxygen:

$$H_2O + \frac{1}{2}O_2 + 2e^- \rightarrow 2OH^-$$

Thus hydroxide ions (OH^-) are released at the cathode and, away from the region of the cell, these combine with Fe^{++} ions:

$$Fe^{++} + 2OH^- \rightarrow Fe(OH)_2$$

Iron(II) hydroxide, $Fe(OH)_2$, so formed quickly oxidises to iron(III) hydroxide, $Fe(OH)_3$, the basis of the reddish-brown substance we call rust. Thus corrosion of the steel is actually accelerated when a coating of some substance which is cathodic to steel becomes damaged. Fortunately, in the case of both tin and nickel, metals commonly used to coat iron and steel, the difference in electrode potential in the cell so formed is small, so that the acceleration of attack is not great. The further apart metals are in the electro-chemical series (Table 21.1) the greater the rate of corrosion of the anode.

(b) Metals which are *anodic* towards iron, such as zinc and aluminium. These metals will go into solution in preference to steel (Fig 21.3B) to which they will therefore offer some chemical, as well as mechanical, protection. Since they protect the steel by going into solution themselves, such protection is generally referred to as 'sacrificial'. Zinc is used in this way for galvanising steel products, whilst aluminium is effective as a protective paint partly for the same reason. It must be emphasised however that the object of galvanising should be to produce a sound continuous coating of zinc on the surface because sacrificial protection will only be temporary should the coating be porous.

21.34   Having dealt briefly with the principles of electrolytic corrosion, we are now in a position to give an adequate explanation of the rusting of mild steel (Fig 21.4). Here electrolytic action takes place between the iron and the oxide film which will be present on the surface—assuming of course that the oxide film is broken and that the gap is covered by an electrolyte (moisture containing dissolved oxygen).

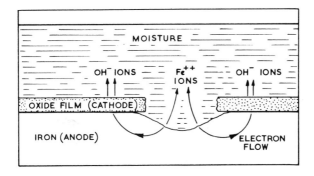

Fig. 21.4   The electrolytic corrosion of iron due to the presence of a surface film of oxide.

Since iron is anodic to its oxide, it will go into solution as ions:

$$Fe \rightarrow Fe^{++} + 2e^-$$

The electrons which are released travel to the cathode and take part in the reaction:

$$H_2O + \tfrac{1}{2}O_2 + 2e^- \rightarrow 2OH^-$$
(from
atmosphere)

As a result, Fe^{++} ions go into solution at the anode and hydroxide ions (OH^-) form at the cathode. When these two different sorts of ions meet *away from the region of electrolytic action* the following reaction occurs:

$$Fe^{++} + 2OH^- \rightarrow Fe(OH)_2$$

This iron(II) hydroxide, $Fe(OH)_2$, is quickly oxidised by atmospheric oxygen to iron(III) hydroxide, $Fe(OH)_3$, which is precipitated as a reddish-brown substance, the main constituent of rust.

21.35   So far we have been dealing only with electrolytic action between some surface coating and the metal supporting it. The same type of corrosion may take place at the surface of a metal in which particles of an impurity are present as part of the microstructure. In Fig 21.3c it is assumed that the particle of impurity is cathodic towards the metallic matrix, thus causing the latter to dissolve. Since impurities are often segregated at crystal boundaries, this will lead to gradual intercrystalline corrosion of the material as more particles of impurity become exposed by electrolytic action.

21.36   Those metallic alloys which have a crystal structure consisting of two different phases existing side by side are also prone to electrolytic corrosion. If one phase is anodic with respect to the other it will tend to dissolve when the surface is wetted by a suitable electrolyte. Patches of pearlite in steel will tend to corrode in this manner (Fig 21.5). Ferrite is

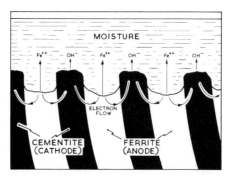

Fig. 21.5   The electrolytic corrosion of pearlite.

anodic to cementite, consequently ferrite goes into solution (as Fe^{++} ions) and the electrolytic action is similar to that which takes place between ferrite and an oxide film during the rusting of mild steel mentioned above. This leaves the brittle cementite platelets standing in relief, but these will ultimately break off as corrosion proceeds and increases their fragility. It is therefore apparent that in order to have a high corrosion-resistance, a metal or alloy should have a structure consisting of one type of phase only, so that no electrolytic action can take place. Similarly, as engineers will know, it is unwise to rivet or weld together two alloys of widely different electrode potentials if they are likely to come into contact with a substance which can act as an electrolyte. In such circumstances the alloy with the lower electrode potential may corrode heavily due to electrolytic action. An example of bad practice of this type is illustrated in Fig. 21.6. Here a steel pipe has been connected to a copper pipe. If the system carries water which is not chemically pure, then electrolytic corrosion of the steel, which is anodic to copper, can be expected. This corrosion will be much more rapid than if the whole pipe were of steel.

Those readers who are householders will probably be aware that the galvanised cold-water tank corrodes more quickly in that region adjacent to the *copper* inlet-valve mechanism, zinc being strongly anodic to copper. This form of plumbing is doubtless very good for trade, particularly as failure of the tank usually occurs when the room beneath has been newly decorated. The obvious remedy is to match the copper plumbing with a copper tank—which is less expensive than it sounds. Fortunately most modern domestic plumbing systems now used 'plastics' (polypropylene) cold-water tanks which are of course impervious to electrolytic corrosion and are also relatively inexpensive.

21.37 The extent to which electrolytic action between two dissimilar electrodes takes place depends not only upon the difference in potential between them but also upon the *chemical nature* of the electrolyte, and in particular the concentration of hydrogen ions (pH). For example, our motor cars rot away even more rapidly as a result of the salt used on roads during winter. When electrovalent compounds dissolve in water molecules

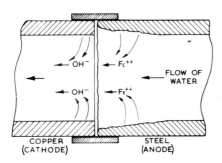

Fig. 21.6 The accelerated corrosion of a steel pipe due to the presence of an adjacent copper pipe.

of the latter tend also to ionise thus increasing the hydrogen ion concentration, pH.

The connection between electrode potential (E_h) and the hydrogen ion concentration (pH) was investigated in Belgium by M. Pourbaix. He derived diagrams which connect these values with the electrochemical reactions between metals and water which was made either acid or alkaline. Both electrode potential, E_h, and hydrogen ion concentration, pH, influence the transfer of electrons in the system so that hydrogen evolution and oxygen absorption depend on these factors and these reactions are the basis of electrolytic corrosion. The effects of E_h and pH on the extent to which water will liberate either hydrogen or oxygen during what we term 'electrolysis' are illustrated in Fig. 21.7. Hydrogen will only be liberated under conditions of E_h and pH which are represented by a point below '*ab*', and oxygen can only be liberated if E_h and pH are represented by a point above '*cd*'. Hence there is a *domain* (between *ab* and *cd*) on the E_h/pH diagram where the electrolysis of water is impossible. As a result the electrolysis of water cannot take place, regardless of the pH of the electrolyte, unless the electrodes have a minimum potential difference of at least 1.2 volts, that is the distance apart of *ab* and *cd* in terms of E_h.

21.38   In Fig. 21.7 the equilibrium of hydrogen and oxygen are represented in terms of E_h and pH. However, equilibrium of a metal in contact

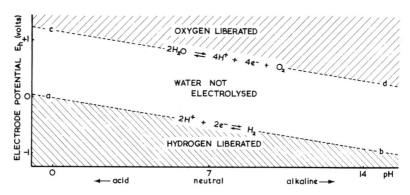

Fig. 21.7   The E_h/pH diagram for water.

with water can also be represented using a Pourbaix diagram. Since the corrosion of iron (or mild steel) is important this is represented as a simplified diagram in Fig. 21.8. The complete Pourbaix diagram for iron is more complex since several chemical reactions are involved. However, a simplified diagram can be drawn based on a study of the following chemical equilibria:

(i) $Fe \rightleftharpoons Fe^{++} + 2e^-$ Iron atoms form *soluble* iron(II) (Fe^{++}) ions.
(ii) $Fe^{++} \rightleftharpoons Fe^{+++} + e^-$ Iron(II) ions are oxidised to iron(III) ions.

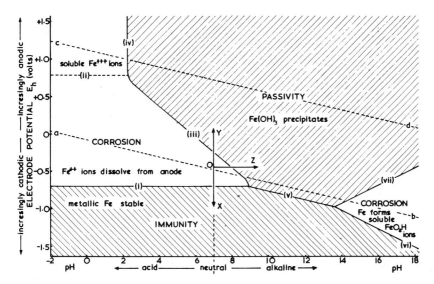

Fig. 21.8 A simplified Pourbaix diagram for the Fe/H$_2$O system.

(iii) $Fe^{++} + 3OH^- \rightleftharpoons Fe(OH)_3 + e^-$ Insoluble iron(III) hydroxide precipitates.

(iv) $Fe^{+++} + 3H_2O \rightleftharpoons Fe(OH)_3 + 3H^+$ Insoluble iron(III) hydroxide precipitates.

(v) $Fe + 3H_2O \rightleftharpoons Fe(OH)_3 + 3H^+ + 3e^-$ Insoluble iron(III) hydroxide precipitates.

(vi) $Fe + 2H_2O \rightleftharpoons FeO_2H^- + 3H^+ + 2e^-$ Ferr*ite* ions go into solution.

(vii) $FeO_2H^- + H_2O \rightleftharpoons Fe(OH)_3 + e^-$ Soluble ferrite ions precipitate as iron(III) hydroxide.

When the electrode potential ($E_h = -0.44$ volts) of iron is measured in pure water (pH = 7) it will be represented on the Pourbaix diagram (Fig. 21.8) by the point O. This indicates that corrosion can take place under such conditions of E_h and pH. The diagram suggests three ways in which the extent of corrosion could be reduced:

(*a*) The potential can be changed in the negative direction (*X*) so that it enters the domain of immunity. This is the normal 'sacrificial' or cathodic protection such as is offered by zinc or aluminium.

(*b*) The potential can be changed in a positive direction (*Y*) into the domain of passivity by applying a suitable external EMF (21.110).

(*c*) The pH of the electrolyte can be increased, ie made more alkaline (*Z*) so that the metal becomes passive. Inhibitors have this effect but need careful control because the passivity domain extends over a very narrow pH range, making constant surveillance of the electrolyte composition necessary.

Electrolytic Action or Wet Corrosion Involving Mechanical Stress

21.40 It has already been shown (21.35 and 21.36) that electrolytic corrosion can occur as a result of different parts of a microstructure being of different compositions and, hence, different electrode potentials. Thus intergranular corrosion can occur in a cast metal since impurities, segregated at grain boundaries, have different electrode potentials to the rest of the structure. Nevertheless a perfectly uniform wrought metal may corrode more rapidly if it is in the cold-worked condition.

As we have seen cold-working causes a movement of dislocations towards crystal boundaries (4.19). This 'pile-up' of dislocations results in the formation of a region of elastic strain in the vicinity of the crystal boundary and the 'locked-up' stresses associated with this strain constitute a region of *higher energy level* than is present in the remainder of the crystal. If an electrolyte is present then energy can dissipate itself by the grain-boundary atoms going into solution as ions. That is the high-energy grain-boundary material is *anodic* to the rest of the crystal.

21.41 This type of intergranular corrosion manifests itself in many instances. The 'season cracking' of 70–30 brass has been mentioned (16.33), but of greater significance in the modern world is the accelerated corrosion which occurs in mild steel which has been cold formed. In many instances it is possible to apply a stress-relieving annealing process to the cold-worked material but in other cases this is inappropriate either because mechanical distortion of the structure would result or because the results of cold-work are necessary to give rigidity and strength. Adequate protection of the surface by some form of coating is then the only way to prevent corrosion.

Consider the cold-formed rim of a motor-car 'wing' (Fig. 21.9). The

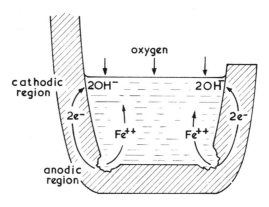

Fig. 21.9 Corrosion due to local cold-work in the rim of a motor car wing.

forming operation will have cold-worked the rim more severely than the rest of the structure making it anodic to its surroundings so that, assuming that the paint work is defective, corrosion will be accelerated there particularly when salt-laden moisture collects during winter. This type of corrosion in metals is generally termed *stress corrosion* and may not necessarily be caused by cold-work. For example, locked-up stresses may be set up near to welded joints as a result of non-uniform cooling. Similar situations may occur during the cooling of a casting. It must be admitted, however, that in both welded joints and castings galvanic corrosion is much more likely to be the result of variations in composition rather than variations in strain energy.

21.42   Corrosion assists in other forms of failure which plague the engineer. Thus *corrosion fatigue* refers to the failure of a component in which the propagation of a fatigue crack is helped by corrosion, or indeed initiated by corrosion. A surface flaw may be accentuated by corrosion and then act as a focal point for the initiation of a fatigue crack (5.65). Once such a fatigue crack has started its progress will be accelerated by corrosion due to the anodic effect of high stress concentrations always present at the root of the crack.

21.43 Impingement corrosion   describes a process involving electrolytic action following mechanical abrasion of the surface. Thus particles of grit or entrained air bubbles carried in a flow of liquid may impinge upon an oxide-coated surface, wearing away the oxide skin. Electrolytic action between the cathodic oxide skin and the anodic metal beneath follows (Fig. 21.10) leading to solution of the metal. This form of attack is prevalent in tubes carrying sea water but also occurs with fresh water, particularly when the flow of water is so rapid that the rate of damage to protective films exceeds the rate of repair by oxidation. Corrosion in this instance is not a result of mechanical working but of removal of an oxide film so that simple galvanic action can occur between the rest of the oxide film and the exposed metal.

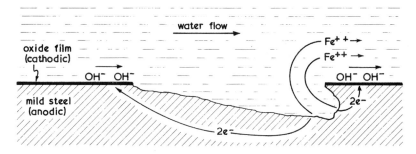

Fig. 21.10 'Impingement corrosion' results from a combination of mechanical abrasion and electrolytic action. The corrosion product will be flushed away.

Electrolytic Action or Wet Corrosion Involving Electrolytes of Non-uniform Composition

21.50 Electrolytic action may occur between two electrodes of *precisely the same composition* if these electrodes are in contact with an electrolyte which is different in composition at the surface of each electrode. This situation can be described by reference to what is termed a concentration cell (Fig. 21.11). Here both electrodes are of pure iron but the electrolyte,

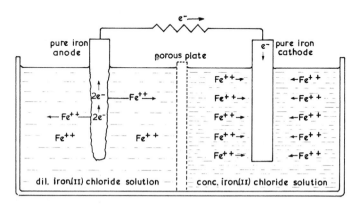

Fig. 21.11 A 'concentration' cell.

iron(II) chloride solution, is more concentrated on the right of the cell than on the left. The cell is divided by a porous plate—similar to that in an old-fashioned Leclanché cell—which prevents the two solutions from mixing freely but is sufficiently porous to allow the passage of ions and so maintain 'electrical contact'. The main reaction in the cell is represented by:

$$Fe \rightleftharpoons Fe^{++} + 2e^-$$

and the direction in which it proceeds at any point in the cell is governed by the concentration of the reactants. Therefore, on the left-hand side, the relatively low concentration of Fe^{++} ions will cause iron atoms to form ions, releasing electrons to the external circuit:

$$Fe \rightarrow Fe^{++} + 2e^-$$

On the right-hand side of the cell the already high concentration of Fe^{++} ions will tend to make the reaction go in the opposite direction as these ions receive electrons from the external circuit:

$$Fe \leftarrow Fe^{++} + 2e^-$$

Since iron atoms are entering solution as Fe^{++} ions at the left-hand electrode this is obviously the anode, whilst the electrode on the right-hand side becomes the cathode as is indicated by the movement of electrons in the external circuit.

21.51 Electrolytic action of this type is the basis of corrosion which occurs in a variety of different instances in engineering practice. For example, so-called 'crevice corrosion' is caused by a non-uniform distribution of dissolved oxygen in water thus producing a 'concentration cell' situation. In Fig. 21.12 two steel plates of similar composition are bolted together and, although the gap between them is small, water will be drawn into the gap by capillary action. If, for a short time, the concentration of dissolved oxygen increases in that water exposed to the atmosphere, then the region '*C*' of the plate will become cathodic to that at '*A*'. The plate in the vicinity of *A* is covered by water containing less oxygen and so becomes anodic. As soon as this happens iron atoms will begin to ionise in the anodic area of the crevice forming Fe^{++} ions, and the electrons so released will travel quickly through the steel plate to the cathodic area to form OH^- ions:

$$2H_2O + O_2 + 4e^- \rightarrow 4OH^-$$

As the concentration of OH^- ions in the cathodic area is thus increased so a persistent concentration cell is set up since oxygen is unable to dissolve in the water trapped in the crevice, whereas oxygen is able to dissolve continuously from the atmosphere in the cathodic area.

Having the greater mobility, Fe^{++} ions travel outwards to the cathodic area more quickly than the OH^- ions travel to the anodic region. Therefore the reaction:

$$Fe^{++} + 2OH^- \rightarrow Fe(OH)_2$$

occurs mainly outside the crevice. $Fe(OH)_2$ is rapidly oxidised to $Fe(OH)_3$, 'rust'. Thus the formation of a concentration cell *accelerates corrosion but it accelerates it where the concentration of oxygen is lower*. This explains

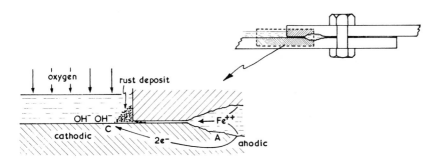

Fig. 21.12   The 'crevice corrosion' of steel plates.

why steel corrodes most in places where it would seem to be better protected from the atmosphere.

21.52 The fairly rapid rusting which occurs in the vicinity of defective paintwork on a motor car is explained in Fig. 21.13. Water begins to penetrate beneath the loose paint film but oxygen does not and so the metal beneath the paint becomes anodic to that which is exposed to the atmosphere. Crevice corrosion then proceeds as above.

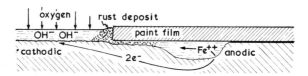

Fig. 21.13 Sub-paint rusting of a motor car body. Another case of crevice corrosion.

One frequently hears a motor-car owner bemoaning the fact that 'rust is eating its way beneath the painting work'. In fact of course the reverse is true—Fe^{++} ions are moving *outwards* from beneath the paint to form rust at the cathodic surface. Some of the Fe^{++} ions will meet up with OH^- ions as these move slowly inwards beneath the paint. As a result some rust forms beneath the paint producing blisters which lift it off.

Accumulations of chemically-inert dirt on the surface of exposed metal will also give rise to this form of corrosion. Such encrustations allow moisture to penetrate and this moisture becomes increasingly deficient in oxygen thus giving rise to anodic spots beneath the dirt. This situation is made worse by the salt used on roads in winter since Cl^- ions will tend to congregate along with OH^- ions outside the dirt encrustations and considerably increase the concentration gradient within the electrolyte. It is far more important to wash away accumulated dirt from the hidden parts of a motor-car—particularly during winter—than it is to waste time polishing the areas which are visible. 'Wings' invariably rust from beneath particularly where 'mud traps' have been incorporated into the design.

21.53 Finally we will consider why a sheet of high-quality mild steel rusts under the action of a simple rain drop (Fig. 21.14). Here the concen-

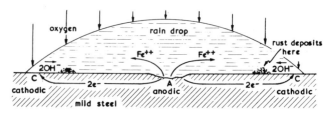

Fig. 21.14 The rusting of mild steel in the presence of a rain drop containing dissolved oxygen.
Note the position of the anulus of rust—near the outer rim because of the lower mobility of the OH^- ions compared with those of Fe^{++}.

tration of dissolved oxygen in the rain drop will be higher at 'C' than at 'A' where the depth of water is greater, so that dissolved oxygen reaches the metal surface at A more slowly. A concentration cell is therefore set up so that atoms of iron at A go into solution as Fe^{++} ions releasing electrons which pass through the 'external circuit' (the mild steel sub-surface) to C where OH^- ions form. A deposit of $Fe(OH)_2$ is produced mainly around the periphery of the rain drop and this oxidises immediately to form $Fe(OH)_3$ or 'rust'.

This type of corrosion would appear to be important on the massive scale with the many off-shore structures associated with oil and gas operations in the North Sea, for as water depth increases so the oxygen content decreases as in the case with the droplet of water described above. However the rate of corrosion is probably influenced more directly by factors such as sea water velocity (impingement corrosion—21.43), chlorine ion $(-Cl^-)$ concentration, marine organisms, temperature and the rise and fall of the tides.

The Prevention of Corrosion

21.60 There are two principal methods by which corrosion may be prevented or minimised. First, the metallic surface can be insulated from the corrosive medium by some form of protective coating. Such coatings include various types of paints and varnishes, metallic films having good corrosion-resistance and artificially thickened oxide films. All of these are generally effective in protecting surfaces from atmospheric corrosion. Zinc coatings are used to protect iron and steel from the rusting action of moist atmospheres and though zinc offers its 'sacrificial protection' (21.33) as a second line of defence, it should be clearly understood that the main objective is to produce a sound continuous film of zinc which will seal off the iron completely from atmospheric action. Sacrificial protection is a temporary phenomenon and is only effective for a limited time since the zinc dissolves quickly once electrolytic action begins.

Tin coatings offer protection against most animal and vegetable juices encountered in the canning industry but due to current very high costs of tin extremely thin coatings of the metal are used on the mild-steel cans. This tin coat is now generally covered by a film of some organic polymer (plastics material) as the main protection, so that modern tin cans tend to rust on the outside rather than the inside.

21.61 In circumstances where corrosive action is severe, or where mechanical abrasion is likely to damage a surface coating, it may be necessary to use a metal or alloy which has an inherent resistance to corrosion. Such corrosion-resistant alloys are relatively expensive, so that their use is limited generally to chemical-engineering plant, marine-engineering equipment and other special applications.

The Use of a Metal or Alloy Which Is Inherently Corrosion-resistant

21.70   The corrosion-resistance of a pure metal or a homogeneous solid solution is generally superior to that of an alloy in which two or more phases are present in the microstructure. As mentioned above, the existence of two phases leads to electrolytic action when the surface of the alloy comes into contact with an electrolyte. The phase with the lower electrode potential will behave anodically and dissolve, leading to pitting of the alloy surface; and the greater the difference in electrode potentials between the phases, the more rapid will be corrosion. Resistance to corrosion will generally be at a minimum when the second phase is segregated at the crystal boundaries, particularly if this phase is anodic to the matrix. In such circumstances serious intercrystalline corrosion will occur.

21.71   Most of the alloys which are used because of their high corrosion-resistance exhibit solid-solution structures. Aluminium–magnesium alloys (17.30) containing up to 7.0 % magnesium fulfil these conditions and are particularly resistant to marine atmospheres. 18–8 stainless steel is completely austenitic when correctly heat-treated but faulty heat-treatment may lead to the precipitation of chromium carbide at the grain boundaries. This gives rise to a zone adjacent to the grain boundary which is impoverished in chromium (Fig. 21.15(i)) so that it becomes anodic to the remainder of the crystal which is still rich in chromium. Consequently in the presence of a strong electrolyte (mineral acids and the like) intergranular corrosion will occur (Fig. 21.15(ii)). 'Weld decay' occurs in 18–8 stainless steels for similar reasons (20.93).

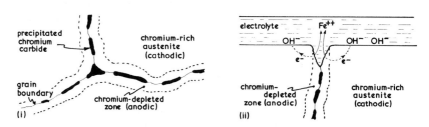

Fig. 21.15   'Weld decay' in austenitic stainless steel.

21.72   As might be expected, coring in a cast alloy of the solid-solution type gives rise to a quicker rate of corrosion than that which obtains when the same solid solution has been cold-worked and annealed. This treatment produces greater homogeneity in the structure and reduces the extent to which electrolytic action will take place between the core and outer fringes of individual crystals. Similarly, the presence of impurities segregated at the crystal boundaries of a metal used in either the alloyed or unalloyed state will accelerate corrosion by setting up electrolytic action with the metal when an electrolyte is present.

INCREASE IN RESISTANCE TO INTERCRYSTALLINE CORROSION

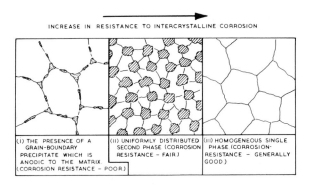

(I) THE PRESENCE OF A GRAIN-BOUNDARY PRECIPITATE WHICH IS ANODIC TO THE MATRIX (CORROSION RESISTANCE – POOR.)

(II) UNIFORMLY DISTRIBUTED SECOND PHASE (CORROSION RESISTANCE – FAIR)

(III) HOMOGENEOUS SINGLE PHASE (CORROSION-RESISTANCE – GENERALLY GOOD)

Fig. 21.16   The relative corrosion-resistance of types of microstructure.

21.73   Although a homogeneous single phase metal or alloy will tend to be less vulnerable to simple electrolytic corrosion, it will of course succumb to attack which is caused by mechanical stress (21.40) or by local variations in electrolyte composition (21.50). Thus stainless steel is susceptible to crevice corrosion since it relies for its corrosion resistance on a supply of oxygen to maintain a dense protective film of chromium oxide on the surface (13.16). For similar reasons stainless steel may suffer impingement corrosion when abrasive particles damage the oxide skin in the presence of strong electrolytes such as acid chloride solutions.

The effects of cold-work in producing regions of high-energy value which accelerate corrosion have already been mentioned. 70–30 brass, normally a reasonably corrosion-resistant single-phase alloy when of high purity, nevertheless suffers 'season cracking' (16.33) as a result of intergranular corrosion associated with deep drawing and the pile-up of dislocations at grain boundaries so produced.

Protection by Metallic Coatings

21.80   Protection afforded by metallic coatings can be either 'direct' or 'sacrificial'. Direct protection depends on an unbroken film of some corrosion-resistant metal covering the article, and if the film becomes broken, corrosion may be accelerated by electrolytic action between the film and the metal beneath. In the case of sacrificial protection, however, the metallic film becomes anodic in the event of a break in the film, and thus dissolves in preference to the cathodic surface beneath. It follows that, when protection is limited to the direct type, as in the case of tin coatings on steel, the quality of the coating is most important, since acceleration, and not inhibition, of corrosion would follow a break in the film. In both cases, of course, protection of the direct type is the fundamental aim of the metal-coating process, and it is only in the possibility of the coating becoming broken that the effects of electrolytic action must be considered.

A number of methods are available for the production of metallic coatings. The most widely used are either electro-plating or dipping the articles to be coated into a bath of the molten metal. Other methods involve spraying or volatilising the protective metal on to the surface whilst mechanical 'cladding' is also used.

21.81 'Cladding' is used chiefly in the production of 'clad' sheet and is employed prior to the final manufacturing stages of a component. The base metal is sandwiched between pieces of the coating metal and the sandwich is then rolled to the required thickness. In some cases the film of protective metal can be sprayed on to the base surface and then rolled on. 'Alclad' is duralumin coated with pure aluminium (up to 0.01 mm thick depending upon the overall thickness of the stock) but other high-strength alloys of aluminium, which have poor corrosion resistance because of their multi-phase microstructures, are also clad. Mild steel clad with stainless steel is also available but it is important to note that with all forms of cladding damage to the protective skin can take place during fabrication.

21.82 Hot-dip Metal Coating   Tin and zinc are the metals most often used to produce metallic coatings in this manner, though use is also made of aluminium. Ideally the coating should adhere to the base surface by solid solution between the two. In some cases however, brittle intermetallic compounds are formed at or near the interface such as $FeSn_2$, $FeSn$, or Fe_2Sn in the case of hot-dip tinning.

(a) *Hot-dip Tinning*. Tinplate is principally used in the manufacture of cans for packaging a wide variety of foodstuffs. The non-toxic properties of tin, combined with its good corrosion-resistance and the ease with which tin-coated articles can be soldered, make it a useful metal for coating articles which come into contact with all types of food.

The mild-steel sheet used for the manufacture of tinplate is usually about 0.25 mm thick, and, after being annealed, it is freed from all traces of oxide by pickling in either dilute sulphuric acid or dilute hydrochloric acid. The tinning unit comprises a thermostatically controlled vessel of molten tin in which is submerged a system of guides and rollers for conducting the pickled sheets down through a layer of molten flux into the tin and then upwards and out of the tin through a layer of palm oil. As they pass through the palm oil, the sheets are subjected to a squeegeeing action by tinned-steel rollers, which serve to control the final thickness of coating on the sheets.

After leaving the tinning machine, the tinned sheets are cooled and then cleaned, usually by washing them in an alkaline detergent. Finally, they are polished in either bran or wood meal.

Other iron and steel items of kitchenware such as mincing machines and mixers, are coated by hot-dipping after fabrication. After being cleaned, degreased and pickled, the component is immersed in a pot of molten tin, on the surface of which is a layer of flux. If the operations have been correctly carried out a brilliant finish is obtained which requires little in the way of after-work.

(b) *Hot-dip Galvanising*. The term galvanising is derived from the name of Luigi Galvani, the Italian physiologist, whose classic observations of dead frogs, in 1786, led to Volta's development of the electric cell. In 1829 Faraday observed that when zinc and iron, in contact with each other, were exposed to the air in the presence of a salt solution the iron did not rust; whereas if the zinc were removed the iron rusted rapidly. Faraday attributed this to 'the effect of chemical action . . . in exciting electricity'. In 1836 a Frenchman, Sorel, took out a patent for a process of coating steel with zinc by hot-dipping, and named the process galvanising. The word was used to indicate the sacrificial electro-chemical protection afforded iron by the zinc.

Before being galvanised, the surface of steel must be thoroughly cleaned, either by shot blasting or, more generally, by pickling in dilute sulphuric acid. In order to prevent contamination of the molten zinc by iron, the work is then washed to remove all traces of iron salts formed by pickling. It is necessary, however, to protect the surface of the work from fresh oxidation, so it is immersed in a flux bath consisting of a solution of zinc chloride and ammonium chloride. This flux is usually dried on to the surface in an oven which is provided with an adequate draught and working at a low temperature, to prevent interaction between the flux and the iron surface.

The work, with its protective flux coating, is then immersed in the molten zinc bath. The flux coating peels away to form a layer on the surrounding bath surface, which in turn is protected from oxidation; and the surface of the work is left clean so that it is immediately wetted by the molten zinc. The thickness of the coating produced depends largely upon the temperature of the bath and the rate at which the work is withdrawn, but good-quality galvanised articles should carry between 0.3 and 0.6 kg/m^2 of zinc on the surface. Thinner and more flexible coatings are obtained by using a zinc bath containing about 0.2% aluminium. This limits the formation of intermetallic compounds of iron and zinc on the steel surface.

In addition to corrugated iron for roofing, the large number of galvanised products includes buckets, barbed wire, water tanks, marine fittings, bolts, and fencing fittings, agricultural equipment and window frames. Many galvanised products such as dust bins have been replaced by plastics mouldings whilst aluminium is becoming increasingly popular for the manufacture of window frames.

(c) *Hot-dip Coating with Aluminium* (*Aluminising*) is used in the protection of iron and steel surfaces under a wide variety of trade names, the best-known of which is 'Aludip'. In all of these processes the surface must be cleaned by grit blasting, de-greasing and/or pickling. It is then carefully dried before immersion in a bath of molten aluminium at about 700°C, generally through a layer of molten flux. In some processes steel sheet is heated in an atmosphere of hydrogen at 1000°C prior to dipping in order to reduce any oxide films present, but in addition to the extra cost of this treatment the quality of the product is often less satisfactory.

In some methods, particularly in the USA, the aluminium-dipped mild-steel sheet is then rolled to give a better surface finish. This product is used to replace tinplate for some containers.

(d) *Hot-dip Coating with Zinc/Aluminium Alloys*. These relatively new commercial processes are said to give a much more corrosion-resistant coating on steel sheet, strip, wire and tube than is obtained by conventional hot-dip galvanising. The high-aluminium coatings *Galvalume* and *Zincalume* use an alloy containing 55Al–45Zn and small amounts of silicon; whilst *Galfan* coatings are based on 95Zn–5Al (with very small additions of 'rare earth' metals). Though *Galfan* coating is less corrosion resistant than those of the 55Al–45Zn alloys it has better formability and retains paint more effectively.

21.83 Coating by Means of a Spray of Molten Metal   Metal spraying consists in projecting so-called 'atomised' particles of molten metal from a special pistol on to a suitably prepared surface. Surface preparation usually involves blasting the surface with an abrasive; steel grit having replaced sharp silica sand for this purpose because of health hazards involved when the latter is used. It is essential that the surface is roughened, rather than peened or polished by the grit in order that good adhesion shall take place. Hence attention to the condition of the grit is important. The metals most commonly used for spraying are zinc and aluminium, though coatings of tin, lead, cadmium, copper, silver and stainless steel can be so deposited.

Early metal spraying processes, originally invented by Dr. Schoop, used an oxyacetylene flame to carry particles of metal on to the work surface. The metal was injected into the flame as a powder. Later modifications of the Schoop process used electrical melting of the metal. An arc was struck between two zinc wires in the pistol and the molten zinc so produced was 'atomised' and carried forward in a blast of compressed air. *Flame impingement* methods are now the most common, using a specially designed flame gun in conjunction with either a continuous rod or a hopper feeding powder to the high-velocity flame. The temperature is generally sufficient to melt the metal, but not necessarily so, and in either case the small particles are projected with high energy at the work surface.

The most important factor is rigid control of the flame conditions. If too high a temperature or too oxidising flame conditions prevail then the particles may be oxidised before they reach the work surface. Nevertheless it is a satisfactory method for structural steel provided that flame conditions are controlled.

Metal spraying has wide application in view of its portability and flexibility; thus, large structures, such as storage tanks, pylons and bridges, can be sprayed on site. Notable recent examples include the Forth Road Bridge and the Volta River Bridge (Ghana), both of which were zinc coated using modern developments of the Schoop process.

21.84 Sherardising   is a cementation process, similar in many respects to carburising, in that zinc is made to combine with an iron or steel surface by heating the work with zinc dust at a temperature *below* the melting point of zinc. The first patent for this process was taken out in London in 1901 by Sherard Cowper-Coles, after whom the process is named. Much thinner coatings can be obtained by sherardising than are possible with hot-dip galvanising, and generally about 0.15 kg/m^2 of zinc is used. Moreover, a more even film is obtained, thus making possible the treatment of such articles as nuts and bolts, the threaded portions of which would undoubtedly become clogged by hot-dipping.

As in all other coating processes, the surface of the work must first be properly prepared. This usually involves degreasing, followed by acid pickling in cold 50% hydrochloric acid or in hot 10% sulphuric acid. Shot blasting is also used, particularly in the case of iron castings, to remove graphite and core sand.

The work is packed into mild-steel drums along with some zinc powder. The drums are heated to 370°C and rotated slowly so that the tumbling action brings all the components into contact with the powdered zinc.

'Calorising' is a process in which steel components are coated with aluminium by a similar method. The components are treated in a heated rotating cylinder containing a mixture of aluminium and aluminium oxide powders. It is used principally to provide a protective coating for heat-treatment pots, ladles and furnace components. Nickel alloys may be protected from attack by sulphur gases by aluminising them. The components are packed into a heat-resistant box along with aluminium powder. Air is then purged from the box by flushing it with argon or nitrogen. The tightly-sealed box is then heated for several hours to promote aluminisation of the surface. 'Chromising' is a similar process involving chromium coating of a mild steel surface in which corrosion resistance comparable with that of a 12% chromium steel can be achieved.

21.85 Electro-plating   The formation of metal coatings by electro-deposition is well known, and a wide variety of metals can be thus used, including copper, nickel, chromium, cadmium, gold and silver. Tin and zinc can also be electro-deposited, and a coating thus formed has advantages over one produced by hot-dipping in respect of flexibility, uniformity and control of thickness of film.

The usual degreasing and pickling processes must be applied to the surface before any attempt is made at electro-plating it. Most instances of peeling and blistering in the case of chromium-plating are due to inadequate preparation of the surface. The author has seen specimens in which the layer of chromium had been deposited on to oxide scale, with the inevitable peeling of the chromium as a result.

In the actual process of electro-plating the article to be plated is made the cathode in an electrolytic cell. Sometimes the metal to be deposited is contained, as a soluble salt, in the electrolyte, in which case the anode is a non-reactive conductor, such as stainless steel, lead or carbon. In most cases, however, the anode consists of a plate of the pure metal which is being deposited, whilst the electrolyte will contain a salt or salts of the same metal. Then, the anode gradually dissolves and maintains the concentration of the metal in the electrolyte as it is deposited on to the articles forming the cathode.

The conditions under which deposition takes place are very important, so that the cell voltage, the current density (measured in amperes per square metre of cathode surface), the ratio of anode area to cathode area and the time of deposition, as well as the composition and temperature of the electrolyte, must all be strictly controlled if a uniform adherent and nonporous film is to be obtained.

21.86 Vacuum Coating and 'Ion Plating' In vacuum-coating processes the work pieces are contained in a vessel from which air is evacuated. The coating metal is electrically heated within the vessel so that, because of the near vacuum conditions, it vaporises at low temperatures and will then condense on any cold surfaces with which it makes contact. Head-light reflectors and mirrors of all types are coated in this way.

With both vacuum coating and electroplating complex shapes are difficult to coat uniformly and to overcome this a technique called 'ion plating' was introduced. Here the article to be plated is made the cathode in a vessel containing low-pressure argon (Fig. 21.17). 'Glow discharge' occurs in the region of the cathode due to ionisation of the argon and the Ar^+ ions are attracted to the cathode, bombarding it and having a cleaning effect on the surface. This is known as 'sputter cleaning' or 'ion scrubbing'.

Metal atoms are present as vapour emitted from the tungsten foil 'boat' and about 1% of these ionise. These M^+ ions are immediately accelerated towards the work-piece comprising the cathode with which they collide. The kinetic energy of the ions is dissipated as heat so that good adhesion is provided because of the tendency of the plating ions to diffuse into the substrate. By adequate control of the evaporation source alloy coatings can be applied as well as pure metals.

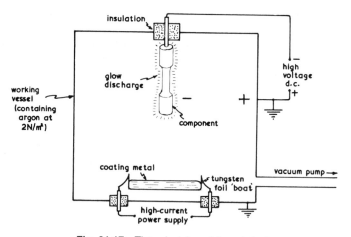

Fig. 21.17 The principles of 'ion-plating'.

Protection by Oxide Coatings

21.90 In some instances the film of oxide which forms on the surface of a metal is very dense and closely adherent. It will then protect the metal surface beneath from oxidation. Stainless steels owe their resistance to corrosion to the presence of a high proportion of chromium, which is one of these elements that form oxide films impervious to oxygen. The blueing of ordinary carbon steel by heating it in air produces an oxide film of such a nature that it affords partial protection from corrosion.

21.91 Anodising Reference has already been made (21.21) to the protection afforded aluminium by the natural film of oxide which forms on its surface. Anodic oxidation, or anodising, is an electrolytic process for thickening this oxide film. This process may be applied for several reasons, such as to provide a key for painting, to provide an insulating coating for an electrical conductor or to provide a surface which may be dyed, as well as to increase the resistance of aluminium to corrosion.

Before being anodised the surface of the article must be chemically clean. Preliminary treatment involves sand-blasting, scratch brushing or barrel polishing, according to the nature of the component. This is followed by degreasing in either the liquid or vapour of trichlorethylene; or by electrolytic cleaning. The latter process is used mainly for highly polished surfaces. The article is made the cathode in an electrolytic bath comprising a 1% solution of sodium hydroxide, and treatment is continued for about half a minute at a voltage of 14 and a current density of 550 A/m^2. The mechanism of this cleaning process is largely one of flotation of the film of grease by hydrogen bubbles given off at the surface of the article which comprises the cathode.

In the actual anodising operation which follows, the aluminium article to be treated is made the anode in an electrolyte containing either chromic, sulphuric or oxalic acid; the cathode being a plate of lead or stainless steel. When an electric current is passed, oxygen is formed at the anode and immediately combines with the aluminium surface of the article. The layer of oxide thus formed grows outwards from the surface of the aluminium. The normal thickness of a satisfactory anodic film produced commercially varies between 0.007 and 0.015 mm, and a film having a thickness within these limits would be formed by anodising a component in a 15% sulphuric acid solution at 20°C for about thirty minutes, using a current density of about 100 A/m^2 at a cell voltage of 15. A longer period of treatment produces a thicker, but soft and spongy, film which would be unsatisfactory in service. The thickness of the natural film produced on an aluminium surface by exposure to air at normal temperatures is of the order of 0.000 013 mm.

The surface produced by anodising can be dyed by immersion in either hot or cold baths of dyestuff, but whether or not the film is dyed, it should always be sealed. The anodic coating is somewhat porous as it leaves the electrolytic tank and is easily stained, but sealing renders the coating impermeable. A simple, but quite effective, sealing process consists in treatment with hot or boiling water, which renders the coating non-absorptive without any visible change in appearance. Other sealing processes include treatment in solutions of 1% nickel or cobalt acetates at 100°C, followed by boiling in water, the application of linseed oil or lanolin to the surface or the application, while the article is warm, of a mixture containing equal parts of turpentine, oleic acid and stearic acid.

21.92 Beryllium Oxide Coatings The addition of small amounts of beryllium to aluminium-base alloys causes a coating of beryllium oxide, BeO, to form on the surface. This is even more protective than the natural skin of aluminium oxide, Al_2O_3, described above. Unfortunately beryllium is an extremely expensive metal. Nevertheless 0.1–0.3% beryllium is added to some high-quality aluminium alloy castings for aerospace purposes.

In addition to providing the protective BeO skin beryllium acts as a 'scavenger' of oxygen and nitrogen in the melt. It also refines the structure of any iron-based intermetallic compounds which may be present, replacing the coarse needles with small equi-axed crystals.

Protection by Other Non-metallic Coatings

21.100   Coatings of this type usually offer only a limited protection against corrosion and are, more often than not, used only as a base for painting.

21.101 Phosphating   A number of commercial processes fall under this heading, but in all of them a coating of phosphate is produced on the surface of steel or zinc-base alloys by treating them in or with a solution containing phosphoric acid and, generally, a metallic phosphate. The phosphate film which forms on the surface is usually grey in colour, about 0.0005 mm thick and offers only limited protection against corrosion but since its surface is rough it provides an excellent 'key' for paint, varnish or lacquer.

In automobile body work it also limits any filiform or scab corrosion resulting from stone chipping by thus preventing any 'crevice corrosion' (21.51) beneath the paint film.

In the automobile industry the spot-welded car body (termed the 'body in white' at this stage) is usually given a preliminary degreasing treatment. The phosphate coating is then achieved by using a phosphoric acid spray in order that the hydrogen so liberated is flushed away:

$$Fe + 2H_3PO_4 \rightarrow Fe(H_2PO_4)_2 + H_2 \uparrow$$

The reaction will then proceed to completion:

$$3Fe(H_2PO_4)_2 \rightarrow Fe_3(PO_4)_2 + 4H_3PO_4$$

Iron phosphate, $Fe_3(PO_4)_2$, forms as an insoluble grey adherent crystalline coating on the surface being treated.

A number of proprietary processes based on phosphating have been introduced since 1903. Some readers may have used the preparation *Kurust* which is available for the small-scale treatment of rusted steelwork, principally the bodywork of decrepit—though not necessarily ancient—motor cars.

21.102 Chromating   Chromate coatings are produced on magnesium-base alloys, and on zinc and its alloys, by immersing the articles in a bath containing potassium bichromate along with various other additions. The colour of the films varies with the bath and alloy, from yellow to grey and black. With steel a chromate film has little permanence but it is sometimes used following phosphating or zinc-coating of steel and as an additive to zinc-based primer paints.

21.103 Electrophoresis This process—also known as 'electro-painting'—is used mainly to apply undercoat paint to steel which has already been phosphated. The chemistry of the process is fairly involved, but briefly, the paint bath consists of colloidal particles of the appropriate 'resin' suspended in a water solution which contains an electrolyte. DC electrodes one of which is the article being painted, are immersed in the bath.

The resin particles, constituting the paint, acquire a charge by adsorbing ions from the electrolyte so that they are then attracted towards the appropriate electrode. Sometimes the ions are provided by one of the reagents in the paint bath. For example, if a colloid is kept in suspension by soap solution the negatively-charged ion of the soap (the 'fatty acid' radical) may be adsorbed by the colloid particles giving them a negative charge so that the combined particle is attracted to the anode which will be the article being painted.

This method produces a very uniform paint film on both flat surfaces and sharp edges. As soon as the colloidal particle is deposited it insulates that portion of the surface and so other charged particles then 'seek out' bare areas which are still charged. For this reason a very complete and uniform film can be deposited and if an ammeter is included in the circuit it will read zero when coating is complete. The fact that the paint is carried in a water solution reduces fire risks but at the same time the resultant film is dense because a phenomenon known as electro-osmosis causes molecules of the liquid to migrate away from the electrode so that the paint film becomes denser. The process is of obvious value in the motor industry where complete bodies can be coated by dipping into an electrophoretic bath.

21.104 The 'Elphal' Process   is similar in principle but is used to coat steel strip with aluminium. Here finely-divided aluminium is carried in a water/alcohol bath which also contains an electrolyte. The aluminium particles are *not* ions and neither this process nor electrophoresis generally should be confused in any way with electroplating. In this case aluminium particles become charged by *attachment* to ions from the electrolyte and consequently migrate to the steel surface (electrode) where they adhere loosely. The solvent is removed by heating and the coating then compacted by cold rolling. Finally the aluminium film is sintered on to the surface of the steel by heat treatment.

21.105 Surface Protection at High Temperatures   The drive towards higher efficiency and performance means that engineering environments are becoming increasingly hostile. The first line of defence against deterioration of any component is its surface. Unfortunately many materials which are designed for strength do not offer a high resistance to corrosion and wear, especially at high temperatures. This is a major problem in the design of gas turbine engines, particularly in respect of the blades.

Many problems of wear and corrosion in gas turbines have been solved by using plasma-sprayed coatings on the surface. The plasma torch (Fig. 21.18) uses the energy in a thermally ionised gas to propel partially molten fine powder particles on to a prepared surface so that they adhere on impact. A DC arc is initiated between the electrodes by a high EMF. The design of the torch is such that the energy involved is concentrated into a very small region so that temperatures up to 30 000°C can be achieved. As the heated argon expands rapidly the particles are accelerated to very high velocities (up to 600 m/s). In order to prevent oxidation of the particles the torch is usually fitted with an inert gas shield.

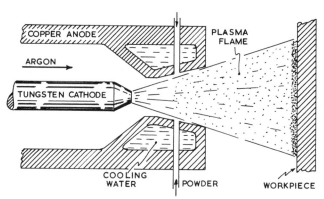

Fig. 21.18 A typical plasma torch.

Most of the coating materials are based on aluminium and chromium, both of which form protective films of tenacious refractory oxide. Such surface films are effectively self-sealing and since the ionic mobilities (21.21) of chromium and aluminium within the film are low, oxidation rates beneath the film are also very low. The surface tenacity of chromia/alumina film is increased by the addition of small amounts of the 'rare-earth' metal yttrium.

These materials are designated $MCrAlY$ coatings (Table 21.2) where M is either nickel or cobalt (or a mixture of both) and are used for turbine corrosion protection in advanced military and civil aero gas turbines.

Table 21.2 *Some UCAR* MCrAlY Alloys.*

UCAR designation	Composition (%)					
	Co	Ni	Mo	Cr	Al	Y
LCO-7	63	—	—	23	13	0.6
LCO-22	39	32	—	21	7.5	0.5
LCO-37	44	23	—	30	3	0.5
LCO-34	0.5	67	0.5	20	11	0.5

*UCAR is a trade mark of the Union Carbide Corporation

Cathodic Protection

21.110 Steel structure will corrode more rapidly if it becomes anodic to its environment. Thus the hull of a ship can be expected to rust more quickly at the stern since it is anodic to the nearby manganese bronze propellers. Moreover salt water acts as a strong electrolyte since it contains large concentrations of Cl^- as well as OH^- ions. The hull can be sacrificially protected in the region of the propellers by bolting slabs of zinc or magnesium to the hull. Since these slabs are more strongly anodic than is the steel hull towards the propellers, then they naturally replace the steel hull as anode. The hull then assumes the rôle of 'external circuit' in the galvanic system so produced, so that it transmits electrons without suffering ionisation (Fig. 21.19).

21.111 Similarly an iron or steel pipe or storage tank situated below ground can be sacrificially protected by burying a slab of zinc or magnesium adjacent to it. Alternatively an *external EMF* may be used to render the steel structure passive to its surroundings (Fig. 21.20) That is the EMF is applied in the direction Y indicated on the Pourbaix diagram (Fig. 21.8). The inert graphite electrode is made the anode by virtue of the applied EMF and so the steel tank becomes cathodic to it. Obviously the value of the EMF necessary to provide passivity will depend on the pH of the soil as well as other factors, and the electrical output will need to be controlled by a feed-back system which automatically adjusts the EMF (E_h) required from the transformer/rectifier system. E_h can be monitored by a high-resistance voltmeter connected to a reference electrode. Where mains electricity—and the necessary qualified maintenance back-up—are available this system will be cheaper to operate than any electro-chemical source, ie sacrificial anodes. This is obvious thermodynamically since the metals of which the sacrificial anodes are made have to be deposited electrolytically in the first place.

21.112 It is also more satisfactory to use a power-impressed EMF to

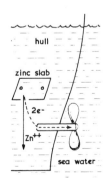

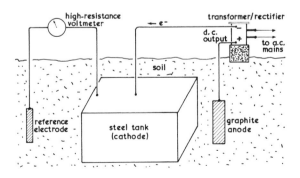

Fig. 21.19 Cathodic protection of a ship's hull.

Fig. 21.20 Cathodic protection of a buried steel tank by means of an 'impressed' EMF and a non-consumable graphite anode.

limit corrosion in a ship's hull since the EMF so provided can be monitored to suit prevailing conditions which will vary with salinity and temperature of the ocean, conditions of paintwork on the hull (area of metal exposed) and the area of the hull in contact with the water (which, in turn, depends upon the load the ship is carrying).

In some cases buried pipe lines can be cathodically protected by using impressed EMF supplied by small windmill-driven dynamos. A similar method is sometimes used in the cathodic protection of off-shore oil or gas rigs, but because frequent inspection is necessary to check for corrosion of electrical contacts the sacrificial anode system is usually preferred. 300 kg anodes of Al–Zn–In alloy are generally distributed uniformly around the structure. A large structure with up to twenty-five years of design life may require up to 500 tonnes of such anode material.

Exercises

1. Explain why: (i) a penknife blade dipped into copper sulphate solution becomes quickly coated with copper; (ii) zinc will dissolve in dilute sulphuric acid whilst silver will not. (Table 21.1)
2. A plate of silver and a plate of cadmium are placed in dilute sulphuric acid. Make a diagram to show the chemical reactions which take place within the simple cell which is set up when the plates are connected electrically, externally to the electrolyte. (Fig. 21.2; Table 21.1)
3. A piece of mild steel has been copper plated. Show to what extent this coating will protect the steel from rain water if the coating becomes scratched. (Table 21.1; 21.33)
4. Outline the principles involved in the electrolytic corrosion of alloys. Explain the meaning of the term sacrificial corrosion and illustrate this principle in the protection of steels by metal coatings. (21.33)
5. Discuss the phenomena which are responsible for the high corrosion-resistance of stainless steels. (21.70)
6. Why does a 0.5% carbon steel rust more quickly in the normalised condition than in the water-quenched state? (21.36)
7. What are the advantages of electro-galvanising over hot-dip galvanising? Give a brief outline of the latter process. What other methods are available for applying a coating of zinc to a steel? (21.82)
8. A broad elastic band is stretched over a flat piece of stainless steel plate. The assembly is then immersed in an acid ferric chloride solution for about two months. Describe the appearance of the steel plate after this treatment and explain what has happened. (21.73)
9. We get the rotten motor cars we deserve! Discuss this statement with respect to corrosion. (13.35; 21.40; 21.52)
10. A mild-steel nail has a cold-forged head. Show how this nail is likely to rust when immersed in rain water. (21.40)
11. Explain how a study of the relevant Pourbaix diagram can help in overcoming some corrosion problems. (21.38)

Bibliography

Ahmed, N. A. G., *Ion Plating Technology*, John Wiley, 1987.

Carter, V. E., *Metallic Coatings for Corrosion Control*, Newnes-Butterworths, 1977.

Dennis, J. K. and Such, T. E., *Nickel and Chromium Plating*, Butterworths, 1972.

Evans, U. R., *The Corrosion and Oxidation of Metals*, Edward Arnold, 1960 (Supplements 1968, 1976.)

Fontana, M. G., *Corrosion Engineering*, McGraw-Hill, 1988.

Friend, W. Z., *Corrosion of Nickel and Nickel-base Alloys*, John Wiley, 1979.

Proskurkin, E. V. and Gorbunov, N. S., *Galvanising, Sherardising and Other Diffusion Coatings*, Technicopy Ltd., 1975.

Ross, T. K., *Metal Corrosion (Engineering Design Guide)*, Oxford University Press, 1978.

Scully, J. C., *The Fundamentals of Corrosion*, Commonwealth and International Library, 1975.

Stewart, D. and Tullock, D. S., *Principles of Corrosion and Protection*, Macmillan, 1968.

Trethewey, K. R. and Chamberlain, J., *Corrosion for Students of Science and Engineering*, Longman, 1989.

West, J. M., *Basic Corrosion and Oxidation*, Ellis Horwood, 1980.

BS 729: 1986 *Hot-dip Galvanized Coatings on Iron and Steel Articles*.

BS 2920: 1973 *Cold-reduced Tinplate and Cold-reduced Blackplate*.

BS 1872: 1984 *Electroplated Coatings of Tin*.

BS 1615: 1987 *Anodic Oxidation Coatings on Aluminium*.

BS 5599: 1978 *Specification for Hard Anodic Oxide Coatings on Aluminium for Engineering Purposes*.

BS 5493: 1977 *Code of Practice for Protective Coating of Iron and Steel Structures against Corrosion*.

BS 5466: 1979/1988 *Methods for Corrosion Testing of Metallic Coatings*.

BS 3189: 1973 *Phosphate Treatment of Iron and Steel*.

ANSWERS TO NUMERICAL PROBLEMS

CHAPTER 1

(1) Al_2O_3; (2) 3.98 g; (3) 4.9 t; (6) $2HNO_3 + MgO = Mg(NO_3)_2 + H_2O$;
(7) 28°C; (11) 65.

CHAPTER 2

(2) 849 N/mm²; (3) (iii); (4) 208.3 kN/mm²; (5) 1264 N/mm²;
(6) (i) 95.5 kN/mm²; (ii) 507 N/mm²; (iii) 691.5 N/mm²; (iv) 5.4%;
(7) (a) 1944 N; (b) 22 mm.

CHAPTER 3

(4) 112

CHAPTER 4

(2) 27.1 N/mm²; (7) 450°C (approx.).

CHAPTER 8

(6)(a) (i) 7/4; (ii) 3/2; (iii) 21/13; (iv) 3/2.

CHAPTER 9

(1)(i) Mg_2Si; (iii) Liq., Si and Mg_2Si; (iv) Si and Mg_2Si are dissolved;
    (v) Solid Si + Liq. (64.5 Si–35.5 Mg); (vi) Liq./Si = 1.29/1;
    (viii) Mg_2Si and Si; (ix) Mg_2Si/Si = 3.73/1
(3)(a) Bi and Sb are close together in Group V of the Periodic Table;
  (b) (i) 577°C; (ii) 96Sb–4Bi; (iii) Liq.: 44Sb–56Bi; solid:
    88.5Sb–11.5Bi; (iv) Liq./solid = 1.4/1; (v) 20Sb–80Bi; (vi) 405°C.
(4) 415°C; Liq.: 22.5Sb–77.5Bi, solid: 72.5Sb–27.5Bi.
(5) (i) 437°C; (iii) Equal amounts at 300°C; at 0°C: β/α = 9/8.
(7) (i) 15.5% (at 450°C); (ii) 328°–550°C;
    (iii) 580°C; (iv) α: 90.8Al–9.2Mg; Liq.: 72Al–28Mg; (v) α/Liq. =

1.76/1; (vi) From 1.5% to 8.3%; (viii) (b)—because all Mg will be in solid solution.
(8) (i) 771°C; (ii) 89.6Ag–10.4Sn; (iii) 685°C; (iv) 76Ag–24Sn; (v) Solid solution β: 83.6Ag–16.4Sn; Liq.: 77.7Ag–22.3Sn; (vi) β/Liq. = 2.7/1.
(9) Approximately 24.5% at 127°C.
(11) (i) X:AuSn₄; Y:AuSn₂; Z:AuSn.

CHAPTER 11

(1) Ferrite/Cemenite = 7.42/1
(4) (i) 783°C; (ii) Pearlite/Ferrite = 1.27/1; (iii) 930°C → 723°C; (iv) Produces brittleness due to networks of primary cementite.
(5) (i) Austenite (0.15%C) only; (ii) Austenite (0.37%C) + Ferrite (0.01%C) Austenite/Ferrite = 1.57/1: (iii) Ferrite (0.02%C) + Cementite (6.69%C), Ferrite/Cementite = 50.3/1 (or Ferrite/Pearlite = 5/1).

CHAPTER 13

(1) 0.65%C; 660°C.
(2) 0.26%C, 1145°C.
(3) 552 N/mm²
(4) Approx. 50% ferrite and 50% austenite.

CHAPTER 16

(1) 800°C; α = 63.5Cu–36.5Zn; β:54Cu–46Zn; β/α = 1.71/1.
(2) δ/α = 1.04/1.
(3) (i) β′ (martensitic); (ii) α + β′.
(4) 1390°C to 1355°C.

CHAPTER 17

(1) α/eutectic = 1.29/1.
(2) (i) α; 99.8Al–0.2Cu; CuAl₂:45.9Al–54.1Cu.
  (ii) α/CuAl₂ = 3.57/1.
  (iii) Between 487°C and 573°C.
  (iv) 530°C, ie mid-way between above temperatures.
  (vi) Very weak and brittle.

CHAPTER 18

(1) (i) 470°C; 410°C; (ii) 480°C; 300°C; (iii) 600°; 560°C.
(2) Al/eutectic = 3.975/1
(3) (i) and (ii) can be hardened by precipitation treatment; (iv) by martensite formation.

CHAPTER 19

(1) 1.97 mm.

CHAPTER 20

(7) Martensitic structure—hence brittle.
(10) A:α/β = 1.67/1; G:β/α = 1.67/1; W:β/α = 7.72/1.

INDEX